Theoretical and Applied Mechanics

Valentin Molotnikov • Antonina Molotnikova

Theoretical and Applied Mechanics

 Springer

Valentin Molotnikov
Don State Technical University
Rostov-on-Don, Russia

Antonina Molotnikova
Rostov Institute for the Defense of
Entrepreneur Protection
Rostov-on-don, Russia

ISBN 978-3-031-09314-2   ISBN 978-3-031-09312-8   (eBook)
https://doi.org/10.1007/978-3-031-09312-8

This Springer imprint is published by the registered company Springer Nature Switzerland AG
The registered company address is: Gewerbestrasse 11, 6330 Cham, Switzerland

Dedicated to the blessed memory of Boris Anisimovich Bondarenko (1923–2017)—an outstanding mathematician and mechanic, a talented teacher and organizer of science, Doctor of Physical and Mathematical Sciences, and academician of the Academy of Sciences of the Republic of Uzbekistan.

Valentin and Antonina

Abstract

The textbook outlines the fundamentals of theoretical and applied mechanics, supplemented by selected sections from the developments of recent decades (drone flight, car variators, etc.). The book is designed to be studied during one to two semesters in the training of bachelors and masters in the direction of "Engineering." The course includes the main sections of the disciplines "Theoretical Mechanics," "Theory of Mechanisms and Machines," "Strength of Materials," "Machine Parts and Design Fundamentals," "Interchangeability, Standardization, and Technical Measurements," and "Introduction to CAD of based AutoCAD – AutoLISP". Such a representative set of disciplines allows the book to be used by students of other technical and technology majors in each of these disciplines. The conciseness makes the textbook a valuable aid in conditions of limited time for the preparation of both during a semester and a session. In addition to being concise, this differs from existing textbooks in its presentation of the state of the art in mechanics, demonstration of engineering applications computer applications (MathCad, COSMOSWorks, Inkscape, AutoCAD, AutoLISP, etc.), presentation of updated data on engineering materials - from composites to the prospects graphene. There are examples of designs of units in modern machines (combine harvester "Don," the car "Infiniti" etc.), and also there is information about national standards in mechanical engineering. Examples are given of both simple and complex engineering calculations. At the end of each chapter are questions for self-checking and multivariate problems for exercises.

Acknowledgements

The authors express their gratitude to Doctor of Physical and Mathematical Sciences, member of the Russian National Committee of IUTAM prof. L. M. Zubov and prof. V. P. Zabrodin for selfless work on reviewing the manuscript. Special thanks to our son Zaur, who inspired us to create this book for his helpful advice on book design and, as a native speaker, for taking on much of the communication with Springer. We also thank Professor V. G. Malinin, Doctor of Physical and Mathematical Sciences, for his help in translating this book into English.

Introduction

Modern scientific and technological progress and the rapid development of high technologies have led to the need to include in the curricula of educational institutions many new disciplines, the knowledge of which is becoming mandatory for a modern specialist. Although the duration of training is limited, the modernization of curricula inevitably leads to a reduction in the number of hours allocated for the development of fundamental general technical disciplines, including theoretical and applied mechanics. The desire to preserve the completeness of the course, and the rigor and clarity of the main provisions, combined with the clarity of the presentation of the material and the demonstration of its application in technology, prompted the authors to write a short but complete textbook.

The concentrated presentation of the course made it necessary to exclude from consideration a number of issues traditionally presented in courses of theoretical mechanics. Some of them seem to be utilitarian (e.g., graphostatics), while others are applications of the main theorems of theoretical mechanics (fluctuations of a material point) and do not contain fundamentally new material compared to what is studied in school physics courses.

At the same time, to solve many problems of modern technology, sometimes the methods of the traditional course of theoretical mechanics are not enough, and it is necessary to use the general methods of analytical mechanics and control theory. In this regard, the author considered it appropriate to move away from the tradition and include in the book material on both classical and the latest variational methods of mechanics. And if the study of this material is still provided for only by a few educational standards for master's programs, then it can be used in optional classes.

In order not to overload the book with complex mathematical constructions, when presenting this material, the authors deliberately never discussed the most important question of the existence and uniqueness of solutions to the formulated problems.

In recent decades, almost all over the world, a wave of reforms in the field of education has unfolded. As a result, most higher education institutions were transformed into universities and academies. This circumstance obliges the authors of textbooks to emphasize the style bordering on monographic. How far the authors succeeded, the readers can judge. At the same time, the authors have retained certain classical elements of the textbook. For example, at the end of each chapter, there are questions for self-control and multivariate tasks for exercises.

The book is based on lectures given by the authors in recent decades in industrial, military, and agricultural universities of the Soviet Union and Russia.

The authors express their gratitude to the Candidate of Technical Sciences, assoc. Veselovsky V. A., who carefully read the manuscript and made a number of valuable comments that were taken into account in the final version of the manuscript.

We are grateful in advance to readers who will send feedback or suggestions to either of the email addresses: v.molotnikov@gmail.com or molotnikova@gmail.com.

Contents

Notation Conventions

Theoretical Mechanics

$\boldsymbol{F},\ \boldsymbol{F}_1,\ \boldsymbol{P}$	Force vector
$F,\ F_1,\ P$	Force module
$\boldsymbol{i},\ \boldsymbol{j}_1,\ \boldsymbol{k}$	Unit vectors of coordinate axes
$F_x,\ F_y,\ F_z$	Force projections on the x, y, and z axes
q	Running load intensity
p	Surface load (pressure)
$\Delta l,\ \Delta s,\ \Delta V$	Element of length, surface, volume
$\varphi,\ \phi,\ \gamma$	Flat angle
t	Time
$\displaystyle\sum_{k=1}^{n}$	Summation symbol
$\boldsymbol{M}$	The vector of the moment of force
$(\boldsymbol{F},\ \boldsymbol{F}')$	A couple of forces
$(\boldsymbol{r} \times \boldsymbol{F})$	Vector product of vectors
$(\boldsymbol{F}_1,\ \boldsymbol{F}_2,\ \boldsymbol{F}_3)$	Force system
$x_C,\ y_C,\ z_C$	Coordinates of the center of mass
ρ	Body density
g	Acceleration of free fall
$S_x,\ S_y$	Static moments of the body relative to x, y
$\boldsymbol{r}$	Point movement vector
$\boldsymbol{v}$	Velocity vector of a point
$\boldsymbol{a}$	Acceleration vector of the point
$\boldsymbol{a}_n$	Normal acceleration vector
$\boldsymbol{a}_\tau$	Tangential acceleration vector
$\boldsymbol{a}_k$	Coriolis acceleration
$\delta\varphi,\ \delta x \ldots$	Possible movements
$Q_1,\ Q_2, \ldots$	Generalized forces
$U,\ T$	Potential (kinetic) energy

Φ Vector of inertia forces
K Kinetic moment of the system
K_x, K_y, K_z Kinetic moments relative to the x, y, z axes
J_x, J_y, J_z Moments of inertia of the system relative to the axes x, y, z
S Impact impulis
u Control vector
ΔG Functional
ψ Absolutely non-zero vector
$\Pi(q_1, q_2, \ldots q_n)$ Potential energy in generalized coordinates

Theory of Mechanisms and Machines

S Number of links
W Number of degrees of freedom
n Number of moving links of the kinematic chain
P_i Number of kinematic pairs of class i
μ_t Time scale
μ_l Length scale
μ_v Speed scale
ω_i Angular velocity of the i-th link
w_{a_i} Acceleration of the point a_i
ε_i Angular acceleration of the i-th link
ξ, η, ζ Normalized kinematic characteristics
G_i Gravity of the links
F_i Inertia forces of the links
M_i Moments of inertia of the links
P_y Balancing force
M_y Balancing moment
m_i Mass of the i-th link
Π Pole of the "Zhukovsky lever"

Resistance of Materials

P_i External forces
N, Q_x, Q_y, M_x, M_y, M_z Internal efforts
p Total voltage at a given point of the site
ε_l Relative linear strain in the direction of l
ε' Relative transverse strain
μ Poisson's ratio
δ Residual elongation
Δl Absolute linear deformation

u	Specific potential energy of deformation
n	Safety margin factor
$[\sigma]$	Allowable stress
P_{per}	Permissible load
$\sigma_1, \sigma_2, \sigma_3$	Principal stresses
ε_x	Linear deformation in the direction of the x axis
γ_{xy}	Shear strain
G	Shear modulus
E	Elastic modulus (Young's)
$\sigma_{max}, \sigma_{min}$	Extreme normal stresses
τ_{max}, τ_{min}	Extreme tangential stresses
Θ	Relative twist angle
J_p	Polar moment of inertia
GJ_p	Torsional rigidity of the timber
W_p	Polar moment of resistance
$[\tau]$	Permissible shear stress
W_x	Moment of resistance about the x axis
σ_{equ}^{IV}	Design stress according to 4 strength theory
EJ	Bending stiffness
i_x, i_y	Radii of gyration
λ	Spring settlement
c	Spring rate
A_{for}	Forced vibration amplitude
k_d	Dynamic coefficient
σ_{-1}	Endurance limit at symmetric stress cycle
n	Fatigue safety factor
$k_{\sigma\sigma}$	Stress concentration factor
P_k	Critical force

Machine Parts and Design Basics

p^m	Contact stress
S	Friction path
$Q(t)$	Probability of failure
$P(t)$	Uptime probability
IT	Size tolerance
W_z	Waviness height
F_t	Circumferential force
P	Power transmitted by a rotating body
T	Torque transmitted by a rotating body
η	Coefficient of performance (efficiency)
i	Drive gear ratio
m	Engagement module

a_w	Center distance
σ_H	Contact stress
ψ_{bm}	Ratio of the width of the ring gear
σ_F	Tooth bending stress
k_t	Heat transfer coefficient
L, L_t	Length and number of chain links
s	Chain safety margin
f	Coefficient of friction
E_p	Reduced modulus of elasticity
t	Durability of the flat belt
H_0	V-belt resource
n_σ, n_τ	Fatigue strength reserve coefficients
L	Nominal bearing life
p	Specific pressure in plain bearing
F	Design force in the centrifugal clutch
$[\tau]_s$	Allowable rivet shear stress
σcru	Shear stress in the keyway

Introduction to CAD Based on AutoCad ~ AutoLisp

CAD	Computer-aided design
$Cloud''CAD$	Cloud CAD systems
$GOST$	Russian State standards
$BHATCH$	Lisp hatch command

Part I
Theoretical Mechanics

Chapter 1
Statics

Abstract Fundamental concepts of theoretical mechanics are introduced, such as force, system of forces, pair of forces, moment of pair, etc. Engineering devices that restrict the freedom of a solid body are considered. These devices are called constraints. The principle of freedom from ties, which allows us to consider the body free, is described. The formulation of Varignon's theorem on the moment of the resultant system of forces and its proof is given. The conditions of equinocity of plane and spatial systems of forces are investigated. Examples of solving problems about the equilibrium of a system of forces are given. As an engineering application, the calculation of farms using a computer is given.

Keywords Forces · Force systems · Forces concentrated · Force distributed · Space Euclidean · Movement mechanical · Equilibrium · Problems theoretical mechanics · Body free · Body not free · Truss

1.1 Introduction

Theoretical mechanics studies the simplest form of motion of matter—mechanical motion. *Mechanical movement* is the change in time of the relative position of material bodies. At the same time, the question of the physical nature of the causes of motion is not considered, which does not interfere with the wide coverage of problems of mechanical displacements and conditions of equilibrium of bodies. Movement takes place in space and in time. The *space* in theoretical mechanics is taken to be the usual three-dimensional Euclidean space. *Time* is assumed to be the same at any point in space and in any frame of reference. It is believed that space and time are not associated with the movement of bodies.

Theoretical mechanics is based on Newton's laws (Nikolai 1952; Zhukovskii 1952; Whittaker 1904) the validity of which has been verified by centuries of practice. At the same time, she makes extensive use of the abstraction method. So, for example, *if the dimensions of a material body are small in relation to other bodies or in comparison with the distances from it to these bodies, then it can be considered as a material point.* The latter differs from the geometric point by the presence of a finite mass. Another useful abstraction adopted in theoretical mechanics is *an absolutely rigid body*, which means a body that does not change its geometric shape under any circumstances.

Newton's laws are basic not only for theoretical mechanics but also for mechanics in a broad sense. Modern mechanics embraces many areas of science and technology, including the theory of elasticity, resistance of materials, mechanics of liquids and gases, theory of mechanisms and machines, celestial mechanics, theory of regulation, etc.

In our time, theoretical mechanics proper is understood as a relatively narrow section of mechanics, which includes the mechanics of a material point and the mechanics of absolutely rigid bodies and their systems. Despite this, theoretical mechanics is the scientific foundation of general mechanics. It is widely used in technology, deriving many problems from specific problems of designing machines, astronautics, the theory of automatic control and regulation, etc.

Theoretical mechanics is one of the most important fundamental general scientific disciplines. It plays an essential role in the training of engineers of all specialties. General engineering disciplines are based on the results of theoretical mechanics: resistance of materials, machine parts, theory of mechanisms and machines, etc. The main task of theoretical mechanics is to study the motion of material bodies under the action of forces. An important particular problem is the study of the balance of bodies under the action of forces.

The foundations of modern theoretical mechanics were created by the great mathematicians, physicists, and engineers like Galileo, Newton, D'Alembert, Euler and many others. An invaluable contribution to its development was made by Russian scientists Ostrogradsky, Zhukovsky, Kovalevskaya, Lyapunov, Tsiolkovsky, James (James 2005), etc.

By the nature of the problems studied, theoretical mechanics is traditionally divided into three sections: statics, kinematics, and dynamics. *Statics* is a branch of theoretical mechanics that studies the conditions of equilibrium of bodies or their systems. By *equilibrium* is meant a state of rest with respect to a certain coordinate system. The laws and methods of statics are widely used in engineering calculations.

Kinematics is the science of the movement of material bodies in space from a geometric point of view, regardless of the reasons that cause or change this movement. Its laws and conclusions provide invaluable service in the analysis and synthesis of various mechanisms.

Dynamics studies the mechanical motion of material objects under the action of forces and in this sense combines the kinematic properties of motion with the properties of the forces that prompted the motion.

To service the problems of creating new technology, nonclassical branches of theoretical mechanics have arisen and are intensively developing in recent decades (Molotnikov 2004). These include the mechanics of bodies of variable mass, the mechanics of gyroscopes, variational problems of mechanics, the dynamics of space flight, the mechanics of the special theory of relativity, as well as mechatronics, which lies at the junction of theoretical mechanics, microelectronics, computer science, and the theory of automatic control.

1.2   The Subject and Tasks of Theoretical Mechanics

Theoretical mechanics studies the simplest form of motion of matter—mechanical motion. *Mechanical movement* is the change in time of the relative position of material bodies. At the same time, the question of the physical nature of the causes of motion is not considered, which does not interfere with the wide coverage of problems of mechanical displacements and conditions of equilibrium of bodies. Movement takes place in space and in time. The *space* in theoretical mechanics is taken to be the usual three-dimensional Euclidean space. *Time* is assumed to be the same at any point in space and in any frame of reference. It is believed that space and time are not associated with the movement of bodies.

Theoretical mechanics is based on Newton's laws, the validity of which has been verified by centuries of practice. At the same time, she makes extensive use of the abstraction method. So, for example, *if the dimensions of a material body are small in relation to other bodies or in comparison with the distances from it to these bodies, then it can be considered as a material point.* The latter differs from the geometric point by the presence of a finite mass. Another useful abstraction adopted in theoretical mechanics is *an absolutely rigid body,* which means a body that does not change its geometric shape under any circumstances.

Newton's laws are basic not only for theoretical mechanics but also for mechanics in a broad sense. Modern mechanics embraces many areas of science and technology, including the theory of elasticity, resistance of materials, mechanics of liquids and gases, theory of mechanisms and machines, celestial mechanics, theory of regulation, etc.

In our time, theoretical mechanics proper is understood as a relatively narrow section of mechanics, which includes the mechanics of a material point and the mechanics of absolutely rigid bodies and their systems. Despite this, theoretical mechanics is the scientific foundation of general mechanics. It is widely used in technology, deriving many problems from specific problems of designing machines, astronautics, the theory of automatic control and regulation, etc.

Theoretical mechanics is one of the most important fundamental general scientific disciplines. It plays an essential role in the training of engineers of all specialties. General engineering disciplines are based on the results of theoretical mechanics: resistance of materials, machine parts, theory of mechanisms and machines, etc. The main task of theoretical mechanics is to study the motion of material bodies under the action of forces. An important particular problem is the study of the balance of bodies under the action of forces.

The foundations of modern theoretical mechanics were created by the great mathematicians, physicists, and engineers like Galileo, Newton, D'Alembert, Euler, and many others. Scientists of the nineteenth–twentieth centuries made an invaluable contribution to its development like Ostrogradsky, Zhukovsky, Kovalevskaya, Lyapunov, Tsiolkovsky, James (James 2005) , etc.

By the nature of the problems studied, theoretical mechanics is traditionally divided into three sections: statics, kinematics, and dynamics. *Statics* is a branch of theoretical mechanics that studies the conditions of equilibrium of bodies or their systems. By *equilibrium* is meant a state of rest with respect to a certain coordinate system. The laws and methods of statics are widely used in engineering calculations.

Kinematics is the science of the movement of material bodies in space from a geometric point of view, regardless of the reasons that cause or change this movement. Its laws and conclusions provide invaluable service in the analysis and synthesis of various mechanisms.

Dynamics studies the mechanical motion of material objects under the action of forces and in this sense combines the kinematic properties of motion with the properties of the forces that prompted the motion.

To service the problems of creating new technology, nonclassical branches of theoretical mechanics have arisen and are intensively developing in recent decades. These include the mechanics of bodies of variable mass, the mechanics of gyroscopes, variational problems of mechanics, the dynamics of space flight, the mechanics of the special theory of relativity, as well as mechatronics, which lies at the junction of theoretical mechanics, microelectronics, computer science, and the theory of automatic control.

1.3   Strength: Force Systems

The fundamental concept in statics is the concept of force. *Force* is defined as a measure of the mechanical interaction of bodies. It is characterized by three elements: an application point, a direction, and a numerical value (modulus).

The first two elements define the line of action of the force. The graphic image of a force is a vector directed along the line of its action, and the length of the vector on a certain scale is equal to the modulus of the force.

Let us agree to denote the forces by capital letters of the bold Latin script, for example, F, F_1, P, etc. (Fig. 1.1). The modulus of force will be denoted by the same letter of the regular bold font: F, F_1, P, and t. P. As is known from the course of physics, the unit of force in the SI system is the newton (N); in the technical system of units MKGSS, the kilogram-force (kg or kgf) (1 kgf $= 9.80665$ N); and in the CGS system, dyne (1 dyn $= 10^{-5}$ N).

Fig. 1.1 Image and
designation of forces

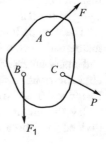

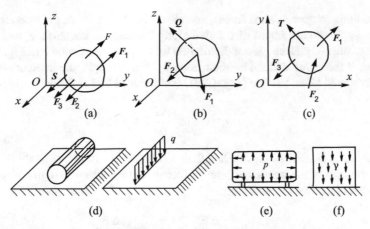

Fig. 1.2 System of forces: (**a**) parallel forces; (**b**) system of converging forces; (**c**) flat system of forces; (**d**) forces distributed along the line; (**e**) forces distributed over the surface; (**f**) forces distributed over the volume of the body

It follows from the definition that the concept of force presupposes the presence of an object to which it is applied, i.e., material body. Let us agree to call the body *free*, if its movements are not limited by anything. Otherwise, the body is said to be *non-free*. To set the force, it is enough to specify three projections of the force on the axis of the rectangular coordinate system x, y, z and the coordinates of the point of application of the force. If the unit vectors (orts) of these axes are denoted by i, j and k, then the force is defined by

$$F = F_x i + F_y j + F_z k, \qquad (1.1)$$

where F_x, F_y, F_z are projections of the force F on the coordinate axis x, y, z.

The set of forces acting on a given body is called the *system of forces*. Figure 1.2 shows systems of forces that are most commonly found in engineering practice.

In the case when the force is the result of direct contact between two bodies and the contact area is very small compared to the size of the bodies, it is said that a concentrated force is applied to each of the interacting bodies. Examples of such forces can be the pressure of the car wheels on the rails, the pressure of the rolling bearing balls on the rings, etc. As mentioned above, the concentrated forces are represented by vectors applied at the corresponding points of the beam or other structural element.

If all the forces of the system are parallel, they form a system of parallel forces. Such a system of forces is shown in Fig. 1.2a. In the case when the lines of action of all forces intersect at one point, we speak of a system of converging forces (Fig. 1.2b). Forces whose vectors are located in the same plane (Fig. 1.2c) form a planar system of forces.

In addition to concentrated forces, so-called distributed forces are often found in technology. There are forces distributed along the line, on the surface, and on the volume of the body. Examples of distributed force systems are shown in Fig. 1.2d–f. Distributed forces are defined by their intensity. For example, the intensity q of the load distributed along the line (Fig. 1.2d) is determined by the formula

$$q = \lim_{\Delta l \to 0} \frac{\Delta F}{\Delta l},$$

where Δl is an element of the line length and ΔF is the force acting on this element.

The concept of the intensity of the surface p and volume γ systems of forces is introduced in exactly the same way:

$$p = \lim_{\Delta s \to 0} \frac{\Delta F}{\Delta s}; \quad \gamma = \lim_{\Delta V \to 0} \frac{\Delta F}{\Delta V};$$

moreover, Δs and ΔV are, respectively, the elementary area and elementary volume allocated in the vicinity of the point where the intensity of distributed forces is determined, and ΔF is still the force acting on this element. Systems of forces are called equivalent, if each of them has the same effect on the kinematic state of the solid. A force equivalent to some system of forces is called *resultant* of this system of forces. A force that coincides in modulus with the resultant, but has the opposite direction on the general line of action, is called *balancing*.

If a system of forces applied to a solid at rest does not bring it out of this state, it is called a *system of mutually balancing forces*.

Forces acting on a mechanical system from material bodies or points of another system are called *external*. Forces of the same interaction between material points or parts of the same mechanical system are called *internal*.

1.4   Axioms of Theoretical Mechanics

The properties of forces and systems of forces, established as a result of numerous observations and experimental verification, form the basis of theoretical mechanics and are formulated in the form of axioms. The axioms of mechanics reflect, in essence, the laws of Galileo-Newton.

Axiom 1   The axiom of the equilibrium of a system of two forces. *Two forces, equal in magnitude and directed in opposite directions along the common line of action, are in equilibrium* (Fig. 1.3a).

Axiom 1 defines the simplest system of forces that is statically equivalent to zero. If two forces F_1 and F_2 are in equilibrium, then naturally they form a system of forces equivalent to zero. Such a system of forces, when applied to a body at rest, will not change its state of rest.

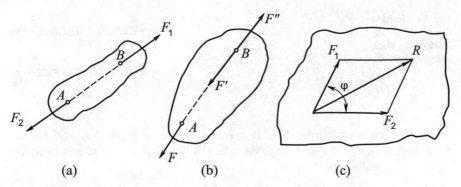

Fig. 1.3 To the axioms of forces: (**a**) equilibrium of two forces; (**b**) three forces equal in modulo; (**c**) parallelogram of forces

Axiom 2 The axiom of joining and excluding balancing forces. *The effect of a system of forces on a solid will not change if a system of forces equivalent to zero is added to or excluded from it.*

Consequence The force can be transferred along the line of its action without changing its modulus and direction.

Proof Let a force F be applied to the body at the point A (Fig. 1.3b). Let us apply at the point B, located on the line of action of the force F, mutually balancing forces F' and F'', directed along the line of action of the force F. We choose the modules of these forces to be the same as the F force module. Based on axiom 2, the system of forces F and F'' can be excluded as statically equivalent to zero. Then a force F' will be applied to the body at B, which is equivalent to the force F at A. □

The proved corollary allows us to consider the force as a *sliding vector in rigid body statics.*

Axiom 3 Axiom of the parallelogram of forces. *It is the resultant of two forces with intersecting lines of action applied at the point where they intersect and is represented by the diagonal of the parallelogram based on these forces.*

The formulated position is known to the reader from the physics course and is expressed by the vector equality (Fig. 1.3c):

$$R = F_1 + F_2. \tag{1.2}$$

The module of the resultant is calculated by the formula

$$R = \sqrt{F_1^2 + F_2^2 + 2F_1 F_2 \cos \varphi}\,, \tag{1.3}$$

where φ is the angle between the vectors F_1 and F_2.

Fig. 1.4 To the axiom of
Solidification: (**a**) axial
tension; (**b**) axial
compression

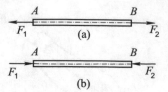

Axiom 4 Axiom of inertia. *If is attached to a material point (body) An ax-iom!inertiaa balanced system of forces, then the point (body) is at rest or moves rectilinearly and evenly.*

This axiom expresses Galileo's law of inertia.

Axiom 5 The axiom of equality of action and reaction. *Every action of one body on another corresponds to an equal and opposite reaction.*

Axiom 5 in physics is known as Newton's third law. It is necessary to remember that the forces of interaction are always applied to different bodies, and therefore it is impossible to talk about their equivalence. It should also be borne in mind that material points (bodies) can interact not only in contact with each other but also at a distance through force fields.

Axiom 6 The axiom of solidification. *Equilibrium of the forces applied to a deformable solid is not disturbed if the body is considered absolutely solid.*

It follows from this axiom that external forces applied to a deformable solid must satisfy the equilibrium conditions written in the assumption of non-deformability of this body. However, for a deformable body, these conditions are necessary, but not always sufficient.

Let's explain this with an example. Figure 1.4 shows a AB thread with two equal and opposite forces F_1 and F_2 applied to it. In the case when the applied forces stretch the thread (Fig. 1.4a), there is an equilibrium. If the forces F_1 and F_2 compress the thread (Fig. 1.4b), the system loses its equilibrium. Therefore, in the example considered, two balancing forces acting on the thread satisfy the equilibrium condition only if there is an additional requirement—the forces must stretch, but not compress the thread.

1.5 Connections and Their Reactions

Bodies that in one way or another restrict the freedom of movement of a non-free body are called *connections*. The action of bonds on a non-free body is carried out by means of forces called *bond reactions*. Unlike specified or active forces that can cause a change in the kinematic state of a body, the reactions of constraints depend on the type of constraints, as well as on the nature of the active forces.

One of the basic provisions of theoretical mechanics is the principle of *being free from ties*. According to this principle, *any non-free solid body can be considered as free, if the connections are discarded, and their action is replaced by the reactions applied to the given body.*

Let's consider the main types of connections that are most often found in mechanical devices.

1. **Smooth surface**. Smooth is an ideal surface that can be ignored by the friction of a given body. Such a surface prevents the body from moving only in the direction of the normal drawn through the point where the body touches the surface. Therefore, the reaction of a smooth surface is directed along the general normal to the surfaces of touching bodies and is applied at the point (K, Fig. 1.5a) of their contact.

   In the case when one of the contact surfaces has an angular point, the reaction is directed along the normal to the smooth contact surface. Examples of such relationships are shown in Fig. 1.5b, c. At the *position in*, the bar rests on a smooth surface at points M and L. At the point M, the reaction F_1 is directed along the normal to the surface of the bar. At the point L, the reactie P reaction of the tip is directed along the normal to the smooth surface of the supporting body at the tip of the tip.

2. **A single terminal**. Figure 1.5d shows a body pivotally connected at O_1 with a weightless single rod, the second end of which is attached to a fixed base by a Ohinge. In this case, the connection for the body is the rod OO_1, which forces the point O_1 to be at a constant distance from the point O. The force of the body's action on the rod should be directed along the straight line OO_1. Consequently, the reaction direction R of a single rod coincides with its axis (Fig. 1.5e). In this case, the reaction force vector R can be directed to the point O_1 or in the opposite direction.

3. **Flexible link (or thread)**. This relationship (Fig. 1.6) is similar to to the single rod discussed above, with the only difference that the reaction T of the flexible thread is directed along the thread to the suspension point A.

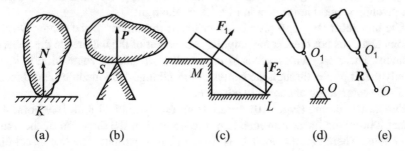

|(a)|(b)|(c)|(d)|(e)|

Fig. 1.5 Type of communication: (**a**) smooth surface; (**b**) point; (**c**) support at two points; (**d, e**) single rod

Fig. 1.6 Flexible connection
(thread)

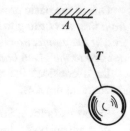

Fig. 1.7 Pivotally fixed
support: (**a**) a flat hinge; (**b**)
planar hinge diagrams

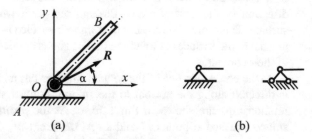

4. **Hinge-fixed support (cylindrical joint).** Devices of this kind are widely used
   in engineering for the mobile connection of two bodies. In Fig. 1.7a, two bodies
   A and B are connected by an axis passing through holes in these bodies. The
   body B attached by the hinge to the support A can rotate around the axis O of
   the hinge in the drawing plane. The O hinge also does not prevent the B body
   from moving in the direction of the hinge axis. Therefore, the reaction R of a
   cylindrical hinge can have any direction in the xOy plane defined by the angle
   α. Schematically, a hinge-fixed support is represented in one of the ways shown
   in Fig. 1.7b on the right.
5. **The articulated support** is a device schematically shown in Fig. 1.8. The lower
   support cage is mounted on rollers that can roll on the base plate A. The link
   only prevents vertical movement of the upper Bclip. Therefore, as in the case of
   a smooth surface, the reaction R of such a support passes through the center of
   the hinge and is directed perpendicular to the rolling plane. Schematic images of
   a pivoting support are shown in Fig. 1.8 on the right.
6. **The ball joint** (Fig. 1.9) prohibits any linear movement of the center O of the
   joint and does not prevent any angular movement of the ball inside the spherical
   cavity. Therefore, with respect to the reaction R, such a connection is known
   only that it passes through the center of the Ohinge. The modulus and direction
   of the vector R are unknown in advance.
7. **The** thrust bearing (Fig. 1.10) differs from the cylindrical joint (see point 4) in
   that it prohibits linear movements of the body B in the direction of the axis of
   the joint. Therefore, the point O is fixed in the pendulum. The line of action of
   the reaction $R$the of the pendulum passes through this point.

Fig. 1.8 Pivotally movable
support

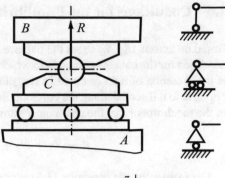

Fig. 1.9 Ball joint

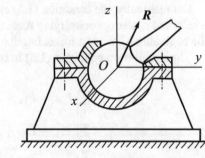

Fig. 1.10 Thrust bearing

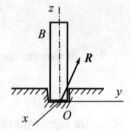

1.5.1   Classification of Constraints

The constraints imposed by connections on the freedom of movement of points in
a mechanical system can be expressed mathematically by equations or inequalities
that connect the coordinates of points, their time derivatives, and time itself. For a
single point, the coupling equation can generally be written in the form

$$f(x,\, y,\, z;\, \dot{x},\, \dot{y},\, \dot{z};\, \ddot{x},\, \ddot{y},\, \ddot{z};\, \ldots ;\, t) \geqslant 0. \tag{1.4}$$

In the case of strict equality, the relation (1.4) is called *holding* or *two-way,*
and in case of inequality—*releasing* or *one-way.* Relationships for which the
expression (1.4) does not explicitly include time are called stationary. If, in addition,
the relationship does not depend on time-derived coordinates, it is called *geometric.*

1.6   Conditions for the Equilibrium of Converging Forces

Based on axiom 1 (p. 8) about the balance of two forces, a necessary and sufficient condition for the balance of a body to which a system of converging forces is applied is the presence of a force equal in magnitude to the resultant R force system and opposite to it directed along the common line of their action. The latter is equivalent to the requirement that the resultant R vanish, that is,

$$R = F_1 + F_2 + \ldots + F_n = 0. \tag{1.5}$$

Geometrically, the condition (1.5) expresses the fact that in the force polygon of a balanced system of converging forces, the end of the last force must coincide with the beginning of the first force. In other words, the power polygon must be closed.

The equilibrium condition (1.5) in coordinate form takes the form

$$
\begin{aligned}
\sum_{i=1}^{n} F_{ix} &= F_{1x} + F_{2x} + \ldots + F_{nx} = 0; \\
\sum_{i=1}^{n} F_{iy} &= F_{1y} + F_{2y} + \ldots + F_{ny} = 0; \\
\sum_{i=1}^{n} F_{iz} &= F_{1z} + F_{2z} + \ldots + F_{nz} = 0.
\end{aligned}
\tag{1.6}
$$

We will also use the abbreviated condition record (1.6) as

$$\sum X_i = 0; \quad \sum Y_i = 0; \quad \sum Z_i = 0, \tag{1.7}$$

where the sums symbolically mean what is expanded written in the formulas (1.6).

Thus, for the equilibrium of the system of converging forces, it is necessary and sufficient that the algebraic sums of the projections of all forces on the coordinate axes are equal to zero.

The equilibrium conditions (1.5) or (1.6) (or (1.7)) allow, on the one hand, to check whether a given system of forces is balanced. On the other hand, if it is known that the body is in equilibrium due to superimposed connections, these conditions allow us to determine the reactions of connections or forces in structural elements.

1.7   A System of Parallel Forces: Power Pairs

Let two parallel forces act on a solid F_1 and F_2. There are two possible cases:

(1) the forces are directed in the same direction (Fig. 1.11a);
(2) the forces are directed in opposite directions (Fig. 1.11b).

Fig. 1.11 System of parallel forces: (**a**) co-directional forces; (**b**) the presence of an opposing force

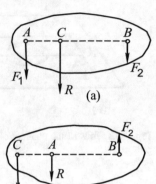

From school physics course known that in the first case, the resultant R forces F_1 and F_2, applied at the points A and B, so is aimed, as the force F_1, F_2. The point C of the application of the resultant divides the segment AB into parts inversely proportional to the modules of forces. Thus, in case 1 (Fig. 1.11a) we have

$$R = F_1 + F_2; \quad \frac{AC}{BC} = \frac{F_2}{F_1}. \tag{1.8}$$

Using the known properties of the proportion, you can get

$$\frac{BC}{F_1} = \frac{AC}{F_2} = \frac{AB}{R}. \tag{1.9}$$

In case 2, when the forces F_1 and F_2 are directed in opposite directions (Fig. 1.11b), the resultant R has the direction of the greater modulo of the forces F_1, F_2. The modulus of the resultant is equal to the difference between the modulus of the greater and lesser of these forces. The point C of the application of the resultant lies on the continuation of the segment AB closer to the larger force and is removed from the points A and B by distances inversely proportional to the force modules. If $F_1 > F_2$, then

$$R = F_1 - F_2; \quad \frac{CA}{CB} = \frac{F_2}{F_1} \text{ or } \frac{CB}{F_1} = \frac{CA}{F_2} = \frac{AB}{R}. \tag{1.10}$$

Let $F_1 = F_2$ be the case in question. Such a system of two equal modulo, parallel and oppositely directed forces, is called *pair of forces* or simply *pair* (Fig. 1.12). We agree to denote the pairs of forces (F, F'), (P, P'), etc.

The plane (Π), in which the forces F and F' that form a pair are located, is called the *plane of action of the pair*. The pair has no resultant, but the forces of the pair are not balanced, since they are not directed in a straight line. The shortest distance between the lines of action of the pair's forces is called the *shoulder of the pair*.

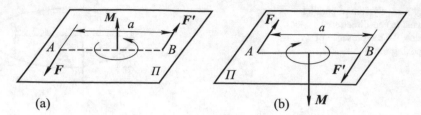

Fig. 1.12 Toward the concept of couple moment: (**a**) a left-handed pair; (**b**) clockwise rotation pair

The pair tends to cause the effect of rotation of the solid to which it is applied. The magnitude of this effect is characterized by the *moment of the pair.* The moment of couple is defined as the vector module equal to the product of one force couple on her shoulder and directed perpendicular to the plane of action of the pair so that the end of this vector to see a couple of forces tending to rotate the plane of its action is opposite to the rotation clockwise (Fig. 1.12).

Thus, for the pair (F, F') shown in Fig. 1.12a having the shoulder a, the moment M is directed perpendicular to the plane Π up, and for the pair shown in Fig. 1.12b, the moment vector m is directed down. In Fig. 1.12, the direction in which the pair tends to rotate the Π plane of its action is marked with arrows.

By definition, the module of the vector M will be

$$M = Fa. \tag{1.11}$$

In the case when only pairs located in the same plane are considered, the moment of a pair of forces is sometimes defined as

$$M = \pm Fa, \tag{1.12}$$

taking the upper sign for the moment shown in Fig. 1.12a, and the lower sign for the moment in the opposite direction (Fig. 1.12b).

Pairs of forces are called *equivalent,* if their moments are numerically equal and have the same sign. From this definition, it follows that a pair of forces, without changing its action on a solid, can be transferred to any place in the plane of its action, turn its shoulder at any angle, or change this shoulder along with the force modules, leaving the magnitude of the moment and the direction of rotation unchanged.

1.8   Moment of Force Relative to the Point and Axis

Consider an arbitrary force F and some point O that we will also call *the center* or *the pole.* Draw the plane Π through the vector F and the center O (Fig. 1.13) and from the point O drop the perpendicular to the line of action of the force F.

Fig. 1.13 To the concept of
the moment of force relative
to a point

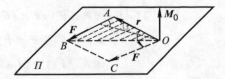

The length a of this perpendicular is called the *arm of the force F relative to the center O.*

The moment of force F relative to the point O is the vector M_0, applied at this point perpendicular to the plane Π containing the force and the point, and directed in the direction from which the rotation performed by the force around the center appears to occur counterclockwise.

The modulus of this vector is equal to the product of the force modulus on the shoulder:

$$M_0 = Fa. \tag{1.13}$$

Denote by A and B, respectively, the beginning and end of the vector F (Fig. 1.13). The vector r drawn from the center of O to the point A is called the radius vector of this point. Considering the triangle OAB, we conclude that the modulus of the vector M_0 is equal to twice the area of the triangle OAB or the area of the parallelogram $OABC$ constructed on the radius vector r and the vector F, moved to the point O. It follows that the moment of force relative to a point can be expressed as a vector product:

$$M_0 = r \times F. \tag{1.14}$$

Let $Oxyz$ be a Cartesian system of rectangular coordinates with orts i, j, k, whose origin coincides with the center of O. Let's denote the coordinates of the point (A) of the force application by x, y, z, and the projection of the force F on the coordinate axis by F_x, F_y, F_z. Applying the rule for calculating the vector product, we can rewrite the previous equality in the form

$$M_0 = \begin{vmatrix} i & j & k \\ x & y & z \\ F_x & F_y & F_z \end{vmatrix} = (yF_z - zF_y)i + (zF_x - xF_z)j + (xF_y - yF_x)k. \tag{1.15}$$

It follows from the expression (1.15) that the projections of the moment of force relative to the point on the coordinate axes are expressed by the formulas

$$M_{0x} = yF_z - zF_y; \quad M_{0y} = zF_x - xF_z; \quad M_{0z} = xF_y - yF_x. \tag{1.16}$$

If F is the resultant of a system of converging forces $(F_1, F_2, \ldots, F_n)$, then using the distributivity property of the vector product, we get

$$M_0(F) = r \times F = r \times (F_1 + F_2 + \ldots \times F_n)$$
$$= r \times F_1 + r \times F_2 + \ldots + r \times F_n$$
$$= M_0(F_1) + M_0(F_2) + \ldots + M_0(F_n). \tag{1.17}$$

The equality (1.17) proves the following statement.

Theorem *The moment of a resultant system of converging forces relative to an arbitrary center is equal to the sum of the moments of the component forces relative to the same center.*

Assume that a force F is applied to a solid at A. Let's draw an arbitrary axis Oz that is not parallel to the force F, as well as a plane Π that is perpendicular to this axis. Projecting the beginning A and end B of the vector F onto the plane Π, we construct the projection F' of the vector F onto the plane Π (Fig. 1.14).

The moment of force F relative to the axis z is the product of the projection module F' of the force F on the plane Π perpendicular to the axis and the shoulder d of this projection relative to the point O of the intersection of the axis with the plane:

$$M_z(F) = \pm F'd. \tag{1.18}$$

The upper sign is taken if, looking from the end of the z axis, the observer sees the projection F' tending to rotate the Π plane counterclockwise. Otherwise, the moment is $M_z(F)$ is considered negative.

Denoting by S the area of the triangle $o'B'$, from the definition (1.18), we get

$$M_z(F) = \pm 2S. \tag{1.19}$$

We prove the following proposition.

Theorem *The projection of the moment of force relative to a point on the axis passing through this point is equal to the moment of force relative to the axis.*

Fig. 1.14 To the theorem on
the moment of the resultant

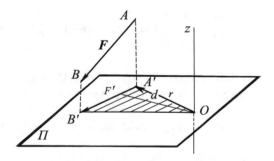

Fig. 1.15 Projection of the
moment of force on the axis

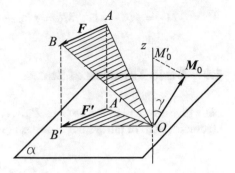

Proof Let's give the force F and an arbitrary axis z passing through a fixed point
O (Fig. 1.15). Draw a plane α perpendicular to the $z$axis through this point. Let's
build a projection of F' of the force F on the plane α and connect the pole O with
the beginning and end of each vector. As a result, we get two triangles OAB and
$OA'B'$. By analogy with the formula (1.19), we can make sure that the modulus of
the moment of force F relative to the center O is equal to twice the area S_{OAB} of
the triangle OAB, and its moment relative to the axis z is expressed as twice the
area $S_{OA'B'}$ of the triangle $OA'B'$, i.e.,

$$M_0 = 2S_{OAB}; \quad M_z = 2S_{OA'B'}. \tag{1.20}$$

But the triangle $o'B'$ is a projection of the triangle OAB on the plane α, and
therefore

$$S_{OA'B'} = S_{OAB} \cdot \cos\gamma, \tag{1.21}$$

where γ is the angle between the planes of the triangles under consideration.

Draw from the center O the vector M_0 of the moment of force F relative to
the point O. Since the vector M_0 is by definition perpendicular to the plane of the
triangle OAB, and the z axis is perpendicular to the plane α by construction, we
conclude that the angle between M_0 and z is equal to the angle γ between the
planes of the triangles OAB and $o'B'$. Therefore, multiplying the equality (1.21) by
two, taking into account the formulas (1.20), we get

$$M_0 \cos\gamma = M_z. \tag{1.22}$$

But the product of the modulus of the vector M_0 by the cosine of the angle γ is
the projection $\overrightarrow{OM_0'}$ of this vector on the z axis. Thus, the theorem is proved. $\square$

The center of O is compatible with the origin of the Cartesian coordinate system
$Oxyz$. Then, based on Theorem 2 and the formulas (1.16), we obtain the following
expressions of the moment of force F relative to the coordinate axes:

$$M_x = yF_z - zF_y; \quad M_y = zF_x - xF_z; \quad M_z = xF_y - yF_x. \qquad (1.23)$$

1.9   Basic Theorem of Statics

Let's give a system of forces $(F_1, F_2, \ldots, F_n)$ that are randomly located in space. Geometric sum of the system of forces: $(F_1, F_2, \ldots, F_n)$

$$F_0 = \sum_{i=1}^{n} F_i$$

It is called its *main vector*. The vector sum of the moments of these forces relative to some pole O (the center of reduction) is called *the main moment* of the system of forces relative to this pole:

$$M_0 = \sum_{i=1}^{n} M_{0i}.$$

Let's consider a solid body with a force F applied at the point A (Fig. 1.16). We apply a system of two forces F' and F'' equivalent to zero at an arbitrary point B of the body, and $F' = F = -F''$. The system of forces (F, F', F'') is equivalent to the force F'и паре (F, F'') with a moment

$$M(F, F'') = \overrightarrow{BA} \times F.$$

But the vector product $\overrightarrow{BA} \times F$ expresses the moment of force F relative to the point B, i.e.,

$$M_B(F) = \overrightarrow{BA} \times F.$$

Thus, the *force applied at any point of a solid body can be transferred parallel to itself to any other point of this body while adding a pair whose moment is equal to the moment of this force relative to the new point of application.*

Fig. 1.16 The lemma of parallel carrying over of force

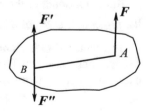

Fig. 1.17 To prove the main
theorem of statics

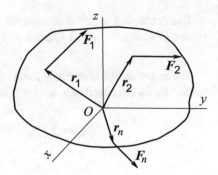

Let's call this proposition *the parallel force transfer lemma*. With this in mind,
we now prove the following.

Theorem *Every system of forces is equivalent to the main vector of the system
applied at some point of the body (the center of reduction) and the main moment of
all forces relative to this point.*

Proof Let an arbitrary spatial system n of forces (F_1, F_2, ..., F_n) acts on a
certain solid. Choose a certain center of the O cast (Fig. 1.17) and match it with
the origin of the rectangular coordinate system $Oxyz$. Radius-vectors of points
of application of forces, denote by r_1, r_2, ..., r_n, and the moments of forces
with respect to point O respectively in M_{01}, ..., M_{0n}, and on the basis of the
formula (1.14):

$$M_{01} = r_1 \times F_1; \ldots; M_{0n} = r_n \times F_n.$$

Let's perform a parallel transfer of the forces F_1, F_2, ..., F_n to the center of
the cast and add these forces as converging. We get one power:

$$F_0 = F_1 + F_2 + \ldots + F_n, \tag{1.24}$$

which by definition is equal to the main vector of the original system of forces F_1,
F_2, ..., F_n.

On the basis of the lemma of parallel carrying over of force during the sequential
transfer of forces F_1, F_2, ..., F_n to the center of the cast, we have to add each
time a match with a moment M_{01}, M_{02}, ..., M_{0n}. Due to the definition of pair
equivalence, the moment of the resulting pair will be

$$M_0 = M_{01} + M_{02} + \ldots + M_{0n} = \sum_{i=1}^{n} M_{0i} = \sum_{i=1}^{n} r_i \times F_i. \tag{1.25}$$

But according to the definition of M_0, there is a main point of the system of forces $(F_1, F_2, \ldots, F_n)$ relative to the center of O, and the theorem is proved.

$\square$

When setting the coordinate forces $F_1, F_2, \ldots, F_n$, the projections of the main vector F_0 on the coordinate axis will be

$$
\begin{aligned}
F_{0x} &= \sum_{i=1}^{n} F_{ix} = F_{1x} + F_{2x} + \ldots + F_{nx}; \\
F_{0y} &= \sum_{i=1}^{n} F_{iy} = F_{1y} + F_{2y} + \ldots + F_{ny}; \\
F_{0z} &= \sum_{i=1}^{n} F_{iz} = F_{1z} + F_{2z} + \ldots + F_{nz},
\end{aligned}
\tag{1.26}
$$

and its modulus and guide cosines are determined by the formulas

$$
F_0 = \sqrt{F_{0x}^2 + F_{0y}^2 + F_{0z}^2};
$$
$$
\cos(\widehat{x, F_0}) = \frac{F_{0x}}{F_0}; \quad \cos(\widehat{y, F_0}) = \frac{F_{0y}}{F_0}; \quad \cos(\widehat{z, F_0}) = \frac{F_{0z}}{F_0}.
\tag{1.27}
$$

For projections of the main moment M_0 by the formulas (1.16), we have

$$
\begin{aligned}
M_{0x} &= \sum_{i=1}^{n} M_{0x}(F_i) = \sum_{i=1}^{n} (y_i F_{iz} - z_i F_{iy}); \\
M_{0y} &= \sum_{i=1}^{n} M_{0y}(F_i) = \sum_{i=1}^{n} (z_i F_{ix} - x_i F_{iz}); \\
M_{0z} &= \sum_{i=1}^{n} M_{0z}(F_i) = \sum_{i=1}^{n} (x_i F_{iy} - y_i F_{ix}).
\end{aligned}
\tag{1.28}
$$

Then the modulus and guide cosines of the principal moment are determined by the formulas

$$
M_0 = \sqrt{M_{0x}^2 + M_{0y}^2 + M_{0z}^2};
$$
$$
\cos(\widehat{x, M_0}) = \frac{M_{0x}}{M_0}; \quad \cos(\widehat{y, M_0}) = \frac{M_{0y}}{M_0}; \quad \cos(\widehat{z, M_0}) = \frac{M_{0z}}{M_0}.
\tag{1.29}
$$

1.10 Equilibrium of an Arbitrary System of Forces

Let an arbitrary (in general, spatial) system of forces be applied to a solid $(F_1, F_2, \ldots, F_n)$. Using the main theorem of statics, we bring it to the center of O. As a result, we get the main vector F_0 and the main moment M_0, which are equivalent to the original system. For the equilibrium of this system, it is necessary and sufficient that the conditions are met simultaneously:

$$F_0 = 0; \quad M_0 = 0. \tag{1.30}$$

The need for conditions (1.30) follows from the fact that if any of them is not met, the source system is reduced either to the resultant (for $M_0 = 0, F_0 \neq 0$) or to a pair (for $M_0 \neq 0, F_0 = 0$). In both cases, the system will be unbalanced. However, the conditions (1.30) are sufficient, since with $F_0 = 0$, the system can only be reduced to a pair of M_0, and since $M_0 = 0$, the system is balanced.

If $Oxyz$ is a Cartesian system of rectangular coordinates with the origin in the center of the reduction O, the vectors of forces and moments are given by their projections:

$$F_0\{x_0, y_0, z_0\}; \quad F_1\{x_1, y_1, z_1\}; \quad \dots; \quad F_n\{x_n, y_n, z_n\};$$
$$M_0\{M_{0x}, M_{0y}, M_{0z}\}; \quad M_{01}\{M_{1x}, M_{1y}, M_{1z}\}; \quad \dots; \quad M_{0n}\{M_{nx}, M_{ny}, M_{nz}\},$$

and then, in projections on the x, y, z axis, the equilibrium equations take the form

$$X_0 = \sum_{i=1}^{n} X_i = X_1 + X_2 + \dots + X_n = 0;$$
$$Y_0 = \sum_{i=1}^{n} Y_i = Y_1 + Y_2 + \dots + Y_n = 0;$$
$$Z_0 = \sum_{i=1}^{n} Z_i = Z_1 + Z_2 + \dots + Z_n = 0;$$
$$M_{0x} = \sum_{i=1}^{n} M_{ix} = M_{1x} + M_{2x} + \dots + M_{nx} = 0;$$
$$M_{0y} = \sum_{i=1}^{n} M_{iy} = M_{1y} + M_{2y} + \dots + Mny = 0; \tag{1.31}$$
$$M_{0z} = \sum_{i=1}^{n} M_{iz} = M_{1z} + M_{2z} + \dots + M_{nz} = 0,$$

where X_0, Y_0, Z_0 are projections of the main vector and M_{0x}, M_{0y}, M_{0z} are projections of the main moment on the x, y, and z axes.

Let us make the following remarks about the conditions (1.31).

Firstly, some of the terms included in these equations may represent projections of previously unknown reactive or active forces or moments. By solving a system of Eqs. (1.31) with respect to these unknowns, we can determine their values, if only the number of unknowns does not exceed six.

Secondly, the equilibrium equations are linear, which does not create problems when solving them. Thirdly, for special cases of systems of forces, the equilibrium equations (1.31) are simplified, since part of them is identically satisfied.

1.11 Solving Problems on the Balance of Systems of Forces

1.11.1 General Comment

Let us consider a number of general provisions applied to solving problems on the equilibrium of a system of forces acting on a single solid body or on a system of bodies.

It is recommended to start solving the problem by freeing the body or system of interacting bodies from the superimposed connections. Based on the principle of release (p. 11), the discarded bonds should be replaced by reactions. The correct replacement of connections by their reactions is one of the main stages of solving the equilibrium problem.

After completing this stage, all active forces must be applied to the body in question. When considering a system of bodies, the forces of interaction between individual bodies do not need to be applied, since these forces are internal to the system as a whole and will not be included in the equilibrium equations of the system of bodies. To determine them, after finding the reactions of bonds imposed on the system of bodies, the given system is divided into separate bodies. At the same time, it is necessary to ensure that the forces of interaction between bodies or groups of articulated bodies at the points of dismemberment are equal in modulus and opposite in direction.

If there are distributed forces, they should be replaced by statically equivalent concentrated forces. This operation is performed according to the following rules. The load with intensity q, evenly distributed over a straight line segment of length a (Fig. 1.18a), is replaced by a force $F = qa$ applied in the middle of the segment a.

With a triangular load distribution (Fig. 1.18b), the statically equivalent force P passes through the center of gravity of the triangle ABC, which is separated from the leg BC at a distance of $b/3$. The modulus of force P is equal to the area of the triangle ABC, i.e., $P = q_0b/2$.

In more complex cases of distributed forces, the equivalent load should be determined using methods of equivalent transformations of force systems defined by

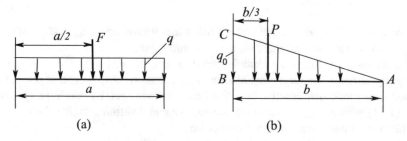

(a)   (b)

Fig. 1.18 Reducing distributed loads to concentrated forces: (**a**) uniform loading; (**b**) triangular load

the main theorem of statics. However, if necessary, the transformation is performed using integration operations.

After identifying all the active and reactive forces, you can start composing the equilibrium equations. Here, first of all, you need to select the coordinate axes and moment points. Their successful choice significantly simplifies the equilibrium equations, reducing the total number of terms included in the equations and the number of unknowns. This can be achieved using the following recommendations.

The coordinate axes should be chosen so that one or two unknown forces are perpendicular to one of the coordinate axes and, therefore, parallel to the other axis. With this choice of axes, only one unknown force will enter the corresponding equilibrium equation of the body. When using moment points, you need to select the intersection point of the two forces you are looking for. This point is usually a cylindrical joint.

We give examples of solving problems on the equilibrium of systems of forces applied to a solid and a system of bodies.

1.11.2 Example 1

The winch is equipped with a ratchet wheel of diameter d_1 (Fig. 1.19a) with a dog A. On a drum with a diameter of d_2, fixed to the wheel, a cable is wound that supports the load Q. Determine the pressure R on the B axis of the dog if given: $Q = 50$ kN, $d_1 = 420$ mm, $d_2 = 240$ mm, $h = 50$ mm, $a = 120$ mm. Ignore the dog's weight.

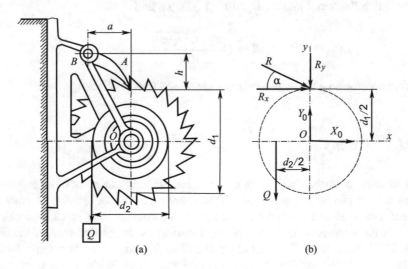

(a)   (b)

Fig. 1.19 For example 1: (a) diagram of the mechanism; (b) force diagram

D e c i s i o n. We dissect the system and consider the balance of the ratchet wheel. The reaction of the dog R (Fig. 1.19b) is equal to the desired pressure on the B axis and is directed in the opposite direction along the straight line BA at an angle α to the horizontal line. Decompose the vector R into two components along the axes x and y. The modules of these components will be

$$R_x = R \cos \alpha; \quad R_y = R \sin \alpha.$$

At the point O, we have a cylindrical hinge. We release the wheel from this connection and apply the reactions X_0 and Y_0. Thus, the ratchet wheel is under the action of a flat system of forces.

Choose a moment point O and write down the condition for the equilibrium of the moments of all forces relative to this point:

$$\sum M_0 = 0 : \quad Q\frac{d_2}{2} - R_x\frac{d_1}{2} = 0,$$

i.e.,

$$Rd_1 \cos \alpha = Qd_2. \tag{1.32}$$

Find the cosine of the angle α. Have

$$\mathrm{tg}\alpha = \frac{h}{a}; \quad \cos\alpha = \frac{1}{\sqrt{\mathrm{tg}^2\alpha + 1}} = \frac{a}{\sqrt{h^2 + a^2}}.$$

Substituting this result into the equality (1.32), we find

$$R = Q\frac{d_2\sqrt{h^2 + a^2}}{ad_1}.$$

After performing the calculation, we get $R = 31$ kN.

1.11.3   *Example 2*

The structure consists of a rigid square and a rod, which at the point C are connected to each other pivotally (Fig. 1.20a). The external links imposed on the structure are a rigid seal at A and a smooth surface at B. The design is affected by a pair of forces with a moment $M = 60$ kN· m, a uniformly distributed load of intensity $q = 20$ kN/m, a force $F_1 = 10$ kN, inclined to the axis of the horizontal section of the square at the angle $\alpha_1 = 60°$, and the force $F_2 = 30$ kN, which is the angle $\alpha_2 = 75°$ with the horizon (Fig. 1.20a).

Determine the bond reactions at points A, B, and C at $a = 0.2$ m.

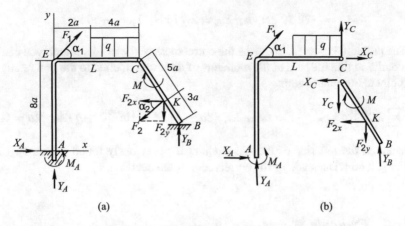

Fig. 1.20 For example 2: (**a**) design diagram; (**b**) dismemberment scheme

D e c i s i o n. Let's free the rod BC from the bonds (Fig. 1.20b) and write down the conditions of its equilibrium:

$$\sum M_C^{(BC)} = 0; \quad -M + Y_B \cdot 8a \cdot \cos \alpha_1 - F_{2x} \cdot 5a \cdot \sin \alpha_1 - F_{2y} \cdot 5a \cdot \cos \alpha_1 = 0;$$

$$\sum Y^{(BC)} = 0; \quad -Y_C - F_{2y} + Y_B = 0;$$

$$\sum X^{(BC)} = 0; \quad -X_C - F_{2x} = 0, \tag{1.33}$$

where

$$F_{2x} = F_2 \cos \alpha_2 = F_2 \cos 75° = 30 \cdot 0,2588 = 7,77 \text{ кН};$$

$$F_{2y} = F_2 \sin \alpha_2 = F_2 \sin 75° = 30 \cdot 0,9659 = 28,98 \text{ кН}.$$

From Eqs. (1.33), we find

$$Y_B = 96.22 \text{ kN}, \quad Y_C = 67.25 \text{ kN}, \quad X_C = -7.77 \text{ kN}.$$

Now we write the equilibrium conditions of the gon AEC:

$$\sum M_A^{(AEC)} = 0; \quad M_A - F_1 \cdot 8a \cdot \cos \alpha_1 - 16qa^2 + Y_C \cdot 6a - X_C \cdot 8a = 0;$$

$$\sum X^{(AEC)} = 0; \quad X_A + F_1 \cos \alpha_1 + X_C = 0;$$

$$\sum Y^{(AEC)} = 0; \quad Y_A + F_1 \sin \alpha_1 - q \cdot 4a + Y_C = 0. \tag{1.34}$$

Solving the system (1.34), we find:

$$M_A = -72.32 \text{ kN} \cdot \text{м}; \ X_A = 2.77 \text{ kN}; \ Y_A = -59.89 \text{ kN}.$$

The problem is solved. To check the correctness of the results found, you can use the condition that the sum of the moments of all forces applied to the gon relative to the point C is equal to zero:

$$\sum M_C^{(AEC)} = 0; \ -Y_A \cdot 6a + X_A \cdot 8a - F_1 \cdot 6a \cdot \sin 60° + q \cdot 4a \cdot 2a = 0.$$

Calculating the left side of the last equality for previously found values Y_A, X_A, and M_A, it convinces you of the correctness of the decision.

1.11.4 Example 3

The *ABDE* tripod, which has the shape of a regular pyramid, is hinged on two cantilever beams (Fig. 1.21). A rope is thrown over the block, fixed at the top of the E of the tripod, which evenly lifts a load of weight P with the help of a winch. From the block to the winch, the cable runs parallel to the console. Determine the embedding reactions of the first console, neglecting its weight and the weight of the tripod. The tripod height is $l/2$ (Fig. 1.21a).

D e c i s i o n. Let us free the node E from the bonds. We obtain the system of converging forces shown in Fig. 1.21b. To determine the forces in the tripod rods, we use the conditions (1.6) of the equilibrium of this system, equating to zero the sum of the projections of all forces on the xyz axis:

$$\sum X^{(E)} = 0, \ \sum Y^{(E)} = 0, \ \sum Z^{(E)} = 0.$$

To write these sums, we define the angles formed by the forces S_{EA}, S_{EB}, S_{ED}, and their projections onto the plane xOy with the axes of the selected coordinate system. Since the triangle ABD is equilateral and BC is its bisector, we get

$$BC = \frac{2}{3} AB \cos \angle ABC = \frac{2}{3} \cdot \frac{l}{2} \cos 30° = \frac{2}{3} \cdot \frac{l}{2} \cdot \frac{\sqrt{3}}{2} = \frac{l}{2\sqrt{3}}.$$

By the condition of the problem, $CE = l/2$. With this in mind, let us denote by α the angles made by the axes of the rods and the vertical force P:

$$\alpha = \angle CEA = \angle CEB = \angle CED.$$

From the triangle BCE, we find:

$$\tan \alpha = \frac{BC}{CE} = \frac{l \cdot 2}{2l\sqrt{3}} = \frac{1}{\sqrt{3}}, \text{ i.e. } \alpha = 30°.$$

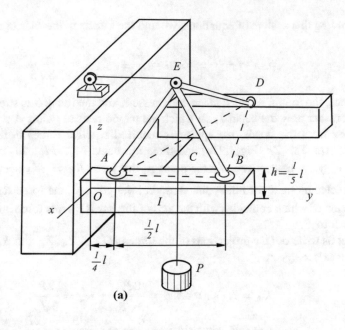

(a)

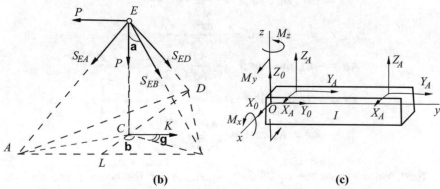

(b)   (c)

Fig. 1.21 Example 3: (**a**) design sketch; (**b**) diagram of forces at node E; (**c**) forces in support bar AB

Further, note that the height DL of the triangle ABD is parallel to the x axis and $\angle BCL = \beta = 60°$. Drawing the line $CK \parallel Oy$ through the point C, we have $\angle BCK = \gamma = 30°$.

Having dealt with the angles, you can start compiling the equilibrium equations for the node E:

(1) $\sum X^{(E)} = 0$: $S_{EA} \sin\alpha \cos\beta + S_{EB} \sin\alpha \cos\beta - S_{ED} \sin\alpha = 0$,

(2) $\sum Y^{(E)} = 0$: $-P - S_{EA} \sin\alpha \sin\beta + S_{EB} \sin\alpha \cos\gamma = 0$,

(3) $\sum Z^{(E)} = 0$: $-P - S_{EA} \cos\alpha - S_{EB} \cos\alpha - S_{ED} \cos\alpha = 0$.

Solving this system of equations, we find the efforts in the rods of the tripod:

$$S_{EA} = -\frac{8P}{3\sqrt{3}}, \quad S_{EB} = \frac{4P}{3\sqrt{3}}, \quad S_{ED} = -\frac{2P}{3\sqrt{3}}.$$

Thus, the rods EA and ED are compressed, and the rod EB is stretched.

Consider now the beam I, on which the tripod rests at points A and B. Freeing the bar from the bonds, we apply six required reactive forces in the seal: three forces X_0, Y_0, Z_0 (Fig. 1.21c) and three moments M_x, M_y, M_z relative to the axes x, y, z, respectively. At the point A, the force $R_A = \frac{8P}{3\sqrt{3}}$ acts on the beam I, directed along EA to the point A, and at the point B, the force $R_B = \frac{4P}{3\sqrt{3}}$, the direction of which coincides with the axis of the bar BE ; this force is directed from point B to E.

Let us write out the projections of the forces $R_A X_A, Y_A, Z_A$ and $R_B X_B, Y_B, Z_B$ on the axis x, y, z:

$$X_A = R_A \sin \alpha \cos \beta = \frac{8P}{3\sqrt{3}} \cdot \frac{1}{2} \cdot \frac{1}{2} = \frac{2P}{3\sqrt{3}};$$

$$Y_A = -R_A \sin \alpha \sin \beta = \frac{8P}{3\sqrt{3}} \cdot \frac{1}{2} \cdot \frac{\sqrt{3}}{2} = -\frac{2P}{3};$$

$$Z_A = -R_A \cos \alpha = -\frac{8P}{3\sqrt{3}} \cdot \frac{\sqrt{3}}{2} = -\frac{4P}{3};$$

$$X_B = -R_B \sin \alpha \cos \beta = -\frac{4P}{3\sqrt{3}} \cdot \frac{1}{2} \cdot \frac{1}{2} = -\frac{P}{3\sqrt{3}};$$

$$Y_B = -R_B \sin \alpha \sin \beta = -\frac{4P}{3\sqrt{3}} \cdot \frac{1}{2} \cdot \frac{\sqrt{3}}{2} = -\frac{1}{3}P;$$

$$Z_B = R_B \cos \alpha = \frac{4P}{3\sqrt{3}} \cdot \frac{\sqrt{3}}{2} = \frac{2}{3}P.$$

Next, we compose the equilibrium equations for the beam I and find the unknown reactive forces:

(1) $\sum X = 0: \ X_0 + X_A + X_B = 0;$

$$X_0 = -X_A - X_B = -\frac{2P}{3\sqrt{3}} + \frac{P}{3\sqrt{3}} = -\frac{P}{3\sqrt{3}};$$

(2) $\sum Y = 0: \ Y_0 + Y_A + Y_B = 0: \ Y_0 = -Y_A - Y_B = \frac{2}{3}P + \frac{1}{3}P = P;$

(3) $\sum Z = 0: \ Z_0 + Z_A + Z_B = 0: \ Z_0 = -Z_A - Z_B = \frac{4P}{3} - \frac{2P}{3} = \frac{2}{3}P;$

(4) $\sum M_x = 0: \ M_x + Z_A \cdot \frac{l}{4} + Z_B \cdot \frac{3l}{4} - Y_A \cdot \frac{l}{10} - Y_B \cdot \frac{l}{10} = 0;$

$$M_x = -\frac{1}{4}lZ_A + \frac{l}{10}(Y_A + Y_B);$$

$$M_x = -\frac{l}{4} \cdot \frac{4P}{3} - \frac{3l}{4} \cdot \frac{2P}{3} + \frac{l}{10}\left(-\frac{2P}{3} - \frac{P}{3}\right) = -\frac{4}{15}Pl;$$

(5) $\sum M_y = 0$: $M_y + X_A \cdot \dfrac{l}{10} + X_B \cdot \dfrac{l}{10} = 0$;

$$M_y = -\frac{l}{10}(X_A + X_B) = -\frac{l}{10}\left(\frac{2P}{3\sqrt{3}} - \frac{P}{3\sqrt{3}}\right) = -\frac{Pl}{30\sqrt{3}};$$

(6) $\sum M_z = 0$: $M_z - X_A \cdot \dfrac{l}{4} - X_B \cdot \dfrac{3l}{4} = 0$;

$$M_z = \frac{1}{4}(X_A + 3X_B) = \frac{l}{4}\left(\frac{2P}{3\sqrt{3}} - \frac{3P}{3\sqrt{3}}\right) = -\frac{Pl}{12\sqrt{3}}.$$

1.12 Truss Calculation

A truss is a geometrically unchanging structure of rectilinear rods connected at the ends by hinges. If the centerlines of all truss members are in the same plane, then the truss is called flat. Otherwise, it is called spatial. The hinge axes where the ends of the rods converge are called truss nodes. In the trusses used for bridges, coverings, and floors of building structures, there are (Fig. 1.22) upper and lower chords, as well as a lattice. The lattice consists of inclined rods, called braces, and vertical elements—struts (the latter may be absent).

The distance between the truss supports is called its span. The truss is divided along the span into panels, usually limited by chord nodes.

If the hinges connecting the truss rods are considered ideal (i.e., without friction), the external loads are applied at the nodes and the self-weight of the rods is neglected, and then all rods experience only tension or compression. Note that in real structures, the truss rods are not hinged but rigidly connected to each other. However, in this case, the hinge-rod design scheme is applicable to them with a sufficient degree of approximation.

Under the calculation of the truss, we here mean the definition of efforts in its rods. To solve this problem, we will use the method already applied in the examples of the previous section. The method consists in the fact that the efforts in the elements of the truss are determined by sequential mental cutting of the nodes

Fig. 1.22 Farm and its elements

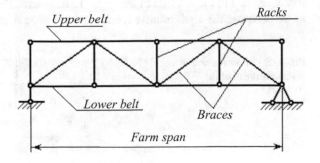

and drawing up the equations of the equilibrium of the forces converging in the considered node of the truss. The sequence of consideration of nodes is usually determined by the condition that the number of unknown forces applied to the node does not exceed the number of equations of its equilibrium (two for a flat and three for a spatial truss). The described calculation algorithm is called the knot cutting method.

When starting the calculation, we do not know in advance which truss rods are stretched and which are compressed. Therefore, we will conditionally assume that all the rods are stretched. This means that the reactions of all the rods are directed away from the nodes. If, as a result of calculations under this assumption, a negative value is obtained for some force, then the corresponding rod is compressed.

The forces in individual truss members can be zero. Such rods are usually called zero. They can be determined without performing a farm calculation. Indeed, the following propositions follow from the equations of equilibrium of knots.

Lemma 1 *If two rods converge at a truss node and there is no external load on the node, then these rods are zero.*

Lemma 2 *If three rods converge at an unloaded node of a truss, two of which are directed along one straight line, and then the third rod is zero.*

Lemma 3 *If an external force is applied to the node of a flat truss, in which two rods converge, directed along the axis of one of the rods, then the force in this rod is equal in magnitude to the applied force, and the second rod is zero.*

In the case when it is required to determine the force in one or more rods, and not in all elements of the truss, it is advisable to use the method of dividing the structure into two parts. With the proper choice of the section, as a rule, three rods are mentally cut. To determine the forces in three dissected rods, the equations of the equilibrium of the forces applied to any of the obtained parts of the truss can be written.

Consider, for example, the farm shown in Fig. 1.23, and suppose that the support reactions R_1 and R_2 have already been found. Let it be required to determine the effort in the belt element 4. Mentally make a cut I-I and consider the balance of forces applied to the left side of the truss. To determine the required force in the rod 4, we compose the equilibrium equation, taking node A as the moment point. From

Fig. 1.23 To the method of dismemberment of the structure

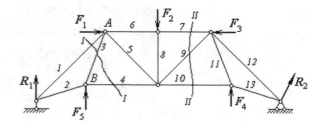

this equation, we will immediately find the force of interest to us in the rod 4, since the other two unknown forces—in the rods 1 and 3—will not enter the equation. Similarly, to determine the force in bar 1, point B should be selected as the moment point. For a known force in bar 4, the force in element 3 can be determined, for example, from the equation of moments of the left side of the truss, taking the axis of the left support hinge as the moment point.

Further, it is necessary to find the effort in the brace 9. Let's draw a mental section II-II and write down the equation of equilibrium of forces applied to the right (or left) part of the truss. It is advisable to take this equation in the form of equality to zero of the sum of the projections of all right or left forces on the vertical axis, since this equation will contain only one unknown unknown effort. The forces in the belt elements 7 and 10 are also easily determined by the method of the moment point.

1.12.1   Calculation of a Triangular Cantilever Truss

Example 1 Determine the forces in the rods of the cantilever truss shown in Fig. 1.24, at $P = 20\,\text{kN}$, $a = 2\,\text{m}$.

S o l u t i o n. The problem can be solved without first defining the support reactions. Let's apply the method of cutting nodes. First, we cut out node A, in which two rods converge. We select the coordinate axes xy and write down the conditions for the equilibrium of forces converging at node A:

$$\sum X = 0; \quad N_2 + N_1 \cos\alpha = 0,$$
$$\sum Y = 0; \quad -P + N_1 \sin\alpha = 0. \tag{1.35}$$

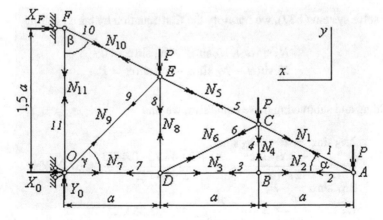

Fig. 1.24 For example 1

Let's define the angles α, β, and γ (Fig. 1.24). We have

$$\tan\alpha = \frac{1.5a}{3a} = 0.5; \quad \alpha = 26°34'; \quad \beta = 90° - \alpha = 63°26';$$
$$\sin\alpha = 0.4472; \quad \cos\alpha = 0.8944;$$
$$\sin\beta = 0.8944; \quad \cos\beta = 0.4472; \quad \gamma = 45°.$$

From the second equation of the system (1.35), we find

$$N_1 = \frac{P}{\sin\alpha} = \frac{20}{0.4472} \approx 44.72 \text{ kN.}$$

For this value of N_1, the first equation of the system (1.35) gives

$$N_2 = -N_1\cos\alpha = -44.82 \cdot 0.8944 \approx -40 \text{ kN.}$$

Next, you can cut out the B node, since three forces converge here, one of which has already been defined. Based on Lemma 2, we conclude that rod 4 is zero, that is, $N_4 = 0$. Then for the equilibrium of the node, it is necessary that

$$N_3 = N_2 = -40 \text{ kN.} \tag{1.36}$$

Next, consider the node C. The forces in bars 1 and 4 are known; therefore, only two forces, N_5 and N_6, need to be determined. We write down the equilibrium conditions of the node:

$$\sum X = 0 : \ N_5\cos\alpha + N_5\cos\alpha - N_1\cos\alpha = 0;$$
$$\sum Y = 0 : \ N_5\sin\alpha - N_6\sin\alpha - N_1\sin\alpha - N_4 - P = 0. \tag{1.37}$$

To solve system (1.37), we multiply the first equation by $\tan\alpha$, getting

$$N_5\sin\alpha + N_6\sin\alpha - N_1\sin\alpha = 0;$$
$$N_5\sin\alpha - N_6\sin\alpha - N_1\sin\alpha = P.$$

Adding and subtracting these equalities, we find

$$2N_5\sin\alpha = P + 2N_1\sin\alpha;$$
$$N_5 = \frac{P + 2N_1\sin\alpha}{2\sin\alpha} = \frac{20 + 2\cdot 44.72 \cdot 0.4472}{2\cdot 0.4472} \approx 67.08 \text{ kN;}$$
$$2N_6\sin\alpha = -P;$$
$$N_6 = -\frac{P}{2\sin\alpha} = -\frac{20}{2\cdot 0.4472} \approx -22.36 \text{ kN.}$$

Now we can consider the equilibrium of node D:

$$\sum X = 0: \ N_7 - N_6 \cos \alpha - N_3 = 0;$$
$$\sum Y = 0: \ N_8 + N_6 \sin \alpha = 0. \tag{1.38}$$

From the system (1.38), we calculate the efforts:

$$N_7 = N_6 \cos \alpha + N_3 = -22.36 \cdot 0.8944 - 40 = -60 \text{ kN};$$
$$N_8 = -N_6 \sin \alpha = 22.36 \cdot 0.4472 = 10 \text{ kN}.$$

Cut out the node E and compose the equilibrium equations for the forces converging here:

$$\sum X = 0: \ N_{10} \cos \alpha + N_9 \cos \gamma - N_5 \cos \alpha = 0;$$
$$\sum Y = 0: \ -P + N_{10} \sin \alpha - N_9 \sin \gamma - N_8 - N_5 \sin \alpha = 0/ \tag{1.39}$$

We express the effort N_{10} from the first equation of the system (1.39):

$$N_{10} = N_5 - N_9 \frac{\cos \gamma}{\cos \alpha} \approx 67.08 - 0.7906 N_9.$$

Substituting this result into the second equation of the system (1.39), we find

$$-20 + (67.08 - 0.7906 N_9) \cdot 0.4472-$$
$$-0.7071 N_9 - 10 - 67.08 \cdot 0.4472 = 0.$$

Next, we find

$$-1.0607 N_9 - 30 = 0; \quad N_9 = -28.28 \text{ kN};$$
$$N_{10} = 67.08 + 28.28 \cdot 0.7906 \approx 89.44 \text{ kN}.$$

Consider the equilibrium of the node F:

$$\sum X = 0: \ -X_y - N_{10} \sin \beta = 0;$$
$$\sum Y = 0: \ -N_{11} - N_{10} \cos \beta = 0. \tag{1.40}$$

From the first equation of the system (1.40), we determine the support reaction X_y:

$$X_y = -N_{10} \sin \beta = -89.44 \cdot 0.8944 \approx -80 \text{ kN}.$$

The second equation of system (1.40) gives

$$N_{11} = -N_{10} \cos \beta \approx -89.44 \cdot 0.4472 \approx -40 \text{ kN}.$$

The forces in all bars are defined. To find the reactions of the support O, consider the equilibrium of forces converging at the node O:

$$\sum X = 0: \ -X_0 - N_7 - N_9 \cos \gamma = 0;$$
$$\sum Y = 0: \ N_{11} + Y_0 + N_9 \sin \gamma = 0. \tag{1.41}$$

Performing calculations, we get

$$X_0 = -N_7 - N_9 \cos \gamma = 60 + 28.28 \cdot 0.7071 \approx 80 \text{ kN};$$
$$Y_0 = -N_{11} - N_9 \sin \gamma = 40 + 28.28 \cdot 0.7071 \approx 60 \text{ kN}.$$

To check the correctness of the results obtained, we compose the equilibrium equations for the truss freed from constraints:

$$\sum X = 0: \ -X_y - X_0 = 0;$$
$$\sum Y = 0; \ Y_0 - 3P = 0;$$
$$\sum M_y = 0: \ X_0 \cdot 1.5a - Pa - P \cdot 2a - P \cdot 3a = 0.$$

It is easy to verify that for the found values of X_y, X_0, Y_0, the written equations are satisfied identically.

1.12.2   Calculation of a Symmetrical Bridge Girder

Example 2 In the bridge girder shown in Fig. 1.25, nodes C, D, and E are loaded with the same load P. The braces make angles of $45°$ with the horizon. Find the forces in bars 1, 2, 3, and 4 caused by the given load.

S o l u t i o n. Due to the symmetry of the truss and the load, we conclude that the support reactions R_A and R_B are equal to each other and amount to half of the entire external load:

$$R_A = R_B = \frac{3P}{2}.$$

Let's draw a through section $I - I$ and consider the equilibrium of the right side of the truss. Taking E as the moment point, we can write

$$\sum M_E = 0: \ N_1 a + R_B \cdot a = 0,$$

i.e.

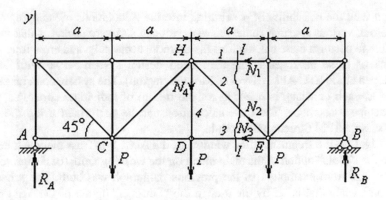

Fig. 1.25 Symmetrical bridge girder

$$N_1 = -R_B = -\frac{3}{2}P.$$

Taking H as the moment point, we similarly get

$$\sum M_B = 0: \ -N_3 a - Pa + R_B \cdot 2a = 0,$$

whence $N_3 = 2P$.

Let us now compose the equilibrium equation for the right side of the farm in the form of the equality to zero of the sum of the projections of all forces on the y axis:

$$\sum Y = 0: \ N_2 \cos 45° - P + R_B = 0.$$

Find

$$N_2 = \frac{P - R_B}{\cos 45°} = \frac{P - \frac{3}{2}P}{\cos 45°} = -\frac{1}{2}P \cdot \frac{2}{\sqrt{2}} = -\frac{P}{\sqrt{2}}.$$

To determine the effort in rack 4, consider the equilibrium of the node D. Composing the equation of the projections of the forces converging at the node D onto the y axis, we obtain $N_4 = P$. The problem has been solved.

1.13 Computer Calculation of Truss

As can be seen from the examples of calculating trusses given in the previous paragraph, the procedure for determining the forces in the rods is, in principle, simple. But with a large number of nodes and rods, it becomes tedious and fraught with the possibility of penetrating into the calculations of annoying errors. Therefore, various research centers have developed software packages that make it possible to perform these calculations faster, more accurately, and error-free.

One of these packages is the automated design system APM WinMachine, developed at the STC APM (Korolev, Moscow region). The system contains several modules, each of which is intended for the design of individual structures, parts, or machine assemblies. The truss calculation can be performed using the APM WinTruss or APM Structure3D module.

In Fig. 1.26 is a fragment of the window of the APM WinTruss module interface. Using the "Tools" option of the main menu on the working field, the design scheme of the farm from example 1 of the previous paragraph was built. The scheme is built on a scale that is set by the user in the "Settings" option of the main menu. In the "Tools" option, loads and supports are selected. For strength calculations, cross-sections and bar materials can also be assigned here.

After the calculation scheme is built, the "Calculation" option is selected and the "Results" are viewed. In Fig. 1.27 is shown the results of the calculation. Comparing the values of the forces in the rods with those found in example 1 (see the previous section), we conclude that the computer results coincide with sufficient accuracy with the data of the "manual" calculation.

In conclusion, note that the APM package is not the only one that allows you to perform these calculations. You can, for example, use the Lira-9 program developed by Ukrainian specialists (NIIASS, Kiev) or other engineering applications of Western production.

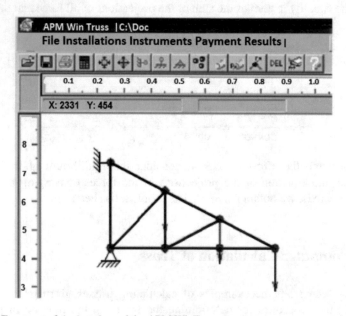

Fig. 1.26 Fragment of the interface of the APM WinTruss program

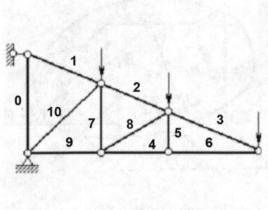

N°	Stress MPa	Load N
0	- 40.00	- 40000.0
1	89.44	89442.7
2	67.08	67082.0
3	44.72	44721.4
4	- 40.00	- 40000.0
5	0.00	0.0
6	- 40.00	- 40000.0
7	10.00	10000.0
8	- 22.36	- 22360.7
9	- 60.00	- 60000.0
10	- 28.28	- 28284.3

Stresses in rods

Fig. 1.27 Results of computer calculations

1.14   Center of Parallel Forces

Consider a system of parallel and equally directed forces:

$$(F_1, F_2, \ldots, F_n),$$

which is not in equilibrium, but is reduced to the resultant R (Fig. 1.28). Obviously, the resultant R is directed in the same way as the forces of the system and is modulo equal to the sum (see the formula (1.8)) modules of applied forces:

$$R = \sum_{i=1}^{n} F_i. \tag{1.42}$$

Rotate each of the forces of the system $(F_1, F_2, \ldots, F_n)$ near the point of its application in the same direction at the same angle. The forces obtained as a result of the specified rotation are denoted by $(F'_1, F'_2, \ldots, F'_n)$. They are shown in Fig. 1.28 with dashed lines.

Since the modules of the initial and rotated forces are respectively equal to $F_1 = F'_1, F_2 = F'_2, \ldots, F_n = F'_n$, then obviously the modules of the resultant R' system of rotated forces and the resultant R equal to each other:

$$R' = R.$$

We show that the line of action of the force R' passes through the point C of the application of the resultant R regardless of the angle of rotation. In fact, first adding up the two forces F_1 and F_2, we find that their resultant R_1 passes through

Fig. 1.28 Center of parallel forces

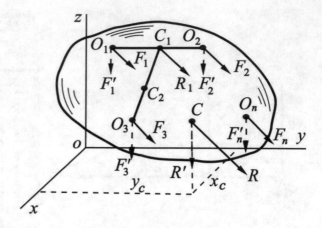

the point C_1 lying on the straight line O_1O_2 connecting the points of application of forces F_1 and f_2. According to the formulas (1.8), the distances of the point C_1 from the points O_1 and O_2 for any identical turns of forces F_1 and F_2 remain unchanged, and the equality $F_2 \cdot C_1 O_2 = F_1 \cdot C_1 O_1$ is satisfied. Adding now the force R_1 with the following force F_3 and continuing this operation until all the forces of the system are summed (F_1, F_2, ..., F_n), we will see that the resultants R and R' pass through the same point C.

The point C through which the line of action of the resultant system of parallel forces passes for any turns equal for all forces near the points of their application is called the *center of parallel forces*.

Determine the position of the center of parallel forces. Choose an arbitrary coordinate system $Oxyz$ (Fig. 1.28) and denote the coordinates of the points of force application in these axes by $O_1(x_1, y_1, z_1)$, ..., ..., $O_n(x_n, y_n, z_n)$. Turning all the forces of the system (F_1, F_2, ..., F_n) so that they become parallel to the Oz axis. Calculate the moment of the resultant R' system of rotated forces (F'_1, F'_2, ..., F'_n) relative to the y axis:

$$M_y((R')) = \sum_{i=1}^{n} M_y((F'_i)). \tag{1.43}$$

But in accordance with Eq. (1.42),

$$M_y((R')) = R' \cdot x_C = R \cdot x_C$$

and besides,

$$M_y(F'_1) = F'_1 x_1; \dots; M_y(F'_1) = F'_n x_n,$$

receive

$$Rx_C = F_1 x_1 + F_2 x_2 + \ldots + F_n x_n.$$

We'll find it from here:

$$x_C = \frac{F_1 x_1 + F_2 x_2 + \ldots + F_n x_n}{F_1 + F_2 + \ldots + F_n} = \frac{\sum F_i x_i}{R}.$$

Turning all the forces of the system $(F_1, F_2, \ldots, F_n)$ parallel to the Ox axis, and then Oy, we get two similar formulas for the coordinates y_C and z_C. Thus, the coordinates of the center of parallel forces are determined by the formulas

$$x_C = \frac{\sum F_i x_i}{R}; \quad y_C = \frac{\sum F_i y_i}{R}; \quad z_C = \frac{\sum F_i z_i}{R}. \tag{1.44}$$

It is easy to verify that the formulas (1.42) and (1.44) remain valid for parallel forces directed in different directions. In this case, they should be considered F_i not modules of forces, but their projections on the axis parallel to the forces and, in addition, should be $R \neq 0$.

1.15   The Center of Gravity of a Rigid Body

All bodies located near the earth's surface are affected by the force of gravity. If the body is divided into separate elementary volumes, then the force of attraction directed to the center of the Earth acts on each small particle. In the case when the size of the body is small compared to the radius of the Earth, the forces of gravity acting on elementary volumes can be considered parallel to each other.

The center of gravity of a body is a point that is the center of parallel gravity forces applied to the elementary volumes of the body. The Center of gravity of an absolutely solid body is a geometric point that is constant relative to the body. It can be located both inside and outside the body.

Consider a body that has a volume of V. We divide this body into elementary parts and denote the weight of each of them by ΔP_i. As a result, we get a system of parallel gravity forces with the center of parallel forces C. The gravity of an elementary part is equal to the product of its mass Δm_i and the acceleration of gravity g, i.e., $\Delta P_i = \Delta m_i g$. If we denote the volume of the elementary part by ΔV_i, and the density of the body by ρ, then

$$\Delta P_i = \rho g \Delta V_i. \tag{1.45}$$

In the case when the body is a material surface S with a surface density ρ, instead of the formula (1.45), we can write

$$\Delta P_i = \rho g \Delta S_i, \tag{1.46}$$

where ΔS is the area of the elementary part of the surface S.

In the case of a material line of length l, we get the same result:

$$\Delta P_i = \rho g \Delta l_i, \tag{1.47}$$

and ρ is the linear density, and Δl_i is the length of the line element.

We can determine the coordinates of the body's center of gravity using the formulas (1.44) for the center of parallel forces ΔP_i. If the body under consideration is homogeneous, then the density of ρ is the same for all elementary parts of the body, and when substituting expressions (1.45)...(1.47) in formulas (1.44), it will be reduced. Then the formulas defining the center of gravity in the limit will be expressed as integrals:

- for volume V:

$$x_C = \frac{1}{V} \iiint\limits_{(V)} x\,dV; \quad y_C = \frac{1}{V} \iiint\limits_{(V)} y\,dV; \quad z_C = \frac{1}{V} \iiint\limits_{(V)} z\,dV; \tag{1.48}$$

- for the S surface:

$$x_C = \frac{1}{S} \iint\limits_{(S)} x\,dS; \quad y_C = \frac{1}{S} \iint\limits_{(S)} y\,dS; \quad z_C = \frac{1}{S} \iint\limits_{(S)} z\,dS; \tag{1.49}$$

- for the material line l:

$$x_C = \frac{1}{l} \int\limits_{(l)} x\,dl; \quad y_C = \frac{1}{l} \int\limits_{(l)} y\,dl; \quad z_C = \frac{1}{l} \int\limits_{(l)} z\,dl. \tag{1.50}$$

Integrals included in formulas (1.48)... (1.50) are called static moments of volume, area, or line relative to coordinate planes. In particular, for a flat figure of area S, the static moments S_x, S_y relative to the axes Ox and Oy will be

$$S_x = \iint\limits_{(S)} y\,dS = y_C S; \quad S_y = \iint\limits_{(S)} x\,dS = x_C S. \tag{1.51}$$

Note 1 If a homogeneous body has a plane, axis, or center of symmetry, then its center of gravity is located in the plane of symmetry, on the axis of symmetry, or in the center of symmetry, respectively.

Note 2 If the body can be divided into a finite number of parts, for each of which the position of the center of gravity is known, then it is advisable to calculate the coordinates of the center of gravity of the entire body directly using the

Fig. 1.29 Plate with a hole

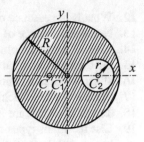

formulas (1.44). In this case, the number of summands in the sums will be equal to the number of parts that the body is divided into.

Note 3 If the body has cavities or notches, the negative weight method can be used to determine the coordinates of its center of gravity.

Let us, for example, determine the center of gravity of a circular plate of radius R with a circular cutout of radius r (Fig. 1.29), the center of which C_2 is removed from the point C_1 by a distance a. From what is said in Remark 1, it follows that the center of gravity of the plate lies on the x axis, which is the axis of symmetry. To find the x_C coordinate, we add the area of the plate to a full circle and apply the method of splitting the body into parts. Let part I be a circle of radius R, $C_1(0; 0)$ be its center of gravity, and $F_1 = \pi R^2$ be the area of the circle. A circle of radius r is taken as part II; its area is taken with the «minus» > sign: $F_2 = -\pi r^2$; and the center of gravity of part II $C_2(a; 0)$. Now, using the first of the formulas (1.44), we find

$$x_C = \frac{F_1 x_1 + F_2 x_2}{F_1 + F_2} = \frac{\pi R^2 \cdot 0 - \pi r^2 \cdot a}{\pi R^2 - \pi r^2} = -\frac{ar^2}{R^2 - r^2}.$$

Self-Test Questions

1. Give a definition of «power».
2. List the characteristics of strength.
3. Name the units of force measurement in the SI, MCGSS, and GHS systems.
4. What is the system of forces?
5. Give examples of concentrated and distributed forces.
6. What is the resultant system of forces?
7. What is the balancing force?
8. Define the external and internal forces.
9. Formulate an axiom about the equilibrium of two forces.
10. What systems of forces are called statically equivalent?
11. Name the simplest system of forces equivalent to zero.

12. What is the essence of the axiom of adding and excluding balancing forces?
13. Name the essence of the axiom of solidification.
14. What does the axiom of inertia express?
15. Give the formulation of the axiom of equality of action and reaction.
16. What is called a connection?
17. What is the feedback response?
18. To which object are the reaction forces applied?
19. List the main types of bonds for which the direction of the reaction force is known in advance.
20. Name the connections for which the point of application of the reaction is known in advance, but not its direction.
21. Give the definition of a system of converging forces.
22. What is the main vector of the system of forces?
23. What is the difference between the main vector and the resultant system of forces?
24. For a system of forces resultant and the main vector of the same?
25. What are the methods for determining the resultant system of converging forces.
26. How are the projections of the resultant system of converging forces expressed through the projections of the forces of this system?
27. What is a force polygon?
28. Write down the equilibrium condition of the system of converging forces in vector form.
29. Formulate the equilibrium conditions of the system of converging forces in coordinate form.
30. What problems allow us to solve the equilibrium conditions of a system of converging forces?
31. Name the necessary and sufficient condition for the equilibrium of a system of converging forces.
32. Which system of forces is called a pair?
33. What is called the pair's shoulder?
34. What is called the moment of force relative to a point? What is the dimension of this quantity?
35. Which plane is called the plane of action of the pair?
36. Which pairs are called equivalent?
37. Write down the vector and scalar dependencies between the elements of the pair.
38. Formulate theorems about the equivalence of pairs.
39. What is the resulting pair?
40. Write down the formula for determining the resulting system of pairs.
41. Name the equilibrium condition of a flat system of pairs.
42. Give a vector record of the equilibrium condition of an arbitrary system of pairs.
43. How to calculate the modulus of the moment of force relative to a point?
44. Formulate a theorem about the moment of the resultant system of converging forces.
45. What is called the moment of force relative to the axis?

46. Write down the formula linking the moment of a force about a point moment of this force about the axis that passes through the same point.
47. Formulate the theorem of moment of resultant of concurrent force system.
48. Is formulated as lemma of parallel carrying over of force?
49. How are the coordinates of the center of parallel forces determined?
50. What is called the center of gravity of the body?
51. How are the properties of symmetry used in determining the centers of gravity of bodies?
52. What is the essence of the method of negative weights?
53. What structure is called a truss?
54. List the main structural elements of the truss.
55. What are the methods of calculating farms known to you?

Control Tasks for the Section "Statics"

T a s k 1. The design consists of two parts. Determine which method of connecting parts of the structure has the smallest reaction modulus specified in Table 1.1, and

Table 1.1 Initial data for options 1–10

Option number according to Fig. 1.23	P_1, kN	P_2, kN	M, kN·m	q, kN/m	The form of a sliding pair	The studied reaction
1	5.0	–	24.0	0.8		X_A
2	5.0	10.0	22.0	1.0		R_A
3	7.0	9.0	20.0	1.2		R_B
4	8.0	–	18.0	1.4		M_A
5	9.0	–	16.0	1.6		R_A
6	10.0	8.0	25.0	1.8		M_A
7	11.0	7.0	20.0	2.0		R_B
8	12.0	6.0	15.0	2.2		M_A
9	13.0	–	10.0	2.4		X_A
10	14.0	–	12.0	2.6		R_A

for this method of connecting, determine the reactions of the supports, as well as the C connection (Tatlor 2005), (Yablonskii et al. 2000).

In Fig. 1.30 is shown the first connection method—using the C hinge. The second method of connection—using a moving pair—is shown in Table 1.1.

Example of Completing a Task

Given: design scheme (Fig. 1.31); $P_1 = 5\ kN$, $P_2 = 7\ kN$, $M = 22\ kN \cdot m$, $q = 2\ kN/m$; $\alpha = 60°$.

Determine the reaction of supports and connections C for the method of joints (hinge or slide pair) at which the modulus of reaction of the Oprah A lesser.

S o l u t i o n 1. *Determination of the reaction of the support A at the hinge joint with the point C.*

Consider a system of balancing forces applied to the entire structure (Fig. 1.32). Let's make an equation of equilibrium of moments of forces relative to point B. To simplify the calculation of the moment of force P_1, we decompose it into horizontal and vertical components: $P_1'' = P_1 \sin 60° = 4.33\ kN$, $P_1' = P_1 \cos 60° = 2.5\ kN$:

$$\sum M_{iB} = 0;\ \ P_1' \cdot 3 + P_1'' \cdot 8 - Q \cdot 1 - Y_A \cdot 5 + X_A \cdot 1 - M+ \\ +p_2\sqrt{1.0^2 + 1.5^2} = 0, \tag{1.52}$$

where $Q = q \cdot 4 = 2 \cdot 4 = 8\ kN$.

After data substitution and calculations, Eq. 16.20 takes the form

$$X_A - 5Y_A = -24.74\ kN. \tag{$*$}$$

The second equation with the unknown X_A, Y_A is obtained by considering the system of balancing forces applied to the part of the structure located to the left of the hinge C (Fig 1.33):

$$\sum M_{iC} = 0;\ \ P_1'' \cdot 6 + Q \cdot 2 + X_A \cdot 4 - Y_A \cdot 3 = 0,$$

шг after calculations

$$4X_A - 3Y_A = -41.98\ kN. \tag{1.53}$$

Solving the system of Eqs. (*) and (16.21), we find

$$X_A = -7.97\ kN,\ \ Y_A = 3,36\ kN.$$

Calculate the reaction modulus of the support a at the hinge joint at the point C:

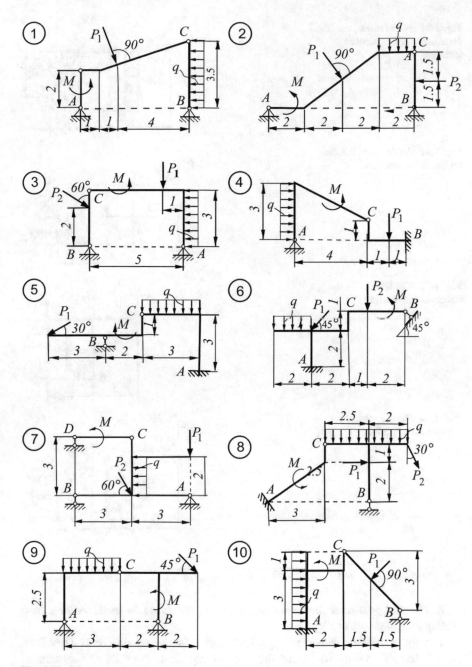

Fig. 1.30 Diagrams of composite structures (for task 1)

Fig. 1.31 For example, the calculation of a composite structure

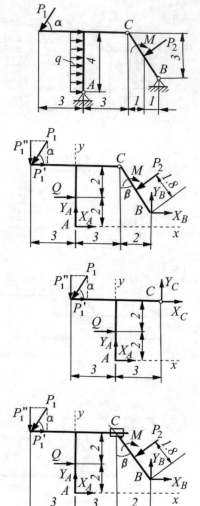

Fig. 1.32 A system of forces on the entire structure

Fig. 1.33 The left part of the structure with sealing

Fig. 1.34 Design scheme for connection by sealing

$$R'_A = \sqrt{X_A^2 + T_A^2} = \sqrt{7.97^2 + 3.36^2} = \sqrt{74.81} = 8.65 \, kN. \quad (**)$$

2. *The design scheme for connecting parts of the structure at the point c by a sliding pair is shown in* Fig. 1.34.

The force systems shown in Figs. 1.32 and 1.34 are no different. Therefore, Eq. (*) remains valid. To obtain the second equation, consider the system of balancing forces applied to the left of the moving pair (Fig. 1.35a).

Let's make an equilibrium equation:

$$\sum X_i = 0; \quad X_A + Q - P'_1 = 0, \qquad (1.54)$$

Fig. 1.35 Sliding seal in the support C: (**a**) the left side of the structure; (**b**) diagram of the right side of the structure

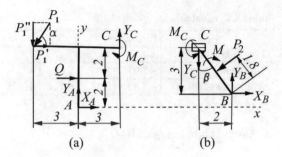

(a)   (b)

where from

$$X_A = 5.5 \, kN$$

and from the equation (*), we find

$$Y_A = 3.85 \, kN.$$

Therefore, the reaction modulus of the support A with the moving pair in the connection C will be

$$R''_A = \sqrt{X_A^2 + Y_A^2} = \sqrt{5.5^2 + 3.85^2} = \sqrt{45.07} = 6.71 \, kN.$$

Comparison of the last result with the value (**) leads to the conclusion that when connecting a sliding pair, the reaction modulus of the support A is less than when connecting a hinge joint. We find the components of the reaction of the support B and the sliding pair.

For the left-hand side of C (Fig. 1.35a)

$$\sum Y_i = 0; \quad -P''_1 + Y_A + Y_C = 0, \tag{1.55}$$

where from

$$Y_C = P''_1 - Y_A = 0.48 \, kN.$$

The components of the reaction of the support B and the moment in the moving pair are found from the equilibrium equations of the right part of the structure (Fig. 1.35b):

$$\sum M_{iB} = 0; \quad M_C + Y_C \cdot 2 - M + P_1 \cdot 1.8 = 0,$$
$$\sum X_i = 0; \quad -P_2 \cos + X_B = 0, \tag{1.56}$$
$$\sum Y_i = 0; \quad -Y_c + Y_B - P_2 \sin = 0.$$

Table 1.2 Initial data for options 1–10

Type of construction	Forces, kN						Moments, kN·m
	X_A	Y_A	R_A	Y_C	X_B	Y_B	M_C
With swivel joint	−7.97	3.36	8.65	–	–	–	–
With a pair of slides	−5.50	3.85	6.71	±0.48	5.82	4.37	±8.44

From a right triangle BCD

$$\sin \beta = BD/BC = 2.0/\sqrt{2^2 + 3^2} = 0.555,$$
$$\cos \beta = CD/BC = 3.0/3.61 = 0.832.$$

Solving Eqs. (1.56) with respect to M_C, X_B, Y_B, we obtain

$$M_C = 8.44 \, kN \cdot m; \quad X_B = 5.82 \, kN; \quad Y_B = 4.37 \, kN.$$

To check the correctness of the determination of reactions, we will make sure that the equation of equilibrium of forces applied to the entire structure, which was not used earlier, is satisfied (see Fig. 1.35(b)):

$$\sum M_{iA} = P_1' \cdot 4 + P_1'' \cdot 3 - Q \cdot 2 - M - P_2 \sin \beta \cdot 4 - P_2 \cos \beta \cdot 2.5 - X_B \cdot 1 + $$
$$+ Y_B \cdot 5 = 2.5 \cdot 5 + 4.33 \cdot 3 - 8 \cdot 2 - 22 - 7 \cdot 0.555 \cdot 4 + 7 \cdot 0.832 \cdot 2.6 - $$
$$-5.82 \cdot 1 + 4.37 \cdot 5 = 59.40 - 59.36 \approx 0.$$

The calculation results are shown in Table 1.2.

T a s k 2. Determine the reactions of the truss supports to a given load, as well as the forces in all of its rods by cutting out nodes. The data required for the calculation are given in the table below. Truss schemes are shown in Fig. 1.36.

Option number	P_1	P_2	P_3	a	h	α
	kN			m		grad
1	4	9	2	2.0	–	30
2	10	3	4	2.5	–	60
3	2	12	6	3.0	–	60
4	10	10	5	4.0	–	60
5	2	4	2	–	2.0	60
6	3	7	5	4.0	3	-
7	4	6	3	4.0	–	60
8	5	7	7	3.2	–	45
9	10	8	2	5.0	–	60
10	3	4	5	4.4	3.3	–

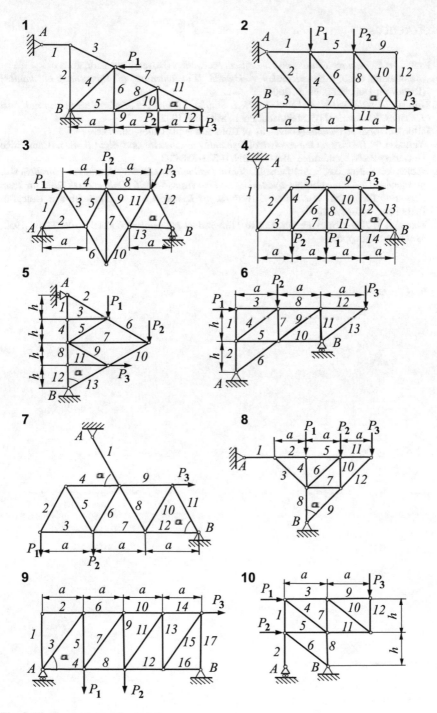

Fig. 1.36 Truss schemes for task 2

References

J. James, *An Elementary Treatise on Theoretical Mechanics* (Dover, New York, 2005)

V. Molotnikov, *Osnovy teoreticheskoi mekhaniki [Fundamentals of Theoretical Mechanics]* (Rostov-on-Don, Fenix Publ., 2004)

E. Nikolai, *Teoreticheskaya mekhanika. CH. 1. Statika. Kinematika [Theoretical Mechanics. Part 1. Statics. Kinematics]* (Gostekhizdat Publ., Moscow, 1952)

J. Tatlor, *Classical Mechanics* (University of Colorado Publ., Colorado, 2005)

E. Whittaker, *A Treatise on the Analytical Dynamics of Particles and Rigid Bodies*. (Cambridge University Press, Cambridge, 1904). ISBN: 0-521-35883-3

A. Yablonskii, S. Noreiko, S. Vol'fson, dr., *Sbornik zadanii dlya kursovykh rabot po teoreticheskoi mekhanike: Uchebnoe posobie dlya tekhnicheskikh vuzov, 5-e izd. [Collection of Tasks for Term Papers on Theoretical Mechanics: A Textbook for Technical Universities]*, 5th edn. (Integral-Press Publ., Moscow, 2000)

N. Zhukovskii, *Teoreticheskaya mekhanika [Theoretical Mechanics]* (Gostekhizdat Publ., 1952, Moscow–Leningrad)

Chapter 2
Kinematics

Abstract Kinematics is a branch of theoretical mechanics that studies the movement of points, bodies, and systems of bodies. The reasons that cause or change the motion of the body are not considered in kinematics. The term comes from the French *cinematique*—movement. To describe motion, kinematics studies the trajectories of points, lines, and other geometric objects in space, as well as some of their properties, such as speed and acceleration. Astrophysics uses kinematics to describe the motion of celestial bodies and systems. Mechanical engineering, robotics, and biomechanics use it to describe the movement of systems consisting of connected parts, such as a motor, robotic arm, human body skeleton, and others.

Keywords Movement · Body count's · Cycloid · Trajectory · Movement methods of specifying · Plan speeds · Acceleration plan · Movement difficult · Addition of velocities · Formula borax · Speed up Corioli

2.1 Ways to Set the Movement

Motion in mechanics is a change in the position of a point or body in space over time. As mentioned above, the reasons for this movement are not considered by Whittaker (1904), Begss (1983), Molotnikov (2004). The section of mechanics that studies the geometric properties of motion without taking into account mass and forces acting on a point or body is called *kinematics*.

When describing the motion of a body (point), we must specify which other body the motion is considered relative to. This object is called the *reference body*. A coordinate system is associated with the reference body, taking it as the *reference system*. Time is considered independent of motion, i.e., it is the same in all reference systems if the same event is selected for the initial moment.

In the process of movement, all points of the body make different movements. For example, when a wheel rolls on a straight rail, the points of its axis move in a straight line, and the points of the rim describe complex curves called cycloids. Therefore, the study of the movement of the body should be preceded by the study of the movement of the point.

When a body moves, every point describes a continuous line, which is called the *trajectory of the point*. If the trajectory of a point is a straight line, then its movement is called *rectilinear*. If the trajectory is a curved line (generally speaking, a spatial one), then the point is said to make *a curved* movement.

The movement of a point relative to the selected reference frame is considered set if the method for determining the position of the point at any time is known. Let's look at ways to set the movement.

Let the point M (Fig. 2.1) move relative to some reference system $Oxyz$. The position of this point at any time t can be determined by specifying a vector r drawn from the origin O to the point M. The vector r is called the *radius vector* of the point M. Over time, the radius vector changes in both magnitude and direction. Therefore, it is a vector function of the scalar argument t, which is symbolically written as

$$r = r(t). \tag{2.1}$$

The equality (2.1) defines *the law of motion of a point in vector form*. This method of setting the point movement is called *vector* method.

The geometric position of the ends of the r vectors when the t argument is changed is called the *hodograph* of this vector. Thus, the hodograph of the radius vector represents the *trajectory* of the moving point.

The position of a point in the reference system $Oxyz$ (Fig. 2.1) can also be determined by its coordinates x, y, and z. If a point moves, i.e., its position changes in time, then the coordinates of the point will be functions of time:

$$x = x(t); \quad y = y(t); \quad z = z(t). \tag{2.2}$$

The dependencies (2.2) are equations of motion of a point in Cartesian rectangular coordinates. At the same time, they are parametric equations of the point trajectory. Specifying motion by equations of the type (2.2) is called *coordinate method for describing the motion of a point*. Of course, you can also set the movement of a point in a polar, spherical, or any other coordinate system.

Fig. 2.1 Radius-vector of the point

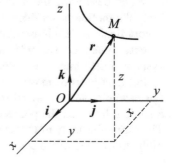

Fig. 2.2 To the natural way
of setting the point movement

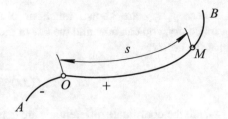

Let the trajectory AB of the point M be known (Fig. 2.2). To set the law of
motion of a point along the trajectory, we select the starting point O corresponding
to the position of the point M at some time $t = t_0$. Distances s in one direction
from the point O along the trajectory AB will be considered positive, and in the
other—negative. Then the position of the point M on the trajectory will be uniquely
determined by the curved coordinate s, which is a function of time:

$$s = s(t). \qquad (2.3)$$

Setting the movement of a point with the described technique is called *natural
way of setting the movement*. In the natural way, setting the motion is equivalent
to setting the trajectory, choosing the starting point, and knowing the law of motion
(2.3) of the point along the trajectory.

The three ways of setting the point movement are interrelated and allow for an
unambiguous transition from one method to another. Let's say, for example, that the
movement is set using the coordinate method (2.2). To go to the vector method, it is
enough to know the law (2.1). Receive

$$r = xi + yj + zk, \qquad (2.4)$$

where i, j, k are the unit vectors of the x, y, z axes. The module of the radius
vector r and its direction are determined by the formulas

$$r = \sqrt{x^2 + y^2 + z^2}; \qquad (2.5)$$

$$\cos(\widehat{x, r}) = \frac{x}{r}; \ \cos(\widehat{y, r}) = \frac{y}{r}; \ \cos(\widehat{z, r}) = \frac{z}{r}. \qquad (2.6)$$

The transition from the coordinate method of setting the movement to the natural
method is also quite simple. Excluding the time t from the dependencies (2.2), we
get the trajectory equation in the form

$$f_1(x, y, z) = 0; \ f_2(x, y, z) = 0, \qquad (2.7)$$

where f_1, f_2 are known functions of the specified arguments. With a known trajectory, you can now find the law of motion (2.3). To do this, write the differential of the arc ds:

$$ds = \pm\sqrt{(dx)^2 + (dy)^2 + (dz)^2},$$

where the coordinate differentials are defined by the formulas

$$dx = \dot{x}dt; \; dy = \dot{y}dt; \; dz = \dot{z}dt, \quad \left(\dot{x} = \frac{dx}{dt}; \; \dot{y} = \frac{dy}{dt}; \; \dot{z} = \frac{dz}{dt}\right). \qquad (2.8)$$

Taking into account the equalities (2.8), the formula for the arc differential can be rewritten as

$$ds = \pm\sqrt{\dot{x}^2 + \dot{y}^2 + \dot{z}^2}\, dt.$$

Integrating this expression in the interval from $t = t_0$ (the starting point) to some point t, we get the law of motion of the point:

$$s = \pm \int_{t_0}^{t} \sqrt{\dot{x}^2 + \dot{y}^2 + \dot{z}^2}\, dt. \qquad (2.9)$$

The upper sign before the integral is selected when the movement occurs in the direction of the positive reference of the arc on the trajectory (2.7). Otherwise, keep the «minus> > sign before the integral.

Example The motion of a point in the plane OHU is given by the equations

$$x = a\cos\omega t; \; y = a\sin\omega t, \qquad (2.10)$$

where a and ω are positive constants. Find the trajectory of the point and the law of its movement along the trajectory.

S o l u t i o n. The trajectory equation in coordinate form is found by excluding time t from the parametric equations (2.10). To do this, we square both equations and add the resulting equalities. Find

$$x^2 + y^2 = a^2.$$

Therefore, the trajectory is a circle of radius a with the center at the origin (Fig. 2.3).

Then, calculating the derivatives of x and y by t, we have

$$\dot{x} = -a\omega\sin\omega t; \; \dot{y} = a\omega\cos\omega t.$$

Fig. 2.3 For example, traffic
tasks

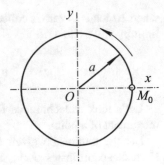

Substituting these expressions into the formula (2.9), we have

$$s = \pm a\omega \int_{t_0}^{t} \sqrt{\sin^2 \omega t + \cos^2 \omega t}\, dt; \quad s = \pm a\omega \cdot (t - t_0). \tag{2.11}$$

Let's assume $t_0 = 0$ for certainty. Then according to Eqs. (2.10), at the initial moment $x = a$, $y = 0$. This means that the origin of M_0 (Fig. 2.3) lies on the Ox axis. Assuming that the point rotates around the circle in the direction opposite to the clockwise rotation, we must keep the upper sign in the formulas (2.11), since only under this condition will the coordinate y increase and x decrease in accordance with Eqs. (2.10). Thus, the law of motion of a point has the form

$$s = a\omega t.$$

2.2   Speed and Acceleration of the Point

Assume that at time t the position of a point is determined by the radius vector $r(t)$, and at time $t + \Delta t$, the radius vector $r(t + \Delta t)$. Vector

$$\Delta r = r(t + \Delta t) - r(t)$$

is *moving* point during time Δt. Vector quantity

$$v_{\text{cp}} = \frac{\Delta r}{\Delta t} \tag{2.12}$$

defines the average speed of a point over time Δt. As can be seen from the definition (2.12), the direction of the average velocity coincides with the direction of the vector of movement of the point for the considered time Δt. The limit of the ratio (2.12) when Δt tends to zero is called *the speed of the point* at a given time. Denoting this limit by v, we can write

$$v = \lim_{\Delta t \to 0} \frac{\Delta r}{\Delta t} = \frac{dr}{dt} = \dot{r}. \tag{2.13}$$

Let's look at techniques for calculating the speed of various ways to set the movement of a point.

Let the motion be given in the coordinate way in the system of rectangular Cartesian coordinates $Oxyz$, i.e., the functions are given:

$$x = x(t); \quad y = y(t); \quad z = z(t).$$

According to the expression (2.4),

$$r = xi + yj + zk.$$

Differentiating the vector r by time, we get

$$v = \frac{dr}{dt} = \frac{dx}{dt}i + \frac{dy}{dt}j + \frac{dz}{dt}k. \tag{2.14}$$

Therefore, the projections v_x, v_y, v_z, and v on the coordinate axis will be

$$v_x = \frac{dx}{dt} = \dot{x}; \quad v_y = \frac{dy}{dt} = \dot{y}; \quad v_z = \frac{dz}{dt} = \dot{z}. \tag{2.15}$$

The modulus and direction of the velocity vector are determined by the formulas

$$v = \sqrt{v_x^2 + v_y^2 + v_z^2} = \sqrt{\dot{x}^2 + \dot{y}^2 + \dot{z}^2}; \tag{2.16}$$

$$\cos(\widehat{x, v}) = \frac{v_x}{v} = \frac{\dot{x}}{\sqrt{\dot{x}^2 + \dot{y}^2 + \dot{z}^2}}; \quad \cos(\widehat{y, v}) = \frac{v_y}{v} = \frac{\dot{y}}{\sqrt{\dot{x}^2 + \dot{y}^2 + \dot{z}^2}};$$

$$\cos(\widehat{z, v}) = \frac{v_z}{v} = \frac{\dot{z}}{\sqrt{\dot{x}^2 + \dot{y}^2 + \dot{z}^2}}. \tag{2.17}$$

In the case of polar coordinates (r, φ), the functions $r = r(t)$ and $\varphi = \varphi(t)$ are set. To calculate the speed, we use the relation between rectangular Cartesian and polar coordinates:

$$x = r \cos \varphi; \quad y = r \sin \varphi.$$

Fig. 2.4 To derive formulas
for the speed and acceleration
of a point

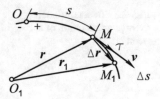

We differentiate these functions by time:

$$\dot{x} = \dot{r} \cos \varphi - r\dot{\varphi} \sin \varphi; \quad \dot{y} = \dot{r} \sin \varphi + r\dot{\varphi} \cos \varphi.$$

Substituting the obtained expressions into the formula (2.16), we find

$$v = \sqrt{\dot{x}^2 + \dot{y}^2} = \sqrt{\dot{r}^2 + r^2\dot{\varphi}^2}. \tag{2.18}$$

When specifying the motion of a point in a natural way, its trajectory and the law of motion along the trajectory are known in the form $s = s(t)$. Calculate the speed of the point. Let's fix some fixed point O_1 that does not belong to the trajectory (Fig. 2.4). The position of a point M at time t is determined, on the one hand, by the arc coordinate s, and, on the other, by the radius vector $\boldsymbol{r}$. Using the speed definition (2.13), we have

$$v = \frac{d\boldsymbol{r}}{dt} = \frac{d\boldsymbol{r}}{ds} \cdot \frac{ds}{dt} = \frac{d\boldsymbol{r}}{ds} \cdot \dot{s}.$$

Let's introduce the designation:

$$\boldsymbol{\tau} = \frac{d\boldsymbol{r}}{ds}.$$

The vector $\boldsymbol{\tau}$ is directed tangentially to the trajectory in the direction of the positive reference of the arc s. The modulus of this vector is equal to one as the limit of the ratio of the chord length Δr to the length of the arc it contracts Δs (Fig. 2.4). Thus,

$$\boldsymbol{v} = \boldsymbol{\tau}\dot{s}. \tag{2.19}$$

Since $\boldsymbol{\tau}$ is a unit vector, the *numerical value of the point velocity at the moment is equal to the time derivative of the curvilinear coordinate s*:

$$v = \dot{s}. \tag{2.20}$$

The value $\dot{s}$ is sometimes called the *algebraic speed* of a point.

Fig. 2.5 To determine the
average acceleration of a
point

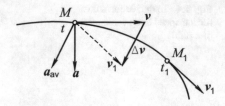

Let a moving point M have a speed v at a given time t. At the moment $t_1 = t + \Delta t$, the point will move to the position M_1, having the speed v_1 (Fig. 2.5). Move the vector v_1 parallel to itself to the point M. Then the speed increment over time Δt is graphically represented by the vector Δv.

Value

$$a_{av} = \frac{\Delta v}{\Delta t} \tag{2.21}$$

called *the average acceleration* of a point over time Δt.

Acceleration a points at time t are called the limit, that the average acceleration tends to when Δt tends to zero:

$$a = \lim_{\Delta t \to 0} a_{av} = \lim_{\Delta t \to 0} \frac{\Delta v}{\Delta t} = \frac{dv}{dt}. \tag{2.22}$$

Using the speed definition (2.13), you can also represent

$$a = \frac{d}{dt}\left(\frac{dr}{dt}\right) = \frac{d^2 r}{dt^2}. \tag{2.23}$$

Note the following. If the trajectory of a point is not a flat curve, then the vector a_{av} lies in a plane that passes through the tangent to the trajectory at M and a straight line parallel to the tangent at M_1. The limit position of this plane when M_1 tends to the point M is called *touching plane.* Therefore, the acceleration vector a lies in the contiguous plane and is directed toward the concavity of the trajectory. For a flat trajectory, the acceleration vector lies in the plane of the curve and is also directed toward its concavity.

Let the movement of a point be set in the coordinate way. Let's determine the modulus and direction of acceleration of a point using the known equations of its motion:

$$x = x(t); \quad y = y(t); \quad z = z(t).$$

Because

$$r = xi + yj + zk,$$

applying the formula (2.23) gives

$$a = \frac{d^2 r}{dt^2} = \frac{d^2 x}{dt^2} i + \frac{d^2 y}{dt^2} j + \frac{d^2 z}{dt^2} k. \tag{2.24}$$

Therefore, the components of the vector a along the axes x, y, and z will be

$$a_x = \frac{d^2 x}{dt^2} = \dot{v}_x = \ddot{x}; \ a_y = \frac{d^2 y}{dt^2} = \dot{v}_y = \ddot{y}; \ a_z = \frac{d^2 z}{dt^2} = \dot{v}_z = \ddot{z}. \tag{2.25}$$

For known acceleration projections on coordinate axes, the modulus and direction of the acceleration vector of a point can be determined by the following formulas:

$$a = \sqrt{a_x^2 + a_y^2 + a_z^2}; \tag{2.26}$$

$$\cos(\widehat{x, a}) = \frac{a_x}{a} = \frac{\dot{v}_x}{\sqrt{\dot{v}_x^2 + \dot{v}_y^2 + \dot{v}_z^2}}; \ \cos(\widehat{y, a}) = \frac{a_y}{a} = \frac{\dot{v}_y}{\sqrt{\dot{v}_x^2 + \dot{v}_y^2 + \dot{v}_z^2}};$$

$$\cos(\widehat{z, a}) = \frac{a_z}{a} = \frac{\dot{v}_z}{\sqrt{\dot{v}_x^2 + \dot{v}_y^2 + \dot{v}_z^2}}.$$

$$\tag{2.27}$$

In the natural way of specifying motion, the position of a point on the trajectory is characterized by the law $s = s(t)$. Determine the acceleration of the point. To do this, we represent the velocity vector using the formula (2.19):

$$v = \tau \frac{ds}{dt}.$$

In accordance with the definition (2.22), we differentiate this relation in time t using the rule for calculating the derivative of the product of two functions. Receive

$$a = \frac{dv}{dt} = \frac{d\tau}{dy} \cdot \frac{ds}{dt} + \tau \cdot \frac{d^2 s}{dt^2} = \frac{d\tau}{ds} \cdot \frac{ds}{dt} \cdot \frac{ds}{dt} + \tau \cdot \frac{d^2 s}{dt^2}. \tag{2.28}$$

But according to the formula (2.20), $ds/dt = \dot{s}$ is the algebraic speed of a point, and its square is in any case a positive value:

$$\frac{ds}{dt} \cdot \frac{ds}{dt} = v^2.$$

The derivative of the ORT τ in the arc coordinate s is (Rashevskii 1950) n/ρ, where n is the unit vector of the main normal to the trajectory at this point and ρ is the radius of curvature of the trajectory at this point. Therefore, the first term in the expression (2.28) represents the component of the acceleration vector directed along the main normal to the center of curvature. Denoting it by a_n, we can write

$$a_n = \frac{v^2}{\rho} \cdot n. \tag{2.29}$$

This component is called *normal acceleration of the point*.

The second component of the acceleration vector (2.28) is denoted by a_τ. It represents the projection of the acceleration vector of a point onto a tangent and is called the *tangent acceleration of a point*. With the notation (2.20), we have

$$a_\tau = \tau \cdot \frac{dv}{dt} = \tau \cdot \frac{d^2 s}{dt^2}. \tag{2.30}$$

Thus, the acceleration of the point in the natural way of setting the movement will be

$$a = a_n + a_\tau, \tag{2.31}$$

where the components a_n and a_τ are defined by the formulas (2.29) and (2.30).

The modulus of the acceleration vector is

$$a = \sqrt{a_n^2 + a_\tau^2} = \sqrt{\left(\frac{v^2}{\rho}\right) + \left(\frac{dv}{dt}\right)^2}, \tag{2.32}$$

and its direction is determined by the formula

$$\cos(\widehat{\tau, a}) = \frac{a_\tau}{a}. \tag{2.33}$$

Note also that taking into account the formulas (2.13) and (2.23) for the curvature (K) of the trajectory, we can obtain

$$K^2 = \frac{1}{\rho^2} = \frac{v^2 a^2 - (va)^2}{v^6}; \quad K = \frac{1}{\rho} = \frac{|v \times a|}{|v|^3}. \tag{2.34}$$

Example 1 (Molotnikov 2012) Crank OA of the mechanism shown in Fig. 2.6 rotates uniformly around the point O, so that the angle φ formed by the crank guide axis changes proportionally to time:

$$\varphi = \omega t, \ (\omega - const).$$

Make an equation of motion of the slider B, and determine its speed and acceleration, if it is known that the length of the crank OA is r, and the length of the connecting rod $AB = l$.

Fig. 2.6 Crank and slide
mechanism

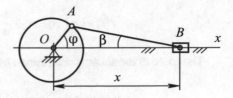

S o l u t i o n. Let's take the slider guide as the x axis, defining the reference
point at O. The equation of motion of the slider B will be known if a function $x(t)$
is found that characterizes the position of the point B at any time t. Have

$$x = r \cos \varphi + l \cos \beta,$$

where β denotes the angle formed by the connecting rod and guide axis (Fig. 2.6).
Let's express this angle in terms of φ.

From the triangle OAB by the sine theorem, we have

$$\frac{OA}{AB} = \frac{\sin \beta}{\sin \varphi},$$

i.e., $\sin \beta = \dfrac{r}{l} \sin \varphi$. Let's denote $r/l = \lambda$. Then

$\sin \beta = \lambda \sin \varphi, \ \cos \beta = \sqrt{1 - \lambda^2 \sin^2 \varphi}$.

Decompose $\cos \beta$ in a series of powers of $\sin \varphi$. Receive

$$\cos \beta = 1 - \frac{1}{2} \lambda^2 \sin^2 \varphi + \frac{\frac{1}{2} \cdot \left(-\frac{1}{2}\right)}{1 \cdot 2} \lambda^4 \sin^4 \varphi + \ldots$$

If $r \ll l$, then the third and subsequent terms of this series can be ignored. In
fact, for example, when $\lambda = \dfrac{1}{5}$ the multiplier is

$$\frac{\frac{1}{2} \cdot \left(-\frac{1}{2}\right)}{1 \cdot 2} \lambda^4 = -\frac{1}{5000}.$$

Thus, approximately we can write

$$\cos \beta = 1 - \frac{1}{2} \lambda^2 \sin^2 \varphi = 1 - \frac{1}{2} \lambda^2 \frac{1 - \cos 2\varphi}{2} = 1 - \frac{\lambda^2}{4} \cos 2\varphi.$$

Substituting this expression into the function x and taking into account that $\varphi = \omega t$, we finally get the following equation of motion of the slider B:

$$x(t) = r \left(\cos \omega t + \frac{\lambda}{4} \cos 2\omega t \right) + l - \lambda r / 4.$$

The speed of the slider is determined by differentiating the equation of its motion:

$$v = v_x = \frac{dx}{dt} = -r\omega \left(\sin \omega t + \frac{\lambda}{2} \sin 2\omega t \right).$$

The resulting formula allows you to calculate the speed of the slider at any time. For example, if $\varphi = \pi/2 \ \omega t = \pi/2, t = \pi/2\omega$, we have

$$v_x \left(\frac{\pi}{2\omega} \right) = -r\omega \left(\sin \frac{\pi}{2} + \frac{\lambda}{2} \sin \pi \right) = -r\omega < 0.$$

Therefore, the velocity vector at the moment under consideration is directed opposite to the x axis.

The acceleration of the slider B is obtained by differentiating the projection of its speed on the x axis:

$$a_x = \frac{dv_x}{dt} = -r\omega^2 (\cos \omega t + \lambda \cos 2\omega t).$$

For example, at $t = \pi/2\omega$, $(\varphi = \pi/2)$, we calculate

$$a_x(\pi/2\omega) = -r\omega^2 \left(\cos \frac{\pi}{2} + \lambda \cos \pi \right) = -r\omega^2 (-\lambda) = \lambda r \omega^2 > 0;$$

in other words, the slider acceleration coincides with the direction of the x axis.

Example 2 (Molotnikov 2017) The equations of motion of a point have the form

$$x = r \cos t^2; \quad y = r \sin t^2; \quad z = bt.$$

Determine the tangent and normal acceleration of the point, as well as the radius of curvature of the trajectory at any time.

D e c i s i o n. Find out the trajectory of the point. Excluding time from the first two equations of motion, we get

$$x^2 + y^2 = r^2.$$

Therefore, the projection of the trajectory on the plane xOy is a circle of radius r. The z coordinate changes according to the equation $z = bt^2$. From this we conclude that the point moves along the surface of a circular cylinder with a generatrix parallel to the z axis (Fig. 2.7). Let's denote $\varphi = t^2$. Then the equations of motion take the form

Fig. 2.7 Moving a point
along a spiral line

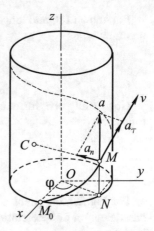

$$x = r\cos\varphi; \quad y = r\sin\varphi; \quad z = b\varphi.$$

Since the coordinate z is proportional to the angle φ (Fig. 2.7) of the rotation of the radius drawn to the point N, the trajectory of the point is a helix.

Determine the velocity projections on the coordinate axis:

$$v_x = \dot{x} = -r \cdot 2t \cdot \sin t^2;$$
$$v_y = \dot{y} = r \cdot 2t \cdot \cos t^2;$$
$$v_z = \dot{z} = 2bt.$$

By differentiating the last formulas, we find the acceleration projections of the point M:

$$a_x = -4rt^2 \cos t^2 - 2r \sin t^2;$$
$$a_y = -4rt^2 \sin t^2 + 2r \cos t^2;$$
$$a_z = 2b.$$

Now we can calculate the velocity and acceleration modules of the point:

$$v = \sqrt{v_x^2 + v_y^2 + v_z^2} = 2t\sqrt{r^2 + b^2};$$
$$a = \sqrt{a_x^2 + a_y^2 + a_z^2} = 2\sqrt{4r^2t^4 + r^2 + b^2}.$$

Calculate the normal and tangent acceleration of the point. The tangent acceleration is determined by the formula (2.30):

$$a_\tau = \frac{dv}{dt} = 2\sqrt{r^2 + b^2}.$$

For known full and tangent accelerations, we find the normal acceleration of the point M:

$$a_n = \sqrt{a^2 - a_\tau^2} = 4rt^2.$$

Now, using the formula (2.29), we calculate the radius of curvature of the trajectory:

$$\rho = \frac{v^2}{a_n} = \frac{1}{r}(r^2 + b^2).$$

For Fig. 2.7 the point C represents the center of curvature; at the point M, a parallelogram of accelerations is constructed and the direction of total acceleration is shown.

2.3   The Simplest Movements of a Solid Body

Every solid body can be considered as a set of points enclosed in its volume. In the case of an absolutely solid body, the points have the property of immutability of the distance between them. We show that the position of such a body in space in the general case is completely determined by setting six independent parameters. To do this, consider three arbitrary points of the body that do not lie on the same straight line. The coordinates of these points are denoted by (x_1, y_1, z_1), (x_2, y_2, z_2) and (x_3, y_3, z_3). Due to the immutability of the distances r_{ij} between ith and jth points, their coordinates must satisfy the equations:

$$\begin{aligned}
(x_1 - x_2)^2 + (y_1 - y_2)^2 + (z_1 - z_2)^2 &= r_{12}^2; \\
(x_2 - x_3)^2 + (y_2 - y_3)^2 + (z_2 - z_3)^2 &= r_{23}^2; \\
(x_3 - x_1)^2 + (y_3 - y_1)^2 + (z_3 - z_1)^2 &= r_{31}^2.
\end{aligned} \qquad (2.35)$$

Therefore, of the nine coordinates of the points under consideration, only six are independent. The randomness of the choice of these three points allows us to conclude that the position of a free solid body relative to any fixed frame of reference is completely determined by six independent parameters. If the body is anchored at any point, its position will be determined by only three independent parameters.

In general, the number of independent parameters that uniquely determine the position of a rigid body in space is called the *number of degrees of freedom of the body.* The motion of a solid body will be set if the law of change in time of the parameters that determine its position relative to the selected reference system is known.

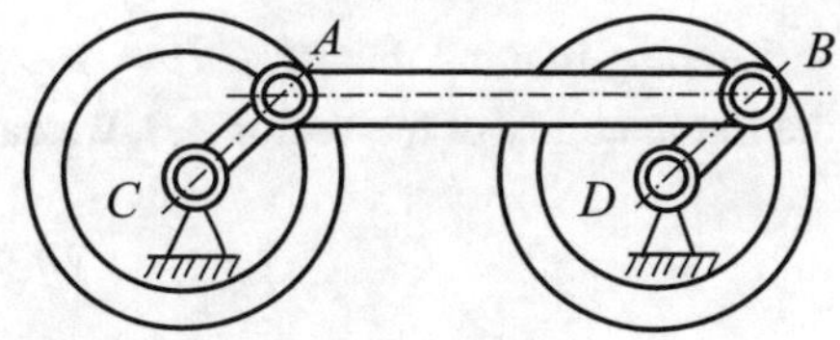

Fig. 2.8 Forward movement
of the partner

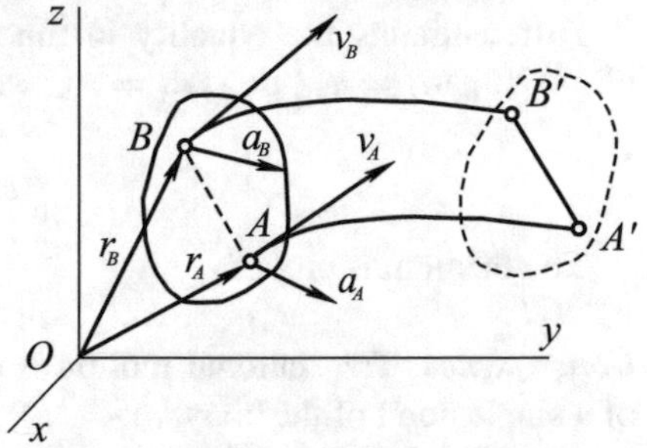

Fig. 2.9 To study the
translational motion of a solid
body

2.3.1 Forward Motion of a Solid Body

Let in the process of motion of a solid body any straight line connecting two points
of the body remains parallel to itself. This movement is called *translational*. The
pistons of an internal combustion engine or compressor, for example, move forward
relative to the cylinders. The trajectories of the piston points are straight lines.
Another example is the translational motion of the AB partner (Fig. 2.8) when the
cranks CA and DB $(CA = DB)$ rotate. The trajectories of the points of the partner
are circles.

In forward motion, the following occurs.

Theorem *For all points of a translationally moving body, the trajectory, velocity,*
and acceleration are the same.

Proof Select two points A and B that belong to the body. The radius vectors of
these points satisfy the condition (Fig. 2.9):

$$r_B = r_A + \overrightarrow{AB}. \tag{2.36}$$

The vector $\overrightarrow{AB}$ is constant modulo, and since the motion is translational, this
vector does not change in direction either.

From Eq. (2.36), it follows that the hodograph of the radius vector of the point
B, which is the trajectory of this point, is shifted relative to the hodograph of the
point A by a constant vector $\overrightarrow{AB}$. By performing this shift, you can combine the
trajectories of points A and B. Therefore, these trajectories are the same:

$$\frac{dr_B}{dt} = \frac{dr_A}{dt} + \frac{d}{dt}(\overrightarrow{AB}).$$

Keeping in mind that $d\mathbf{r}_B/dt = \mathbf{v}_B$, $d\mathbf{r}_A/dt = \mathbf{v}_A$, and $d(\overrightarrow{AB})/dt = 0$ due to the immutability of the vector $\overrightarrow{AB}$, we can write

$$\mathbf{v}_B = \mathbf{v}_A. \tag{2.37}$$

Thus, the velocities of the points of the body in translational motion are the same. Differentiating the equality in time (2.37) and considering that by definition (2.22) $d\mathbf{v}_B/dt = \mathbf{a}_B$, $d\mathbf{v}_A/dt = \mathbf{a}_A$, we get

$$\mathbf{a}_B = \mathbf{a}_A. \tag{2.38}$$

The theorem is proved.   □

Consequence Translational motion is completely characterized by the movement of a single point of the body.

To set the translational motion of the body, it is enough to know the coordinates (x, y, z) any of its points as a function of time, i.e.,

$$x = x(t); \quad y = y(t); \quad z = z(t). \tag{2.39}$$

In general, the motion of a single point of a solid body during translational motion has no restrictions. Therefore, a translationally moving body has three degrees of freedom and equality (2.39) is *equations of translational motion of a solid.*

2.3.2   Rotation of a Solid Body Around a Fixed Axis

Consider the motion of a rigid body with two fixed points A and B (Fig. 2.10). From the condition of immutability of distances between any points of the body, it follows that all points line AB remains stationary. The straight line AB is called the axis of rotation, and the motion of the body under these conditions is called *rotational.* It is easy to see that any point C (Fig. 2.10) of the body describes a circle centered on the axis of rotation.

We fix a certain moment t_0 and draw two half-planes I and II through the axis of rotation AB, which coincide with each other at $t = t_0$.

Next, we fix the half-plane I, and connect the half-plane II to the body, so that it rotates with it. Then the position of the body at any time $t \geqslant t_0$ will be uniquely determined by the angle φ taken with a certain sign (Fig. 2.10), which is called *the angle of rotation of the body.*

Function

$$\varphi = \varphi(t) \tag{2.40}$$

is called the *equation of rotation of a solid body around a fixed axis.*

Fig. 2.10 Rotational motion
of a solid body

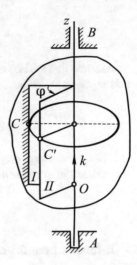

The Oz axis is compatible with the axis of rotation and we will choose the starting point O and the positive direction on it at our discretion. The unit vector of the Oz axis is denoted by k. The angle φ will be considered positive if it is set against the direction of clockwise rotation for an observer looking from the positive direction of the Oz axis, and negative otherwise.

If in a short period of time $\Delta t = t_1 - t$, the body rotates by an angle $\Delta\varphi = \varphi_1 - \varphi$, and then the value ω_{cp}, defined by the formula

$$\omega_{cp} = \frac{\Delta\varphi}{\Delta t}; \tag{2.41}$$

it is called the average angular velocity of the body rm over the time interval Δt. Aiming this interval to zero, we get *the angular velocity of the body* at the moment t:

$$\omega_z = \lim_{\Delta t \to 0} \frac{\Delta\varphi}{\Delta t} = \frac{d\varphi}{dt} = \dot{\varphi}. \tag{2.42}$$

According to the formulas (2.42), angular velocity can be either positive or negative. If the angle φ is measured in radians and the time t is measured in seconds, then the unit of angular velocity is rad/s or s^{-1}.

We will also use the vector representation of the angular velocity:

$$\boldsymbol{\omega} = \omega k, \tag{2.43}$$

where $\omega = |\omega_z|$ is the absolute value of the angular velocity. In engineering, often instead of angular velocity with uniform rotation of the body, the concept of rotation frequency is used, which is taken as the number of revolutions n performed in one minute. It is easy to establish that

$$\omega = \frac{2\pi n}{60} = \frac{\pi n}{30} \, c^{-1} \, (n - \text{в об/мин}). \tag{2.44}$$

Assume that at time t the angular velocity of the body is $\omega(t)$, and at time $t + \Delta t$ it is $\omega(t + \Delta t)$. Value

$$\varepsilon_{cp} = \frac{\omega(t + \Delta t) - \omega(t)}{\Delta t} = \frac{\Delta \omega}{\Delta t} \tag{2.45}$$

is called *the average angular acceleration* over time Δt. Ratio limit (2.45) at $\Delta t \to 0$ is called *the angular acceleration of the body* at a given time:

$$\varepsilon = \lim_{\Delta t \to 0} \frac{\Delta \omega}{\Delta t} = \frac{d\omega}{dt} = \dot{\omega} = \ddot{\varphi}. \tag{2.46}$$

It follows from the definition (2.46) that the SI unit of angular acceleration is c^{-2}.

The angular acceleration vector is the time derivative of the angular velocity vector. Using the formula (2.43), we get

$$\boldsymbol{\varepsilon} = \varepsilon \boldsymbol{k}. \tag{2.47}$$

From this we conclude that the angular acceleration vector, as well as the angular velocity vector, is directed along the axis of rotation.

Consider a point M of a solid body located at a distance h from the axis of rotation OO_1 (Fig. 2.11a). Let r be the radius vector of this point drawn from an

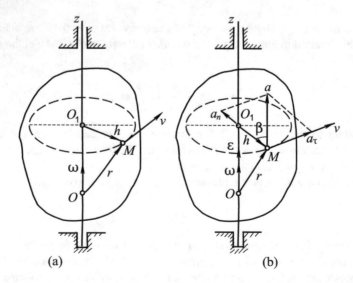

Fig. 2.11 Kinematics of rotational motion of the body: (**a**) point speed; (**b**) point acceleration

arbitrary point O on the axis of rotation. Let's denote by h the vector drawn to the point M from the center O_1 of its circular trajectory.

If during dt an elementary rotation of the body by an angle $d\varphi$ occurs, then the point M will move ds along its trajectory, which is easy to calculate: $ds = hd\varphi$. Then the speed of the point will be

$$v = \frac{ds}{dt} = h\frac{d\varphi}{dt} = h\omega. \tag{2.48}$$

Let's make sure that the vector representation of the velocity can be expressed by the formula

$$v = \omega \times r. \tag{2.49}$$

In fact, the vector $\omega \times r$ is perpendicular to the plane in which the vectors ω and r lie. In the direction, it is parallel to the velocity v, directed tangentially to the trajectory of the point M. The vector product module $\omega \times r$ will be

$$|\omega \times r| = \omega r \sin(\widehat{\omega, r}) = \omega h = v,$$

which proves the formula (2.49). Given this result and the definition (2.45), we calculate the acceleration of the point M:

$$a = \frac{dv}{dt} = \frac{d}{dt}(\omega \times r) = \frac{d\omega}{dt} \times r + \omega \times \frac{dr}{dt}.$$

Since $d\omega/dt = \varepsilon$ and $dr/dt = v$, we get

$$a = \varepsilon \times r + \omega \times v. \tag{2.50}$$

We show that the first term in the formula (2.50) is the tangent acceleration of the point, and the second is its normal acceleration. The vector $\varepsilon \times r$ is directed tangentially to the trajectory at M. Its modulus according to the formula (2.30) will be

$$a_\tau = \frac{d^2s}{dt^2} = h\frac{d\omega}{dt} = h\varepsilon = \varepsilon r \sin(\widehat{\varepsilon, r}) = |\varepsilon \times r|.$$

For the vector product $\omega \times v$, we have (Fig. 2.11b)

$$|\omega \times v| = \omega v \sin(\widehat{\omega, v}) = \omega v = \frac{v^2}{h}.$$

Since the vectors ω and v are orthogonal, the vector $\omega \times v$ is directed to the axis of rotation, and since its modulus is v^2/h, the formula (2.29) implies that the vector $\omega \times V$ represents the normal acceleration of a point in a rotating body:

$$a_n = |\boldsymbol{\omega} \times \boldsymbol{v}| = \frac{v^2}{h} = \omega^2 h.$$

Then the total acceleration of the point M will be

$$a = \sqrt{a_\tau^2 + a_n^2} = \sqrt{h^2 \varepsilon^2 + h^2 \omega^4} = h\sqrt{\varepsilon^2 + \omega^4}. \tag{2.51}$$

The angle β (Fig. 2.11b) formed by the total acceleration with the radius of the circumscribed point of the circle is determined from the formula

$$\operatorname{tg} \beta = \frac{a_\tau}{a_n} = \frac{\varepsilon}{\omega^2}. \tag{2.52}$$

Example An inextensible thread is wound on a pulley with a radius of $R = 20$ cm, on which the load B hangs. Moving vertically down from the rest state according to the equation $x = 4t^2$ (where x is the distance from the fixed horizontal axis mn, in m), the load rotates the pulley. Find the law of rotational motion, the angular velocity and angular acceleration of the pulley, as well as the total acceleration of the wheel rim point (Fig. 2.12).

S o l u t i o n. The structure consists of three bodies: a pulley that performs a rotational movement and a thread and a load that move translationally. The point A belongs to two bodies simultaneously—the pulley and the thread. Since the point B also belongs to the thread, we have

$$v_A = v_B = \frac{dx}{dt} = 8t.$$

For an inextensible thread, the velocities of the points (A) belonging to the pulley and the thread coincide. Therefore,

$$v_A = \omega R = 0,2\omega \ (\text{м/с}).$$

Тогда

$$0,2\omega = 8t; \quad \omega = 40t \ \text{с}^{-1}.$$

The angular acceleration of the pulley is determined by differentiating the angular velocity over time:

$$\varepsilon = \frac{d\omega}{dt} = 40 \ \text{с}^{-2}.$$

According to the problem condition, the movement starts from a state of rest. Therefore, the initial conditions will be

Fig. 2.12 For example, the
rotational movement of the
body

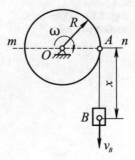

$$t = t_0 = 0 : \ \omega(t_0) = 0, \ \varphi(t_0) = 0. \tag{2.53}$$

The law of rotational motion of the pulley is found by integrating the equation

$$\frac{d\varphi}{dt} = \omega = 40t$$

under the initial condition (2.53):

$$\varphi = 40 \cdot \frac{t^2}{2} + C \ (C - const).$$

From the condition $\varphi \mid_{t=t_0} = 0$, it follows that $C = 0$. Then the law of rotational motion of the pulley will be

$$\varphi = 20t^2.$$

The tangent acceleration of the point A is determined by the formula

$$a_\tau = \varepsilon R = 40 \cdot 0.2 = 8 \text{ m/s}^{-2}.$$

The normal acceleration of the point A will be

$$a_n = \omega^2 R = (40t)^2 \cdot 0.2 = 320t^2 \text{ m} \cdot \text{s}^{-2}.$$

The total acceleration of the point A is determined by the formula (2.32)

$$a = \sqrt{a_\tau^2 + a_n^2} = \sqrt{8^2 + (320t^2)^2} = 8\sqrt{1 + 1600t^4}.$$

Fig. 2.13 Complicated
motion of a point

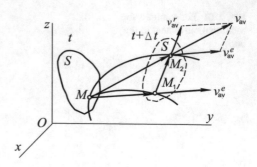

Fig. 2.13 Complicated motion of a point

2.4 Complicated Motion of a Point

Let the body S be associated with an immutable reference system that performs a
certain motion in the fixed coordinate system $Oxyz$ (Fig. 2.13). Consider a point M
that moves relative to a moving immutable system. The motion of a point M relative
to a fixed coordinate system $Oxyz$ is called *absolute or complex,* and the kinematic
characteristics of such motion (trajectory, speed, acceleration) are called absolute.

The motion of a point relative to a movable, unchangeable frame of reference
associated with the body S is called *relative.* Kinematic parameters of such motion
are also called relative. We agree to mark them with indexes r (from *English
relative*—relative). If a point M is fixed in a moving frame of reference at some
point in time, it will move only as a point of the moving system. This movement of
the point is called *portable.* We will use the index e (from *eng.* external-external).

Theorem *The absolute velocity of a material point is equal to the geometric sum
of its portable and relative velocities:*

$$v = v_e + v_r.$$

Proof Consider the positions of an immutable mobile system S at two close time
points t and $t + \Delta t$. The movements of a point M in absolute, relative, and figurative
motions (Fig. 2.13) along some (generally curved) trajectories are represented by the
vectors $\overrightarrow{MM_2}$, $\overrightarrow{M_1M_2}$, and $\overrightarrow{MM_1}$, respectively. But the vector $\overrightarrow{MM_2}$ is equal to the
geometric sum of the vectors $\overrightarrow{M_1M_2}$ and $\overrightarrow{MM_1}$ (Fig. 2.13):

$$\overrightarrow{MM_2} = \overrightarrow{MM_1} + \overrightarrow{M_1M_2}. \tag{2.54}$$

Average absolute speed the point M in time Δt, the expression (2.12), is equal
to the ratio of the displacement vector $\overrightarrow{MM_2}$ in time Δt, i.e.,

$$v_{av} = \frac{\overrightarrow{MM_2}}{\Delta t}.$$

Fig. 2.14 For example 1

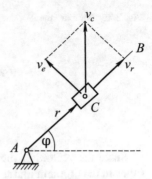

Similarly, we can write down expressions for the average velocity of relative (v_{av}^e) and portable (v_{av}^r) movements. Thus, dividing both parts of the equality (2.54) by Δt, we have

$$v_{av} = v_{av}^e + v_{av}^r.$$

In the limit at $\Delta t \to 0$, we get

$$v = v_e + v_r, \tag{2.55}$$

and that's what I needed to prove.   □

The proved theorem is extremely important in mechanics. Let's look at examples of its application.

Example 1 The backstage AB rotates in the plane around a fixed point A with an angular velocity ω. The backstage stone C slides along the backstage at a speed of v. Find its absolute speed (Fig. 2.14).

S o l u t i o n. The point C makes a complex movement. It moves along with the stage and, in addition, moves along the stage. Relative to the stage, the point C makes a rectilinear movement at the speed of v. Taking the backstage AB as a mobile reference system, we get the following value of the relative velocity of the point C:

$$v_r = v = \frac{dr}{dt}.$$

The rotation of the backstage AB is a portable movement for the point C, and therefore the portable speed of this point will be

$$v_e = \omega r = r \frac{d\varphi}{dt}.$$

Fig. 2.15 For example 2

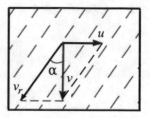

Since the vectors of the transport and relative velocities are orthogonal to each other in the problem under consideration, we will have

$$v_c^2 = v_e^2 + v_r^2 = v^2 + r^2\omega^2.$$

Example 2 A raindrop falls vertically down at a speed of v relative to the ground (Fig. 2.15). Determine the trajectory and speed of the relative movement of the drop relative to the car moving in a straight line at a speed of u on a horizontal road surface.

D e c i s i o n. We consider the movement of a drop as complex, consisting of a portable with the car and relative to the car. By the speed addition theorem,

$$v = v_e + v_r.$$

In this problem, $v_e = u$ and $v \perp u$. Therefore,

$$\text{tg}\,\alpha = \frac{u}{v}; \quad v = \sqrt{u^2 + v^2}.$$

Thus, the trajectories of raindrops are inclined to the vertical at an angle of α. For Fig. 2.15 dotted lines represent the traces of raindrops, i.e., the relative movement of the drops on the side window of the car, if it is located in a vertical plane.

2.5   Flat Solid Motion

The motion of a body is called flat if any of its points always moves in the same plane.

From this definition, it follows that the planes in which the individual points of the body move are parallel to each other and parallel to the same fixed plane. For this reason, the plane motion of the body is also called plane-parallel.

Flat motion is most common in engineering. Most links of mechanisms and machines make a flat or rotational movement. In many cases, the rotation of the body is a special case of flat motion.

All points of the body lying on a straight line MM' drawn perpendicular to the planes α and β move translationally. Therefore, to study the motion of the points of the line under consideration, it is sufficient to study the motion of only one point of this line, for example, the point M belonging to the section S.

A similar argument can be made for any other straight line perpendicular to the α plane. Therefore, to study the plane motion of a solid body, it is sufficient to study the motion of a flat figure S in its plane β, parallel to the fixed plane α.

2.5.1 Equations of Plane Motion of a Solid Body

In the plane β parallel to the fixed plane α (Fig. 2.16), we introduce the fixed coordinate system Oxy. Combine the Oxy plane with the drawing plane (Fig. 2.17), and we will investigate the motion of the S section in this plane.

To set the position of the figure S in the fixed axes ox, it is enough to know the position of some segment AB whose ends belong to S. In turn, the position of the segment AB is completely determined by the coordinates of the point $A(x_A, y_A)$ and the angle φ that the segment AB forms with the axis Ox (Fig. 2.17).

The point A moving with the body, which is chosen at random to determine the position of the section S, is called the *pole*. The position of the body in space at any time will be determined if the dependencies are set:

$$x_A = f_1(t); \quad y_A = f_2(t); \quad \varphi = f_3(t), \tag{2.56}$$

Fig. 2.16 Flat body movement

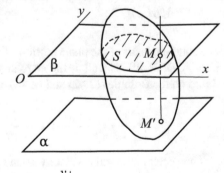

Fig. 2.17 Components of flat motion

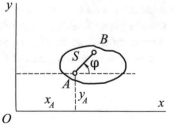

Fig. 2.18 Velocities of body
points in flat motion ıo

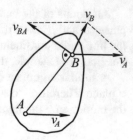

where f_1, f_2, f_3 are known time functions.

Dependencies (2.56) are *equations of plane motion of a solid.* The first two of them determine the translational motion of the pole and simultaneously give the equation of the pole trajectory in parametric form. The third equation (2.56) describes the rotational part of a body's motion around a movable axis passing through a pole.

Thus, the plane motion of a rigid body is a complex motion consisting of translational motion of the figure S and rotational motion around an axis passing through the pole perpendicular to the plane S. In this case, the translational motion of the figure S for any of its points is portable, and the rotation of the section S around an arbitrary pole A represents the relative movement of the points of the body.

2.5.2 Angular Velocity and Angular Acceleration in Plane Motion

The rotational part of the plane motion of a solid is characterized by the dependence $\varphi = f_3(t)$, (2.56), as well as the angular velocity ω and the angular acceleration ε. Based on the formulas (2.42) and (2.46), we have

$$\omega = \frac{d\varphi}{dt} = \frac{df_3(t)}{dt}; \; \varepsilon = \frac{d\omega}{dt} = \frac{d^2\varphi}{dt^2} = \frac{d^2 f_3(t)}{dt^2}. \tag{2.57}$$

The vectors of angular velocity $\boldsymbol{\omega}$ and angular acceleration $\boldsymbol{\varepsilon}$ are directed along a movable axis of rotation passing through the pole A perpendicular to the cross-section plane S. Since the rotational part of the plane motion of the body does not depend on the choice of pole, the angular velocity and angular acceleration vectors can be applied to any point of the figure S, i.e., $\boldsymbol{\omega}$ and $\boldsymbol{\varepsilon}$ are *free vectors.*

2.5.3   *Speed Points in Planar Movement of the Body*

Let's choose an arbitrary point A of the section S as the pole (Fig. 2.18) moving at the speed $\boldsymbol{v}_A$. To determine the velocity of a point B, we apply the velocity addition theorem (2.55). Receive

$$\boldsymbol{v}_B = \boldsymbol{v}_{Be} + \boldsymbol{v}_{Br}, \tag{2.58}$$

where $\boldsymbol{v}_B$ is the absolute velocity of a point B, $\boldsymbol{v}_{Be}$ the speed of point B in a figurative translational movement of the shape S, and $\boldsymbol{v}_{Br}$ the speed of point B in relative motion, which is a rotation of the plane figure S around the pole A with angular velocity ω.

Since the translational motion of the figure together with the point A is chosen for the portable one, then all points of the figure S have the same portable velocities and coincide with the absolute speed of the pole A, i.e.,

$$\boldsymbol{v}_{Be} = \boldsymbol{v}_A.$$

The relative velocity vector $\boldsymbol{v}_{Br}$ is directed perpendicular to AB in the direction of rotation, and its modulus is

$$v_{Br} = \omega \cdot AB.$$

The relative velocity of a point B can also be expressed as a vector product:

$$\boldsymbol{v}_{Br} = \boldsymbol{\omega} \times \overrightarrow{AB}.$$

To emphasize the fact that the speed of relative motion of a point B is obtained from the rotation of a flat shape around a movable axis passing through the pole A, it is convenient to denote the vector $\boldsymbol{v}_{Br}$ through $\boldsymbol{v}_{BA}$ and present the formula (2.58) as

$$\boldsymbol{v}_B = \boldsymbol{v}_A + \boldsymbol{v}_{BA}, \tag{2.59}$$

where

$$v_{BA} = \omega \cdot AB. \tag{2.60}$$

Thus, *the speed of any point of the figure in plane motion is equal to the geometric sum of the speed of the pole and the relative speed of this point in rotational motion around the pole.*

Fig. 2.19 To the concept of
an instantaneous center of
velocity

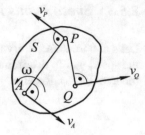

2.5.4 Instant Center of Velocity

We show that for a plane motion of a body at any time, there is a point whose velocity is zero. This point is called the *instantaneous velocity center.* Let S, as before, be a flat figure representing the cross-section of the body of any of the planes parallel to some fixed plane α (Fig. 2.16). Let's assume that we know the speed v_A of some point A (Fig. 2.19) of a flat shape at a fixed time, as well as the angular speed ω of the shape at this moment. Let's take the point A as the pole. Then the formula (2.59) the velocity of any point P figure is equal to the geometric sum of the velocity pole of the v_A and angular velocity v_{BA} points surrounding the pole. From the point A, draw a perpendicular to v_A so that the rotation of the vector v_A to this perpendicular coincided with the direction of rotation of the figure around the pole A (Fig. 2.19). Let's find a point P, on this perpendicular whose rotational velocity is equal to the absolute velocity of the pole, i.e., $v_{PA} = v_A$. Since the velocity directions v_A and v_{PA} are opposite, then $v_{PA} = -v_P$, and the point velocity P will be

$$v_P = v_A + v_{PA} = 0.$$

Therefore, the point P chosen by us at the time under consideration is the instantaneous center of velocities. Its position determines the segment AP, the length of which according to the formula (2.60) will be

$$AP = \frac{v_A}{\omega}. \tag{2.61}$$

At each instant of time, the instantaneous velocity center is the only point whose velocity is zero. Taking this point as a pole and considering that the speed of the pole in this case is zero, it is easy to determine the speed of any point Q (Fig. 2.19) of a flat figure. In fact, according to (2.59) and (2.60) for the point Q, we have

$$v_Q = v_{QP} = \omega \cdot PQ, \tag{2.62}$$

Fig. 2.20 For example 1

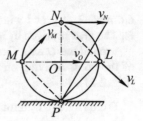

Fig. 2.21 For example 2

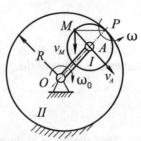

where PQ is the distance from the point Q to the instantaneous velocity center P. The vector v_Q is perpendicular to the segment PQ and is directed toward the rotation of the shape around the pole.

It follows from the above that to determine the position of the instantaneous velocity center at known velocities of two points of the figure (e.g., points A and Q, Fig. 2.19), it is sufficient to find the intersection point of the perpendiculars to the velocity vectors of these points. In some cases, you can immediately specify the point of a flat shape, the speed of which is zero at the moment under consideration. Thus, in the case of rolling without sliding one body over the surface of another stationary body, the point of contact of the bodies is the instantaneous center of velocity (point P, Fig. 2.20).

Example 1 A wheel with a radius of R (Fig. 2.20) rolls without sliding along a fixed straight line, having a speed of v_0. Determine the speed of the points M, N, and L of the wheel rim at a given time.

D e c i s i o n. The instantaneous center of velocity in the problem under consideration is located at the point P where the wheel meets the straight line. The angular velocity of the wheel is determined from the formula (2.61)

$$\omega = \frac{v_0}{OP} = \frac{v_0}{R}.$$

The speeds of the points specified in the condition are determined by the formula (2.62). Given that $MP = LP = R\sqrt{2}$, we get

$$v_N = \omega \cdot PN = \omega \cdot 2R = 2v_0; \quad v_M = v_L = \omega \cdot MP = v_0\sqrt{2};$$
$$v_N \perp PN; \quad v_M \perp MP; \quad v_L \perp LP.$$

Example 2 A gear I with radius r is engaged with a fixed gear wheel II with radius R. The gear is driven by a crank OA that rotates uniformly with an angular velocity ω_0 around the O axis. Determine the angular velocity of the gear I and the velocity of the point M lying on the diameter of the gear perpendicular to the crank OA (Fig. 2.21).

D e c i s i o n. Determine the speed of the point A of the crank OA:

$$v_A = \omega_0 \cdot OA = \omega_0(R - r).$$

The position of the instantaneous velocity center P is known. Calculate the angular velocity of the gear I:

$$\omega = \frac{v_A}{PA} = \frac{\omega_0(R - r)}{r}.$$

Then the speed of the point M will be

$$v_M = \omega_0 \cdot PM = \omega_0\sqrt{2}(R - r).$$

The vector v_M is directed along the straight line MP (Fig. 2.21).

2.5.5   Speed Plan

Let the velocities v_A, v_B, and v_C of points A, B, C be known for the plane movement of the figure S (Fig. 2.22a).

Set aside from some center O (Fig. 2.22b) in the selected scale μ_v segments:

$$Oa = \frac{v_A}{\mu_v}; \quad Ob = \frac{v_B}{\mu_v}; \quad Oc = \frac{v_C}{\mu_v}$$

in the direction of the velocity vectors of points A, B, and C, respectively. Connect the points a, b and c. The resulting diagram is called *speed plan*.

Let's set some properties of the speed plan. By the formulas (2.59) and (2.60), we have

$$v_B = v_A + v_{BA};$$
$$v_{BA} = \omega \cdot AB; \quad v_{BA} \perp \overrightarrow{AB}. \tag{2.63}$$

From the triangle Oab, we note that

$$\overrightarrow{Ob} = \overrightarrow{Oa} + \overrightarrow{ab} \text{ or}$$
$$v_B = v_A + \mu_v \cdot \overrightarrow{ab}.$$

Fig. 2.22 Building a speed
Plan: (**a**) point speeds; (**b**)
speed plan

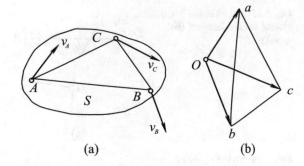

(a)   (b)

Comparing this result with the first of the equalities (2.63), we conclude that
$\mu_v \cdot \overrightarrow{ab} = \boldsymbol{v}_{BA}$. Similarly, we find $\mu_v \cdot \overrightarrow{ac}$, etc. in Addition, from the formulas (2.63)
it follows that $ab \perp AB$, $ac \perp AC$, etc., as well as $ab = \omega \cdot AB$, $ac = \omega \cdot AC$,
from which we get

$$\frac{ab}{AB} = \frac{ac}{AC} = \frac{bc}{BC} = \ldots = \omega.$$

Thus, the segments connecting the ends of the velocity vectors on the velocity
plane are perpendicular to the segments connecting the corresponding points of the
body and are proportional to them. In other words, the figures indicated by the same
symbols on the cross-section S and the velocity plan are similar and rotated relative
to each other by $90°$.

The velocity plan of the mechanism is constructed as a set of velocity plans of
the bodies that make up the mechanism, and all vectors of absolute velocities are
deferred from the common center O.

Example 3 Plot the speed of the mechanism (Fig. 2.23a) for the position shown in
the figure, if the speed v_A of the end of the crank $O_1 A$ is known. The connecting rod
ABC is made in the form of a rigid triangular plate. The crank $O_2 D$ is connected at
the point D by a hinge to the middle of the rod CE ($CD = DE$).

S o l u t i o n. Select the scale of length μ_l (e.g., $\mu_l = 0.01$ m/mm) and draw the
plan of the mechanism at the selected scale with the specified location of its links
(Fig. 2.23a). Set the speed scale μ_v (e.g., $\mu_v = 0.2 \frac{\text{m/s}}{\text{mm}}$) and the position of the
center O (Fig. 2.23b). From the center of O, draw a ray perpendicular to $O_1 A$, and
put the segment $Oa = v_A/\mu_v$ on it. From the same center O, we draw a straight
line Ob parallel to the vector $\boldsymbol{v}_B$, which is directed along $O_1 B$, and from the point
a—a straight line $ab \perp AB$ to the intersection with the line Ob. Then the point b of
the intersection of these lines will determine the end of the vector $\overrightarrow{Ob}$, representing
the scale μ_v speed of the point B. Find $v_B = \mu_v \cdot Ob$.

To construct the point c of the velocity plan, we use the similarity property of the
velocity plan and the mechanism plan, but keep in mind that similar figures of these
plans are rotated relative to each other by $90°$. From the point a, we draw a straight

Fig. 2.23 For example 3: (**a**) plan of the mechanism; (**b**) speed plan

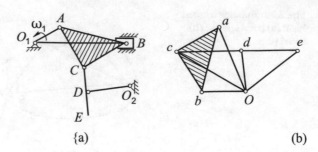

{a}   (b)

Fig. 2.24 To the study of accelerations in planar movement of the body

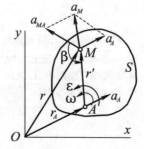

line $ac \perp AC$, and from the point b—a straight line $bc \perp BC$. The intersection of these lines will define a point c representing the end of the velocity vector of the point C. By measuring the length of the segment Oc, we find $v_C = \mu_v \cdot Oc$.

The direction of the absolute velocity vector of the point D is known ($v_D \perp O_2D$). To construct a point d on the velocity plan (Fig. 2.23b), draw a straight line Od parallel to v_D from the center of O, and a line perpendicular to CD from the point c. The intersection of these lines gives the point d. Connecting the straight points O and d, we get the vector $\overrightarrow{Od}$, representing the speed of the point d.

The modulus of the vector $\boldsymbol{v}_D$ is determined by multiplying the length of the segment Od by the scale of the speed plan:

$$v_D = Od \cdot \mu_v.$$

We also find the point e of the velocity plan from the similarity property. Since $DC = DE$, we postpone $de = cd$ on the continuation of cd. Connecting the center of O with the point e, we finish building the plan. The vector $\overrightarrow{Oe}$ represents the absolute velocity of the point E. After measuring the length of the segment Oe, we find

$$v_E = Oe \cdot \mu_v.$$

2.5.6 Determination of Accelerations of Points of a Body in Flat Motion

The position of a point M of a flat figure S (Fig. 2.24) with respect to a fixed frame of reference ox is determined by the radius vector $r = r_A + r'$, where r_A is the radius vector of the pole A, $R' = \overrightarrow{am}$. According to the expression (2.23), the acceleration of the point M is calculated as the second derivative of the radius vector in time, i.e.,

$$a_M = \frac{d^2 r}{dt^2} = \frac{d^2 r_A}{dt^2} + \frac{d^2 r'}{dt^2}.$$

Here, the first term defines the acceleration of the pole a_A, and the second is equal to the acceleration of the point M in rotational relative motion around the pole. Therefore,

$$a_M = a_A + a_{MA}. \tag{2.64}$$

The formulas (2.51) and (2.52) are valid for the acceleration of a_{MA}, i.e.,

$$a_{MA} = MA \cdot \sqrt{\varepsilon^2 + \omega^4}; \quad \operatorname{tg} \beta = \frac{\varepsilon}{\omega^2}, \tag{2.65}$$

where ω and ε are the angular velocity and angular acceleration of the body, respectively, and the angle β between the acceleration direction a_{MA} and the segment AM determines the direction of the vector a_{MA}.

Thus, the acceleration of any point of a body in plane motion is geometrically composed of the acceleration of some other point taken as a pole and the acceleration of a point in its rotational motion around this pole.

When solving problems, it is often more convenient to replace the relative acceleration vector a_{MA} with its tangent a_{MA}^τ and normal a_{MA}^n components whose modules are defined by formulas (see p. 71)

$$a_{MA}^\tau = AM \cdot \varepsilon; \quad a_{MA}^n = AM \cdot \omega^2, \tag{2.66}$$

The vector a_{MA}^n is always directed from the point M to the pole A, and the tangent component a_{MA}^τ it is directed perpendicular to AM in the direction of rotation, if it is accelerated, and against rotation, if it is slowed down.

Example 4 The wheel rolls without sliding in a vertical plane on an inclined straight path. Find the acceleration of the ends of two mutually perpendicular wheel diameters, one of which is parallel to the rail, if at the given time the speed of the wheel center $v_0 = 1$ m/s, the acceleration of the wheel center $a_0 = 3$ m/s^2, and the wheel radius $R = 0.5$ m (Fig. 2.25).

Fig. 2.25 For example 4

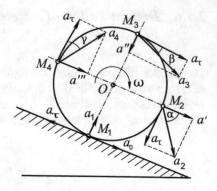

D e c i s i o n. Determine the angular velocity and angular acceleration of the wheel:

$$\omega = \frac{v_0}{R} = \frac{1}{0.5} = 2\,C^{-1}; \quad \varepsilon = \frac{a_0}{R} = \frac{3}{0.5} = 6\,C^{-2}.$$

Using the formulas (2.65), we calculate the modules of normal and tangential acceleration of wheel rim points in relative motion:

$$a_n = \omega^2 R = 2^2 \cdot 0.5 = 2\,\text{м/с}^2;$$
$$a_\tau = \varepsilon R = 6 \cdot 0.5 = 3\,\text{м} \cdot c^{-2}.$$

Now we calculate the acceleration of points.

1. Point M_1:

$$a_1 = a_0 + a^n_{M_1} + a^\tau_{M_1}.$$

The vectors a_0 and $a^\tau_{M_1}$ are equal in modulus and opposite in direction. Therefore,

$$a_0 + a^\tau_{M_1} = 0,$$

and the acceleration of the point M_1 will be

$$a_1 = a_n = 2\,\text{m} \cdot c^{-2}.$$

The direction of the vector a_1 coincides with the direction of the normal acceleration of the point M_1 (from the point M_1 to the point O).

2. Point of M_2:

$$a_2 = a_0 + a^n_{M_2} + a^\tau_{M_2}.$$

Denote

$$a' = a_0 + a^n_{M_2}.$$

The modulus of this vector will be $a' = a_0 - a_n = 1 \text{ m} \cdot c^{-2}$, and the direction is shown in Fig. 2.25. Using these results, we calculate the modulus and direction of the acceleration vector of the point M_2:

$$a_2 = \sqrt{a'^2 + a^{\tau 2}_{M_2}} = \sqrt{1 + 3^2} = \sqrt{10} \approx 3,36 \text{ м} \cdot c^{-2}; \quad \text{tg}\,\alpha = \frac{a^\tau_{M_2}}{a'} = 3.$$

3. Point M_3:

$$a_3 = a_0 + a^n_{M_3} + a^\tau_{M_3}.$$

The vectors a_0 and $a^\tau_{M_3}$ coincide in direction. Denoting their sum by a'', we get

$$a'' = a_0 + a^\tau_{M_3} = 3 + 3 = 6 \text{ м} \cdot c^{-2};$$
$$a_3 = \sqrt{a''^2 + a^{n2}_{M_3}} = \sqrt{6^2 + 2^2} = \sqrt{40} \approx 6,32 \text{ м} \cdot c^{-2};$$
$$\text{tg}\,\beta = \frac{a^n_{M_3}}{a''} = \frac{2}{6} = \frac{1}{3}.$$

4. Point M_4:

$$a_4 = a_0 + a^n_{M_4} + a^\tau_{M_4}.$$

Denoting $\tilde{a} = a_0 + a^n_{M_4}$, we get $\tilde{a} = 3 + 2 = 5 \text{ m} \cdot c^{-2}$, and the direction of this vector coincides with the direction of a_0 (Fig. 2.25). Now calculate

$$a_4 = \sqrt{\tilde{a}^2 + a^{\tau 2}_{M_4}} = \sqrt{5^2 + 3^2} = \sqrt{34} \approx 5,83 \text{ м} \cdot c^{-2}; \quad \text{tg}\,\gamma = \frac{\tilde{a}}{a^\tau_{M_4}} = \frac{5}{3}.$$

The angles α, β, and γ are shown in Fig. 2.25.

2.6   Complex Motion of Points in the General Case

When determining the kinematic characteristics of a complex point motion, it is necessary to consider the change in vector quantities over time in relation to reference systems moving relative to each other. Therefore, it is necessary to calculate the derivatives of vector quantities in moving relative to each other reference systems. Let's get acquainted with the procedure of such calculations.

2.6.1 Absolute and Relative Derivatives of a Vector Function of a Scalar Argument

Consider an arbitrary vector $b(t)$ that depends on time t. Its time derivative in a fixed frame of reference is called *full or absolute derivative* and is denoted by db/dt. The time derivative when taking into account the change in the vector $b(t)$ relative to the mobile reference system is called *relative or local derivative* and denotes $\tilde{d}b/dt$. We get the relationship between these derivatives. Decompose the vector b into components along the axes of the mobile coordinate system $Oxyz$ with orts i, j, k:

$$b = b_x i + b_y j + b_z k. \tag{2.67}$$

In absolute motion of a point, a change in the vector b consists of changing its projections b_x, b_y, b_x on the moving coordinate axes and changing the orts i, j, k due to the movement of the moving coordinate system relative to the stationary one. Therefore, the full derivative of the vector b is calculated using the formula

$$\frac{db}{dt} = \frac{db_x}{dt} i + \frac{db_y}{dt} j + \frac{db_z}{dt} k + b_x \frac{di}{dt} + b_y \frac{dj}{dt} + b_z \frac{dk}{dt}. \tag{2.68}$$

The first three terms of the formula (2.68) are calculated under the assumption that the orts i, j, k are invariant and by definition form the relative derivative of the vector b, i.e.,

$$\frac{\tilde{d}b}{dt} = \frac{db_x}{dt} i + \frac{db_y}{dt} j + \frac{db_z}{dt} k. \tag{2.69}$$

To calculate the derivatives of unit vectors i, j, k by the argument t, we use the formula (Rashevskii 1950), which connects the curvature vector K at an arbitrary point of the trajectory with the tangent vector τ and normals n:

$$K = \frac{d\tau}{ds} = n \cdot \frac{1}{\rho}.$$

Assuming here $\tau = i$, $n = j$, we have

$$\frac{di}{ds} = \frac{1}{\rho} j.$$

But

$$\frac{di}{dt} = \frac{di}{ds} \cdot \frac{ds}{dt}.$$

Since the vector i is single, $\rho = 1$, $ds = d\varphi$. Taking these intermediate results into account, we obtain the following formula for the derivative of the unit vector i:

$$\frac{di}{dt} = \frac{d\varphi}{dt} \cdot j.$$

Remembering now that the time derivative of the rotation angle is the angular velocity ω, we can write

$$\frac{di}{dt} = \omega \times i; \quad \frac{dj}{dt} = \omega \times j; \quad \frac{dk}{dt} = \omega \times k, \tag{2.70}$$

where the last two formulas are written by analogy with the first one.

The vector ω is the angular velocity of the rotational part of the motion around the point O of the mobile coordinate system $Oxyz$ (Fig. 2.26) relative to the stationary reference system $O_1x_1y_1z_1$. It should be directed perpendicular to the plane in which the vectors i and j are located, and so that from the end of its arrow, you can see the rotation of the vector i to j in this plane at an angle $90°$ counterclockwise.

Substituting the expressions (2.69) and (2.70) into the formula (2.68) and putting ω in parentheses, we get

$$\frac{db}{dt} = \frac{\tilde{d}b}{dt} + \omega \times (b_x i + b_y j + b_z k)$$

or

$$\frac{db}{dt} = \frac{\tilde{d}b}{dt} + \omega \times b. \tag{2.71}$$

The relationship (2.71) between the full and local derivatives of the vector is called the *Bour formula*.

Fig. 2.26 To the formula of Bour

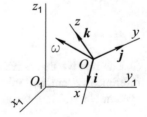

Fig. 2.27 The composition
of velocities

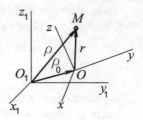

2.6.2   Addition of Velocities

Let a point M move relative to a moving coordinate system $Oxyz$ (Fig. 2.27),
whose motion relative to the stationary system $Ox_1y_1z_1$ is characterized by the
translational velocity v_0 and the angular velocity ω when rotating around the pole O.
We draw the vectors r and ρ, which determine the position of the point in question
in the mobile and stationary systems. We also construct the radius vector of the point
O, indicated in Fig. 2.27 with ρ_0. At any given time, these three vectors are linked
by a dependency

$$\rho = \rho_0 + r.$$

We differentiate this dependence in time, taking into account the changes of
vectors relative to fixed axes, i.e., we calculate the full derivatives of vectors.
Receive

$$\frac{d\rho}{dt} = \frac{d\rho_0}{dt} + \frac{dr}{dt}. \tag{2.72}$$

According to the formula (2.68), $d\rho/dt = v$ is the absolute speed of the point M,
and $d\rho_o/dt = v_o$ is the absolute speed of the point O. To calculate the derivative
dr/dt, use the Boer formula (2.71). We have

$$\frac{dr}{dt} = \frac{\tilde{d}r}{dt} + \omega \times r.$$

The relative derivative $\tilde{d}r/dt$ is the relative velocity of a point M in a mobile
reference system, and the vector ω is the angular velocity of rotation of the mobile
system $Oxyz$. Thus, we get

$$v = v_0 + \omega \times r + v_r. \tag{2.73}$$

The sum of $v_e = v_0 + \omega \times r$ is the velocity of the point of a free rigid body
attached to a moving coordinate system that the point M in absolute motion of the
body currently coincides with. This is the portable speed of v_e points M (p. 74).
Hence, the formula (2.73) again proves the velocity addition theorem:

$$v = v_e + v_r, \tag{2.74}$$

i.e., *the speed of absolute motion of a point is equal to the vector sum of the relative and portable speeds.*

2.6.3   Acceleration of a Point in the General Case of Portable Motion

To determine the absolute acceleration of a point, we calculate the full derivative of its absolute velocity (2.73). Have

$$a = \frac{dv}{dt} = \frac{d}{dt}(v_0 + \boldsymbol{\omega} \times r + v_r).$$

Applying the Bour formula (2.71), for the complete derivatives of the vectors r and v_r, we get

$$\frac{dr}{dt} = \frac{\tilde{d}r}{dt} + \boldsymbol{\omega} \times r; \quad \frac{dv_r}{dt} = \frac{\tilde{d}v_r}{dt} + \boldsymbol{\omega} \times v_r.$$

Even taking into account that

$$\frac{dv_0}{dt} = a_0; \quad \frac{d\boldsymbol{\omega}}{dt} = \boldsymbol{\varepsilon}; \quad \frac{\tilde{d}r}{dt} = v_r;] \frac{\tilde{d}v_r}{dt} = a_r,$$

for the absolute acceleration of a point, we obtain the following formula:

$$a = a_0 + \boldsymbol{\varepsilon} \times r + \boldsymbol{\omega} \times (\boldsymbol{\omega} \times r) + a_r + 2 \cdot (\boldsymbol{\omega} \times v_r), \tag{2.75}$$

where a_0 is acceleration of the point O; $\boldsymbol{\varepsilon} \times r$ is rotational acceleration of the point M, assuming that it moves only together with the moving coordinate system, without currently having relative motion; and $\boldsymbol{\omega} \times (\boldsymbol{\omega} \times R)$ the axial acceleration of the point m under the same assumption. In sum, the first three terms of the formula (2.75) give the portable acceleration a_e of a free point in the general case of its motion together with a mobile coordinate system relative to a fixed reference system. Therefore, the formula (2.75) can be rewritten as

$$a = a_e + a_r + a_k, \tag{2.76}$$

where indicated

$$a_k = 2(\boldsymbol{\omega} \times v_r). \tag{2.77}$$

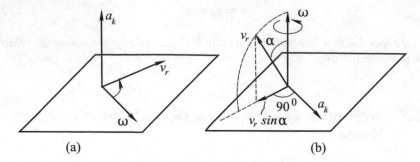

Fig. 2.28 To the Coriolis theorem: (**a**) direction of velocities and Coriolis acceleration; (**b**) to the calculation of the Coriolis acceleration

The acceleration a_k defined by the formula (2.77) is called *Coriolis acceleration, or rotational (incremental) acceleration*. The formula (2.76) expresses **the Coriolis Theorem** *The absolute acceleration of a moving point is equal to the vector sum of the portable, relative, and Coriolis accelerations.*

Let's take a closer look at Coriolis acceleration. The modulus of this acceleration is obviously equal to

$$a_k = 2\omega v_r \sin(\widehat{\boldsymbol{\omega}, \boldsymbol{v_r}}). \tag{2.78}$$

The direction of the vector $\boldsymbol{a}_k$ is determined by the rules of the vector product. This means that the Coriolis acceleration will be directed perpendicular to the plane passing through the vectors $\boldsymbol{\omega}$ and $\boldsymbol{v}_r$ in the direction from which the shortest turn $\boldsymbol{\omega}$ to $\boldsymbol{v}_r$ is seen to occur counterclockwise (Fig. 2.28a). If the vectors $\boldsymbol{\omega}$ and $\boldsymbol{v}_r$ do not lie in the same plane, the vector $\boldsymbol{\omega}$ should be moved parallel to itself to the beginning of the vector $\boldsymbol{v}_r$. This operation is legal, since $\boldsymbol{\omega}$ is a free vector.

To calculate the Coriolis acceleration, you can also use the *rule of N.E. Zhukovsky*, graphically presented in Fig. 2.28b: *the projection of the relative velocity $\boldsymbol{v}_r$ on a plane perpendicular to the vector $\boldsymbol{\omega}$ and equal to $v_r \sin \alpha$ should be multiplied by 2ω and rotated by $90°$ around the vector $\boldsymbol{\omega}$ in the direction of rotation.* is a vector equal in modulus to $2\omega v_r$ and having the direction found by turning the relative velocity projection, and will be the desired Coriolis acceleration.

It follows from the formula (2.78) that the Coriolis acceleration is zero if:

- $\omega = 0$, which is the case, for example, when a moving frame of reference moves forward;
- $\boldsymbol{\omega} \parallel \boldsymbol{v}_r$, i.e., the angular velocity vector is parallel to the relative velocity vector;
- at times when the speed of a point in relative motion vanishes: $v_r = 0$.

Example A The compressor disk with curved channels (Fig. 2.29) rotates uniformly with an angular velocity ω around the O axis perpendicular to the drawing plane. Air flows through channels with a constant relative velocity v_r. Find the projections of

Fig. 2.29 Kinematics of air
particles in the compressor

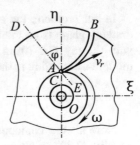

absolute velocity and acceleration on the coordinate axis for an air particle located
at the point C of the channel AB, if the radius of curvature of the channel at the
point C is ρ, and the angle between the normal to the curve AB at the point C and
the radius OC is φ. The length of the radius OC is r.

S o l u t i o n. By the velocity addition theorem (2.74) for the velocity of an air
particle at the point C, we have

$$v = v_e + v_r,$$

and the portable speed modulus is $v_e = \omega r$. For a given direction of rotation of the
disk (Fig. 2.29), the vector of the transport velocity v_e is co-directed with the axis
$O\xi$. Projections of the absolute velocity v on the coordinate axes will be

$$v_\xi = \omega r + v_r \cos\varphi; \quad v_\eta = v_r \sin\varphi.$$

The acceleration of the air particle at the point C is calculated using the formula
(2.76)

$$a = a_e + a_r + a_k. \tag{2.79}$$

When the disk rotates uniformly, the tangential component of the transport
acceleration is zero. Therefore, the transport acceleration coincides with the normal
acceleration of the air particle $a_e^n = \omega^2 r$ and is directed from the point C to the O
axis.

By the condition of the problem, the air particle moves inside the channel AB
with a constant relative velocity. Therefore, the tangential component of the relative
acceleration is zero, and the vector a_r coincides with the normal acceleration of the
point C in relative motion. The modulus of this vector will be $a_r = a_r^n = v_r^2/\rho$. The
vector a_r is directed along the normal CD to the curve AB at the point C.

Calculate the Coriolis acceleration a_k using the formula (2.77)

$$a_k = 2(\omega \times v_r).$$

In our problem, the angular velocity vector $\boldsymbol{\omega}$ is directed perpendicular to the drawing plane from the reader, so the direction of Coriolis acceleration coincides with the ray CE (from the point C to the point E). The Coriolis acceleration modulus, according to the formula (2.78), will be

$$a_k = 2\omega r \sin 90° = 2\omega r.$$

Now all the components of absolute acceleration are known. Projecting the equality (2.79) on the coordinate axes ξ and η, we get

$$a_\xi = \left(2v_r\omega - \frac{v_r^2}{\rho}\right) \sin \varphi; \quad a_\eta = -\left[\omega^2 r + \left(2v_r\omega - \frac{v_r^2}{\rho}\right) \cos \varphi\right].$$

Self-Test Questions

1. What is the difference between the reference body and the reference system?
2. List the main ways to set the point's motion.
3. How is the average speed of the point directed over a certain period of time?
4. Write down the formulas that determine the modulus and direction of the point's velocity when the coordinate method is used to set its motion.
5. How is the speed of a point expressed in terms of a curved coordinate in the natural way of setting the movement?
6. Give the definition of average acceleration points for some time.
7. How are the modulus and guide cosines of the acceleration vector of a point expressed in terms of acceleration projections on rectangular coordinate axes?
8. Write down the formulas for the normal and tangent accelerations of the point in the natural way of setting the motion.
9. What determines the number of degrees of freedom of a solid?
10. Why can't the velocities and accelerations of its points be different when a body is moving forward?
11. Give definitions of the angular velocity and angular acceleration of the body.
12. How are the angular velocity and angular acceleration vectors directed when a body rotates around a fixed axis?
13. How to calculate the speed of a point of a body rotating around a fixed axis? Explain where the velocity vector is directed.
14. Write down the formulas for the normal and tangential accelerations of a point of a body rotating around a fixed axis.
15. Give a definition of the absolute (complex) motion of a point.
16. Which motion of a point is called relative, and which is portable?
17. Formulate a theorem on the addition of velocities.
18. Why is the motion of a solid body called flat? Give examples of links of mechanisms that perform flat motion.

19. What are the simple movements that make up the flat motion of a solid?
20. Why is the motion of a solid called flat? Give examples of links of mechanisms that perform flat motion.
21. What are the simple movements that make up the plane motion of a solid body?
22. How is the speed of an arbitrary point of a body determined in flat motion?
23. What is an instant center of velocity? How is the magnitude and direction of the velocity of an arbitrary point of a body determined at a known position of the instantaneous center of velocity and angular velocity?
24. What are the components of the acceleration of a point in flat motion?
25. Write down the formulas for calculating the tangent and normal components of the relative acceleration of a point in the plane motion of the body.
26. Give the definition of the instantaneous center of acceleration.
27. When a body moves flat at some point in time, it turns out that its points A and B are separated from the instantaneous center of acceleration at distances of 5 and 10 cm. What is the acceleration modulus of a point B if the acceleration modulus of a point A is 3 m/c^2?
28. What is an instant center of velocity? How is the magnitude and direction of the velocity of an arbitrary point of a body determined at a known position of the instantaneous center of velocity and angular velocity?
29. What are the components of the acceleration of a point in flat motion?
30. Write down the formulas for calculating the tangent and normal components of the relative acceleration of a point in flat motion of the body.
31. What is the difference between the absolute and relative derivatives of a vector function of a scalar argument?
32. That expresses the formula of Borax?
33. How is the absolute velocity vector of a point expressed in the general case of its motion?
34. Name the components of the acceleration vector for complex point movement.
35. How are the modulus and direction of Coriolis acceleration determined?
36. In what complex motion of a point is the Coriolis acceleration equal to zero?
37. Formulate the definitions of absolute, relative, and translational motion of a solid.
38. How to determine the speed of an arbitrary point of a body that performs translational relative and translational motion?

Control Tasks for the Section "Kinematics"

Task. The crank O1A rotates at a constant angular velocity $\omega_{O_1A} = 2\ s^{-1}$ (Yablonskii et al. 2000). Determine for a given position of the mechanism the following:

(1) The velocities of points $A, B, C, \ldots$ the mechanism and the angular velocities of all its links using the velocity plan

(2) The velocities of the same points of the mechanism and the angular velocities of its links using instantaneous velocity centers
(3) Accelerations of points A and B and angular acceleration of link AB
(4) Position of the instantaneous acceleration center of the AB link
(5) Acceleration of the point M dividing the link AB in half.

Schemes of mechanisms are shown in Figs. 2.30 and 2.31, and the data required for the calculation are given in Tables 2.1 and 2.2.

Example of Completing a Task

Given: a diagram of the mechanism in a given position (Fig. 2.32); initial data (Table. 2.3). Take the length of the leading link $O_1A = 12\ cm$, and its position is determined by the angle $\varphi = 52°$.

S o l u t i o n.

1. *Determination of the velocities of points and angular accelerations of the links of the mechanism using the velocity plan.*

   (a) Determine the velocities of the points. To do this, we build a diagram of the mechanism at the selected scale (Fig. 2.32). The speed of the point A of the crank $O1A$ will be

   $$v_A = \omega_{O_1A} \cdot O_1A = 2 \cdot 12 = 24\ cm/s.$$

   The vector $\vec{v}_A$ is perpendicular to O_1A and is directed in the direction of rotation of the crank. Building a speed plan. From an arbitrarily selected pole O, we draw a segment Oa that represents the speed of the point A at the selected scale. Determine the velocity of point B through the pole O hold, and parallel velocity v_B, and point a—straight line perpendicular to AB. We get the point b; the segment Ob determines the speed of the point B. We measure the length of the segment Ob and, using the velocity scale, find $v_B = 17.5$ cm/s.

   To determine the velocity of a point C, divide the segment ab of the velocity plan with respect to $ac/cb = AC/CB$.

   The segment Oc represents the velocity of the point C. Using the velocity scale, we get $v_C = 17.5\ cm/s$.

   Continuing the construction of the speed plan, we find v_A, v_B, v_C, v_D, v_E, v_F, v_G, v_F (Table 2.4). In the drawing of the mechanism, the ends of the vectors of points of a straight link (e.g., points A, B, C) are on one straight line (Fig. 2.33).

   (b) *Determine the angular velocities of the links of the mechanism.* The segment ab of the velocity plan expresses the rotational velocity of point B around point A:

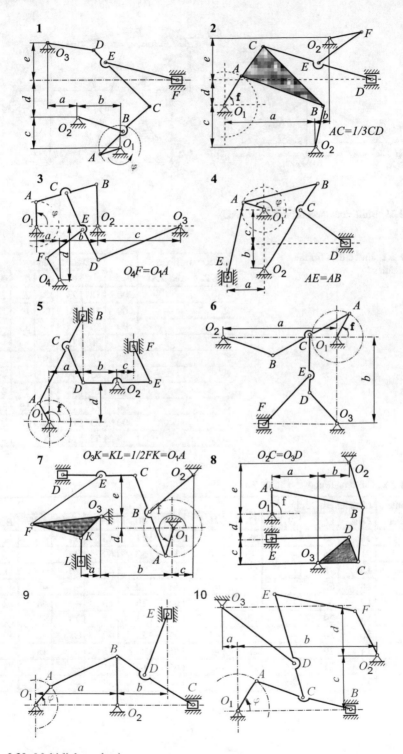

Fig. 2.30 Multi-link mechanisms

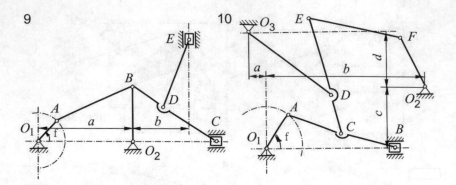

Fig. 2.31 Multi-link mechanisms (continued)

Table 2.1 Initial data for the "kinematics"

Option number,		Distances, cm				
Fig. 2.30 and 2.31	φ, °	a	b	c	d	e
1	200	18	23	18	22	23
2	60	56	10	26	16	25
3	90	15	25	54	35	–
4	115	26	15	23	–	–
5	125	19	19	19	–	
6	60	65	49	–	–	–
7	259	11	42	11	7	24
8	90	27	18	14	15	30
9	200	23	19	20	28	21
10	20	55	21	25	–	–

Table 2.2 Continuation of the Table 2.1

Scheme number	Link lengths, cm									
	O_1A	O_2B	O_2D	O_3D	O_3F	AB	BC	CD	CE	DE	EF
1	14	28	–	28	–	21	21	48	38	–	42
2	21	25	–	–	20	54	52	69	35	–	32
3	15	28	–	58	–	42	21	47	26	–	31
4	15	65	–	–	–	51	22	38	–	–	–
5	12	–	19	–	–	55	19	23	–	38	22
6	15	29	–	24	–	50	25	32	23	–	39
7	16	34	–	–	41	25	25	42	21	–	49
8	14	29	–	23	–	55	32	15	–	45	–
9	21	31	–	25	–	65	62	31	–	11	29
10	15	–	24	–	–	70	35	33	–	17	12

Fig. 2.32 For example, the analysis of a multi-link mechanism

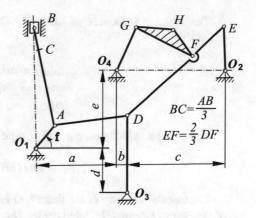

$$BC = \frac{AB}{3}$$

$$EF = \frac{2}{3} DF$$

Table 2.3 Data for analysis of the multi-link mechanism

a	b	c	d	e	AB	AD	O_2D	DE	O_3E	FG	GH	FH	O_4G
32	4	39	19	32	46	29	32	53	18	25	14	14	20

Table 2.4 Mechanism point speeds

Method of determination	Speed of points, cm/s							
	v_A	v_B	v_C	v_D	v_E	v_F	v_G	v_H
According to the speed plan	24	17.5	17.5	17.5	17.5	17.5	14.8	14.4
Using instantaneous velocity centers	24	17.3	17.5	17.4	17.4	17.4	14.6	14.1

Fig. 2.33 Building a speed plan

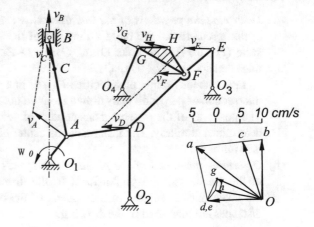

$$ad - v_{AB} = \omega_{AB} \cdot AB;$$

hence, the angular velocity of the link AB will be

$$\omega_{AB} = ab/AB = 19.5/46 = 0.424 \, rad/s. \qquad (2.80)$$

The angular velocities of the AD, DE, and FGH links are determined similarly:

$$\omega_{AD} = ad/AD;$$
$$\omega_{DE} = de/DE;$$
$$\omega_{FGH} = fg/FG.$$

The angular velocity ω_{FGH} can also be determined from the relations

$$\omega_{FGH} = gh/GH = fh/FH.$$

The angular velocity of the link $O_2 D$ is determined by the rotational velocity of the point D around the fixed center O_2:

$$\omega_{O_2 D} = v_D/O_2 D.$$

Similarly, the angular velocities of the links $O_3 E$, $O_4 G$ are determined:

$$\omega_{O_3 E} = v_E/O_3 E; \quad \omega_{O_4 G} = v_G/O_4 G.$$

The angular velocities calculated from these formulas are shown in Table 2.5.

2. *Determination of speeds of points and angular speeds of links using instant centers of speeds.*

(a) *Determine the positions of the instantaneous centers of speeds of the links of the mechanism.* We build a diagram of the mechanism in the selected scale (Fig. 2.34). The links $O_1 A$, $O_2 D$, $O_3 E$, $O_4 G$ revolve around the fixed centers O_1, O_2, O_3, O_4.

Instantaneous center of velocities P_{AB} of link AB is found as the point of intersection of perpendiculars drawn from points A and B to their velocities. The positions of the instantaneous centers of velocities P_{AD} and P_{FGH} are determined similarly. The instantaneous center of velocities of the DE link is at infinity.

(b) *Determine the speed of the points.* The speeds of the points of the links of the mechanism are proportional to the distances from these points to the instantaneous centers of the speeds of the corresponding links. These distances are measured in the drawing.

Table 2.5 Mechanism point speeds

Method of determination	Angular velocities of links, rad/s						
	AB	AD	DE	$O_2 D$	$O_3 E$	FGH	$O_4 G$
According to the speed plan	0.424	0.5	0	0.547	0.972	0.272	0.740
Using instantaneous velocity centers	0.421	0.505	0	0.544	0.967	0.278	0.730

Fig. 2.34 To the definition of instantaneous velocity centers

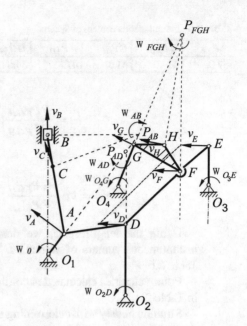

To determine the velocities of the points B and C of the AB link, we have the proportions

$$\frac{v_A}{v_B} = \frac{AP_{AB}}{BP_{AB}}; \quad \frac{v_A}{v_C} = \frac{AP_{AB}}{CP_{AB}}.$$

Hence,

$$v_B = v_A \cdot \frac{BP_{AB}}{AP_{AB}}; \quad v_C = v_A \cdot \frac{CP_{AB}}{AP_{AB}}.$$

Similarly, for the point D of the link AD, we can write

$$\frac{v_A}{v_D} = \frac{AP_{AD}}{DP_{AD}},$$

from where

$$v_D = v_A \cdot \frac{DP_{AD}}{AP_{AD}}.$$

Since the instantaneous center of velocities of the link DE is at infinity, then $v_E = v_F = v_D$.

To determine the velocities of the points G and H, we have the proportions

Table 2.6 Instantaneous centers of speeds

AP_{AB}	BP_{AB}	CP_{AB}	AP_{AD}	DP_{AD}	FP_{FGH}	GP_{FGH}	HP_{FGH}
57	41	41.5	47.5	34.5	62.7	52.8	50.8

$$\frac{v_F}{v_G} = \frac{FP_{FGH}}{GP_{FGH}}; \ \frac{v_F}{v_H} = \frac{FP_{FGH}}{HP_{FGH}}.$$

Hence,

$$v_G = v_F \cdot \frac{GP_{FGH}}{FP_{FGH}}; \ v_H = v_F \cdot \frac{HP_{FGH}}{FP_{FGH}}.$$

Using the length scale, we determine the distances from points to instantaneous centers of velocities. These distances (in cm) are given in Table 2.6.

Point velocities calculated according to the indicated formulas are shown in Table 2.4.

Simultaneously with determining the modules of the speeds of the points, we find the direction of the speeds and the direction of rotation of the links of the mechanism. For example, in the direction of the velocity of the point A and the position of the instantaneous center of velocities P_{AB}, we establish that the rotation of the link AB is clockwise. Therefore, the speed of point B at a given position of the mechanism is directed upward. Similarly, we determine the direction of rotation of the remaining links of the mechanism and the direction of the speeds of its points (Fig. 2.34).

(c) *Determine the angular speeds of the links of the mechanism.* The speed of any point of the link is equal to the product of the angular velocity of the link by the distance from the point to the instantaneous center of speeds:

$$v_A \omega_{AB} \cdot AP_{BA} = \omega_{AD} \cdot AP_{AD}.$$

From here we determine the angular velocities of the links AB and AD:

$$\omega_{AB} = \frac{v_A}{AP_{AB}}; \ \omega_{AD} = \frac{v_A}{AP_{AD}}.$$

The angular velocity of the link $O_2 D$ is determined by the velocity of the point D:

$$\omega_{O_2 D} = \frac{v_D}{O_2 D}.$$

Fig. 2.35 Determining the
acceleration of the point B

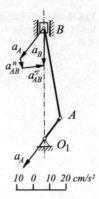

The angular velocity of the link DE at the given position of the mechanism
is equal to zero because the instantaneous center of velocities in this case is at
infinity: $\omega_{DE} = 0$.

Similarly, we determine the angular velocities of the remaining links of the
mechanism:

$$\omega_{O_3E} = \frac{v_E}{O_3E};$$
$$\omega_{FGH} = \frac{v_F}{FP_{FGH}};$$
$$\omega_{O_4G} = \frac{v_G}{O_4G}.$$

The angular velocities of the links calculated according to the indicated ratios
are given in Table 2.5.

3. *Determination of accelerations of points A, B, D and angular accelerations of
links AB and BD.*

(a) *Determine the accelerations $\vec{a}_A$, $\vec{a}_B$ and ε_{AB} (Fig. 2.35).* Using the theorem
on the acceleration of points of a plane figure, we determine the acceleration
of the point B:

$$a_B = a_A + a_{AB}^n + a_{AB}^\tau.$$

Since the crank O_1A rotates uniformly, the acceleration of the point A is
directed to the center of O_1 and is equal to

$$a_A = a_A^n = O_1A \cdot \omega_{O_1A}^2 = 12 \cdot 2^2 = 48 \; cm/s^2.$$

The centripetal acceleration of the point B in the rotational motion of the
connecting rod AB around the pole A is directed from the point B to the
point A and is equal to

Fig. 2.36 Determination of
the acceleration of the link
AD

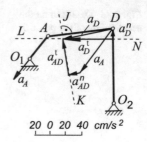

20 0 20 40 cm/s^2

$$a^n_{BA} = AB \cdot \omega^2_{AB} = 46 \cdot 0.4^2 = 7.36 \; cm/s^2.$$

We postpone the pole acceleration from the point B on the appropriate
scale a_A. From the end of the vector a_A, we construct the vector a^n_{AB},
drawing it parallel to BA. Through the end of the vector a^n_{AB}, we draw a
straight line perpendicular to BA, i.e., parallel to rotational acceleration a^τ_{AB}.
The intersection point of this straight line with the straight line along which
the slider acceleration vector B is directed determines the ends of the vectors
a_B and a^τ_{AB}.

By measuring in the drawing, we get

$$a_B = 39 cm/s^2; \; a^\tau_{AB} = 30 \; cm/s^2.$$

Since $a^\tau_{AB} = AB \cdot \varepsilon_{AB}$, then the angular acceleration of the link AB is

$$\varepsilon_{AB} = \frac{a^\tau_{AB}}{AB} = \frac{30}{46} = 0.652 \; s^{-2}.$$

(b) We determine the a_D and ε_{AD} (Fig 2.36).

Point D belongs to two links: AD and O_2D. Taking the point A as the pole,
we obtain

$$a_D = a_A + a^n_{AD} + a^\tau_{AD}.$$

Acceleration of point A is found above: $a_A = 48 \; cm/s^2$.
The centripetal acceleration of the point D in the rotational motion of the link
AD around the pole A is directed from the point D to the point A and is equal to

$$a^n_{AD} = AD \cdot \omega^2_{AD} = 28.5 \cdot 0.5^2 = 7.1 \; cm/s^2.$$

We postpone the acceleration of the pole a_A from the point D on the
appropriate scale. From the end of the a_A vector, we construct a vector a^n_{AD},
drawing it parallel to DA. Draw the straight line JK through the end of the
a^n_{AD} vector perpendicular to DA, i.e., parallel to the tangential acceleration of

the a_{AD}^{τ}. However, it is not possible to determine a_D acceleration with this build because its direction is unknown.

To find the acceleration of the point D, it is necessary to perform the second construction, considering D as the point of the link O_2D. In this case

$$a_D = a_D^n + a_D^{\tau}.$$

The centripetal (normal) acceleration of the point D will be

$$a_D^n = O_2D \cdot \omega_{O_2D}^2 = 8 \ cm/s^2.$$

We postpone the vector a_D^n from the point D, directing it to the center O_2. Draw a straight line LN through the end of the vector a_D^n perpendicular to O_2D, i.e., parallel to the rotational acceleration a_D^{τ}.

The intersection point of this line with JK defines the ends of the vectors a_D, a_{AD}^{τ}, and a_D^{τ}. By measuring in the drawing, we get $a_D = 42 \ cm/s^2$; $a_D^{\tau} = 30 \ cm/s^2$.

Because $a_{AD}^{\tau} = AD \cdot \varepsilon_{AD}$, then the angular acceleration of the AD link will be

$$\varepsilon_{AD} = \frac{a_{AD}^{\tau}}{AD} = \frac{30}{29} = 1.03 \ s^{-2}.$$

4. *Determination of the position of the instantaneous center of acceleration of the link AB* (Fig. 2.37). Let's take the point A as a pole. Then the acceleration of the point B will be

$$a_B = a_A + a_{AB}.$$

Build a parallelogram of accelerations at the point B along the a_B diagonal and the a_A side. The side of the a_{AB} parallelogram expresses the acceleration of the point B in the rotation of AB around the pole A.

Fig. 2.37 Instantaneous
acceleration center of the link
AB

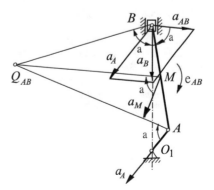

The acceleration a_{AB} makes an angle α with the segment AB, which can be measured in the drawing.

The direction of the a_D^τ vector relative to the pole A allows you to determine the ε_{AB}, in this case clockwise. Putting the angle α from the vectors a_A and a_B in this direction and drawing two beams, we find the point of their intersection Q_{AB}—the instantaneous center of acceleration of the link AB.

5. Determination of the acceleration of the point M. Find the acceleration of the point M using the instantaneous acceleration center. The accelerations of the points of a planar figure are proportional to the distances to their instantaneous acceleration center:

$$\frac{a_M}{a_A} = \frac{MQ_{AB}}{AQ_{AB}}. \tag{2.81}$$

According to the drawing, we define $MQ_{AB} = 67.5\,cm$, $AQ_{AB} = 77\,cm$. Using these results, we find the acceleration of the point M:

$$a_M = a_A \cdot \frac{MQ_{AB}}{AQ_{AB}} = 48 \cdot \frac{67.5}{77} = 42.1\,cm/s^2. \tag{2.82}$$

The acceleration of a_M is equal to the angle α with the straight line MQ_{AB}; the direction of this vector corresponds to the angular acceleration ε_{AB}.

References

J. Begss, *Kinematics* (Taylor and Francis Publ., 1983), p. 1. ISBN 0-89116-355-7

V. Molotnikov, *Osnovy teoreticheskoi mekhaniki [Fundamentals of Theoretical Mechanics]* (Fenix Publ., Rostov-on-Don, 2004)

V. Molotnikov, *Mekhanika konstruktsiiyu Teoreticheskaya mekhanika. Soprotivlenie materialov [Construction mechanics and Theoretical Mechanics. Mechanics of Materials]* (Lan' Publ., Sankt-Peterburg, 2012)

V. Molotnikov, *Tekhnicheskaya mekhanika [Technical Mechanics]* (Lan' Publ., Sankt-Peterburg, 2017)

P. Rashevskii, *Differentsial'naya geometriya [Differential Geometry]* (GITTL, Moscow–Lenigrad, 1950)

E. Whittaker, *A Treatise on the Analytical Dynamics of Particles and Rigid Bodies* (Cambridge University Press, Cambridge, 1904). ISBN 0-521-35883-3

A. Yablonskii, S. Noreiko, S. Vol'fson, dr., *Sbornik zadanii dlya kursovykh rabot po teoreticheskoi mekhanike: Uchebnoe posobie dlya tekhnicheskikh vuzov, 5-e izd. [Collection of Tasks for Term Papers on Theoretical Mechanics: A Textbook for Technical Universities]*, 5th edn. (Integral-Press Publ., Moscow, 2000)

Chapter 3
Dynamics

Abstract The motion of material bodies is studied depending on the forces acting on them. Forces in the dynamics of the variables may depend on time, position of the body, and its velocity. You can come to the concept of inertia if you want to compare the results of the action of one force with different bodies. The property of bodies to change their speed faster or slower under the action of applied forces is called inertia. A quantitative measure of inertia is the mass of a body (a measure of the gravitational properties of a body). The mass of this body will be considered constant (with the exception of specially discussed cases). To distract from the shape of the body, the concept of a material point is introduced. A material point is a material body that has a mass, the size of which can be ignored when studying motion. In dynamics, a body moving translationally can be considered a material point. Dynamics are based on the laws that Newton laid out in 1687. These three laws are also called the Galileo-Newton axioms, since the first law was discovered by Galileo in 1638.

Keywords Dynamics of point · Basic law of dynamics · Law of independence of forces · Problems of dynamics · The theorem of Coriolis · Energy of system · Operation and power · The amount of motion · Kinetic energy · Center of mass · The principle of d'Alembert

3.1 Point Dynamics

Dynamics studies the laws of motion of material bodies under the action of forces. The simplest material object is the *material point*. We have already spoken (p. 5) so-called model material body of any form, the size of which in the considered task can be neglected and to take a body for a geometric point mass equal to body mass. More complex material objects—solid bodies and mechanical systems of bodies—are considered to consist of material points (Yablonskii 1971; Molotnikov 2012; Berezkin 1974).

In dynamics, in contrast to kinematics, when studying the motion of bodies, both the acting forces and the *inertia* of the material bodies themselves are taken into account. Inertia is the property of material bodies to change the speed of their movement faster or slower under the action of applied forces. A quantitative measure of a body's inertia is its mass.

All positions of dynamics are derived from its axioms, also called *laws of classical mechanics*. We have already seen two of them—the axiom of inertia and the axiom of equality of action and reaction—in the section «Statics». Let us formulate two other axioms.

Basic Law of Dynamics   *The acceleration of a material point relative to an inertial frame of reference is proportional to the force applied to the point and in the direction coincides with the force vector.*   If applied to a point force denoted by F, its acceleration relative to the inertial reference system—using a—and mass, using m, then the fundamental law of dynamics expressed by the formula

$$ma = F. \tag{3.1}$$

The mass m included in the formulation of the basic law of dynamics (3.1) is called the *inert mass* of the point. In contrast to the inert mass, the values m and M are included in Newton's law of gravitation:

$$F = G\frac{mM}{r^2}, \ (G\ -const), and$$

they are called *gravitational masses* of material points that are distant from each other at a distance of r. It is experimentally established that with a high degree of accuracy, the inert and gravitational masses are equivalent.

The Law of Independence of the Action of Forces   *When several forces simultaneously act on a material point, the acceleration of a point in an inertial frame of reference from the action of each individual force does not depend on the presence of other forces applied to the point, and the total acceleration of the point is equal to the vector sum of the accelerations from the action of individual forces.*

Let a system of forces be applied to a material point $(F_1, F_2, \ldots, F_n)$, which we will also symbolically denote by (F) for brevity. According to the law of independence of the action of forces, the acceleration of a point from the action of each of the forces of the system is determined by the formula (3.1)

$$ma_1 = F_1, \ ma_2 = F_2, \ \ldots, \ ma_n = F_n. \tag{3.2}$$

With all the forces of the system acting simultaneously (F), the acceleration of a point is equal to the vector sum of the accelerations caused by individual forces:

$$a = \sum_{i=1}^{n} a_i. \tag{3.3}$$

Summing up the equalities (3.2) and using (3.3), we obtain *the basic equation of point dynamics:*

$$m\boldsymbol{a} = \sum_{i=1}^{n} \boldsymbol{F}_i. \tag{3.4}$$

Equation (3.4) remains valid for a non-free material point. In this case, the number of forces applied should also include the forces of bond reactions.

System of Units   The basic law of dynamics (3.1) relates the acceleration, force, and mass of a material point. The dimension of acceleration depends on the choice of units of length and time. Due to the law (3.1), the units of length, time, mass, and force cannot be selected independently. Only three of these values are independent, but the dimension of the fourth is determined by the law (3.1).

In the international system SI per unit of time adopted is the second (s); length, meter (m); and mass, kilogram (kg). Per unit of power adopted here is the Newton (N), which is equal to the force that tells the body mass of 1 kg an acceleration of 1 m/s^2.

In the technical system of units of MCGSS, independent values are taken: for time-second, for length-meter, and for force-kilogram-force (kgf). The unit of mass is derived from these units and is determined from the basic law of dynamics (3.1). It is called the *technical unit of mass* (i.e., m). By definition, one technical unit of mass has a body that under the action of a force of 1 kgf receives an acceleration of 1 m/s^2. In particular, if in the formula (3.1) F is the force of gravity, and the acceleration $a = g$ is the acceleration of gravity, then it follows from the basic law of dynamics that the mass in 1, i.e., m, has a body whose gravity is 9.81 kgf.

In the GHS system, the following independent values are taken as independent values: unit of time , second; unit of mass, gram (g); and unit of length, centimeter (cm). The unit of force—Dina—is derived and has the dimension g· cm/s^2. Therefore, 1 Dina is a force that gives a body with a mass of 1 g an acceleration of 1 cm/s^2. It is easy to calculate that

$$1 \text{ Dina} = 10^{-3} \text{ kg} \cdot 10^{-2} \text{ m/s}^2 = 10^{-5} \text{ N}.$$

Differential Equations of Motion of a Material Point   It is known from kinematics that the acceleration $\boldsymbol{a}$ is expressed in terms of the radius vector $\boldsymbol{r}$ of a point by the formula

$$\boldsymbol{a} = \frac{d^2\boldsymbol{r}}{dt^2}.$$

Substituting this dependence in the formula (3.4), we obtain the differential equation of motion of a material point in vector form:

$$m\frac{d^2\mathbf{r}}{dt^2} = \sum_{i=1}^{n} \mathbf{F}_i. \tag{3.5}$$

Denoting the resultant of all given forces and bond reaction forces by $\mathbf{F}$, we write Eq. (3.5) as

$$m\frac{d^2\mathbf{r}}{dt^2} = \mathbf{F}. \tag{3.6}$$

If we project both parts of Eq. (3.6) on the axis of the Cartesian rectangular coordinate system $Oxyz$, we get

$$m\frac{d^2x}{dt^2} = F_x; \ m\frac{d^2y}{dt^2} = F_y; \ m\frac{d^2z}{dt^2} = F_z, \tag{3.7}$$

where x, y, z are projections of the radius vector $\mathbf{r}$ on the x, y, and z axes, respectively, and F_x, F_y, F_z are projections of the resultant $\mathbf{F}$ on the specified axes. Equalities (3.7) represent differential equations of motion of a material point in a rectangular Cartesian coordinate system.

If a point moves only in one plane Oxy, and then assuming $z \equiv 0$, $F_z = 0$, we get

$$m\frac{d^2x}{dt^2} = F_x; \ m\frac{d^2y}{dt^2} = F_y. \tag{3.8}$$

In another special case, when the point moves in a straight line Ox, we obtain one differential equation for the rectilinear motion of the point:

$$m\frac{d^2x}{dt^2} = F_x. \tag{3.9}$$

Sometimes it is more convenient to consider the movement of a material point in natural moving coordinate axes. Projecting both parts of the formula (3.1) onto natural axes, we obtain

$$ma_\tau = F_\tau; \ ma_n = F_n; \ ma_b = F_b, \tag{3.10}$$

where a_τ, a_n, a_b, F_τ, F_n, F_b are, respectively, the projection of acceleration and resultant forces on the tangent, main normal, and binormal to the trajectory of a point at a given time. In kinematics, it was found that in natural axes, the components of the acceleration vector are expressed in terms of the arc coordinate s, the velocity of the point v, and the radius of curvature of the trajectory ρ by the formulas

$$a_\tau = \frac{d^2s}{dt^2}; \ a_n = \frac{v^2}{\rho}; \ a_b = 0.$$

Substituting these dependencies into formulas (3.10), we obtain differential equations of motion of a material point in projections on natural coordinate axes:

$$m\frac{d^2 s}{dt^2} = F_\tau; \quad m\frac{v^2}{\rho}; \quad 0 = F_b. \tag{3.11}$$

Two Main Problems of Point Dynamics Differential equations of the dynamics of a material point in one or another coordinate system allow us to solve two main problems.

T h e   f i r s t   t a s k   o f   d y n a m i c s. The mass of a point and the law of its motion are known. Find the force acting on the point.

If, for example, the equations of motion are given in a Cartesian coordinate system

$$x = f_1(t); \quad y = f_2(t); \quad z = f_3(t),$$

then from Eqs. (3.7), the projections of the resultant F on the coordinate axis are determined:

$$F_x = m\frac{d^2 x}{dt^2} = m\frac{d^2 f_1}{dt^2}; \quad F_y = m\frac{d^2 y}{dt^2} = m\frac{d^2 f_2}{dt^2}; \quad F_z = m\frac{d^2 z}{dt^2} = m\frac{d^2 f_3}{dt^2}.$$

With the known projections F_x, F_y, F_z of the force F, it is easy to determine its modulus and direction.

T h e   s e c o n d   t a s k   o f   d y n a m i c s. The mass of the point and the resultant of the forces applied to it are known. Find the law of motion of this point.

Let's consider the solution of the problem in a rectangular Cartesian coordinate system. The force F, generally speaking, may depend on time, coordinates of moving point, its speed, acceleration, etc. For simplicity, restrict ourselves to the case when the force F depends only on time, position, and speed point. Under this assumption, the differential equations (3.7) can be written as

$$m\frac{d^2 x}{dt^2} = F_x(t;\ x, y, z;\ \dot{x}, \dot{y}, \dot{z}); \quad m\frac{d^2 y}{dt^2} = F_y(t;\ x, y, z;\ \dot{x}, \dot{y}, \dot{z});$$
$$m\frac{d^2 z}{dt^2} = F_z(t;\ x, y, z;\ \dot{x}, \dot{y}, \dot{z}).$$

Thus, the problem is reduced to integrating a system of three second-order ordinary differential equations. The general solution of each of these equations depends on two arbitrary constants, and the general integral of the system will contain six arbitrary constants C_1, C_2, $\ldots$, C_6. Let's write the general solution of the problem in the form

$$x = f_1(t; C_1, C_2, \ldots, C_6);$$
$$y = f_2(t; C_1, C_2, \ldots, C_6);$$
$$z = f_3(t; C_1, C_2, \ldots, C_6), \tag{3.12}$$

where f_1, f_2, f_3 are known functions of the specified arguments. Differentiating the solution (3.12) by time, we get the velocity projections on the coordinate axes:

$$v_x = \dot{x} = f_1'(t; C_1, C_2, \ldots, C_6);$$
$$v_y = \dot{y} = f_2'(t; C_1, C_2, \ldots, C_6);$$
$$v_z = \dot{z} = f_3'(t; C_1, C_2, \ldots, C_6). \tag{3.13}$$

To isolate a specific movement from the six-parameter trajectory equation (3.12), you must specify the position of the point and its speed at some fixed point in time, for example, at $t = 0$. Substituting (3.12) and (3.13) for the coordinate and velocity values at $t = 0$, we obtain the following six equations for determining six arbitrary constants:

$$x_0 = f_1(0; C_1, C_2, \ldots, C_6); \quad v_{ox} = f_1'(0; C_1, C_2, \ldots, C_6);$$
$$y_0 = f_2(0; C_1, C_2, \ldots, C_6); \quad v_{oy} = f_2'(0; C_1, C_2, \ldots, C_6);$$
$$z_0 = f_3(0; C_1, C_2, \ldots, C_6); \quad v_{oz} = f_3'(0; C_1, C_2, \ldots, C_6), \tag{3.14}$$

where indicated:

$$x_0 = x\,|_{t=0}, \quad y_0 = y\,|_{t=0}, \quad z_0 = z\,|_{t=0};$$
$$v_{ox} = v_z\,|_{t=0}, \quad v_{oy} = v_y\,|_{t=0}, \quad v_{oz} = v_z\,|_{t=0}. \tag{3.15}$$

When a point moves in a plane, we will have only two differential equations, and in the case of rectilinear motion, we will have one equation. In these particular cases, the number of arbitrary constants in the solution (3.12) and initial conditions is reduced to four and two, respectively.

Example 1 The motion of a material point of mass 0.2 kg is expressed by the equations $x = 3\cos 2\pi t$ cm, $y = 4\sin \pi t$ cm (t in seconds). Determine the projection of the force acting on the point, depending on its coordinates.

D e c i s i o n. Write down the differential equations of the point motion under study:

$$m\frac{d^2x}{dt^2} = F_x; \quad m\frac{d^2y}{dt^2} = F_y.$$

Calculate the required derivatives:

$$\frac{dx}{dt} = -6\pi \sin 2\pi t; \quad \frac{d^2x}{dt^2} = -12\pi^2 \cos 2\pi t = -4\pi^2 x;$$

$$\frac{dy}{dt} = 4\pi \cos \pi t; \quad \frac{d^2y}{dt^2} = -4\pi^2 \sin \pi t = -\pi^2 y.$$

Now you can define the desired force projections:

$$F_x = m\ddot{x} = -0,2 \cdot 4 \cdot 3,14^2 \cdot 0,01x = -0,0789x \ (H);$$

$$F_y = m\ddot{y} = -0,2 \cdot 3,14^2 \cdot 0,01y = -0,0197y \ (H),$$

and x and y must be expressed in centimeters.

Example 2 A material point of mass m makes a rectilinear motion under the action of a force that changes according to the law $F = F_0 \cos \omega t$, where F_0 and ω are constant values. At the initial moment, the point had a speed of $\dot{x} = v_0$. Find the equation of motion of the point.

D e c i s i o n. In the case of rectilinear motion of a point, the differential equation of motion has the form

$$m\frac{d^2x}{dt^2} = F,$$

and the initial conditions in our problem are

$$t = 0 : \ x = x_0 = 0; \ v = \frac{dx}{dt} = v_0.$$

Have

$$m\frac{d^2x}{dt^2} = F_0 \cos \omega t;$$

$$m\frac{dx}{dt} = \frac{1}{\omega} F_0 \sin \omega t + C_1 \ (C_1 - const);$$

$$mx = -\frac{F_0}{\omega^2} \cos \omega t + C_1 t + C_2 \ (C_2 - const).$$

We determine the constants C_1 and C_2 from the initial conditions:

$$0 = -\frac{F_0}{\omega^2} + C_2, \ C_2 = \frac{F_0}{\omega^2}; \ C_1 = mv_0.$$

Substituting the found values of constants in the general solution of the differential equation, we find the law of motion of the point:

$$x = -\frac{F_0}{m\omega^2}\cos\omega t + v_0 t + \frac{F_0}{m\omega^2}$$

or

$$x = \frac{F_0}{m\omega^2}(1 - \cos\omega t) + v_0 t.$$

3.2   Dynamic Coriolis Theorem

Let a material point of mass m move in a mobile, generally non-inertial, frame of reference $Oxyz$ (Fig. 3.1). We introduce an inertial frame of reference $O_1x_1y_1z_1$. The resultant of the active forces applied to the point is denoted by F, and the resultant of the reaction forces of bonds is denoted by N. In the inertial frame of reference $O_1x_1y_1z_1$, the equation of motion of a point in vector form will be

$$m a = F + N, \tag{3.16}$$

and a is the absolute acceleration of the point. Applying the acceleration addition theorem (p. 92), we have

$$a = a_e + a_r + a_k, \tag{3.17}$$

where a_e, a_r, a_k are the portable, relative, and Coriolis accelerations of the point, respectively.

Substituting the absolute acceleration a by the formula (3.17) into the equation of motion (3.16), we get

$$m a_r = F + N + \mathbf{\Phi}_e + \mathbf{\Phi}_k, \tag{3.18}$$

where indicated:

Fig. 3.1 To the Coriolis theorem

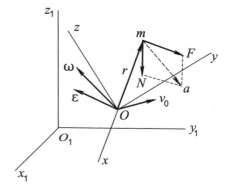

$$\Phi_e = -ma_e; \quad \Phi_k = -ma_k. \tag{3.19}$$

The forces Φ_e and Φ_k defined by the formulas (3.19) are called *portable* and *Coriolis* inertial forces, respectively.

The formula (3.18), which represents the equation of relative motion of a point in vector form, expresses **the dynamic Coriolis theorem**:

Theorem *A material point moves relative to a non-inertial frame of reference in the same way as it does relative to an inertial one, and only to the applied active forces and bond reactions, one should add the transport and Coriolis forces of inertia.*

The inertial forces Φ_e and Φ_k are corrections for the non-inertiality of the reference frame. If the mobile reference system $Oxyz$ moves relative to the main inertial system translationally, uniformly, and rectilinearly ($\omega = 0$, $\varepsilon = 0$), then the inertial forces Φ_e and Φ_k vanish. In this case, the equation of relative motion of the point (3.18) takes the form

$$ma_r = F + N, \tag{3.20}$$

which coincides with Eq. (3.16) of motion relative to the inertial frame of reference. The result reflects the *principle of relativity* of classical —the principle of Galileo–Newton, which can be formulated in the form as follows: *all mechanical phenomena in different inertial reference systems proceed in the same way,* or *no mechanical experiments can detect the inertial motion of the reference system, participating together with it in this movement.*

Example A body weighing $G = 2\,\text{H}$ is placed on a smooth face of a three-sided prism, the other face of which lies on the horizontal plane. What horizontal acceleration should the prism have so that the body does not move relative to the prism, and what pressure does the body produce on the prism in this case, if $\alpha = \text{arctg}(3/4)$ (Fig. 3.2)?

D e c i s i o n. We consider the body as a material point M, which is affected by the body weight G, the normal reaction force N, and the inertial force Φ_e (Fig. 3.2). By the condition, there is no acceleration of the point M in relative motion: $a_r = 0$, so from the formula (3.18), we get

$$G + N + \Phi_e = 0.$$

Projecting this vector equality on the x and y axes moving with the prism, we get

$$G \sin \alpha - \Phi_e \cos \alpha = 0;$$
$$N - G \cos \alpha - \Phi_e \sin \alpha = 0. \tag{3.21}$$

In the problem under consideration,

Fig. 3.2 Accounting for
inertial forces

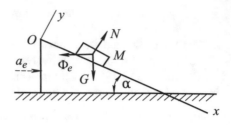

$$G = mg; \quad \Phi_e = ma_e,$$

where m is the mass of the point M, g is the acceleration of gravity, and a_e is the desired acceleration of the prism.

From the first equation of the system (3.21), we find

$$mg \sin \alpha - ma_e \cos \alpha = 0; \quad a_e = g \cdot \mathrm{tg}\, \alpha = \frac{3}{4}g = 7,35 \text{ M/c}^2.$$

Then

$$\Phi_e = ma_e = mg \cdot \mathrm{tg}\, \alpha = G \cdot \mathrm{tg}\, \alpha.$$

From the second equation (3.21), we determine the modulus of the prism reaction force:

$$N = G \cos \alpha + \Phi_e \sin \alpha = G \cos \alpha + G \cdot \mathrm{tg}\, \alpha \sin \alpha = G \cos \alpha (1 + \mathrm{tg}\,^2\alpha);$$

$$N = G \frac{1 + \mathrm{tg}\,^2\alpha}{\sqrt{1 + \mathrm{tg}\,^2\alpha}} = G\sqrt{1 + \mathrm{tg}\,^2\alpha};$$

$$N = 2\sqrt{1 + (3/4)^2} = 2,5 \text{ H}.$$

3.3 General Theorems of Dynamics of a Point

3.3.1 Basic Concepts and Definitions

Many engineering tasks do not require complete knowledge of the motion characteristics of a material point. Individual, practically important aspects of the phenomenon can be studied using general theorems that follow from the basic law of dynamics. In this case, such motion characteristics as the amount of motion, kinetic energy, force momentum, etc. are widely used. We will give definitions of the characteristics that we will need in the following presentation.

The amount of motion of a material point is called the vector $\boldsymbol{q}$, which is equal to the product of the mass of the point m and its speed $\boldsymbol{v}$, i.e.,

$$\boldsymbol{q} = m\boldsymbol{v}. \tag{3.22}$$

In physics, the amount of motion is often called the momentum of a material point. We will use both of these names. The vector of the amount of movement of a point is considered to be applied to the most moving point.

The kinetic energy of a point is the scalar value T, equal to half the product of the mass of the point and the scalar square of its velocity:

$$T = \frac{m\boldsymbol{v}^2}{2} = \frac{mv^2}{2}. \tag{3.23}$$

This value is often also called *living force.*

The effect of a force on a material point depends not only on the magnitude of the force but also on the duration of its action. To characterize the action exerted by a force over a certain period of time, the concept of *force momentum is introduced.*
An elementary force pulse $\boldsymbol{F}$ is a vector quantity $d\boldsymbol{S}$, equal to the product of the force by the elementary time interval of its action dt:

$$d\boldsymbol{S} = \boldsymbol{F}dt. \tag{3.24}$$

As follows from the definition, the elementary impulse is directed along the line of action of the force. Summing up the elementary pulses (3.24), we get the force pulse for a finite time period t_1:

$$\boldsymbol{S} = \int\limits_{0}^{t_1} \boldsymbol{F}dt. \tag{3.25}$$

If the force $\boldsymbol{F}$ remains constant for the entire time from 0 to t_1 both in modulus and direction, we have $\boldsymbol{S} = \boldsymbol{F}t_1$, and the modulus of the force momentum will be $S = Ft_1$. In general, the modulus of the force pulse can be calculated through its projections. In particular, when using the rectangular coordinate system $Oxyz$, the projections of the force pulse $\boldsymbol{F}$ for the time interval from 0 to t_1 on the coordinate axes will be

$$S_x = \int\limits_{0}^{t_1} F_x dt; \; S_y = \int\limits_{0}^{t_1} F_y dt; \; S_z = \int\limits_{0}^{t_1} F_z dt. \tag{3.26}$$

The energy characteristics of the force action when the point moves are *work* and *power.* Let's define these concepts. Let the point M (Fig. 3.3) move along the path $M_0 M_1$ under the force $\boldsymbol{F}$. The elementary work of the force $\boldsymbol{F}$ is a scalar value

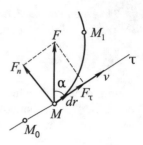

dA equal to the product of the projection F_τ of the force F on the tangent τ to the track, directed in the direction of moving the point, on an infinitesimal arc ds in the direction of the tangent, i.e.,

$$dA = F_\tau ds, \tag{3.27}$$

or, since $F_\tau = F \cos \alpha$,

$$dA = F \cos \alpha ds. \tag{3.28}$$

If F_x, F_y, F_z are projections of the force F on the Cartesian axes x, y, and z, and dx, dy, dz are projections of ds on the same axes, then the formula (3.27) can also be represented as

$$dA = F_x dx + F_y dy + F_z dz. \tag{3.29}$$

When a point is finally moved from the position M_0 to the position M_1 (Fig. 3.3), the work of the force F is calculated as the integral sum of the elementary works, i.e.,

$$A = \int_{(M_0)}^{(M_1)} F_\tau ds$$

or

$$A = \int_{(M_0)}^{(M_1)} (F_x dx + F_y dy + F_z dz). \tag{3.30}$$

Power W of force is the ratio of the work dA to the infinitesimal time interval dt for which this work is completed, i.e.,

$$W = \frac{dA}{dt}. \tag{3.31}$$

If the force is constant in time, it follows that

$$W = \frac{A}{t_1},$$
(3.32)

where t_1 is the time during which A was completed.

Using the formula (3.27), you can get

$$W = \frac{dA}{dt} = \frac{F_\tau ds}{dt} = F_\tau v.$$
(3.33)

3.3.2   Theorem on Changing the Amount of Motion Material Point

Let the mass of the point remain constant. Since $a = dv/dt$, the basic law of dynamics (3.4) of a point moving under the action of n forces can be represented as

$$\frac{d}{dt}(mv) = \sum_{i=1}^{n} F_i \text{ или } \frac{dq}{dt} = \sum_{i=1}^{n} F_i.$$
(3.34)

The formulas (3.34) express **the theorem on changing the amount of motion of a material point in differential form.**

Theorem *The time derivative of the amount of motion of a point is equal to the vector sum of the forces acting on the point.*

Integrating the second of the formulas (3.34) in the range from $t = 0$ to $t = t_1$, we get

$$\int_0^{t_1} dq = \sum_{i=1}^{n} \int_0^{t_1} F_i dt \text{ or } q(t_1) - q(0) = \sum_{i=1}^{n} \int_0^{t_1} F_i dt.$$

Denote $q(t_1) = q_1, q(0) = q_0$ and apply the formula (3.25). Find

$$q_1 - q_0 = \sum_{i=1}^{n} S_i,$$
(3.35)

where S_i is the pulse of the ith force for the time interval from 0 to t_1. Formula (3.35) expresses theorem on change of momentum of a point in the final form.

Theorem *The change of momentum of a material point over a period of time is equal to the vector sum of the momenta of all attached to the point of force for the same period of time.*

The vector equality (3.35) in Cartesian coordinates is represented by three scalar expressions:

$$q_{1x} - q_{0x} = \sum_{i=1}^{n} S_{ix}; \quad q_{1y} - q_{0y} = \sum_{i=1}^{n} S_{iy}; \quad q_{1z} - q_{0z} = \sum_{i=1}^{n} S_{iz}. \tag{3.36}$$

In the case of rectilinear motion of a point, only one of the three equalities (3.36) remains, and when moving along a plane, only two remain.

3.3.3 Theorem on Changing the Kinetic Energy of a Point

Let's write down the basic law of the dynamics of a material point in the form

$$m \frac{d\boldsymbol{v}}{dt} = \boldsymbol{F},$$

where $\boldsymbol{F}$ is the resultant of the forces applied to the point. Multiply both parts of this relation scalar by the differential of the radius vector of the point $d\boldsymbol{r}$. Have

$$m d\boldsymbol{v} \cdot \frac{d\boldsymbol{r}}{dt} = \boldsymbol{F} d\boldsymbol{r}.$$

Given that by definition $d\boldsymbol{r}/dt = \boldsymbol{v}$, we get

$$m \boldsymbol{v} d\boldsymbol{v} = \boldsymbol{F} d\boldsymbol{r}.$$

The left side of this equality can be represented as

$$m \boldsymbol{v} d\boldsymbol{v} = d(m\boldsymbol{v}^2/2) = d(mv^2/2),$$

and the scalar product on its right side is the differential of the dA force $\boldsymbol{F}$. In fact (see Fig. 3.3)

$$\boldsymbol{F} d\boldsymbol{r} = F d(r \cos\alpha) = F_\tau ds = dA.$$

Thus, we finally get

$$d(mv^2/2) = dA. \tag{3.37}$$

The formula (3.37) in differential form expresses the theorem on the change in the kinetic energy of a material point:

Theorem *The differential of the kinetic energy of a point is equal to the elementary work of the force applied to the point.*

Divide both parts of Eq. (3.37) by dt. Then, given that $dA/dt = W$ is the power of F, we get another expression of the theorem:

$$\frac{d}{dt}\left(\frac{mv^2}{2}\right) = W, \tag{3.38}$$

that is, the *derivative of the kinetic energy of a point in time is equal to the power applied to the point of force.*

Let the speed of a point at position M_0 (Fig. 3.3) be v_0, and at position $M_1 - v_1$. Integrating both parts of Eq. (3.37) from the point M_0 to the point M_1, we obtain the mathematical expression of the theorem on changing the kinetic energy of a point in the final form

$$\frac{mv_1^2}{2} - \frac{mv_0^2}{2} = A, \tag{3.39}$$

i.e., *the change in the kinetic energy of a point at the final displacement is equal to the work of the force acting on the point at the same displacement.*

3.3.4 Theorem on Changing the Moment of the Amount of Motion of a Point

According to the expression (3.22), the amount of motion q of a material point is a vector quantity. Let's define the moments of this vector relative to the center and relative to the axis. *For moment of the amount of movement* $q = mv$, the points M relative to the center O (Fig. 3.4a) are the vector L_0, equal to the vector product of the radius vector r of the point M drawn from the center O and the vector q:

$$L_0 = r \times q = r \times mv. \tag{3.40}$$

The modulus of the vector L_0 is equal to the product of the value mv on the shoulder h of the vector mv relative to the center O (Fig. 3.4a):

$$L_0 = mvh. \tag{3.41}$$

It follows that the module L_0 is equal to twice the area of the shaded area in Fig. 3.4a of a triangle with base mv and height h.

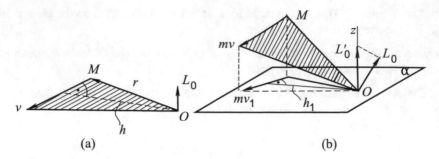

Fig. 3.4 Graphical representation of the moment of the amount of movement

Moment (L_z) of the *momentum* (mv) of the material point M with respect to the z axis (Fig. 3.4b) is the product of the projection (mv_1), taken with a "plus" sign or "minus" of the vector mv onto the plane α, perpendicular to z, on the arm h_1 of this projection with respect to points (O) of intersection of the z-axis and the α-plane:

$$L_z = \pm mv_1 h_1; \tag{3.42}$$

moreover, $L_z > 0$ if the vector $m\ mv_1$ is visible from the end of the z axis in a counterclockwise direction, and $L_z < 0$ otherwise.

Just as in statics (Ch. 1), the moments of the amount of motion relative to the center O and relative to the axis z passing through this center are related by the dependence

$$L_z = L_0 \cos(\widehat{L_0, z}). \tag{3.43}$$

Denoting the coordinates of a moving point M by x, y, z, and projecting the velocity of this point on the coordinate axis by v_x, v_y, v_z, by analogy with the formula (1.23), you can get

$$L_x = m(yv_z - zv_y); \; L_y = m(zv_x - xv_z); \; L_z = m(xv_y - yv_x). \tag{3.44}$$

We prove the following proposition.

Theorem *The time derivative of the moment of the amount of motion of a material point relative to an arbitrarily selected fixed center is equal to the vector sum of the moments of forces acting on the point relative to the same center.*

Proof Let the point M move along the trajectory ab under the force P (Fig. 3.5). Choose an arbitrary center O and draw the radius vector r from the point O to the point M. The moment of force P relative to the center of O will be

$$M_0 = r \times P.$$

Fig. 3.5 On the derivative
theorem

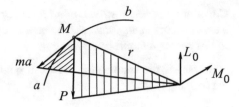

According to the formula (3.40), the moment of the amount of motion of a point M relative to the center O is defined as

$$L_0 = r \times mv.$$

We differentiate the last formula by time, taking into account that $dr/dt = и$ $dv/dt = a$. Receive

$$\frac{dL_0}{dt} = v \times mv + r \times ma.$$

Since the vectors v and mv are collinear, their vector product is zero. Therefore,

$$\frac{dL_0}{dt} = r \times ma = r \times P$$

or

$$\frac{dL_0}{dt} = M_0. \tag{3.45}$$

When multiple forces act on a material point, $(P_1, P_2, \ldots, P_n)$ the moment M_0 of the resultant P is equal to the vector sum of the moments of the component forces, and the formula (3.45) takes the form

$$\frac{dL_0}{dt} = M_{10} + M_{20} + \ldots + M_{n0} = \sum_{i=1}^{n} M_{i0}. \tag{3.46}$$

The theorem is proved. □

Projecting the vector equality (3.46) on the x, y, z axis, we can write

$$\frac{dL_x}{dt} = \sum_{i=1}^{n} M_{ix}; \quad \frac{dL_y}{dt} = \sum_{i=1}^{n} M_{iy}; \quad \frac{dL_z}{dt} = \sum_{i=1}^{n} M_{iz}. \tag{3.47}$$

Corollary 1 *If the resultant of all forces applied to a material point passes through a certain fixed center, then the moment of the amount of motion of the point relative to this center remains constant.*

In fact, if the resultant P passes through the center of O, then $M_0 = 0$ and by the formula (3.45)

$$\frac{d L_0}{dt} = 0, \qquad (3.48)$$

i.e., $L_0 = const$.

Corollary 2 *If the moment of the resultant of all forces applied to a material point relative to some axis is zero, then the moment of the amount of movement of the point relative to this axis remains constant.*

From the relations (3.47), it follows that, for example,

$$\text{if } \sum_{i=1}^{n} M_{ix} = 0, \text{ then } \frac{d L_x}{dt} = 0,$$

which means that $L_x = const$.

Example A ball of weight p attached to an inextensible thread slides along a smooth horizontal plane (Fig. 3.6). The other end of the thread is drawn at a constant speed C into a hole made on the plane. Determine the movement of the ball and the thread tension T, if it is known that at the initial moment the thread is located in a straight line Ox, the distance between the ball and the hole is R, and the projection of the initial velocity of the ball perpendicular to the direction of the thread is v_0.

D e c i s i o n. We introduce the polar coordinates r, φ, which determine the position of the ball M on the plane α. Then the initial conditions will be

$$t_0 = 0: \ r_0 = R, \ \varphi_0 = 0. \qquad (3.49)$$

Three forces act on the ball: the weight of the ball is p, the reaction force of the plane is N, and the reaction of the thread is T. Since the moment of each of these forces relative to the vertical axis Oz is zero, then based on corollary 2, the moment

Fig. 3.6 To study the
movement of the ball

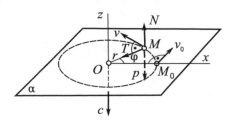

of the amount of motion relative to this axis must be constant, i.e.,

$$L_z = const.$$ (3.50)

By the condition of the problem, the coordinate r changes according to the equation

$$r(t) = R - ct.$$ (3.51)

To find the function $\varphi(t)$, use the condition (3.50). By definition we have

$$L_z = mvr = mr \cdot (\omega r) = mr^2 \dot{\varphi},$$

where m is the mass of the ball.

At the initial moment, $L_{z0} = mv_0 R$. Then, by virtue of the condition (3.50), we get

$$mr^2 \dot{\varphi} = mv_0 R,$$

where from

$$\dot{\varphi} = \frac{v_0 R}{r^2} = \frac{v_0 R}{(R - ct)^2}.$$

Let's integrate this equation in time:

$$\varphi = \frac{v_0 R}{c(R - ct)} + C \quad (C - const).$$

We define the integration constant C from the initial condition (3.49):

$$t_0 = 0 : \quad \varphi_0 = 0; \quad 0 = \frac{v_0 R}{cR} + C; \quad C = -\frac{v_0}{c}.$$

Substituting the found value C in the general solution φ, we get the second of the parametric equations of motion of the ball:

$$\varphi(t) = \frac{v_0 t}{R - ct}.$$ (3.52)

Thus, the movement of the ball is described by a system of Eqs. (3.51) and (3.52). Excluding time from this system, we obtain the equation of the ball's trajectory in polar coordinates:

$$\varphi = \frac{v_0}{c} \left(\frac{R}{r} - 1 \right).$$ (3.53)

Note that the condition of constancy of the angular momentum of the ball ($L_z = mr^2\dot{\varphi} = const$) relative to the z axis implies that as the value of r decreases, the angular velocity of the thread rotation increases.

To determine the thread tension T, we write the projection of the vector expression of the basic law of point dynamics on the thread direction:

$$ma_n = T, \tag{3.54}$$

where $m = p/g$ is the mass of the ball, $a_n = v^2/\rho$ is the normal acceleration of the ball when moving along the trajectory (3.53), and v is the absolute speed, and ρ is the radius of curvature of the trajectory at a point determined by the moment of time t.

The ball makes a complex movement. The rectilinear movement of the ball with the thread to the center O with a constant speed $v_r = c$ is taken as relative, and the rotation of the ball with the stretched thread with a speed $v_e = \omega r$ is taken as portable. Then, by the theorem of addition of velocities, we find

$$v = \sqrt{v_r^2 + v + e^2} = \sqrt{c^2 + r^2\dot{\varphi}^2}.$$

Substituting the derivative of the function φ according to the formula (3.52) and performing calculations, we find

$$v = \sqrt{c^2 + \frac{v_0^2 R^2}{r^2}}.$$

The curvature of a plane trajectory (3.53) in polar coordinates r, φ is calculated by the formula (2.34), which can be given the form in polar coordinates:

$$\frac{1}{\rho} = \frac{|r^2 + 2r'^2 - rr''|}{(r^2 + r'^2)^{3/2}}, \quad \left(r' = \frac{dr}{d\varphi}; \; r'' = \frac{d^2r}{d\varphi^2}\right). \tag{3.55}$$

Using the trajectory equation (3.53), we calculate

$$r = \frac{R}{(c/v_0)\varphi + 1}; \; r' = -\frac{cr^2}{Rv_0}; \; r'' = \frac{2r^3c^2}{v_0^2 R^2}.$$

Substituting these results into the formula (3.55), we find

$$\frac{1}{\rho} = r^{-1}\left(1 + \frac{c^2r^2}{v_0^2 R^2}\right)^{-3/2}.$$

Now you can calculate the normal acceleration of the ball:

$$a_n = \frac{v^2}{\rho} = \frac{v_0^3 R^3}{r^3 \sqrt{v_0^2 R^2 + c^2 r^2}}.$$

Taking this result into account, we finally obtain the desired thread tension from the formula (3.54) :

$$T = \frac{p v_0^3 R^3}{g r^3 \sqrt{v_0^2 R^2 + c^2 r^2}}.$$

3.4 System Dynamics

3.4.1 The Geometry of the Masses

Consider a mechanical system consisting of a finite number of N material points with masses m_1, m_2, ..., m_n, whose radius vectors drawn from the same point O (Fig. 3.7) are denoted by r_1, r_2, ..., R_n. *The center of mass of the system* is a geometric point C whose radius vector is defined by the formula

$$r_C = \sum_{i=1}^{N} \frac{m_i r_i}{M}, \qquad (3.56)$$

where $M = \sum_{i=1}^{N} m_i$ is the mass of the system.

Denoting the Cartesian coordinates of the material points that make up the system by (x_1, y_1, z_1), (x_2, y_2, z_2), ..., (x_n, y_n, z_n) and projecting the vector

Fig. 3.7 To the mass geometry

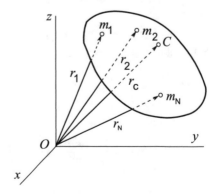

equality (3.56) on the coordinate axis, we obtain formulas for determining the coordinates of the center of mass:

$$x_C = \sum_{i=1}^{N} \frac{m_i x_i}{M}; \quad y_C = \sum_{i=1}^{N} \frac{m_i y_i}{M};$$

$$z_C = \sum_{i=1}^{N} \frac{m_i z_i}{M}. \tag{3.57}$$

From the definition (3.56), it follows that the center of mass is not a material point, but a geometric point that may not coincide with any material point in the system (e.g., in the case of a ring).

Vector

$$S_O = \sum_{i=1}^{N} m_i r_i$$

it is called the *static moment of the* system relative to the O point.

Scalar value:

$$S_{xy} = \sum_{i=1}^{N} m_i z_i; \quad S_{yz} = \sum_{i=1}^{N} m_i x_i; \quad S_{xz} = \sum_{i=1}^{N} m_i y_i \tag{3.58}$$

They are called *static moments of the* system relative to the Oxy, Oyz, and Oxz planes, respectively.

The radius vector and coordinates of the center of mass of the system are expressed in terms of static moments using the formulas

$$r_C = \frac{S_0}{M}; \quad x_C = \frac{S_{yz}}{M}; \quad y_C = \frac{S_{xz}}{M}; \quad z_C = \frac{S_{xy}}{M}. \tag{3.59}$$

In the case of a solid body, summation in formulas (3.56)...(3.58) is replaced by integration:

$$r_C = \int_{(M)} \frac{r\,dm}{M}; \tag{3.60}$$

$$x_C = \int_{(M)} \frac{x\,dm}{M}; \quad y_C = \int_{(M)} \frac{y\,dm}{M}; \quad z_C = \int_{(M)} \frac{z\,dm}{M}. \tag{3.61}$$

To describe the rotational movements of a body in dynamics, special geometric characteristics of the mass distribution are used. Here are the definitions of these characteristics.

The moment of inertia of a body relative to a point O is the value defined by the formula

$$J_O = \int\limits_{(M)} r^2 dm. \tag{3.62}$$

The moment of inertia relative to a point is often also called the *polar moment of inertia.*

The moment of inertia of the body relative to the l axis is called the integral:

$$J_l = \int\limits_{(M)} d^2 dm, \tag{3.63}$$

where d is the distance from the point of the body to the l axis, and integration is performed over all body masses.

The centrifugal moments of inertia of a body relative to any pair of Cartesian coordinate axes are called magnitudes:

$$J_{xy} = \int\limits_{(M)} xy \, dm; \quad J_{yz} = \int\limits_{(M)} yz \, dm; \quad J_{zx} = \int\limits_{(M)} zx \, dm. \tag{3.64}$$

The axes with respect to which the centrifugal moments of the body turn to zero are called *main axes of inertia.* The main axes passing through the center of mass of bodies are called *the main central axes.*

3.4.2 Theorem on the Motion of the Center of Mass of the System

Consider a mechanical system consisting of N material points. Let some k-th point of the system with mass m_k be affected by external (active and reactive) forces with the resultant F_k^e and internal forces, the resultant of which is denoted by F_k^i.

According to the basic law of dynamics, the acceleration of the point a_k is related to the acting forces by the dependence:

$$m_k a_k = F_k^e + F_k^i, \quad (k = 1, 2, \ldots, N),$$

i.e.,

$$m_1 a_1 = F_1^e + F_1^i,$$

$$m_2 a_2 = F_2^e + F_2^i,$$

$$\dots\dots\dots\dots,$$

$$m_N a_N = F_N^e + F_N^i. \tag{3.65}$$

Adding the left and right parts of the equalities (3.65), we have

$$\sum_{k=1}^{N} m_k a_k = \sum_{k=1}^{N} m_k F_k^e + \sum_{k=1}^{N} m_k F_k^i. \tag{3.66}$$

Let's analyze the obtained equality. First, using the formula (3.56), we can write

$$\sum_{k=1}^{N} m_k r_k = M r_C,$$

where r_k is a radius-vector k-th point of the system, M is the mass, and r_C is a radius-vector of center of mass. By differentiating this equality twice in time, we find

$$\sum_{k=1}^{N} m_k \frac{d^2 r_k}{dt^2} = M \frac{d^2 r_C}{dt^2},$$

or

$$\sum_{k=1}^{N} m_k a_k = M a_C, \tag{3.67}$$

and a_C is the acceleration of the center of mass of the system.

Second, according to Newton's third law (p. 8), any two points j and k of the system interact with equal modulo and oppositely directed forces F_{jk}^i and F_{kj}^i, the sum of which is zero. Therefore, *for all points of the system, the resultant of internal forces vanishes,* i.e.,

$$\sum_{k=1}^{N} a_k^i = 0. \tag{3.68}$$

Substituting the results (3.67) and (3.68) into the equality (3.66), we get

$$M a_C = \sum_{k=1}^{N} F_k^e. \tag{3.69}$$

The formula (3.69) expresses the theorem on the motion of the center of mass of the system.

Theorem *The product of the mass of a mechanical system and the acceleration of its center of mass is equal to the vector sum of all external forces applied to the system.*

Corollary 1 *If the resultant of all external forces applied to a system is zero, then the center of mass of this system moves uniformly and rectilinearly.*

This is known as the law of conservation of motion of the center of mass. In fact, given $\sum_{k=1}^{N} \mathbf{F}_k^e = 0$, the formula (3.69) implies that $\mathbf{a}_C = 0$ or $\mathbf{v}_C = const$, which means that the velocity vector remains unchanged in both magnitude and direction. As you know from the physics course, this movement is called uniform and straight.

Corollary 2 *If the sum of the projections of all forces acting on the system on some axis is zero, then the projection of the velocity of the center of mass of the system on the same axis is a constant value.*

Proof For the proof, we write the vector equality (3.69) in projections on the axis of the Cartesian system of rectangular coordinates $Oxyz$. We get three scalar equalities:

$$M\frac{d^2 x_C}{dt^2} = \sum_{k=1}^{N} F_{kx}^e; \quad M\frac{d^2 y_C}{dt^2} = \sum_{k=1}^{N} F_{ky}^e; \quad M\frac{d^2 z_C}{dt^2} = \sum_{k=1}^{N} F_{kz}^e, \tag{3.70}$$

where x_C, y_C, and z_C are coordinates of the center of mass of the system and F_{kx}, F_{ky}, F_{kz} are projections of the kth force on the corresponding axis.

Let, for example, $\sum_{k=1}^{N} F_{kx}^e = 0$. Then the first of Eqs. (3.70) gives

$$\frac{d^2 x_C}{dt^2} = 0 \text{ или } \frac{dx_C}{dt} = v_{Cx} = const, \tag{3.71}$$

which proves corollary 2. $\square$

Example Determine the movement of a floating crane lifting a load weighing $P_1 = 2$ t when the crane boom is rotated by $30°$ to the vertical position. The weight of the crane $P_2 = 20$ t, and the length of the boom OA is 8 m (Fig. 3.8). Ignore the water resistance and weight of the arrow.

D e c i s i o n. The initial speed of a system consisting of a crane weighing P_2 and a load P_1 is zero: $v_{Cx} = 0$. Therefore, according to the second of the formulas (3.71), $x_c = const$.

Fig. 3.8 Floating crane

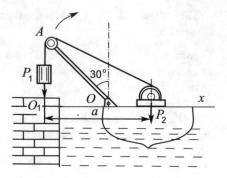

Choose the beginning of the axis x at the point O_1 under the weight P_1; shift to the left due to the crane lifting P_1 we denote by ξ, and the coordinate power P_2 before lifting— through a.

Thus, at the initial moment, we have $x_1 = 0$, $x_2 = a$, and the coordinate of the center of mass of the system will be

$$x_C = \frac{P_1 x_1 + P_2 x_2}{P_1 + P_2} = \frac{P_2 a}{P_1 + P_2}.$$

After lifting the load,

$$x_1' = x_1 + OA \cdot \sin 30° - \xi;$$

$$x_2' = a - \xi;$$

$$x_C' = \frac{P_1 x_1' + P_2 x_2'}{P_1 + P_2} = \frac{P_1(OA \cdot \sin 30° - \xi) + P_2(a - \xi)}{P_1 + P_2}.$$

Since $x_C = x_c'$, we have

$$P_2 a = P_1(OA \cdot \sin 30° - \xi) + P_2(a - \xi).$$

From here we find

$$\xi = \frac{P_1 \cdot OA \cdot \sin 30°}{P_1 + P_2} = \frac{2 \cdot 8 \cdot 0,5}{2 + 20} =$$
$$= 0,36 \text{ м.}$$

3.4.3   Theorem on Changing the Amount of Motion of a System

The quantity of motion of a mechanical system consisting of N material points is the vector quantity Q, which is equal to the geometric sum of the quantities of motion of all points in the system:

$$Q = \sum_{i=1}^{N} q_i = \sum_{i=1}^{N} m_i v_i. \tag{3.72}$$

Write the equality (3.56) as

$$\sum_{i=1}^{N} m_i r_i = M r_C.$$

Differentiating both parts of this formula at the time, you will receive at a constant mass:

$$\sum_{i=1}^{N} m_i \frac{dr_i}{dt} = M \frac{dr_C}{dt} \text{ или } \sum_{i=1}^{N} m_i v_i = M v_C,$$

i.e.,

$$Q = M v_C. \tag{3.73}$$

Thus, *the amount of motion of the system is equal to the product of the mass of the entire system and the speed of its center of mass.* It follows from this result that for a fixed center of mass, the amount of motion of the system is zero. In particular, when a solid body rotates around a fixed axis passing through the center of mass, the amount of motion of the body is zero. For complex body movement, the value Q characterizes only the translational part of the movement. For example, for a rolling wheel, $Q = m v_C$ and does not depend on whether the wheel rotates around the center of mass C or not.

We write the differential equation of motion of a system consisting of N material points. According to the law (3.66), we have

$$\sum_{k=1}^{N} m_k a_k = \sum_{k=1}^{N} F_k^e + \sum_{k=1}^{N} F_k^i.$$

By the property (3.68), the sum of internal forces vanishes: $\sum_{k=1}^{N} F_k^i = 0.$ In addition,

$$\sum_{k=1}^{N} m_k a_k = \frac{d}{dt} \sum_{k=1}^{N} m_k v_k = \frac{dQ}{dt}.$$

Thus, we finally have

$$\frac{d\boldsymbol{Q}}{dt} = \sum_{k=1}^{N} \boldsymbol{F}_k^e. \tag{3.74}$$

Equation (3.74) in differential form expresses the theorem on changing the amount of motion of the system.

Theorem *The time derivative of the amount of motion of the system is equal to the main vector of all external forces acting on the system.*

In projections on Cartesian rectangular axes $Oxyz$, the vector equation (3.74) gives

$$\frac{dQ_x}{dt} = \sum_{k=1}^{N} F_{kx}^e; \quad \frac{dQ_y}{dt} = \sum_{k=1}^{N} F_{ky}^e; \quad \frac{dQ_z}{dt} = \sum_{k=1}^{N} F_{kz}^e. \tag{3.75}$$

We get another expression of the theorem. Let at some point in time $t_0 = 0$ the amount of movement of the system is $\boldsymbol{Q}_0$, and at the moment $t_1 > t_0$—$\boldsymbol{Q}_1$. Integrating Eq. (3.74) in the range from $t_0 = 0$ to t_1, we get

$$\boldsymbol{Q}_1 - \boldsymbol{Q}_0 = \sum_{k=1}^{N} \int_0^{t_1} \boldsymbol{F}_k dt.$$

But, according to the expression (3.69), the integrals on the right side represent the impulses S_k^e of external forces. Therefore,

$$\boldsymbol{Q}_1 - \boldsymbol{Q}_0 = \sum_{k=1}^{N} S_k^e. \tag{3.76}$$

The obtained equality expresses the theorem on changing the amount of motion of the system in integral form.

Theorem *The change in the amount of motion of the system over a finite period of time is equal to the sum of the impulses applied to the system by external forces over the same period of time.*

In projections on coordinate axes, the vector equality (3.76) is equivalent to three scalar equations:

$$Q_{1x} - Q_{0x} = \sum_{k=1}^{N} S_{kx}^e; \quad Q_{1y} - Q_{0y} = \sum_{k=1}^{N} S_{ky}^e; \quad Q_{1z} - Q_{0z} = \sum_{k=1}^{N} S_{kz}^e. \tag{3.77}$$

Corollary 1 *Let the main vector of external forces acting on the system be zero:*

$$\sum_{k=1}^{N} \boldsymbol{F}_k^e = 0.$$

Then we get from Eq. (3.74) that $\boldsymbol{Q} = const$.

Thus, *if the vector sum of all external forces applied to the system is zero, then the vector of the amount of motion of the system does not change either modulo or in direction.*

Corollary 2 *If the sum of the projections of all the external forces of the system on some axis (e.g., Ox) is zero, according to the formulas (3.75), we get*

$$Q_x = const \ at \ \sum_{k=1}^{N} F_{kx}^e = 0.$$

The results expressed by corollaries 1 and 2 are called *the law of conservation of the amount of motion of the system,* which can also be formulated as follows: *internal forces cannot change the amount of motion of a mechanical system.*

Example In the mechanism shown in Fig. 3.9, a moving wheel of radius r weighs p and has a center of gravity at O_1; a rectilinear rod AB weighs k times more than a moving wheel and has a center of gravity in its middle. The crank OO_1 rotates around the O axis with a constant angular velocity ω. Calculate the amount of movement of the system, ignoring the mass of the crank.

S o l u t i o n. Select the coordinate axes $Oxyz$ and the direction of rotation of the crank OO_1, as shown in the figure. The point O_1 is the center of mass of the

Fig. 3.9 On the quantity of
motion theorem

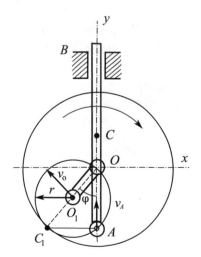

moving wheel, and its velocity vector is directed perpendicular to O_1 and modulo $\omega \cdot r$.

The velocity projections on the coordinate axis will be

$$v_{O_1 x} = -\omega r \cos \varphi = -\omega r \cos \omega t;$$

$$v_{O_1 y} = \omega r \sin \varphi = \omega r \sin \omega t.$$

The vector $\boldsymbol{v}_A$ of the absolute velocity of the point A is directed along the axis of the rod AB. Using the instantaneous velocity center C_1, we find the module of the vector $\boldsymbol{v}_A$:

$$v_A = \omega \cdot C_1 A = \omega \cdot 2r \sin \varphi = 2\omega r \sin \omega t.$$

Given that $v_{Ax} = 0$, $v_{Ay} = v_A$, we calculate the projections of the amount of movement on the coordinate axes:

$$Q_{O_1 x} = \frac{p}{g} v_{O)1x} = -\frac{p}{g} \omega r \cos \omega t;$$

$$Q_{O_1 y} = \frac{p}{g} v_{O_1 y} = \frac{p}{g} \omega r \sin \omega t;$$

$$Q_{ABx} = 0; \quad Q_{ABy} = \frac{kp}{g} \cdot 2\omega r \sin \omega t.$$

Summing up the corresponding projections for the entire system, we have

$$Q_x = -\frac{p}{g} \omega r \cos \omega t; \quad Q_y = \frac{p}{g} \omega r (1 + 2k) \sin \omega t.$$

3.4.4 Theorem on Changing the Moment of Quantity of Motion of the System

Earlier (p. 122), the definition of the moment of the amount of movement of a material point was given. For mechanical systems, *main angular momentum* relative to any point O (or *kinetic moment system*) is called the vector sum of the moments of momentum of all points of this system which are taken with respect to point O. Denoting this value by $\boldsymbol{K}_O$, by definition we have

$$\boldsymbol{K}_O = \sum \boldsymbol{M}_O (m_i \boldsymbol{v}_i) = \sum (\boldsymbol{r}_i \times m_i \boldsymbol{v}_i). \tag{3.78}$$

Here and further, the summation extends to all N points of the system.

Using the formulas (3.44), we write the equality (3.78) in projections on Cartesian rectangular axes:

Fig. 3.10 To calculate the kinetic moment of the system

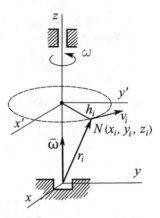

$$K_x = \sum M_x(m_i \boldsymbol{v}_i) = \sum m_i(y_i v_{xi} - z_i v_{yi});$$

$$K_y = \sum M_y(m_i \boldsymbol{v}_i) = \sum m_i(z_i v_{xi} - x_i v_{zi});$$

$$K_z = \sum M_z(m_i \boldsymbol{v}_i) = \sum m_i(x_i v_{yi} - y_i v_{xi});$$

$$(i = 1, 2, \ldots, N), \qquad (3.79)$$

where K_x, K_y, K_z are the kinetic moments of the system relative to the coordinate axes; x_i, y_i, z_i and v_{xi}, v_{yi}, v_{zi} are the coordinates and projections of the velocity vector of the ith point on the coordinate axes, respectively.

To understand the mechanical meaning of the kinetic moment of the system, we calculate its value for a solid body rotating around a fixed axis Oz with an angular velocity ω (Fig. 3.10). By the formulas (3.73), we have

$$K_z = \sum M_z(m_i \boldsymbol{v}_i).$$

But when the body rotates around the Oz axis, the velocity modulus of any $i$point is $v_i = h_i \omega$, and the vector of the amount of motion $m_i \boldsymbol{v}_i$ is perpendicular to the segment h_i omitted from the point N_i to the axis of rotation. Vector $m_i \boldsymbol{v}_i$ lies in the plane normal to Oz. Therefore, for any point of the body, the moment of the amount of movement relative to the Oz axis will be

$$M_z(m_i \boldsymbol{v}_i) = h_i m_i v_i = m_i h_i^2 \omega.$$

Then for the whole body, we get

$$K_z = \sum m_i h_i^2 \omega = \omega \sum m_i h_i^2.$$

According to the formula (3.63), the sum in this formula is the moment of inertia J_z of the body relative to the Oz axis. Thus, we finally have

$$K_z = J_z\omega, \tag{3.80}$$

i.e., *kinetic momentum of a rotating body about a fixed axis equals the moment of inertia of the body about this axis at the angular velocity of the body.*

We now show that the theorem on the change in the moment of the amount of motion for a single material point, proved in 3.3.4, will also be valid for a mechanical system.

If all external and internal forces are applied to the points of the system and we denote the resultant of external and internal forces applied to the kth point, respectively, by F_k^e and F_k^i, then based on the theorem (3.45) for the kth point, we can write

$$\frac{d}{dt}[M_O(m_k v_k)] = M_O(F_k^e) + M_O(F_k^i).$$

Making such equations for all points of the system and adding their left and right parts, we get

$$\frac{d}{dt}\left[\sum M_O(m_k v_k)\right] = \sum M_O(F_k^e) + \sum M_O(F_k^i).$$

But by the property of internal forces (3.68), the last sum is zero. So finally we have

$$\frac{dK_O}{dt} = \sum M_O(F_k^e). \tag{3.81}$$

The result expresses the following theorem of moments:

Theorem *The time derivative of the kinetic moment of the system relative to a point is equal to the geometric sum of the moments of external forces acting on the system relative to the same point.*

In projections on the axis of a Cartesian rectangular coordinate system $Oxyz$, the equality (3.81) has the form

$$\frac{dK_x}{dt} = \sum M_x(F_k^e); \quad \frac{dK_y}{dt} = \sum M_y(F_k^e); \quad \frac{dK_z}{dt} = \sum M_z(F_k^e). \tag{3.82}$$

Corollary 1 *Let the sum of the moments of all external forces relative to the point O be zero, i.e.,*

$$M_O(F_k^e) = 0.$$

In this case, from theorem (3.81), it implies that $K_O = const$.

Thus, *if the sum of moments of all applied to a system of external forces relative to a fixed point is equal to zero, the angular momentum of the system about that point remains unchanged in magnitude and direction.*

Corollary 2 *Let the sum of the moments of all external forces relative to some axis, for example, Ox, vanish. Then it follows from the formulas (3.82) that $K_x = const$, i.e., if the sum of the moments of all external forces acting on the system relative to any axis is zero, then the kinetic moment of the system relative to the same axis is a constant value.*

The formulated consequences express the *law of conservation of the kinetic moment of the system*. According to this law, *internal forces cannot change the main moment of the amount of motion of the system.*

To illustrate the law, consider a mechanical system rotating around a fixed axis Oz. If in this case $\sum M_z(F_k^e) = 0$, then by corollary 2 and the formula (3.80), $J_z\omega = const$. For an immutable system (absolutely solid), $J_z = const$; hence, $\omega = const$. If the system is variable, then, generally speaking, $J_z \neq const$ and the angular velocity can change along with the change in the moment of inertia J_z. Since $J_z\omega = const$, the angular velocity decreases with increasing moment of inertia J_z, and increases with decreasing moment of inertia. The results obtained explain the known effects in experiments with the Zhukovsky platform.

The theorem on changing the kinetic moment allows us to obtain a differential equation of rotation of a body around a fixed axis. Let, for example, Oz be the axis of rotation of the body. By the formulas (3.82) and (3.80), we have

$$\frac{dK_z}{dt} = \sum M_z(F_k^e) \text{ and } K_z = J_z\omega.$$

From

here, for an immutable system, we get

$$J_z\frac{d\omega}{dt} = \sum M_z(F_k^e).$$

Expressing the angular velocity in terms of the rotation angle φ, we have

$$J_z\ddot{\varphi} = \sum M_z(F_k^e). \tag{3.83}$$

The dependence (3.83) is a differential equation of the rotational motion of a body around a fixed axis.

Example 1 A uniform round disk with a weight of $P = 50$ kgf and a radius of $R = 30$ cm rolls without sliding along a horizontal plane, making 60 rpm around its axis. Calculate the main moment of the amount of disk movement relative to (1) the axis passing through the center of the disk perpendicular to the plane of motion and (2) relative to the instantaneous axis (Fig. 3.11).

Fig. 3.11 For example 1

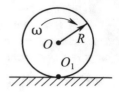

D e c i s i o n. The kinetic moment of the disk relative to the Oz axis passing through the point O perpendicular to the plane of motion of the disk is calculated by the formula (3.80)

$$K_z = J_z \omega.$$

Calculating the integral in the formula (3.63) for a solid disk, we obtain

$$J_z = \frac{MR^2}{2},$$

where $M = P/g$ is the mass of the disk and g is the acceleration of gravity. Given that the angular velocity ω is related to the rotational speed n (rpm) by the dependence $\omega = \pi n/30 (c^{-1})$, we get

$$K_z = \frac{P}{g} \frac{R^2}{2} \frac{\pi n}{30} = \frac{\pi n P R^2}{60 g}.$$

Having in mind that 1 kgf = 9,81 N, we calculated

$$K_z = \frac{3.14 \cdot 60 \cdot 50 \cdot 9.81 \cdot 0.3^2}{60 \cdot 9.81} = 14.13 \text{ kg} \cdot \text{m}^2 \cdot c^{-1}.$$

To calculate the kinetic moment of the disk relative to the $O_1 z'$ axis parallel to Oz, calculate the moment of inertia of the disk $J_{z'}$ relative to the $O_1 z'$ axis. To do this, we use Steiner's theorem (see, e.g., Molotnikov 2004, p. 243), according to which the moment of inertia relative to the new axis $O_1 z'$, parallel to the central axis Oz, is calculated by the formula

$$J_{z'} = J_z + MR^2 = \frac{MR^2}{2} + MR^2 = \frac{3}{2} MR^2.$$

Next, we calculate

$$K_{z'} = J_{z'}\omega = \frac{3}{2} MR^2 \frac{\pi n}{30} = \frac{\pi n P R^2}{20 g};$$

$$K_{z'} = \frac{50 \cdot 9.81 \cdot 3.14 \cdot 60 \cdot 0.3^2}{20 \cdot 9.81} = 42.4 \text{ kg} \cdot \text{m}^2 \cdot c^{-1}.$$

Example 2 For rapid braking of large flywheels, an electric brake is used, consisting of two diametrically located poles that carry a winding fed by a direct current. The currents induced in the flywheel body as it moves past the poles create a braking moment M_1 proportional to the speed v on the flywheel rim: $M_1 = kv$, where k is a coefficient depending on the magnetic flux and the size of the flywheel. The moment M_2 from the friction in the bearings can be considered constant. The diameter of the flywheel is D, and its moment of inertia relative to the axis of rotation is J. Find the time interval after which the flywheel rotating at the angular velocity ω_o will stop.

D e c i s i o n. Write down the differential equation of the rotational motion of the flywheel:

$$J\frac{d\omega}{dt} = \sum M(F_k^e).$$

In the right part, we must substitute with a negative sign the sum of the moments M_1 and M_2, since these moments slow down the rotation of the flywheel, i.e.,

$$J\frac{d\omega}{dt} = -M_1 - M_2 - -kv - M_2.$$

This is a differential equation with separable variables. Its general integral has the form

$$\frac{2J}{kD}\ln\left(\frac{kD}{2}\omega + M_2\right) = -t + C \quad (C - const).$$

We find the constant C from the initial condition: $\omega(0) = \omega_0$. From here we get

$$C = \frac{2J}{kD}\ln\left(\frac{kD}{2}\omega_o + M_2\right).$$

Thus, the equation of motion of the flywheel will be

$$t = \frac{2J}{kD}\ln\frac{\dfrac{kD}{2}\omega_o + M_2}{\dfrac{kD}{2}\omega + M_2}.$$

The flywheel will stop at the moment T when the angular velocity turns to zero. Find

$$T = t\,|_{\omega=0} = \frac{2J}{kD}\ln\left(1 + \frac{kD\omega_o}{2M_2}\right).$$

3.4.5   Theorem on Changing the Kinetic Energy of a System

The kinetic energy (T) *of a mechanical system* is the sum of the kinetic energies of all its points, i.e.,

$$T = \sum \frac{m_k v_k^2}{2},\tag{3.84}$$

where m_k and v_k are the mass and velocity of the kth point, respectively, and the summation extends to all points of the system. It follows from the definition (3.84) that the kinetic energy of a system can be zero only if all its points are at rest.

We calculate the kinetic energy for various cases of system motion.

For *translational motion* , all points of the system, including the center of mass C, move at the same speeds, so that $v_k = v_C$ for all k and from the formula (3.84), we get

$$T_{post} = \sum \frac{m_k v_C^2}{2} = \frac{1}{2} v_C^2 \sum m_k$$

or

$$T_{const} = \frac{1}{2} M v_C^2,\tag{3.85}$$

where M is the mass of the system.

In the case of *rotational motion* of the system around some axis Oz with angular velocity ω, we have $v_k = \omega h_k$, where h_k is the distance of the point from the axis of rotation. Then for the kinetic energy (T_{BP}), receive

$$T_{BP} = \sum \frac{m_k h_k \omega^2}{2} = \frac{1}{2} \omega^2 \sum m_k h_k^2$$

or, with the formula (3.63)

$$T_{Bp} = \frac{1}{2} J_z \omega^2.\tag{3.86}$$

In the general case of motion, the velocity of an arbitrary point in the system can be represented as a sum: the speed of the center of mass $\boldsymbol{v}_C$ is taken for the pole and $\boldsymbol{v}_k'$ of k-th point in the rotational motion with angular velocity ω around the instantaneous axis CP, i.e.,

$$\boldsymbol{v}_k = \boldsymbol{v}_C + \boldsymbol{v}_k',$$

and the module of the vector $\boldsymbol{v}_k'$ is equal to $V_k' = \omega h_k$. Then

$$v_k^2 = v_k^2 = (v_C + v_k')^2 = v_C^2 + v_k'^2 + 2v_C v_k'.$$

Substituting this result into the formula (3.84), we get

$$T = \frac{1}{2} v_C^2 \sum m_k + \frac{1}{2} \omega^2 \sum n_k h_k^2 + v_C \sum m_k v_k'.$$

Given that the first sum is the mass of the system M, the second sum is the moment of inertia of the system relative to the instantaneous axis CP, and the third sum, which is the amount of motion of the system when rotating around the axis CP passing through the center of mass, is zero, we obtain in general the motion of the system:

$$T = \frac{1}{2} M v_C^2 + \frac{1}{2} J_{CP} \omega^2. \tag{3.87}$$

Earlier, we proved a theorem about the change in the kinetic energy of a point. Considering it as any k-th point of the system, based on the formula (3.38), we can record

$$d\left(\frac{m_k v_k^2}{2}\right) = dA_k^e + dA_k^i,$$

where dA_k^e, dA_k^i are elementary operations of external and internal forces applied to the k point. Composing such relations for all points of the system and adding them together, we get

$$d\left(\sum \frac{m_k v_k^2}{2}\right) = \sum dA_k^e + \sum dA_k i$$

or

$$dT = \sum dA_k^e + \sum dA_k^i. \tag{3.88}$$

The resulting equality expresses the theorem on the change in the kinetic energy of the system in differential form:

Theorem *The differential of the kinetic energy of the system is equal to the sum of the elementary work of all external and internal forces acting on the system.*

Integrating the equality (3.88) between two positions of the system—the initial and final—in which the kinetic energy is equal to T_0 and T, respectively, we get

$$T - T_0 = \sum A_k^e + \sum A_k^i. \tag{3.89}$$

Formula (3.89) is expressed in a finite (or integral) form of the theorem on change of kinetic energy of the system.

Theorem *The change of kinetic energy of the system when it moves from one position to another is equal to the sum of all external and internal forces applied to the system at the relevant displacement points of the system at the same time it moves.*

In the case of an immutable system (an absolutely solid body), the sum of the internal forces is zero:

$$\sum A_k^i = 0.$$

Therefore, for such a system, the formula (3.89) can be represented as

$$T - T_0 = \sum A_k^e, \tag{3.90}$$

that is, the change in the kinetic energy of an absolutely solid body during any movement is equal to the sum of the work of all external forces acting on the body on the movements of their application points.

Example 1 Calculate the kinetic energy of the rocker mechanism if the moment of inertia of the crank OA (Fig. 3.12) relative to the axis of rotation perpendicular to the drawing plane is J_O, the length of the crank is a, the mass of the rocker is m, and the mass of the rocker stone is ignored. The crank OA rotates at an angular velocity of ω.

At what positions of the mechanism does the kinetic energy reach the highest and lowest values?

D e c i s i o n. The kinetic energy of the mechanism consists of the kinetic energy T_1 of the crank performing a rotational motion with a constant angular velocity ω, and the kinetic energy T_2 of the backstage in its translational motion at a speed of v. Have

Fig. 3.12 For example 1

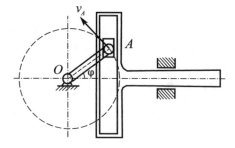

$$T_1 = \frac{1}{2} J_O \omega^2;$$

$$T_2 = \frac{1}{2} mv^2 = \frac{1}{2} m(v_A \sin \varphi)^2 = \frac{1}{2} m(\omega a \sin \varphi)^2.$$

Adding T_1 and T_2, we find the kinetic energy of the entire system:

$$T = \frac{\omega^2}{2} (J_O + ma^2 \sin^2 \varphi).$$

From the result obtained, it is obvious that the function T has a minimum value for $\varphi = k\pi$, $(k = 0, 1, 2, \ldots)$, and the maximum is when $\varphi = (\pi/2) + k\pi$.

Example 2 A constant torque M is applied to a gate drum with a radius of r and a weight of P_1. A load P_2 is attached to the end of the A cable wound on the drum, which rises along an inclined plane located at an angle α to the horizon. What angular velocity will the gate drum acquire by turning at the angle φ? The coefficient of friction of sliding the load on an inclined plane is f. The weight of the cable is ignored, and the drum is considered a uniform round cylinder. At the initial moment, the system was at rest (Fig. 3.13).

D e c i s i o n. Let's use the theorem on the change in the kinetic energy of the system. By the formula (3.89), we have

$$T - T_0 = \sum A_k^e + \sum A_k^i.$$

According to the condition, the system was at rest at the initial moment. Therefore, $T_0 = 0$. Let's calculate the kinetic energy T at the moment when the drum turns at the angle φ. The system consists of two bodies: a drum with kinetic energy T_1 and a load, whose kinetic energy is denoted by T_2. Have

$$T_1 = \frac{1}{2} J\omega^2 = \frac{1}{2} \cdot \frac{P_1}{g} \cdot \frac{r^2}{2} \cdot 2 = \frac{P_1 r^2 \omega^2}{4g};$$

$$T_2 = \frac{1}{2} m_2 v^2 = \frac{1}{2} \cdot \frac{P_2}{g} \cdot (\omega r)^2 = \frac{P_2}{2g} \omega^2 r^2.$$

Fig. 3.13 For example 2

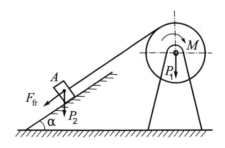

Adding these values, we get

$$T = T_1 + T_2 = \frac{P_1 r^2 \omega^2}{4g} + \frac{P_2 r^2 \omega^2}{2g} = \frac{r^2 \omega^2}{4g}(P_1 + 2P_2). \tag{3.91}$$

Let us now calculate the sum of the work of all the forces acting on the system. When the drum is rotated through an angle φ, in addition to the moment M, the work is done by gravity P_2 and friction force F_{fr}, and the work of the last two forces is negative because they move in the opposite direction to the force. We have

$$\sum A_k = M\varphi - P_2 r \varphi \sin \alpha - f P_2 r \varphi \cos \alpha, \tag{3.92}$$

where $r\varphi$ is the movement of a point A parallel to the inclined plane; $r\varphi \sin \alpha$ is the projection of this movement on the direction of the force P_2; $f P_2 \cos \alpha = F_{fr}$ is the friction force.

Equating the right-hand sides of the formulas (3.91) and (3.92), we get

$$\frac{r^2 \omega^2}{4g}(P_1 + 2P_2) = \varphi [M - P_2 r(\sin \alpha + f \cos \alpha)].$$

We'll find it from here

$$\omega = \frac{2}{r}\sqrt{\frac{M - P_2 r(\sin \alpha + f \cos \alpha)}{P_1 + 2P_2} g\varphi}.$$

3.5   The d'Alembert Principle

The d'Alembert principle is a general method by which dynamic equations are given the form of static equilibrium equations.

Consider a material point M of mass m that is in motion under the action of forces $(F_1, F_2, \ldots, F_n)$ with acceleration a (Fig. 3.14). According to the formula (3.1), the basic equation of point dynamics has the form

Fig. 3.14 On the formulation
of the d'Alembert principle

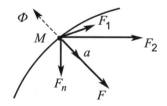

$$ma = F_1 + F_2 + \ldots + F_n.$$

Let's denote $\Phi = -ma$ and move all the terms in the last formula to the right side. Receive

$$F_1 + F_2 + \ldots + F_n + \Phi = 0. \tag{3.93}$$

The force Φ, equal to the product of the mass of a point and its acceleration and directed opposite to the acceleration, is called the force of inertia.

Equation (3.93) can be interpreted as follows: *if at each moment of time the inertial force is added to the forces actually acting on the point, then the resulting system of forces will be balanced.*

This proposition, which is equivalent to Newton's second law, expresses the *principle of d'Alembert.*

Repeating the above reasoning for each of the points of the mechanical system, we come to the *principle of d'Alembert for the system: if at each moment of time to all points of the system, except for the actual external and internal forces, the corresponding forces of inertia are applied, then the resulting system of forces will be balanced and it will be possible to apply all the equations of statics.*

It should be borne in mind that in reality the forces of inertia are applied to the body that gives the system (or point) acceleration. The application of inertial forces to a system (point) is only a formal technique that reduces the problem of dynamics in the form of a solution to the problem of statics.

It is known from statics (Sect. 1.10) that for an arbitrary system of forces to be balanced, the geometric sum of all forces acting on the system must vanish, as well as the sum of their moments relative to an arbitrary center O. Then based on the d'Alembert principle, it should be

$$\sum (F_k^e + F_k^i + \Phi_k) = 0;$$
$$\sum \left[M_O(F_k^e) + M_O(F_k^i) + M_O(\Phi_k) \right] = 0. \tag{3.94}$$

Denote

$$R^\Phi = \sum \Phi_k; \quad M_O^\Phi = \sum M_O(\Phi_k). \tag{3.95}$$

Values R^Φ and M_O^Φ are called, respectively, the main vector and the main moment of inertial forces relative to the center O. Given that for the internal forces of the system, the main vector and the main moment relative to the center O vanish, from the equalities (3.94), we obtain

$$\sum F_k^e + R^\Phi = 0; \quad \sum M_O(F_k^e) + M_O^\Phi = 0. \tag{3.96}$$

Fig. 3.15 For example, on the principle of d'Alembert

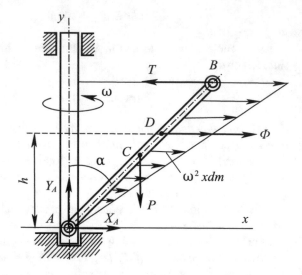

Equations (3.96) do not contain internal forces, which in many cases simplifies the process of solving problems.

Example. A uniform rod AB of length l and weight P is attached by a hinge to a vertical shaft rotating at an angular velocity ω (Fig. 3.15). Find the tension T of the horizontal thread holding the rod at an angle α to the shaft axis.

S o l u t i o n. The AB rod is affected by the active forces P and T, as well as the reactive forces X_A, Y_A. In accordance with the principle of d'Alembert, we add the forces of inertia to this system. For each rod element with a mass of dm, the centrifugal force of inertia is $\omega^2 x dm$, where x is the distance from the element dm to the axis of rotation Ay.

Thus, the inertial forces are distributed along the rod AB according to a linear law. Their resultant passes through the center of gravity of the triangle representing the law of distribution of forces of inertia (Fig. 3.15) and is equal in modulus to the area of this triangle. Therefore, the modulus of the main vector of inertial forces will be

$$\Phi = \frac{P}{g}\omega^2 x_C = \frac{P}{g}\omega^2 \cdot \frac{l}{2}\cdot \sin\alpha,$$

where x_C is the coordinate of the center of mass of the rod AB.

Now let's make an equilibrium equation in the form of equal to zero sum of the moments of all forces applied to the rod, relative to the point A:

$$\sum M_A(F_k) = 0:$$

$$T \cdot l \cos\alpha - \Phi h - P \cdot \frac{l}{2}\cdot \sin\alpha = 0,$$

where $h = (2/3)l \cdot \cos\alpha$—removing the center of gravity of the triangle of inertial forces of the rod AB from the axis Ax. From the written equation of moments, we determine the thread tension:

$$T = P\left(\frac{\omega^2 l}{3g}\sin\alpha + \frac{1}{2}\operatorname{tg}\alpha\right).$$

Self-Test Questions

1. What does dynamics study?
2. What is understood by the material point?
3. What property of material bodies is called inertia?
4. What is the measure of inertia of the material body?
5. How is the basic law of point dynamics formulated?
6. Write down the mathematical expression of the basic law of point dynamics.
7. Give the formulation of the law of independence of the action of forces.
8. Name the basic and derived mechanical quantities in the SI, MCGSS, and CGS systems.
9. Write down the differential equations of motion of a material point in Cartesian coordinates.
10. Write down the differential equations of motion of a material point in projections on natural coordinate axes.
11. Formulate the first problem of point dynamics.
12. What is the second problem of point dynamics?
13. What is an inertial frame of reference?
14. How is the principle of relativity of Galileo-Newton formulated?
15. What is called the amount of movement of a material point?
16. Define kinetic energy of a particle.
17. What is meant by an elementary force impulse?
18. What are the energy characteristics of the force during the motion of a point.
19. Write the theorem on change of momentum of a material point in the differential form.
20. Write the theorem on change of kinetic energy of a particle in a differential form.
21. Give the definition of the moment of momentum of a material point relative to the axis.
22. Give the formulation of the dynamic Coriolis theorem.
23. Name the movements in which the Coriolis force of inertia is zero.
24. At what displacements is the work of gravity (a) positive, (b) negative, and (C) equal to zero?
25. Give a definition of the static moment of a system of material points relative to a certain point.

26. Write down the expression of the static moment of a system of material points relative to the plane.
27. How are the radius vector and coordinates of the center of mass of the system expressed in terms of static moments?
28. What is called the moment of inertia of a body relative to a point?
29. Write down the formula for calculating the moment of inertia of the body relative to the axis.
30. What is called the center of mass of the system?
31. Can the center of mass of a solid body be located outside this body?
32. Write down the formulas for calculating the coordinates of the center of mass in three-dimensional space.
33. Give a mathematical expression of the theorem on the motion of the center of mass of the system.
34. Under what conditions center of mass of a system moves uniformly and rectilinearly?
35. What is called the polar moment of inertia of the body?
36. Give the definition of the axial moment of inertia of the system of material points.
37. What physical law implies that the resultant of the internal forces of the system is zero?
38. Write down the mathematical expression of the theorem on the motion of the center of mass in coordinate form.
39. What is called the amount of motion of a mechanical system?
40. How is the theorem about changing the amount of motion of the system formulated?
41. What is called the main vector and the main moment of inertial forces of the system?
42. What is called the kinetic moment of a mechanical system? What is its dimension?
43. Formulate the d'Alembert principle for a mechanical system.

Control Tasks for the Section "Dynamics"

T a s k. Determine the value of the constant force $\vec{P}$, under the action of which the rolling without sliding of the wheel mass m has a boundary character, i.e., the wheel's adhesion to the base is on the verge of failure. Find also for this case the equation of motion of the center of mass of the wheel C, if at the initial moment of time its coordinate $x_{C_0} = 0$ and the speed $v_{C_0} = 0$. The task options are shown in Fig. 3.16, and the data required for the solution is given in Table 3.1 (Yablonskii et al. 2000).

The following designations are used in the task: i_c, the radius of inertia of the wheel relative to the central axis perpendicular to its plane; R and r, the radii of the

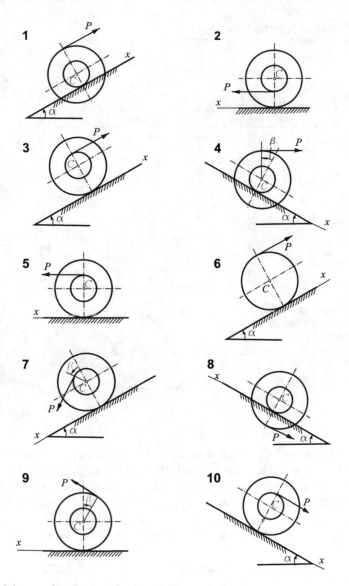

Fig. 3.16 Schemes of mechanisms for the task "Dynamics"

large and small circles; f, the coefficient of adhesion (coefficient of friction at rest); and δ, the coefficient of rolling friction.

N o t e. Wheels for which the radii of inertia are not specified are considered solid homogeneous disks.

Example of Completing a Task Given: $m = 200$ kg; $R = 60$ cm; $r = 10$ cm; $i_c = 50$ cm; $\alpha = 15°$; $\beta = 30°$; $f_{fr} = 0.1$; $\delta = 0$ (Fig. 3.17a).

Table 3.1 Source data for the dynamics task

Scheme number	The parameters of the mechanism							
	m, kg	i_C, cm	R, cm	r, cm	α, grad	β, grad	f, cm	δ, cm
1	300	50	80	40	20	–	0.35	0
2	200	40	60	30	–	–	0.2	0.8
3	180	50	60	20	30	–	0.1	0
4	220	30	70	25	30	30	0.2	0
5	240	40	60	15	–	–	0.1	1.0
6	200	–	50	–	15	–	0.2	0
7	200	45	60	25	30	15	0.25	0
8	150	40	70	25	15	–	0.5	0
9	250	–	–	–	–	30	0.15	0
10	150	40	50	15	20	–	0.3	0.7

Fig. 3.17 The scheme of the mechanism, for example, on the dynamics

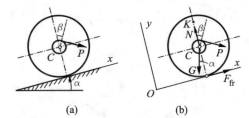

(a)   (b)

S o l u t i o n. The following forces act on the wheel: the gravity of the wheel $\vec{G}$, the normal reaction $\vec{N}$, the force $\vec{P}$, and the friction force $\vec{F}_{fr}$ (Fig. 3.17b) (at $\delta \neq 0$, you need to show the moment of the pair of forces of resistance to friction).

The coupling force $\vec{F}_{fr}$ is conditionally directed toward the positive direction of the x axis.

Write down the differential equations of the plane motion of the wheel:

$$m\ddot{x}_C = \sum X_i^e; \ m\ddot{y}_C = \sum Y_i^e; \ J_C\ddot{\varphi} = \sum M_{Ci}^e$$

or in this case

$$m\ddot{x}_C = P\cos\beta - G\sin\alpha + F_{fr}; \tag{3.97}$$

$$m\ddot{y}_C = N + G\cos\alpha - P\sin\beta; \tag{3.98}$$

$$J_C\ddot{\varphi} = Pr - F_{fr}R. \tag{3.99}$$

The positive direction of the angle of rotation of the wheel is taken to be clockwise, which corresponds to the movement of the wheel in the positive direction of the x axis.

In accordance with this, the clockwise direction is also assumed to be positive when determining the signs of the moments of external forces in the equation of moments (3.99).

To the differential equations (3.97)–(3.99), we add the relation equations:

$$Y_C = R = const; \tag{3.100}$$

$$\omega = \dot{\varphi} = \frac{v_C}{R} = \frac{\dot{x}_C}{R}. \tag{3.101}$$

The last equation expresses the rolling condition of the wheel without sliding.

The condition (3.100) implies that

$$\ddot{y} = 0. \tag{3.102}$$

Differentiating Eqs. (3.101) in time, we obtain

$$\ddot{\varphi} = \ddot{x}_C / R. \tag{3.103}$$

Substituting (3.102) and (3.103) into Eqs. (3.98) and (3.99) and considering that $G = mg$, $J_C = mi_c^2$, we obtain

$$N = P \sin + mg \cos\alpha, \; mi_C^2 \ddot{x}_C / R = Pr - F_{fr} R.$$

Excluding $\ddot{x}_C$ from the last equation and the equality (3.97), we find

$$F_{fr} = \frac{P(Rr - i_C^2 \cos\beta) + i_C^2 mg \sin\alpha}{R^2 + i_C^2}$$

or based on the original data

$$F_{fr} = -0.257P + 208. \tag{3.104}$$

The dependency graph (3.104) is shown in Fig. 3.18. The graph intersects the P axis at the point $P_0 = 809 \; N$

At $0 \leqslant P < P_0 \; F_{fr} > 0$—the coupling force is directed, as shown in the figure, in the positive direction of the x axis.

At $P > 0 \; F_{fr} < 0$, the coupling force is directed in the opposite direction.

The module of the clutch force, which ensures rolling of the wheel without sliding, is subject to the following restrictions:

Fig. 3.18 Dependence of the coupling force on the active force

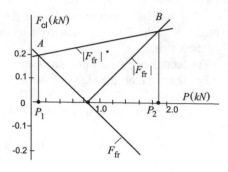

$$|F_{fr}| \leqslant Nf_{(fr)}, \qquad (3.105)$$

(meaning that always $N > 0$).

The limit value of the coupling force modulus will be

$$|F_{fr}|* = Nf_{fr} = (P \sin + mg \cos \alpha) f_{fr}.$$

or based on the original data

$$|F_{fr}|* = 0.05P + 190. \qquad (3.106)$$

In Fig. 3.18 is shown a graph of the dependence of $|F_{fr}|$ on P, which is a polyline. There is also a straight line (3.106). It intersects the graph F_{fr} at points A and B with abscissas P_1 and P_2.

Range of force values P at which the wheel rolls without sliding ($|F_{fr}| \leqslant Nf_{fr}$) is

$$P_1 \leqslant P \leqslant P_2.$$

The boundary values of the force P are found using (3.104) and (3.106), from the conditions

$$F_{fr} = |F_{fr}|*è - F_{fr} = |F_{fr}|*.$$

As a result of calculations, we get

$$P_1 = 58.6 \ N; \quad P_2 = 1923 \ N.$$

The differential equation of motion of the center of mass of the wheel based on (3.97) and further dependencies takes the form

Fig. 3.19 Dependence of the acceleration of the center of mass on the active force

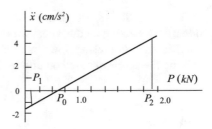

$$\ddot{x}_C = \frac{R[P(R\cos\beta + r) - mg\sin\alpha \cdot R]}{m(R^2 + i_C^2)}$$

or based on the original data

$$\ddot{x}_C = 0.003047 P - 1.498. \tag{3.107}$$

The dependency graph (3.107) is shown in Fig. 3.19: $\ddot{x}_c = 0$ for $P = 492 \, N$. For $P = P_1$, we have $\ddot{x}_C = -1.32 \, m/s^2$.

Integrating this differential equation twice and keeping in mind the zero initial conditions, we find

$$x_C = -0.66t^2;$$

the wheel rolls down an inclined plane.

When $P = P_2$

$$\ddot{x}_C = 4.36 \, m/s^2, \quad x_C = 2.18t^2.$$

The wheel rolls in the direction of the positive direction of the x axis up the inclined plane.

N o t e. The resultant of all forces applied to a wheel rolling without sliding passes through a point K located on a diameter perpendicular to the velocity of the center of mass, at a distance $CK = i_C^2/R$ from it.

In this example,

$$CK = 50^2/60 = 41.7 \, cm.$$

The point K is shown in Fig. 3.17b.

The sum of the moments of all forces applied to the wheel, relative to the axis passing through the point K and perpendicular to the plane of motion, is zero:

$$\sum M_{iK} = 0. \tag{3.108}$$

The condition (3.108) allows you to find the dependence of the coupling force on other forces acting on the wheel without composing differential equations.

This condition also allows us to check the results of calculations performed in solving the problem based on differential equations of plane motion of a solid.

References

E. Berezkin, *Kurs teoreticheskoi mekhaniki, tt. I, II. [The Course of Theoretical Mechanics]* (MGU Publ., Moscow, 1974)

V. Molotnikov, *Osnovy teoreticheskoi mekhaniki [Fundamentals of Theoretical Mechanics]* (Fenix Publ., Rostov-on-Don, 2004)

V. Molotnikov, *Mekhanika konstruktsiiyu Teoreticheskaya mekhanika. Soprotivlenie materialov [Construction Mechanics and Theoretical Mechanics. Mechanics of Materials]* (Lan' Publ., Sankt-Peterburg, 2012)

A. Yablonskii, *Kurs teoreticheskoi mekhaniki. Ch. II. [The Course of Theoretical Mechanics]* (Vysshaya shkola Publ., Moscow, 1971)

A. Yablonskii, S. Noreiko, S. Vol'fson, dr., *Sbornik zadanii dlya kursovykh rabot po teoreticheskoi mekhanike: Uchebnoe posobie dlya tekhnicheskikh vuzov, 5-e izd. [Collection of Tasks for Term Papers on Theoretical Mechanics: A Textbook for Technical Universities]*, 5th edn. (Integral-Press Publ., Moscow, 2000)

Chapter 4
Theory of Impact

Abstract So far we have considered problems concerning the statics and dynamics of discrete (lumped) and continuous material systems when forces of action and reaction act upon these systems in a continuous fashion for the entire duration of a process. This chapter examines the case where changes in the system's momentum that lead to a change in velocity are related to the action of a force or moment of force over a finite and often very short period of time. A phenomenon is called an impact if we observe a sudden (instantaneous) change in the velocity of a particle caused by the action of instantaneous forces. If two bodies collide and the time of the collision process is very short, then we observe a continuous change in the velocity of the body, and because the collision usually lasts for a very short time, it is associated with the generation of relatively large forces.

Keywords Impact · Impulse · Kelvin's theorem · Direct blow · Oblique blow · Central blow · Carnot's theorem · Shock impulse · Spinning impact

4.1 Impact Phenomenon and Its Main Characteristics

So far, we have considered the motion of bodies under the action of ordinary forces that cause a continuous change in the modules and velocity directions of points in the system. In this case, each infinitesimal time interval corresponds to an infinitesimal increment of speed. However, in nature and technology, there are cases when the velocities of the points of the body for a negligible period of time get finite increments (Hagedorn 2008; Krodkiewski 2006; Molotnikov 2012). The reader interested in the physics of impact phenomena is advised to refer to the excellent monograph (Zel'dovich and Raizer 1963).

A phenomenon in which the velocities of all points of the system or parts of these points change by finite values over a negligible period of time is called an impact. Examples of such phenomena can be the impact of a hammer on a workpiece when forging a part, the impact of a ball on a wall, the impact of a copra woman on a pile, etc. The duration of the impact is usually a fraction of a second.

Fig. 4.1 Possible movement
of the hinge lever

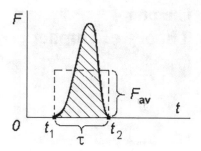

The final change in the velocities of the points of the body for a small period of time becomes possible because the modules of forces that develop during impact are very large. As a result, the impulses of such forces during the impact are finite values. Forces that have these properties are called instantaneous or shock. The characteristic graph of change of impact force F during impact between the beginning and end of the phenomena is shown in Fig. 4.1. As can be seen from the figure, the impact force during shock increases rapidly according to some rule from zero to the maximum value, and then quickly decreases to zero, generally speaking, different law.

In many cases, knowledge of the laws of increasing and decreasing impact force is not required. It is enough to know the integral characteristic of this rapidly changing force during the impact. This characteristic is a shock pulse, the quantitative measure of which is a vector quantity :

$$S = \int_0^\tau F\,dt. \tag{4.1}$$

The modulus of the shock pulse is equal to the area shaded in the figure. A constant force during the impact, which during the impact gives the same modulus of momentum as the variable impact force, is called the average impact force. In Fig. 4.1, the modulus of this force is denoted by. It can be determined from the relation:

$$F_{av}\tau = S.$$

In order for the impact pulse modulus to be a finite value, the impact force must be on the order of $1/\tau$. It follows that for a small τ, the impact force is a large quantity. In contrast, the pulse modulus of a nonimpact force, for example, the elastic force of a spring, during the impact has a strand of magnitude τ, i.e., it is a small value compared to the shock pulse modulus. For this reason, when analyzing the impact phenomenon, the impulses of nonimpact forces can be neglected in comparison with shock impulses.

The impact is accompanied by deformation of bodies. The movements of material points of colliding bodies caused by their deformation during the impact are small. Therefore, we can assume that during the impact, the points of the system do not change their position, and, consequently, neither their radius vectors nor their coordinates change. Let's explain this with an example. Let the body fall on the end of the coil spring. In this case, the movement of the body during the impact is equal to the spring draft during this time. This movement can be neglected in comparison, for example, with the movement of the body during the time from the beginning of the impact to the moment when the spring reaches the maximum draft. For this reason, when approximating the movement of a body during impact, the spring can be considered a solid.

4.2 Basic Theorems of the Theory of Impact

Consider the motion of a material point M (Fig. 4.2) with mass m along the arc of the trajectory AM. Denote by v the velocity of the point at the moment immediately before the impact. Suppose that under the action of shock F and nonimpact F^* forces, the point changed its speed, which immediately after the impact became equal to u, and the movement of the point continues along the trajectory of MB. By the theorem on the change in the amount of motion of a point during the impact τ, we have

$$m\boldsymbol{u} - m\boldsymbol{v} = \int_0^\tau \boldsymbol{F}(\zeta)d\zeta + \int_0^\tau \boldsymbol{F}^*(\zeta)d\zeta.$$

Using the notation (4.1) for the shock force pulse and neglecting the shock force pulse during the impact, we can write

$$m\boldsymbol{u} - m\boldsymbol{v} = \boldsymbol{S}. \qquad (4.2)$$

Formula (4.2) expresses the theorem on change of momentum of a point for the strike.

Theorem *The change of momentum points during shock equal to the shock impulse applied at a point.*

Fig. 4.2 On the proof of Kelvin's theorem

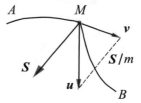

The proved theorem implies in particular that the change in the velocity of a point on impact is equal to

$$u - v = S/m,$$

that is, the specified change is parallel to the shock pulse, as shown in Fig. 4.2 by the dashed line.

In projections on the axes of the Cartesian coordinate system, the Oxyz vector equality (4.2) gives

$$mu_x - mv_x = S_x; \quad mu_y - mv_y = S_y; \quad mu_z - mv_z = S_z. \tag{4.3}$$

Multiply scalar equality (4.2) sequentially by u and v, and then add the resulting equalities. After dividing both parts by 2, we get

$$\frac{mu^2}{2} - \frac{mv^2}{2} = \frac{1}{2} S(u + v). \tag{4.4}$$

But according to theorem on the change in the kinetic energy of a point, the left part of the formula (4.4) is equal to the work of the force F F applied to the point. Therefore,

$$A = \frac{1}{2} S(u + v). \tag{4.5}$$

Thus, the following is proved.

Theorem (Kelvin's) *The work of the force applied to a point over a period of time is equal to the scalar product of the force momentum over the same period of time by the half-sum of the initial and final velocities of the point.*

Let us now consider a mechanical system consisting of N material points. We apply the theorem on the change in the amount of motion on impact to each point of the system:

$$m_k u_k - m_k v_k = S_k^{(e)} + S_k^{(i)}, \quad k = 1, 2, \ldots, N,$$

where $S_k^{(e)}$ and $S_k^{(i)}$ are the external and internal shock pulses tested by the k-point. Here, as before, we neglect the impulses of nonimpact forces during the impact. Summing up the written equalities for all points of the system, we get

$$\sum_k m_k u_k - \sum_k m_k v_k = \sum_k m_k S_k^{(e)} + \sum_k m_k S_k^{(i)}.$$

Denoting the first sum on the left side of the last equality by Q, and the second by S_0, and considering, in addition, that by the property of internal forces $\sum_k m_k S_k^{(i)} = 0$, we have

$$Q - Q_0 = \sum_k m_k S_k^{(e)}. \tag{4.6}$$

Thus, the following proposition is proved.

Theorem *The change in the amount of motion of a mechanical system during the impact is equal to the vector sum of all external shock pulses applied to the points of the system.*

This result expresses a theorem about the change in the amount of motion of the system on impact. In projections on coordinate axes, the Eq. (4.6) looks like

$$Q_x - Q_{0x} = \sum_k S_{kx}^{(e)}; \quad Q_y - Q_{0y} = \sum_k S_{ky}^{(e)}; \quad Q_z - Q_{0z} = \sum_k S_{kz}^{(e)}. \tag{4.7}$$

The result (4.6) can be represented even differently. In fact, according to the formulas (3.72), (3.73)

$$Q = M u_C; \quad Q_0 = M v_C,$$

where M is the mass of the system and u_C and v_C are the velocities of the center of mass after and before impact. Then from the formula (4.6), we get the *theorem about the motion of the center of mass on impact:*

$$M(u_C - v_C) = \sum_k S_k^{(e)}. \tag{4.8}$$

In projections on coordinate axes, the last formula is written as

$$M(u_{Cx} - v_{Cx}) = \sum_k S_{kx}^{(e)}; \quad M(u_{Cy} - v_{Cy}) = \sum_k S_{ky}^{(e)};$$
$$M(u_{Cz} - v_{Cz}) = \sum_k S_{kz}^{(e)}. \tag{4.9}$$

Let $\sum_k m_k S_k^{(e)} = 0$. Then it follows from theorems (4.6) and (4.8) that

$$Q = Q_0; \quad u_C = v_C. \tag{4.10}$$

This means that the *amount of motion of the system and the speed of its center of mass do not change if the vector sum of all external shock pulses acting on the points of the system is zero.*

Similarly, if $\sum_k m_k S_k^{(e)} = 0$, then from formulas (4.7) and (4.9) follows

$$Q_x = Q_{0x}; \ u_{Cx} = v_{Cx}. \tag{4.11}$$

The results obtained express the laws *of conservation of the projection of the amount of motion* and the center of mass of the system on impact.

Let r be the radius vector of the point, and the same one both before and after the impact. Multiply it by a vector equal to (4.2). Receive

$$\boldsymbol{r} \times m\boldsymbol{u} - \boldsymbol{r} \times m\boldsymbol{v} = \boldsymbol{r} \times \boldsymbol{S}. \tag{4.12}$$

Let us now write the formula (4.12) for the k-th of N points in the system:

$$\boldsymbol{r}_k \times m_k \boldsymbol{u}_k - \boldsymbol{r}_k \times m_k \boldsymbol{v}_k = \boldsymbol{r}_k \times \boldsymbol{S}_k^{(e)} + \boldsymbol{r}_k \times \boldsymbol{S}_k^{(i)}, \ k = 1, 2, \ldots, N,$$

and $\boldsymbol{S}_k^{(e)}$ and $\boldsymbol{S}_k^{(i)}$—respectively external and internal shock pulses applied to the k-th point of the system. Summing up such equalities over all N points and recalling the definition (3.78) of the kinetic moment of the system relative to the origin O of the vectors $\boldsymbol{r}_k$, we obtain

$$\boldsymbol{K}_O - \boldsymbol{K}_O^0 = \sum_k \boldsymbol{M}_O(\boldsymbol{S}_k^{(e)}), \tag{4.13}$$

where $\boldsymbol{K}_O$ and $\boldsymbol{K}_O^0$ are kinetic moments of the system, respectively, after and before the impact, $\sum_k \boldsymbol{M}_O(\boldsymbol{S}_k^{(e)}) = \sum_k \boldsymbol{r}_k \times \boldsymbol{S}_k^{(e)}$ is the kinetic moment of all external shock pulses relative to the center O, and it is taken into account that by the property of internal forces

$$\sum_k \boldsymbol{M}_O(\boldsymbol{S}_k^{(i)}) = \sum_k \boldsymbol{r}_k \times \boldsymbol{S}_k^{(i)} = 0.$$

Thus, equality (4.13) proves the following.

Theorem *The change in the kinetic moment of the system relative to a certain center during the impact is equal to the vector sum of the moments of all external shock pulses applied to the points of the system relative to the same center.*

In projections on coordinate axes, the formula (4.13) takes the form

$$K_x - K_x^{(0)} = \sum_k M_x(S_x^{(e)}); \ K_y - K_y^{(0)} = \sum_k M_y(S_y^{(e)});$$
$$K_z - K_z^{(0)} = \sum_k M_z(S_z^{(e)}). \tag{4.14}$$

In the special case when the impact is applied to a solid body rotating around a fixed axis Oz and having a moment of inertia relative to the axis of rotation J_z, the angular velocities before and after the impact are ω_0 and ω, respectively:

$$K_z = J_z\omega; \quad K_z^{(0)} = J_z\omega_0.$$

Then according to (4.14), we get the following change in the angular velocity at impact:

$$\omega - \omega_0 = \frac{1}{J_z}\sum_k M_{kz}(S_k^{(e)}). \tag{4.15}$$

Note that the formula (4.15) does not include the moments of shock pulses from the reactions of fixed points of the axis of rotation, unless there are shock pulses from friction forces at the points of attachment. The following propositions follow from the theorem on the change in the moment of the amount of motion on impact.

Corollary 1 *If the vector sum of the moments of external shock pulses relative to a certain point is zero, then the kinetic moment of the system relative to this point does not change, i.e.,*

$$\boldsymbol{K}_O - \boldsymbol{K}_O^{(0)} = const \ \ for \ \ \sum_k \boldsymbol{M}_O(S_k^{(e)}) = 0. \tag{4.16}$$

The last equality expresses the *law of conservation of the kinetic moment of the system relative to the point at impact.*

Corollary 2 *If the sum of the moments of external shock pulses is zero relative to a certain axis, for example, ox, then the kinetic moment of the system relative to this axis remains constant during impact, i.e.,*

$$K_z - K_z^{(0)} = const \ \ for \ \ \sum_k M_z(S_{kz}^{(e)}) = 0. \tag{4.17}$$

Corollary 2 expresses the *law of conservation of the kinetic moment of the system relative to the axis at impact.*

4.3 Impact of a Point on a Stationary Surface

We will assume that the size of the body can be ignored and consider it as a material point. A point is called a straight point if its velocity v before the impact is directed along the normal to the surface at the point of impact A (Fig. 5.3). After such an impact, the material point will separate from the surface, and its velocity u will be

directed along the same normal to the surface at the point of impact. To characterize
the impact properties of the surface and body, the coefficient is found

$$k = \frac{|\boldsymbol{u}|}{|\boldsymbol{v}|}, \tag{4.18}$$

called *the recovery factor*. When $k = 1$ the impact is called *perfectly elastic*. In this
case, $u = v$. Therefore, there is only a change in the direction of the speed of the
point to the opposite. The second special case is $k = 0$. In this case, $u = 0$, and the
impact is called *absolutely inelastic*. In all other cases, $0 < k < 1$, and say that the
impact is just elastic (or partially elastic).

The phenomenon of a point hitting a stationary surface can be divided into two
phases. The first phase of deflection τ_1 is called the *deformation phase*. At the end
of this phase, the velocity of the point vanishes. We apply to this phase of impact
the theorem on the change in the amount of motion of a point in the projection on
the external normal n (Fig. 4.3) to the surface. Have

$$0 - (-mv) = S_1; \tag{4.19}$$

moreover, the shock pulse S_1 can be calculated as the sum of the elementary
impulses of the reaction force N for the first phase, i.e.,

$$S_1 = \int_0^{\tau_1} N(t)dt. \tag{4.20}$$

This is followed by a recovery phase of duration τ_2. During this phase, the body,
from the moment of the greatest deformation to its separation from the surface,
partially (and with an absolutely elastic impact—completely) restores its original
shape. It turns out that the total impact time $\tau = \tau_1 + \tau_2$. For the recovery phase,
according to the theorem about changing the momentum of a point, we can write

$$mu - 0 = S_2, \tag{4.21}$$

where by analogy with the formula (4.20)

Fig. 4.3 Direct impact

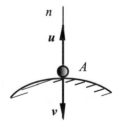

Fig. 4.4 Oblique impact

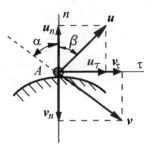

$$S_2 = \int_{\tau_1}^{\tau_2} N(t)dt. \tag{4.22}$$

Expressing u and v from formulas (4.19) and (4.21) and substituting the resulting definition (4.18), we find

$$k + \frac{u}{v} = \frac{S_2}{S_1}. \tag{4.23}$$

Thus, *the recovery coefficient for a direct impact of a point on a fixed surface is equal to the ratio of the numerical values of the shock pulses for the first and second phases of the impact.*

The total shock pulse S is equal to the sum of the shock pulses for the entire time of the impact, i.e.,

$$S = S_1 + S_2 = mv \left(1 + \frac{u}{v}\right) = mv(1 + k). \tag{4.24}$$

From the last formula, it follows that at $k = 1$ $S = 2mv$; at $k = 0$ $S = mv$: *the shock pulse for an absolutely inelastic shock is twice less than the shock pulse for an absolutely elastic direct impact.*

Let us now consider the case when the velocity of the point before impact is an angle α with to the surface at the point of impact (Fig. 4.4). Such a blow is called *oblique* or *indirect*. The angle α is called the *angle of incidence,* and the angle β made up of the velocity direction of the point after impact with the normal to the surface is *the angle of reflection.*

Decompose the vectors u and v into components along the axes n and τ that coincide with the direction of the normal and tangent at the point of impact:

$$v = v_n + v_\tau; \quad u = u_n + u_\tau.$$

The recovery coefficient for an oblique impact is the ratio of the numerical values of the normal components of the velocity of the point after and before the impact, i.e.,

$$k = \frac{|\boldsymbol{u}_n|}{|\boldsymbol{v}_n|} = \frac{u_n}{v_n}. \tag{4.25}$$

Applying the theorem on changing the amount of motion of a point in projections to the normal n, by analogy with the formula (4.23), we obtain

$$k = \frac{u_n}{v_n} = \frac{S_{2n}}{S_{1n}},$$

where S_{2n}, S_{1n} are the projections of shock pulses on the surface normal for the second and first phases of the impact. For a perfectly smooth surface, the tangent components of the velocities $\boldsymbol{u}$ and $\boldsymbol{v}$ coincide, i.e., $u_\tau = v_\tau$. On this condition,

$$\tan \beta = \frac{u_\tau}{u_n} = \frac{v_\tau}{v_n}; \quad \tan \alpha = \frac{v_\tau}{v_n},$$

i.e.,

$$\tan \alpha = k \tan \beta. \tag{4.26}$$

The formula (4.26) gives the relationship between the angle of incidence and the angle of reflection at different recovery coefficients for a perfectly smooth impact surface.

4.4 Experimental Determination of the Recovery Factor

The value of the impact recovery coefficient for various bodies can be determined experimentally by measuring the height h_2 to which the body rises after a direct impact with the surface when falling from a given height h_1 (Fig. 4.5) without initial velocity.

As a test body in experiments, a ball is usually used. When falling from a height of h_1, the speed of the ball immediately before impact will be $v = \sqrt{2gh_1}$. At the moment after its velocity and can be expressed in terms of the height of the lift above the impacted surface by the formula $u = \sqrt{2gh_2}$. Then for the recovery coefficient according to (4.18), we have

$$k = \frac{u}{v} = \sqrt{\frac{h_2}{h_1}}. \tag{4.27}$$

Thus, by setting the drop height h_1 and measuring the lift height h_2, you can get the value of the recovery coefficient for various combinations of ball materials and the impacted surface. More subtle measurements show that the values of the

Fig. 4.5 To the experimental determination of the recovery coefficient

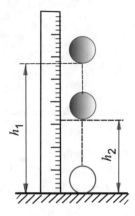

Table 4.1 The values of the recovery factor due to a direct blow

Of the impacting body	Coefficient of restitution k
Wood on wood	1/2
Steel by steel	5/9
Ivory on ivory	8/9
Glass on glass	15/16

recovery coefficients obtained in the described way are only a rough approximation to reality.

It turns out that the recovery coefficient depends not only on the materials of the colliding bodies but also on their masses, shape, collision speeds, and other factors. Table 4.1 shows the values of the recovery coefficient for some materials determined at a collision velocity of about 3 m/s.

4.5 Impact of Two Bodies

Let two bodies with masses m_1 and m_2 move translationally before and after the impact. Let's denote the velocities of their centers of mass before impact through v_1 and v_2, and after impact through u_1 and u_2, respectively. Impact friction will be ignored. In this case, the shock pulses are directed along the general normal at the point of contact, which is also called the *impact line*. If the impact line passes through the centers of mass C_1 and C_2 (Fig. 4.6) colliding bodies, then there is a *central impact*.

Draw the coordinate axis Ox through the centers of mass C_1 and C_2, which we agree to always direct from C_1 to C_2 (Fig. 4.6). For an impact to occur, it must be $v_{1x} > v_{2x}$. Since the body that is hit cannot outstrip the body that is being hit, the condition must also be met $u_{1x} < u_{2x}$. Let the masses m_1, m_2, the velocities of colliding bodies before impact v_{1x}, v_{2x}, and the recovery coefficient k be known. It is required to determine the velocities of bodies after impact u_{1x} and u_{2x}. To solve

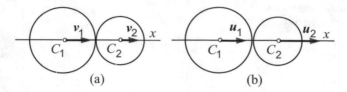

Fig. 4.6 To the experimental determination of the recovery coefficient

this problem, we apply the theorem on changing the amount of motion of colliding bodies. Considering them for the time of impact as one system and ignoring the impulses of internal forces, we get

$$m_1 u_{1x} + m_2 u_{2x} = m_1 v_{1x} + m_1 v_{2x}. \tag{4.28}$$

When two bodies strike, the shock pulse does not depend on the absolute values of the velocities of the colliding bodies, but on how much the velocity of the striking body exceeds the velocity of the impacted one, i.e., on the difference $v_{1x} - v_{2x}$. Therefore, according to the definition (4.18), we have

$$k = \left| \frac{u_{1x} - u_{2x}}{v_{1x} - v_{2x}} \right| = - \frac{u_{1x} - u_{2x}}{v_{1x} - v_{2x}}. \tag{4.29}$$

The last formula can also be written as

$$u_{1x} - u_{2x} = -k(v_{1x} - v_{2x}). \tag{4.30}$$

The joint solution of Eq. (4.28) and any of the equalities (4.29) or (4.30) solves the problem.

S p e c i a l   c a s e:

1. For an absolutely inelastic impact, $k = 0$. In this case, from Eqs. (4.28)–(4.29) follows

$$u_{1x} = u_{2x} = \frac{m_1 v_{1x} + m_2 v_{2x}}{m_1 + m_2}. \tag{4.31}$$

This means that both bodies move at the same speed after impact. The shock impulses acting on the bodies at the moment of impact will be

$$S_{2x} = -S_{1x} = (v_{1x} - v_{2x}) \frac{m_1 m_2}{m_1 + m_2}. \tag{4.32}$$

2. For an absolutely elastic shock $k = 0$ and from Eqs. (4.28) and (4.29), we have

$$u_{1x} = v_{1x} - \frac{2m_2(v_{1x} - v_{2x})}{m_1 + m_2};$$
$$u_{2x} = v_{2x} + \frac{2m_1(v_{1x} - v_{2x})}{m_1 + m_2}. \tag{4.33}$$

The shock pulses tested by the bodies are determined by the formula

$$S_{2x} = -S_{1x} = (v_{1x} - v_{2x})\frac{2m_1 m_2}{m_1 + m_2}. \tag{4.34}$$

Comparing the last result with the formula (4.31), we conclude that for an absolutely elastic shock, the shock impulses of interacting bodies are twice as large as for an absolutely inelastic shock. Note also that if the masses m_1 and m_2 are equal, we obtain from the relations (4.33) $u_{1x} = v_{2x}$, $u_{2x} = v_{1x}$, i.e., *bodies with equal masses exchange velocities at an absolutely elastic impact.*

Example Two balls of mass m_1 and m_2 are suspended on inextensible threads as shown in Fig. 4.7. The left ball is deflected by an angle α and released without initial velocity. After the impact, the right ball is deflected by an angle β. Find the coefficient of restitution of the ball at impact.

D e c i s i o n. We apply the theorem on the change in the kinetic energy of the left ball when moving $A_1 A$:

$$\frac{1}{2}m_1 v_1^2 = m_1 gl(1 - \cos\alpha),$$

where g is the acceleration of gravity. From here we find

$$v_1 = 2\sqrt{gl}\sin\frac{\alpha}{2}.$$

Similarly, we get

$$u_2 = 2\sqrt{gl}\sin\frac{\beta}{2}. \tag{4.35}$$

Fig. 4.7 Example of a direct hit of two balls

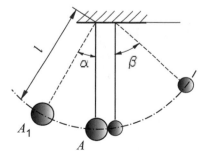

In our case, $v_2 = 0$, and dependencies (4.28) and (4.29) give

$$m_1 u_{1x} + m_2 u_{2x} = m_1 v_{1x}; \quad u_{2x} - u_{1x} = k v_{1x}. \tag{4.36}$$

Bearing in mind that in position A $v_{1x} = v_1$, $u_{2x} = u_2$, from the formulas (4.36), we get

$$k = \frac{(m_1 + m_2) u_2}{m_1 v_1} - 1 = \frac{(m_1 + m_2) \sin \frac{\beta}{2}}{m_2 \sin \frac{\alpha}{2}} - 1. \tag{4.37}$$

4.6   Carnot's Theorem

For an absolutely elastic impact of a point on a perfectly smooth stationary surface, the recovery coefficient $k = 1$, and from the formulas (4.18) and (4.25), it follows that the velocity of the point at the moment of impact changes only in the direction. Since the numerical value of the velocity remains constant, the kinetic energy of the body does not change during the impact.

The situation is different with elastic and absolutely inelastic impacts. In these cases, the numerical value of the velocity during the impact does not remain constant, so the kinetic energy of the body changes. Consider, for example, an absolutely inelastic shock. By theorem (4.2) on changing the amount of motion of a point, we have

$$m\boldsymbol{u} - m\boldsymbol{v} = \boldsymbol{S}, \tag{4.38}$$

where, as before, m is the mass of the point, $\boldsymbol{u}$ and $\boldsymbol{v}$ are its velocities before and after impact, and $\boldsymbol{S}$ is the shock pulse.

In the absence of impact friction, the vector $\boldsymbol{S}$ is directed along the normal to the impacted surface (Fig. 4.8) at the point of impact A. The velocity of the point after impact will in this case be directed tangentially to the surface, i.e., the vectors $\boldsymbol{S}$ and $\boldsymbol{u}$ are orthogonal: $\boldsymbol{S} \perp \boldsymbol{u}$. It follows that the scalar product of these vectors is zero: $\boldsymbol{S} \cdot \boldsymbol{u} = 0$. Taking this result into account, we multiply both parts of equality (4.38) scalar by the vector $\boldsymbol{u}$. We get

Fig. 4.8 To Carnot's theorem

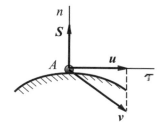

$$mu^2 - mu \cdot v = 0.$$

With this in mind, we record the change in the kinetic energy of a point under an absolutely inelastic impact:

$$\frac{mv^2}{2} - \frac{mu^2}{2} = \frac{mv^2}{2} - \frac{mu^2}{2} + (mu^2 = mu \cdot v) =$$
$$\frac{mv^2}{2} + \frac{mu^2}{2} - mu \cdot v = \frac{m}{2}(v - u)^2.$$

Omitting the intermediate calculations, we can write

$$\frac{mv^2}{2} - \frac{mu^2}{2} = \frac{m}{2}(v - u)^2. \tag{4.39}$$

The vector difference $(v - u)$ is called the *lost velocity.*
Formula (4.39) expresses Carnot's theorem for a material point.

Theorem *The loss of the kinetic energy of a point in an absolutely inelastic impact on a perfectly smooth surface is equal to the kinetic energy of the lost velocity.*

Applying the formula (4.39) for the k-th point of the system and summing over all its points, we obtain

$$T_0 - T = \frac{1}{2} \sum_k m_k (v_k - u_k)^2, \tag{4.40}$$

where indicated

$$T_0 = \sum_k \frac{m_k v_k^2}{2}; \quad T = \sum_k \frac{m_k u_k^2}{2}.$$

Thus, **Carnot's theorem** was obtained for the system:

Theorem *The loss of the kinetic energy of the system during an absolutely inelastic impact in the absence of shock friction is equal to the kinetic energy from the lost velocities of the points of the system.*

4.7 Rotating Body Blow

Let a rigid body of mass m rotate around a fixed axis z (Fig. 4.9). At the moment when it had an angular velocity ω_0, a shock impulse S is applied to the body. The

Fig. 4.9 Action of the shock
pulse S on a rotating body

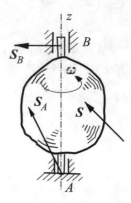

change in the angular velocity of the body as a result of this action based on the last
of formulas (4.15) will be

$$\omega - \omega_0 = \frac{M_z(S)}{J_z},\tag{4.41}$$

where ω is the angular velocity of the body after impact, J_z is the moment of inertia
of the body relative to the axis of rotation, and $M_z(S)$ means the moment of the
external shock pulse relative to the z axis.

The S pulse causes reactive shock pulses S_A and S_B to appear at the reference
points, and the supports experience shock pressures.

Let's find out the conditions under which there are no shock reference pulses.
Choose the following system of coordinate axes: the z axis is directed along the axis
of rotation in the same direction as the angular velocity vector ω_0, and the yoz plane
contains the center of mass of the body C (Fig. 4.10). The impulsive reactions S_A
and S_B in supports A and B are decomposed into components along the axes of the
selected coordinate system (Fig. 4.10a). Denote the distance between the supports
$AB = b$, and the distance from the axis of rotation to the center of mass of the body
through a.

Apply now to study the motion of the body considered to the theorem of change
of momentum of the system at impact and the change in its kinetic moment. Based
on the formulas (4.9) and (4.14), we can write

$$-ma(\omega - \omega_0) = S_{Ax} + S_{Bx} + S_x;$$
$$0 = S_{Ay} + S_{By} + S_y;$$
$$0 = S_{Az} + S_z;$$
$$-J_{xz}(\omega - \omega_0) = -S_{By}b + M_x(S);$$
$$-J_{yz}(\omega - \omega_0) = S_{Bx}b + N_y(S),\tag{4.42}$$

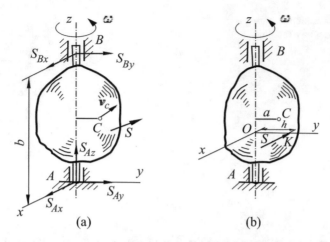

Fig. 4.10 Investigation of the result of the impact pulse: (**a**) decomposition of impulse reactions into components; (**b**) to simplify the calculation of reactions

where S_x, S_y, S_z are the components of the external shock pulse S in the xyz axes and $M_x(S)$, $M_y(S)$ are the moments of this pulse relative to the x and y coordinate axes, respectively.

Equations (4.41) and (4.42) make it possible to determine the reactive impulses in the supports A and B. At the same time, they determine the conditions under which the impact on a rotating body does not cause shock reactions in the bearings, i.e., when all the components of the vectors S_A and S_B turn to zero.

Let's get these conditions. When $S_A = S_B = 0$, in accordance with the second and third of formulas (4.42) should be $S_y = S_z = 0$. This means that the vector S must be orthogonal to the plane yAz (Fig. 4.10a) passing through the center of mass of the body. Since at $S_A = S_B = 0$ equations of system (4.42) do not depend on the choice of the origin on the Az axis. Therefore, in order to simplify further calculations, it is possible to draw the Oxy plane so that the vector S lies in this plane (Fig. 4.10b). In this case, $M_x(S) = M_y(S) = 0$, and the last two equations of system (4.42) give $J_{xz} = J_{yz} = 0$. Physically, this means that the z axis is the main axis of inertia of the body.

Let us now turn to the first equation of system (4.42). For $S_A = S_B = 0$ and $S_x = -S$, it takes the form

$$ma(\omega - \omega_0) = S. \tag{4.43}$$

Since in the case $M_z(S) = Sh$ under consideration, relation (4.38) gives

$$J_z(\omega = \omega_0) = Sh. \tag{4.44}$$

Eliminating the difference from equalities (4.43) and (4.44), we obtain the shoulder $h = OK$ (Fig. 4.10b) of the shock impulse S:

Fig. 4.11 For example, research impact on a rotating body

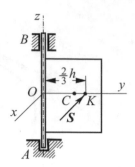

$$h = \frac{J_z}{ma}. \tag{4.45}$$

The point K of intersection of the line of action of the shock impulse with the plane passing through the axis of rotation and the center of mass of the body, in the absence of shock supporting impulses, is called the *center of impact*.

Thus, in order for no reactive impulses to arise at the fixing points of this axis during an impact on a body rotating around the z axis, it is required to satisfy the conditions:

(1) The shock pulse should be located in the Oxy plane, perpendicular to the axis of rotation and passing through O, for which the z axis is the main axis of inertia.
(2) The shock pulse should be directed perpendicular to the plane passing through the axis of rotation and the center of mass of the body.
(3) The shock impulse should be located at a distance $h = J_z/ma$ from the axis, counted in the direction where the center of mass of the body is located.

Example The door has the shape of a homogeneous rectangular plate with a width of h (Fig. 4.11). The axis of rotation of the door is fixed at point A with a thrust bearing, and at point B—with a bearing. Determine the position of the center of impact of the door if it is opened by the impact of the impact impulse.

D e c i s i o n. In accordance with the results (1) and (2) (see above), the shock impulse S must be perpendicular to the plane passing through the axis of rotation and the center of mass of the door C, i.e., perpendicular to the plane of the door itself. The plane in which the vector S lies must pass through the point O on the axis of rotation so that the plane Oxy is the axis of symmetry of the plate.

The distance OK to the center of impact K is determined by the formula (4.45)

$$OK = J_z/ma = J_z/(m \cdot h/2).$$

Considering that the moment of inertia of a rectangular door is

$$J_z = mh^2/3,$$

finally we get

$$OK = \frac{mh^2/3}{mh/2} = \frac{2}{3}h.$$

Consequently, any shock impulse in modulus, the line of action of which is perpendicular to the door plane and passes through the center of impact K at a distance of $2h/3$ from the axis of rotation, does not cause reactive impulses in supports A and B.

Self-Test Questions

1. What phenomenon is called a blow?
2. What forces are called shock forces?
3. What value serves as a quantitative measure of the impact force?
4. What is the effect of the impact force on a material point?
5. Formulate a theorem about the change in the momentum of a material point upon impact. Write down the formula expressions of this theorem in vector and scalar forms.
6. Formulate Kelvin's theorem for a material point upon impact.
7. Write down the formula expressions of the theorem about the change in the momentum of a mechanical system upon impact in vector form and in projections onto the coordinate axes.
8. Can internal shock impulses change the momentum of a mechanical system?
9. Give the formulation of the theorem on the motion of the center of mass of the system upon impact.
10. How is the theorem formulated about the change in the angular momentum of the system relative to a certain center during the impact? Write down the formula representations of this theorem in vector and scalar forms.
11. Formulate the law of conservation of the angular momentum of the system with respect to axis upon impact.
12. What is the impact recovery factor?
13. What impact is called absolutely elastic? Absolutely inelastic?
14. What physical phenomena accompany the deformation phase and the recovery phase upon impact?
15. What is the peculiarity of an absolutely elastic impact?
16. What is the difference between direct and oblique strikes?
17. What is the relationship between the angles of incidence and reflection when the ball hits an ideally smooth stationary surface?
18. What is the lost point speed?
19. How is Carnot's theorem formulated for a material point with an absolutely inelastic impact?

20. Write down the mathematical expression of Carnot's theorem for a mechanical system with absolutely inelastic impact.
21. Do internal shock impulses change the angular momentum of the system?
22. What parameters of motion of a rigid body during rotation around a fixed axis change as a result of impact forces acting on the body?
23. Under what conditions do the shock support impulses of a rotating body vanish?
24. What is called the center of impact?
25. How are the coordinates of the center of impact determined?

Control Tasks for the Section "Theory of Impact"

Variant 1

Cart 1 with a total mass of $m_1 = 6000\,\mathrm{kg}$, moving at a speed of $v_1 = 2.5\,\mathrm{m/s}$ on a horizontal rectilinear path, collides with a stationary cart 2, which together with the container has a mass of $m_2 = 4000\,\mathrm{kg}$, Fig. 4.12. At the end of the collision, the cart2 acquires the speed $u_2 = 2\,\mathrm{m/s}$, and the container—the angular speed of rotation around the edge A, fixed by a thrust bar. Consider a container with a mass of $m_0 = 500\,\mathrm{kg}$ as a homogeneous rectangular parallelepiped ($a = 0.8\,\mathrm{m}, h = 1.5\,\mathrm{m}$). Vertical planes of collision of trolleys are assumed to be smooth. The surface of the rails is absolutely rough, i.e., prevents wheels from slipping when bogies collide. The moments of inertia of the wheels relative to their axes are negligible (Yablonskii et al. 2000).

Determine the speed of the cart 1 at the end of the collision with the cart 2, as well as the impact impulse perceived by the thrust bar.

Variant 2

A weight of $m_0 = 500\,\mathrm{kg}$ falls from a height of $h = 1\,\mathrm{m}$ to the point D of an absolutely rigid beam having a hinge-fixed support A and an elastic support B, the stiffness coefficient of which $c = 20{,}000\,\mathrm{N/cm}$; the impact of the load on the beam is inelastic (Fig. 4.13).

A weight of $m_0 = 500\,\mathrm{kg}$ falls from a height of $h = 1\,\mathrm{m}$ to the point D of an absolutely rigid beam having a hinge-fixed support A and an elastic support B, the stiffness coefficient of which $c = 20{,}000\,\mathrm{N/cm}$; the impact of the load on the beam

Fig. 4.12 Variant 1

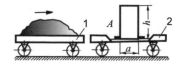

Fig. 4.13 Variant 2

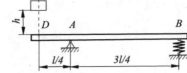

Fig. 4.14 Variant 3

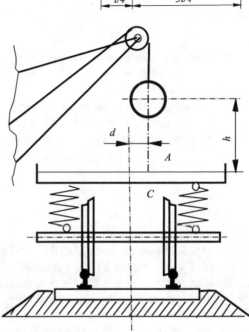

is inelastic (Fig. 4.13). The mass of the beam $m = 6000\,\text{kg}$, and its length $l = 4\,\text{m}$. The horizontal position of the beam shown in the drawing corresponds to the static deformation of the elastic support under the action of the weight of the beam. Take the beam for a thin homogeneous rod, and the load for a material point.

Determine the shock pulse perceived by the beam at the point D, as well as the greatest deformation of the elastic support, assuming that the movement of the point B occurs in a straight line.

Variant 3

Due to the break of the holding cable, the load with a mass of $m_0 = 500\,\text{kg}$ falls from a height of $h = m$ to the platform, which rests on identical and symmetrically located springs. The point A at which the load falls is located in the vertical transverse plane of symmetry of the platform and is separated from the center of gravity C of the platform at a distance $d = 0.6\,\text{m}$, Fig. 4.14

Fig. 4.15 Variant 4

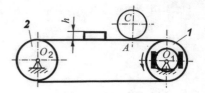

The impact of the load on the platform is inelastic. The mass of the platform $m = 5000$ kg, and its radius of inertia relative to the horizontal longitudinal axis of symmetry $i_c = 0.5$ m.

Taking the platform as an absolutely solid body and considering the load as a material point, determine the speed of the center of gravity and the angular velocity of the platform at the end of the impact. Determine also the shock pulse at the point A.

Variant 4

The load is a homogeneous solid cylinder with a mass of $m = 500$ kg and a radius of $r = 0.2$ m—is moved by the transporter. The conveyor belt is horizontal, and its constant speed $v = 0.6$ m/s; there is no sliding of the belt on pulleys 1 and 2, Fig. 4.15. At some point in time, the movement of the conveyor suddenly stops. Since the surface of the conveyor is absolutely rough, i.e., it does not allow the body to slip under impact, the cylinder will roll along the belt due to a sudden stop of the conveyor. Rolling resistance is negligible.

Determine the shock pulse perceived by the absolutely rough surface of the belt when the conveyor suddenly stops.

Check the speed of the center of gravity (or angular velocity) of the cylinder found for this by Carnot's theorem.

Determine the impact impulse perceived by a thrust step with a height of $h = 0.03$ m, which the cylinder hits after passing a certain distance, if the cylinder does not break off when hitting the step and its slippage occurs.

Variant 5

The transported loads roll from the position A without initial velocity along the inclined plane, which is the angle $\alpha = 15°$ with the horizon, passing along it the distance $s_1 = 3$ m, and continue to roll along the horizontal plane. There is no slip, and the rolling friction coefficient is $\delta = 0.8$ cm, Fig. 4.16.

Fig. 4.16 Variant 5

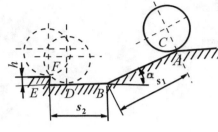

Fig. 4.17 Variant 6

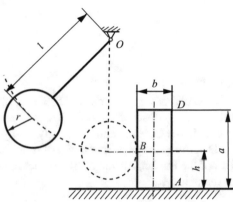

Determine at what distance s_2 should be placed a stop step with a height of $h = 0.2$ m, so that the loads, hitting the edge F of the step, only ascend it, without moving further than the edge F.

The calculation is made for a load—a homogeneous solid cylinder with a mass of $m = 500$ kg and a radius of $r = 0.5$ m.

Assume that there is no separation of the cylinder from the step, and the surface of the step is absolutely rough, i.e., it prevents the cylinder from sliding under impact.

Determine also the horizontal and vertical components of the shock pulse perceived by the cylinder from the side of the step, under the specified conditions.

Variant 6

The pendulum consists of a rod of length $l = 1.2$ m and a homogeneous circular disk with radius $r = 0.1$ m. The mass of the rod is negligible; the mass of the disk $m_0 = 5$ kg, Fig. 4.17.

The pendulum, deflected from the position of stable equilibrium, falls under its own weight, rotating around the fixed axis O; in the vertical position, having an angular velocity $\omega = 3$ rad/s, the pendulum hits the point B of the side face of the body D—a homogeneous rectangular parallelepiped with a mass $m = 6m_0$, ($a = 0.8$ m, $b = 0.4$ m, $c = 0.2$ m) (Fig. 4.17).

Fig. 4.18 Variant 7

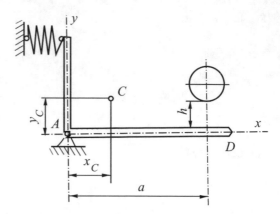

Impact recovery factor $k = 0.5$. The surfaces of the pendulum and the body D at the point of impact are smooth. The plane on which the body D rests is absolutely rough, i.e., it does not allow the body to slip under impact.

Determine the angular velocity of rotation of the body D around the edge A at the end of the impact, as well as the impact momentum perceived by the rough surface at the point A.

Variant 7

The lever consists of two absolutely rigid rods AB and AD, connected at right angles. The lever has a fixed horizontal axis of rotation A and is held at the point B by a spring; $AB = a = 1.5$ m (рис. 4.18).

To the point D of the horizontal rod of the lever, which is at rest, from a height of $h = 0.5$ m, a weight of $m_0 = 100$ kg falls. The mass of the lever is $m = 1000$ kg, and the radius of its inertia relative to the axis of rotation is $i_a = 0.5$ m. The position of the center of gravity C of the lever is determined by the coordinates $x_c = 0.4$ m and $y_c = 0.3$ m. Consider the load as a material point, and take the impact of the load on the lever as inelastic.

Determine the shock pulse tested by the load, as well as the horizontal and vertical components of the shock pulse perceived by the support A.

Variant 8

On the trolley 1 lies a load—a thin-walled homogeneous cylinder with a mass of $m_0 = 500$ kg and a radius of $r = 0.4$ m, which is kept from possible movement along the trolley by a step and an inclined plane that makes up the angle $\alpha = 60^o$ with the horizon (рис. 4.19).

Fig. 4.19 Variant 8

Trolley 1, which together with the load has a mass of $m_1 = 3000\,\text{kg}$, moving along a horizontal rectilinear path, collides with a speed of $v_1 = 3\,\text{m/s}$ on a stationary trolley 2 with a total mass of $m_2 = 6000\,\text{kg}$. At the end of the collision, the trolley 1 stops, and the cylinder, hitting the inclined plane, begins to roll along it.

The cylinder does not detach when it hits the inclined plane; the absolute roughness of the inclined plane prevents the cylinder from sliding under impact. Consider the vertical collision planes of the trolley and trolley as smooth. The surface of the rails is absolutely rough, i.e., it prevents the wheels from slipping when the bogies collide. The moments of inertia of the wheels relative to their axes are negligible.

Determine the angular velocity of the cylinder at the end of the impact on the inclined plane; check the found expression of the angular velocity of the cylinder by Carnot's theorem.

Determine the speed of the trolley 2 at the end of the collision with the trolley 1.

Variant 9

A body D of mass m_0, moving translationally along a horizontal plane, hits the node C of a stationary truss at a speed $v_0 = 3\,\text{m/s}$. The surfaces of the body D and the node C at the point of impact are smooth; the impact recovery coefficient $k = 0.5$. An absolutely rigid truss has a hinge-fixed support O and an elastic support A; $BC = a = 2\,\text{m}$. The mass of the truss $m = 20m_0$, and the radius of its inertia relative to the horizontal axis of rotation O $i_o = 1\,\text{m}$ (Fig. 4.20).

Determine the angular velocity of the truss at the end of the impact and test it by Carnot's theorem.

Determine what speed of translational motion on a smooth horizontal plane will the body D receive after impact.

Variant 10

Deflected by an angle $\alpha = 60°$ from the position of stable equilibrium, the pendulum falls without initial velocity under the action of its own weight, rotating

Fig. 4.20 Variant 9

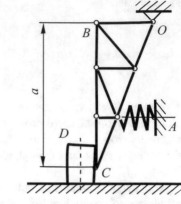

Fig. 4.21 Variant 10

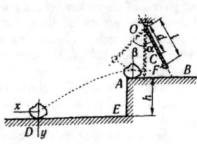

around the fixed axis O. In the vertical position, the pendulum strikes the point F on the resting body in the position A (Fig. 4.21).

The distance from the point O of the intersection of the axis of rotation by the vertical plane of symmetry of the pendulum to its center of gravity C and to the point F located in the same plane is $OC = d = 0.9$ m and $OF = l = 1.1$ m.

The mass of the pendulum is $m = 18$ kg, and the radius of its inertia relative to the axis of rotation is $i_o = 1$ m. The body has a mass of $m_0 = 6$ kg and can be taken as a material point. The coefficient of restitution upon impact of the pendulum on the body $k = 0.2$.

Due to the impact, the body falls from the point A of the plane AB to the point D of the smooth horizontal plane DE. The plane DE is located below the plane AB by $h = 1$ m. The impact of the body at the point D can be considered inelastic ($k = 0$).

Determine the shock pulse at the point D and the equation of motion of the body after this impact, using the coordinate system xDy.

Also determine the angle β of deflection of the pendulum after hitting the body at the A point.

Fig. 4.22 Mechanical
system, for example, on
impact

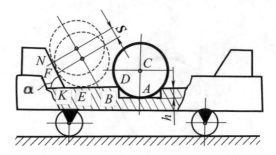

Example of Completing a Task

The load is a homogeneous solid cylinder with a mass of $m = 500\,\text{kg}$ and a radius
of $r = 0.5\,\text{m}$—lies on a moving platform and is kept from possible movement on
the platform by stops—steps (Fig. 4.22).

When the platform suddenly stops, the cylinder hits the edge D of the step BD
with a height of $h = 0.1\,\text{m}$ and goes up this step. Next, the cylinder rolls along the
section DE of the horizontal platform DK and, hitting another stop—the inclined
plane KN, which is the angle $alpha = 60°$ with the horizon—passes along it
the distance $FN = s = 0.1\,\text{m}$. The rolling of the cylinder is not accompanied by
sliding; the rolling resistance is negligible.

There is no separation of the cylinder when hitting the step and the inclined plane;
the absolute roughness of the step and the inclined plane eliminates the sliding of
the cylinder during impact.

Determine the speed of the platform before it stops, as well as the shock pulses
experienced by the cylinder from the side of the step and the inclined one.

Check the found expressions of the angular velocities of the cylinder after hitting
the step and the inclined plane with the help of Carnot's theorem.

S o l u t i o n. When the platform suddenly stops, the forward motion of the
cylinder instantly turns into a rotational motion around the edge D of the step BD,
i.e., the cylinder experiences a shock.

Let's make an equation that expresses the theorem about the change in the kinetic
moment of a mechanical system at impact. For the axis of moments, we take the
horizontal axis passing along the edge D (positions I and II, corresponding to the
beginning and end of the impact on the edge D of the step BD, coincide, Fig. 4.23a):

$$L_{IID} - L_{ID} = \sum M_D(\vec{S}_i^e).$$

The sum of the moments of external shock pulses applied to the cylinder with
respect to the D axis is $sum\, M_D(\vec{S}_i^e = 0$, since the shock pulse crosses the D axis
and therefore

$$L_{IID} = L_{ID}. \tag{4.46}$$

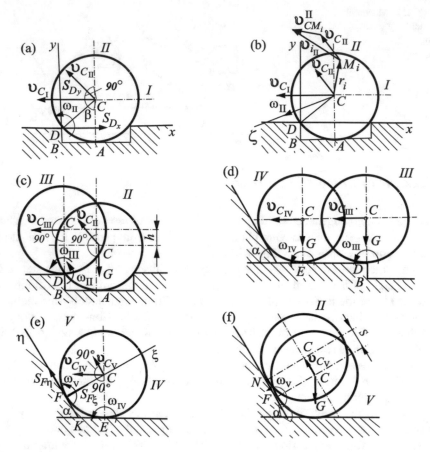

Fig. 4.23 The analysis of the possible positions of the cylinder: (**a**) the positions of the start (I) and end (II) of the stroke; (**b**) kinematics of lost velocities; (**c**) lifting the cylinder onto the BD step; (**d**) rolling the cylinder in section DE from position III to position IV; (**e**) the coincidence of the positions of the beginning (IV) and end (V) of the impact; (**f**) rolling the cylinder on an inclined plane from position (II) to position (V)

Kinetic moment of the cylinder relative to the axis D at the beginning of the impact

$$L_{ID} = mv_{C_1}(r - h),$$

where $v_{C_1} = v$ is the speed of the center of gravity of the cylinder at the beginning of the impact, equal to the speed of the platform before the sudden stop.

Kinetic moment of the cylinder relative to the D axis at the end of the impact

$$L_{IID} = L_D \omega_{II}$$

where J_D is the moment of inertia of the cylinder relative to the axis D at the end of the impact and ω_{II} the angular velocity of the cylinder at the end of the impact

$$L_{IID} = (J_C + mr^2)\omega_{II} = \left(\frac{mr^2}{2} + mr^2\right)\omega_{II} = \frac{3}{2}mr^2\omega_{II},$$

or, taking into account the formula (4.46):

$$\omega_{II} = \frac{2v}{3r^2}(r - h). \tag{4.47}$$

Let us check the expression (4.47) by Carnot's theorem:

$$T_I - T_{II} = T^*,$$

where T_I is kinetic energy of a system of material points before the impact and T_{II} is the same, but at the end of the strike, T^* is kinetic energy corresponding to the lost velocity of the points system.

By definition, we have

$$T_I = \frac{1}{2}mv^2; \quad T_{II} = \frac{1}{2}(mv_C^2 + J_C\omega_{II}^2).$$

Now we derive the formula for the kinetic energy corresponding to the lost velocities Δv_i of points M_i of a solid body in plane motion:

$$\Delta v_i = v_{iI} - v_{iII},$$

where

$$v_{iI} = v_{CI} + v_{CM_i}^I, \quad v_{iII} = v_{CII} + v_{CM_i}^{II};$$

moreover, the rotational velocities around the C axis have the expressions

$$v_{CM_i}^I = \omega_I \times r_i, \quad v_{CM_i}^{II} = \omega_{II} \times r_i.$$

The kinetic energy of the body corresponding to the lost velocities of its points will be

$$T^* = \frac{1}{2}\sum m_i\left[(v_{CI} - v_{CII}) + (v_{CM_i}^I - v_{CM_i}^{II})\right]^2,$$

i.e.,

$$T^* = \frac{1}{2}\left[\sum m_i(v_{CI} - v_{CII})^2 + 2\sum m_i(v_{CI} - v_{CII})(v_{CM_i}^I - v_{CM_i}^{II}) + \sum m_i(v_{CM_i}^I - v_{CM_i}^{II})^2\right].$$

$$\tag{4.48}$$

By simple transformations, the formula (4.48) can be reduced to the following general expression of the kinetic energy of the lost velocities of points of a solid body in plane motion:

$$T^* = \frac{1}{2}m\left[(v_{CI_x} - v_{CII_x})^2 + (v_{CI_y} - v_{CII_y})^2\right] +$$
$$+\frac{1}{2}J_C(\omega_I - \omega_{II})^2. \tag{4.49}$$

For the cylinder in the problem under consideration (Fig. 4.23b)

$$T^* = \frac{1}{2}m\left[(v_{CI} - v_{CII}\cos\beta)^2 + v_{CII}^2\sin\beta\right] + \frac{1}{2}J_C\omega_{II}^2. \tag{4.50}$$

Bearing in mind that

$$v_{CI} = v; \ v_{CII} = \omega r; \ J_C = mr^2/2; \ T^* = T_I - T_{II}; \ \cos = (r - h)/r,$$

we get

$$\omega_{II} = 2(r - h)v/(3r^2). \tag{4.51}$$

Let us now form an equation expressing the theorem on the change in the kinetic energy of the system when the cylinder moves from position II to position III (Fig. 4.23c):

$$T_{III} - T_{II} = \sum A_i^e,$$

where $\sum A_i^e$—is the sum of the work of external forces, equal to $-Gh$ in our case. We have

$$\frac{1}{2}J_D\omega_{III}^2 - \frac{1}{2}J_D\omega_{II}^2 = -Gh.$$

From here

$$\omega_{III} = \sqrt{\omega_{II}^2 - 4gh/(3r^2)}.$$

Considering the movement of the cylinder from position III to IV ($\sum A_i^e = 0$), (Fig. 4.23d), so

$$v_{CIV} = v_{CIII} \ \text{and} \ \omega_{IV} = \omega_{III}. \tag{4.52}$$

We now apply the theorem on the change in the kinetic moment of the system during the transition of the cylinder from the position IV to V. For the axis of moments, we take the horizontal axis that coincides with the generatrix F of the cylinder. In this case, the positions of IV and V corresponding to the beginning and end of the stroke coincide (Fig. 4.23e):

$$L_{VF} - L_{IVF} = \sum M_F(S_i^e).$$

But $\sum M_F(S_i^e) = 0$, since the shock pulse S_F applied to the cylinder intersects the F axis. Therefore,

$$L_{VF} = L_{IVF}.$$

When calculating the kinetic moments of the cylinder L_{IVF} and L_{VF}, we use the theorem on the kinetic moment of the system in the general case of its motion.

Kinetic moment relative to the F axis at the beginning of the impact is

$$L_{IVF} = m v_{CIV} r \cos \alpha + J_C \omega_{IV},$$

where $v_{CIV} = \omega_{IV} \cdot EC$, since the instantaneous center of velocities at the point E is

$$L_{IVF} = m \omega_{IV} r^2 \cos \alpha + (mr^2/2)\omega_{IV} = m \omega_{IV} r^2 (\cos +1/2).$$

Kinetic moment relative to the F axis at the end of the impact is

$$L_{VF} = m v_{CV} r + J_C \omega_V,$$

where $v_{CV} = \omega_v cf$, since the instantaneous velocity center is at F:

$$L_{VE} = m v_V r^2 + (mr^2/2)\omega_V = (3/2)mr^2 \omega_V.$$

But $L_{VF} = L_{IVF}$, so

$$\omega_V = \omega_{IV}(2 \cos \alpha + 1)/3. \tag{4.53}$$

Check the expression (4.53) by Carnot's theorem:

$$T_{IV} - T_V = T^*.$$

Kinetic energy of the cylinder at the beginning of impact on an inclined plane is

$$T_{IV} = \frac{1}{2}(m v_{CIV}^2 + J_C \omega_{IV}^2).$$

Kinetic energy of the cylinder at the end of impact on an inclined plane is

$$T_V = \frac{1}{2}(mv_{CV}^2 + J_C\omega_V^2).$$

The kinetic energy corresponding to the lost velocities of the points of the cylinder is found by the formula (4.49), Fig. 4.23e:

$$T^* = \frac{1}{2}[(v_{CIV} - c_{CV}\cos\alpha)^2 + v_{CV}^2\sin^2\alpha] +$$
$$+ \frac{1}{2}J_C(\omega_{IV} - \omega_V)^2. \tag{4.54}$$

Substituting the written values into the equality (4.7) results in the equality (4.53) after easy transformations.

We make an equation expressing the theorem on the change in the kinetic energy of the system and corresponding to the rolling of the cylinder along an inclined plane from position V to position VI at a distance of s (Fig. 4.23f):

$$T_{VI} - T_V = \sum A_i^e.$$

Since $T_{VI} = 0$, then

$$\frac{mv_C V^2}{2} + \frac{J_C\omega_V^2}{2} = Gs\sin;$$

from where we get

$$\omega_V = \frac{2}{r}\sqrt{\frac{gs\sin\alpha}{3}}. \tag{4.55}$$

From the formulas (4.53), (4.52) and (4.55), find

$$\omega_{II} = \frac{2}{r}\sqrt{\frac{3gs\sin\alpha}{(2\cos\alpha + 1)^2} + \frac{gh}{3}}.$$

The speed of the platform is determined by using the formula (19.1):

$$v = \frac{3r}{r - h}\sqrt{\frac{3gs\sin\alpha}{(2\cos\alpha + 1)^2} + \frac{gh}{3}},$$

i.e.,

$$v = 3.68 \text{ m/s.}$$

Find the shock pulse transmitted to the cylinder from the side of the step. To this end, we compose equations that express the equations of the theorem on the change in the amount of motion of the system on impact in projections on the x and y axes (Fig. 4.23a):

$$m v_{CIIx} + m v_{CIy} = \sum S_{ix}^e; \quad m v_{CIIy} + m v_{CIy} = \sum S_{iy}^e,$$

or

$$= m v_{CII} \cos \beta - (-m v_{CI}) = S_{Dx}; \quad m v_{CII} \sin \beta = S_{Dy};$$

where from

$$S_{Dx} = m v_{CI} - m v_{CII} \cos \beta = m v - m \omega_{II} r (r - h)/r = m v \left[1 - \frac{2}{3} \frac{(r - h)^2}{r^2} \right],$$

i.e., $S_{Dx} = 1055 \, \text{N} \cdot \text{s}$.

$$S_{Dy} = m v_{CII} \sin \beta = m \omega_{II} r \sqrt{1 - \cos^2 \beta} = 589 \, \text{N} \cdot \text{s}.$$

The shock pulse perceived by the cylinder from the side of the step will be

$$S_D = \sqrt{S_{Dx}^2 + S_{Dy}^2} = 1208 \, \text{N} \cdot \text{s}.$$

We find the shock pulse perceived by the cylinder from the side of the inclined plane, for which we compose equations expressing the theorem on the change in the amount of motion of the system during impact in projections on the axis ξ and η (Fig. 4.23e):

$$m v_{CV\xi} - m v_{CIV\xi} = \sum S_{i\xi}^e; \quad m v_{CV\eta} - m v_{CIV\eta} = \sum S_{i\eta}^e,$$

or

$$m v_{CIV} \sin \alpha = S_{F\xi}; \quad m v_{CV} - m v_{CIV} \cos \alpha = S_{F\eta},$$

where from

$$S_{F\xi} = m v_{CIV} \sin \alpha = m \omega_{IV} r \sin \alpha = \frac{3 m \omega_V r \sin \alpha}{2 \cos \alpha + 1},$$

$$S_{F\eta} = m v_{CV} - m v_{CIV} \cos \alpha = m \omega_V r - m \omega_{IV} r \cos \alpha = m \omega_V r - \frac{3 m \omega_V r \cos \alpha}{2 \cos \alpha + 1},$$

from where, using (4.53), we find

$$S_{F\xi} = 691 \, \text{N} \cdot \text{s}; \quad S_{F\eta} = 133 \, \text{N} \cdot \text{s}; \quad S_F = \sqrt{S_{F\xi}^2 + S_{F\eta}^2} = 704 \, \text{N} \cdot \text{s}.$$

References

P. Hagedorn, *Technische Mechanik: Dynamik* (Verlag Harri Deutsch, Frankfurt, 2008)

J. Krodkiewski, *Dynamics of Mechanical Systems* (The University of Melbourne Press, Melbourne, 2006)

V. Molotnikov, *Mekhanika konstruktsiiyu Teoreticheskaya mekhanika. Soprotivlenie materialov [Construction Mechanics and Theoretical Mechanics. Mechanics of Materials]*, (Lan' Publications, Sankt-Peterburg, 2012)

A. Yablonskii, S. Noreiko, S. Vol'fson, dr., *Sbornik zadanii dlya kursovykh rabot po teoreticheskoi mekhanike: Uchebnoe posobie dlya tekhnicheskikh vuzov, 5-e izd. [Collection of Tasks for Term Papers on Theoretical Mechanics: A Textbook for Technical Universities. – 5th edn.]* (Integral-Press Publication, Moscow, 2000)

Y. Zel'dovich, Y. Raizer, *Fizika udarnykh voln i vysokotemperaturnykh gidrodinamicheskikh yavlenii [Physics of shock waves and high-temperature hydrodynamic phenomena]* (Fizmatgiz Publication, Moscow, 1963)

Chapter 5
Elements of Analytic Mechanics

Abstract As you may have noticed, the previous chapters used a vector approach to mechanics based on Newton's laws, which describe motion using vector quantities such as force, speed, and acceleration. These quantities characterize the motion of a body, which is idealized as a "material point" or "particle," understood as the only point to which mass is attached. In contrast, analytical mechanics uses scalar properties of motion that represent the system as a whole—usually its total kinetic energy and potential energy—rather than the Newton vector forces of individual particles. A scalar is a quantity, and a vector is a quantity and direction. These equations of motion are derived from a scalar quantity. At the same time, such artificial techniques as the principle of possible displacements, generalized coordinates, and generalized forces are used. In various fields of modern technology, complex problems arise, for the solution of which it is desirable to have a universal analytical apparatus based on the general principles of mechanics. The development of such an apparatus, the presentation of the general principles of mechanics, the derivation of the differential equations of motion from them, and the study of the equations themselves and methods of their integration constitute the main content of analytical mechanics.

Keywords Possible displacements · The principle of possible displacements · Ideal connections · Nonideal connections · Generalized coordinates · Generalized forces · Conservative forces · Nonconservative forces · Equilibrium in generalized coordinates · General equation of dynamics · Generalized equations of motion

5.1 The Principle of Possible Movements

Possible (or virtual) displacement of a non-free mechanical system is any imaginary infinitesimal displacement allowed at a given time by the connections imposed on the system. It follows from this definition that the possible displacements of points in a mechanical system are considered to be infinitesimal of the first order of smallness (Hand and Finch 1933).

Fig. 5.1 Toward the concept
of possible displacement

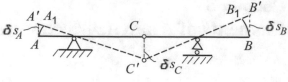

Fig. 5.2 Possible movements
of the crank-slide mechanism

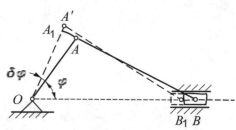

For example, a possible movement of the hinge arm ABC (Fig. 5.1) is the rotation of the links AC and CB as a result of an infinitesimal displacement δS_c of the hinge C in the vertical direction. This offset causes the points A and B to move along the arcs AA_1 and BB_1. However, up to the first order of smallness, these displacements are replaced by possible displacements $\delta S_A = \overline{AA'}$ and $\delta S_B = \overline{BB'}$ perpendicular to AB.

The possible movement of the crank-slider mechanism shown in Fig. 5.2 is the rotation of the crank OA from a given position, determined by the angle φ, to an infinitely small angle $\delta\varphi$ around the axis O. The possible movement δs_A of point A is a segment of the tangent AA' to the arc AA_1, equal in magnitude $\delta s_A = OA \cdot \delta\varphi$. A possible displacement δs_B of point B of the slider is an infinitesimal segment BB_1 of the straight line trajectory of point B.

Actual displacements of points of a non-free mechanical system, performed by the action of applied forces, are always among the possible displacements of points of the system only if the communications imposed on the system do not change over time (Molotnikov 2012). Such connections are called *stationary*.

We will consider a non-free mechanical system of n material points. Let us denote by F_i the resultant of the given active forces applied to the i-th point; by R_i, the resultant of the bond reaction forces at this point; and by δs_i, its possible displacement. *The connections for which the sum of the work of reactions on any possible displacement of the system is equal to zero are called ideal.* By definition, for systems with ideal connections ,

$$\sum R_i \delta s_i \cos(\boldsymbol{R}, \hat{\delta s_i}) = 0, \quad (i = 1, 2, \ldots, n). \tag{5.1}$$

If the system is in equilibrium, then for each of n points, $F_i + R = 0$, that is, $F_i = -R_i$. Since the forces F_i and R_i are equal in magnitude and opposite in direction, the $\cos(\boldsymbol{R}, \hat{\delta s_i})$ work of these forces on the possible displacement of the point δs_i is equal in magnitude, but opposite in sign. Therefore, the sum of the elementary work of all forces applied to the i-th point on any possible displacement is equal to zero:

$$F_i \delta s_i \cos(\boldsymbol{R}, \widehat{\delta s_i}) + R_i \delta s_i \cos(\boldsymbol{R}, \widehat{\delta s_i}) = 0, \ (i = 1, 2, \ldots, n).$$

Summing up such equalities over all points of the system, we get

$$\sum F_i \delta s_i \cos(\boldsymbol{R}, \widehat{\delta s_i}) + \sum R_i \delta s_i \cos(\boldsymbol{R}, \widehat{\delta s_i}) = 0, \ (i = 1, 2, \ldots, n).$$

Hence, in the case of a system with ideal constraints, taking into account conditions (5.35), we have

$$\sum F_i \delta s_i \cos(\boldsymbol{R}, \widehat{\delta s_i}) = 0, \quad (i = 1, 2, \ldots, n). \tag{5.2}$$

Equality (5.36) expresses the principle of possible displacements: *for equilibrium of a mechanical system with ideal stationary constraints, it is necessary and sufficient that the sum of elementary work of all active forces applied to it vanishes on any possible displacement of the system.*

The principle of possible displacements allows us to write the condition of equilibrium of the specified forces for non-free systems consisting of any number of bodies, and the equation of work expressing this principle does not contain reactive forces. Applying the static equations to such complex systems would require finding a large number of bond reactions. Although equality (5.36) does not include the forces of bond reactions, it can be used to determine unknown reactive forces. To do this, the bond whose reaction is to be determined is discarded, and its action is replaced by the reaction force, which is included in the number of active forces. At the same time, of course, the remaining connections must be perfect. Then writing the Eq. (5.36) for the resulting system, we arrive at an equation with respect to the unknown reaction force of the dropped bond.

The described method of determining reactions is also applicable to systems with nonideal connections. For example, if the bond is a rough surface, you can replace it with a perfectly smooth surface by adding a sliding friction force or a pair of forces that prevent rolling to the active forces.

Example 1 A composite beam AD consists of two beams AC and CD, pivotally connected to each other at point C. The end of beam D is embedded in the wall (Fig. 5.3a). Determine the moment of embedment M_D, if equal vertical forces $F_1 = F_2 = F_3 = F$ act on the beam, as well as the moment M of the pair of forces. The dimensions are shown in the figure. Neglect the gravity of the beams.

D e c i s i o n. The bond seal can be replaced with fixed flat hinge, adding the time of the incorporation of M_D (Fig. 5.3b). Neglecting friction at joints C, D and expansion roller B in communications system would be ideal.

We apply the principle of possible displacements to a composite beam with transformed connections. The only possible movement of the system allowed by superimposed connections in the transformed system is the vertical movement δs_C

Fig. 5.3 For example 1:
(a) the initial position of the
beam; (b) possible
displacement of point C

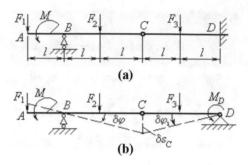

of the hinge C. In this case, the beam will take the position shown in Fig. 6.3b, dotted. The roller support B moves in the horizontal direction, but the applied forces on this movement do not perform work.

For the specified geometric dimensions, $\delta\varphi = \delta\varphi_1$. Denoting the vertical movements of the points of application of forces F_1, F_2, F_3, respectively, through $\delta s + 1$, δs_2, δs_3, according to the formula (5.36), we can write

$$- F_1 \delta s_1 - M \delta\varphi + F_2 \delta s_2 + F_3 \delta s_3 + M_D \delta\varphi_1 = 0. \tag{5.3}$$

Here, with negative signs, the work of forces is written for which the possible movements of the points of application are opposite to the direction of their action. It follows from the geometric features of the system that $\delta s_1 = \delta s_2 = l\delta\varphi$, $\delta s_3 = l\delta\varphi_1$. Substituting these expressions into equality (5.37) and taking into account that $F_1 = F_2 = F_3 = F$, we get

$$\delta\varphi \cdot (-Fl - M + Fl + Fl + M_D) = 0.$$

Since $\delta\varphi = 0$, then we equate to zero the amount in brackets: $Fl - M + M_D = 0$. From here we find

$$M_D = M - Fl.$$

Example 2 In the mechanism, the diagram of which is shown in Fig. 5.4, the crank OA can rotate about a horizontal axis passing through the point O. Along the rod OA, the slider B can move, pivotally connected to the rod BD, which slides in vertical guides. A pair of forces with a moment M is applied to the OA crank; $OD = l$.

Determine at equilibrium the vertical force F applied to the bar BD, depending on the angle φ. Neglect the forces of friction and the gravity of the links of the mechanism.

D e c i s i o n. Let's apply the principle of possible displacements to the mechanism. The active forces are the pair M and the force F. Let us give the system a possible displacement, mentally rotating the link OA at an infinitely small angle $\delta\varphi$

Fig. 5.4 For example 2

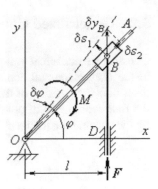

toward the increasing angle φ. The amount of virtual works will be

$$- M\delta\varphi + F\delta y_D = 0, \tag{5.4}$$

where δy_D is the possible displacement of the point of application of the force F in the direction of its action. The BD rod is absolutely solid; therefore, the displacements of points D and B are equal, i.e., $\delta y_d = \delta y_B$. Let's calculate these displacements. Let's decompose the displacement δy_B into components $\delta s_1 \perp OA$ and $\delta s_2 \parallel OA$, (Fig. 5.4). We

$$\delta s_1 = OB \cdot \delta\varphi;$$

but $OB = l/\cos\varphi$ and $\delta y_B = \delta s_1/\cos\varphi$.

From here we find

$$\delta y_B = \delta y_D = \frac{l}{\cos^2\varphi}\delta\varphi.$$

Substituting this expression into the Eq. (5.38), we get

$$\left(-M + \frac{lF}{\cos^2\varphi}\right)\delta\varphi = 0.$$

Thus, we equate the expression in parentheses to zero:

$$-M + \frac{lF}{\cos^2\varphi} = 0,$$

i.e., $F = \frac{M}{l}\cos^2\varphi$.

5.2   Generalized Coordinates: Generalized Forces

Consider a mechanical system consisting of two material points M_1 and M_2 located in the Oxy plane and connected by an inextensible rod of length l (Fig. 5.5). To determine the position of the system, we can arbitrarily set the values only three of the four coordinates of point M_1 and M_2, since the fourth coordinate is determined by the constraint equation

$$(x_1 - x_2)^2 + (y_1 - y_2)^2 - l^2 = 0. \tag{5.5}$$

Constraints of the type (5.39), which impose restrictions on the positions of the points of the system in space, but not on their velocity, are called *geometric* or *holonomic*.

The position of the system under study in the plane can be determined differently. Let's introduce the parameters $q_1 = x_1$, $q_2 = y_1$, $q_3 = \varphi$, (Fig. 5.5). Let us show that setting these parameters determines the position of both points. Indeed, q_1 and q_2 define the point M_1, and for the point M_2, we have

$$x_2 = x_1 + l \cos \varphi = q_1 + l \cos q_3; \;\; y_2 = y_1 + l \sin \varphi = q_2 + l \sin q_3,$$

and the constraint equation is automatically satisfied.

Comparing both ways of setting the position of the system, we come to the conclusion that the Cartesian coordinates of the points M_1 and M_2 are not independent—they are related by Eq. (5.39), and the parameters q_i, $(i = 1, 2, 3)$ are introduced taking into account the constraint equation and each of them can change independently of the others.

In the general case, for a mechanical system consisting of n material points, on which l geometric constraints are imposed, one can choose $s = (3n - l)$ independent parameters $q_1, q_2, \ldots, q_s$. These *independent parameters, which uniquely determine the position of the system in space, are called generalized coordinates of the system. The number (s) of generalized coordinates is called the number of degrees of freedom of the mechanical system.* So, for example, the crank-slider mechanism shown in Fig. 5.2 has one degree of freedom, and a double pendulum (Fig. 5.6), moving in a plane, has two degrees of freedom; angles φ and ψ can be taken as generalized coordinates q_1 and q_2.

Fig. 5.5 Holonomic links

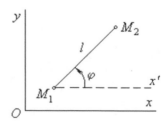

Fig. 5.6 Double pendulum

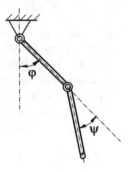

Suppose that the force F_i acts on the i-th point ($i = 1, 2, \ldots, n$) of the system. Let us inform the system of such a possible displacement at which the generalized coordinate q_1 gains an increment δq_1, and the other coordinates do not change. Let δr_{i1} denote the elementary increment of the radius vector of the i-th point caused by the change in the coordinate q_1. Since all other generalized coordinates do not change in this case, the value δr_{i1} is calculated as the partial differential of the radius vector r_i, i.e.,

$$\delta r_{i1} = \frac{\partial r_i}{\partial q_1} \delta q_1. \tag{5.6}$$

Let's calculate now the amount of elementary work of all acting forces on the considered possible displacement. For this purpose, we turn to formula (3.28), which we represent in the form

$$dA = F \cos \alpha \cdot ds = F dr.$$

We get

$$\delta A_1 = F_1 \cdot \delta r_{11} + F_2 \cdot \delta r_{21} + \ldots + F_n \cdot \delta r_{n1}$$

or, using equality (5.36),

$$\delta A_1 = F_1 \frac{\partial r_1}{\partial q_1} \delta q_1 + F_2 \frac{\partial r_2}{\partial q_1} \delta q_1 + \ldots + F_n \frac{\partial r_n}{\partial q_1} \delta q_1. \tag{5.7}$$

Let us introduce the notation

$$Q_1 = \sum F_i \frac{\partial r_i}{\partial q_1}, \tag{5.8}$$

where summation is performed over all points of the system ($i = 1, 2, \ldots, n$). In this case, equality (5.42) can be written in the form

$$\delta A_1 = \boldsymbol{Q}_1 \delta \boldsymbol{q}_1. \tag{5.9}$$

By analogy with formula (3.27), which determines the elementary work of the force $\boldsymbol{F}$, the quantity $\boldsymbol{Q}_1$ is called the *generalized force corresponding to the coordinate* q_1.

Telling the system another independent possible displacement δq_k ($k = 1, 2, \ldots, s$), at which only the generalized coordinate q_k changes, by analogy with equalities (5.42) and (5.43), we obtain

$$\delta A_k = Q_k \delta q_k, \quad (k = 1, 2, \ldots, s); \tag{5.10}$$

$$Q_k = \sum F_i \cdot \frac{\partial \boldsymbol{r}_i}{\partial q_k}, \quad (i = 1, 2, \ldots, n). \tag{5.11}$$

The quantity Q_k is the generalized force corresponding to the q_k coordinate.

If the system is informed of such a possible displacement at which all generalized coordinates change simultaneously, then on the basis of the principle of independence of the action of forces, the elementary work of the system of forces $(\boldsymbol{F}_1, \boldsymbol{F}_2, \ldots, \boldsymbol{F}_n)$ will be defined as the sum of elementary works (5.44), i.e.,

$$\sum \delta A_k = Q_1 \delta q_1 + Q_2 \delta q_2 + \cdots + Q_s \delta q_s. \tag{5.12}$$

Formula (5.46) gives an expression for the complete elementary work of all forces acting on the system in generalized coordinates. Note that the dimension of the generalized force depends on the dimension of the corresponding generalized coordinate and, based on formula (5.44), is determined from the expression

$$[Q] = \frac{[A]}{[q]}, \tag{5.13}$$

where the dimension of the corresponding quantity is symbolically indicated in square brackets. From formula (5.47), it can be seen that if q is a linear quantity, then Q has the dimension of force; if q is the angle (in radians), then the generalized force has the dimension of the moment, and so on.

Example 1 A centrifugal regulator rotates around a vertical axis (Fig. 5.7). The weight of each ball of the governor is G, and the weight of the other parts is not included. The lengths of the rods are equal to l. Taking as the generalized coordinates the angle α formed by the control rods with the vertical, and the angle of rotation of the controller φ around the vertical axis, find the generalized forces corresponding to these generalized coordinates.

D e c i s i o n. Let us inform the generalized coordinate α the increment $\delta \alpha$, leaving the angle φ unchanged. In this case, the points of application of the gravity forces of the balls C_1 and C_2 will receive displacements δs_1 and δs_2, directed

Fig. 5.7 For example 1

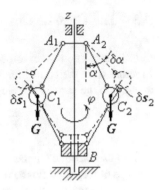

perpendicular to the rods A_1C_1 and A_2C_2. We have

$$\delta s_1 = \delta s_2 = l \cdot \delta\alpha.$$

Let us calculate the sum of the work of the given forces of gravity of the balls on the displacements δs_1 and δs_2 caused by the increment of the generalized coordinate α:

$$\delta A_\alpha = -G\delta s_1 \sin\alpha - G\delta s_2 \sin = -2Gl \cdot \delta\alpha \sin.$$

Using formula (5.46), we determine the generalized force Q_α corresponding to α:

$$Q_\alpha = \frac{\delta A}{\delta\alpha} = -2Gl \sin\alpha.$$

To calculate the generalized force Q_φ corresponding to the generalized coordinate φ, let the angle углу φ increment $\delta\varphi$, leaving the angle α unchanged. In this case, obviously, the points of application of the gravity forces of the balls will move in the plane perpendicular to the z axis of the controller. The work of the given forces G on such a displacement is zero:

$$\delta A_\varphi = 0.$$

Then, by formula (5.10), the generalized force Q_φ corresponding to the coordinate φ is also is equal to zero:

$$Q_\varphi = \frac{\delta A_\varphi}{\delta\varphi} = 0.$$

5.3   The Case of Conservative Forces

Let all forces acting on a mechanical system be conservative. Then there is a force function U of the coordinates of points of the system, the total differential of which is equal to the sum of the elementary work of all forces:

$$\sum \delta A_i = \delta U, \ (i = 1, 2, \ldots, n).$$

Let us introduce generalized coordinates $q_1, q_2, \ldots, q_s$ for a system with s degrees of freedom. Based on formulas (3.89), the generalized forces corresponding to these coordinates are related to the force $Q_1, Q_2, \ldots, Q_s$ function U and potential Π by the dependences

$$Q_1 = \frac{\partial U}{\partial q_1}; \ Q_2 = \frac{\partial U}{\partial q_2}; \ \ldots; \ Q_s = \frac{\partial U}{\partial q_s}; \qquad (5.14)$$

$$Q_1 = -\frac{\partial \Pi}{\partial q_1}; \ Q_2 = -\frac{\partial \Pi}{\partial q_2}; \ \ldots; \ Q_s = -\frac{\partial \Pi}{\partial q_s}. \qquad (5.15)$$

Thus, *if all forces acting on the system are conservative, then the generalized forces are equal to the partial derivatives of the force function (or partial derivatives of the potential energy taken with negative signs) with respect to the corresponding generalized coordinates.*

Example 2 A homogeneous rod AB, having a length l and a weight P (Fig. 5.8), can rotate around the A axis in a vertical plane. A ball M weighing p is threaded onto the rod and connected to a spring. The length of the spring in the unstressed state is a, and the stiffness is c. Disregarding friction, find the generalized forces of the system.

D e c i s i o n. It is obvious that the independent movements of the system are the rotation of the rod around the A axis and the movement of the ball along the rod. Therefore, the system has two degrees of freedom. Let us choose the angle φ

Fig. 5.8 For example 2: generalized coordinates of the system

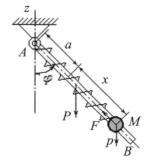

Fig. 5.9 Graphical representation of elastic force work

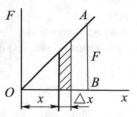

(Fig. 5.8) and the displacement x of the ball from the end of the spring, released from the ball, as the generalized coordinates.

Three active forces are applied to the system: gravity forces P and p and elastic force F. All these forces are conservative. Let's calculate the potential energy of the system in an arbitrary position determined by the generalized coordinates φ and x. We have

$$\Pi == P \cdot \frac{l}{2} \cos - p(a + x) \cos + \frac{1}{2} cx^2.$$

Here, the first term represents the work of gravity P of the rod AB as a result of its deviation from the vertical axis $\grave{A}z$ by an angle φ with the constant x coordinate. The «minus» sign takes into account that the point of application of the force P with the specified deviation moves opposite to the direction of the force. Literally, the same is the second term, which is equal to the work of the force p (the weight of the ball) on the movement $-(a + x) \cos \varphi$ in the direction of this force.

The last term in the expression for the function Π takes into account the work of the spring force $F = cx$ on displacement x. Let us explain the origin of the coefficient $1/2$. When the spring is deformed, the force F does not remain constant, but changes according to the law $F = cx$ (Fig. 5.9). If in position x there is an infinitely small increment of displacement by the value Δx, then the magnitude of the force F with such a displacement can be considered unchanged, and the work of the force on the path Δx will be equal to the product $F \cdot \Delta x$. Graphically, this work is depicted by the area of strip shaded in Fig. 5.9. Therefore, the work of the spring force when it changes from zero to F will be equal to the area of triangle OAB. Using formulas (5.48), we find the generalized forces:

$$Q_1 = -\frac{\partial \Pi}{\partial \varphi} == -\left[\frac{Pl}{2} + p(a + x)\right] \sin \varphi; \quad Q_2 = -\frac{\partial \Pi}{\partial x} = p \cos \varphi - cx.$$

5.4 Equilibrium Conditions of the System in Generalized Coordinates

Consider a mechanical system with ideal stationary constraints. We already know (Sect. 5.2) that the principle of possible displacements is applicable to such systems.

According to this principle, a necessary and sufficient condition for equilibrium is the vanishing of the sum of elementary work of all active forces applied to the system on any possible displacement. We write this condition in the following form:

$$\sum \delta A_i = \sum (F_i \cdot \delta (r_i = 0)),$$

where summation is over all forces. It is easy to verify that such a notation is equivalent to formula (5.36). Taking into account equality (5.46), the written formula can also be represented in the form

$$\sum F_i \cdot \delta (r_i = Q_1 \delta q_1 + Q_2 \delta q_2 + \cdots + Q_s \delta q_s = 0,$$

where s is the number of degrees of freedom of the system.

Therefore, a necessary and sufficient condition for the equilibrium of the considered mechanical system is the equality:

$$Q_1 \delta q_1 + Q_2 \delta q_2 + \cdots + Q_s \delta q_s = 0. \tag{5.16}$$

But the generalized coordinates are independent; therefore, the possible displacements $\delta q_1, \delta q_2, \ldots, \delta qs$ are also independent, moreover arbitrary and infinitesimal values. Assuming $\delta q_1 = 0$, and all the others,$\delta q_i \neq 0$, $(i = 2, 3, \ldots, s)$, from the formula (5.50), we obtain $Q_1 = 0$. Quite similarly, assuming $\delta q_2 \neq 0$ and $\delta q_1 = \delta q_3 = \ldots = \delta q_s = 0$, we will have $Q_2 = 0$, etc. As a result, we come *to the conclusion that for the equilibrium of a mechanical system with ideal, stationary geometric constraints, it is necessary and sufficient that all generalized forces be equal to zero*, i.e.,

$$Q_1 = 0; \; Q_2 = 0; \; \ldots; \; Q_s = 0. \tag{5.17}$$

If the system is under the action of conservative forces, then the equilibrium condition (5.17) together with formulas (5.50) gives the following dependences for the force function U:

$$Q_1 = \frac{\partial U}{\partial q_1} = 0; \; Q_2 = \frac{\partial U}{\partial q_2} = 0; \; \ldots; \; Q_s = \frac{\partial U}{\partial q_s} = 0. \tag{5.18}$$

As is known, the vanishing of all partial derivatives of the strength function with respect to generalized coordinates expresses the necessary condition for the existence of an extremum. Remembering also that the strength function is equal to the potential energy with a negative sign ($U = -\Pi$), we can conclude that *in equilibrium of a mechanical system under the influence of conservative forces, potential energy (and the strength function) can reach an extremum.*

Example Spring AB holds a homogeneous rod OB of length l and weight P at an angle φ to the horizon. The end of the spring A is attached to the horizontal plane

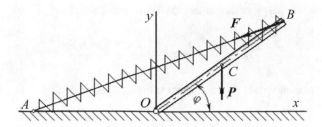

Fig. 5.10 For example 1: (**a**) the initial position of the beam; (**b**) possible displacement of point C

at a distance $OA = l$, (Fig. 5.10). Determine the hardness c spring if the free length of the spring is known to be L.

D e c i s i o n. The system is subject to the action of two active forces: the force of gravity P and the elastic force of the spring F. Both forces are conservative; therefore, to solve the problem, we will use the condition of equilibrium of conservative forces (5.52).

We will take the angle φ of inclination of the rod to the horizon as the generalized coordinate. Let us draw the coordinate axes Ox and Oy through the center of the hinge O and compose an expression for the function Π—the potential energy of the system.

Let us take the horizontal plane AO as a surface with zero gravity potential. Then the potential energy of the force P will be

$$\Pi_P = P y_C = P \frac{l}{2} \sin \varphi.$$

To calculate the potential energy Π_e of elastic deformation of the spring, we calculate its elongation ΔL in the position shown in the figure:

$$\Delta L = AB - L.$$

From an isosceles triangle OAB by the sine theorem, we have

$$\frac{AB}{\sin(\pi - \varphi)} = \frac{l}{\sin \frac{[\pi - (\pi - \varphi)]}{2}}.$$

From here we find

$$AB = 2l \cos \frac{\varphi}{2}.$$

Then

$$\Delta L = AB - L = 2l \cos \frac{\varphi}{2} - L,$$

and the potential energy of deformation of the spring will be

$$\Pi_e = \frac{c(\Delta L)^2}{2} = \frac{c\left(2l\cos\frac{\varphi}{2} - L\right)^2}{2}.$$

The potential energy Π of the system is equal to the sum $\Pi_P + \Pi_e$, i.e.,

$$\Pi = \frac{Pl}{2}\sin\varphi + \frac{c\left(2l\cos\frac{\varphi}{2} - L\right)^2}{2}.$$

Let us find the first derivative of the potential energy of the system with respect to the generalized coordinate φ:

$$\frac{\partial\Pi}{\partial\varphi} = \frac{Pl}{2}\cos\varphi - cl^2\sin + cLl\sin\frac{\varphi}{2}.$$

In equilibrium, this derivative should vanish. From here we find

$$c = \frac{P\cos\varphi}{2\left(l\sin - L\sin\frac{\varphi}{2}\right)}.$$

5.5   General Equation of System Dynamics

In accordance with the d'Alembert principle (Sect. 3.5), for each of the N points of the mechanical system, the set of active forces, forces of reactions of bonds, and forces of inertia satisfies the equations of equilibrium:

$$\boldsymbol{F}_i + \boldsymbol{R}_i + \boldsymbol{\Phi}_i = 0, \quad (i = 1, 2, \ldots, N), \tag{5.19}$$

where $\boldsymbol{F}_i$ is the main vector of active forces, $\boldsymbol{R}_i$ is the main vector of bond reaction forces at a point, and $\boldsymbol{\Phi}_i$ is the point's inertial force. We scalarly multiply each of the equations (5.53) by the possible displacement of a point δr_i and sum over all points of the system. We get

$$\sum \boldsymbol{F}_i\delta r_i + \sum \boldsymbol{R}_i\delta r_i + \sum \boldsymbol{\Phi}_i\delta r_i = 0. \tag{5.20}$$

Formula (5.54) expresses the general equation of the dynamics of a system with any constraints. In the case of a system with ideal constraints, as we know (Sec. 3.5), the second term in (5.54l) vanishes. Remembering also that the force of inertia is expressed through the acceleration $\boldsymbol{a}_i$ of the point relative to the inertial frame of reference by the formula

$$\Phi_i = m_i a_i = -m_i \frac{d^2 r_i}{dt^2},$$

we come to the conclusion that *for systems with ideal constraints, the general equation of dynamics* takes one of the forms

$$\sum (F_i + \Phi_i)\delta r_i = 0; \quad \sum (F_i - m a_i)\delta r_i = 0$$

or

$$\sum \left(F_i - m_i \frac{d^2 r_i}{dt^2} \right) \delta r_i = 0, \tag{5.21}$$

where r_i is the radius vector of the i-th point.

Thus, *at any moment of motion of a mechanical system with ideal connections, the sum of elementary work of all active forces and forces of inertia of points of the system at any possible displacement is equal to zero.*

Let a system with ideal constraints have s degrees of freedom and its position in space is determined by generalized coordinates q_1, q_2, ..., q_s. In the general case of non-stationary connections, the radius vector of each point of the system depends on the generalized coordinates and time, i.e., $r_k = r_k(q_1, q_2, \ldots, q_s; t)$. For a possible movement δr_k, we have

$$\delta r_k = \sum_{k=1}^{s} \frac{\partial r_k}{\partial q_i} \delta q_i, \tag{5.22}$$

since the time is considered unchanged. Substitution of formula (5.22) into the general equation of dynamics (5.21) gives

$$\sum_{i=1}^{s} \left[\sum_{k=1}^{N} F_k \frac{\partial r_k}{\partial q_i} + \sum_{k=1}^{N} \Phi_k \frac{\partial r_k}{\partial q_i} \right] \delta q_i = 0. \tag{5.23}$$

We denote

$$Q_i = \sum_{k=1}^{N} F_k \cdot \frac{\partial r_k}{\partial q_i};$$
$$Q_i^{(\Phi)} = \sum_{k=1}^{N} \Phi_k \cdot \frac{\partial r_k}{\partial q_i}. \tag{5.24}$$

Let's call these values the generalized forces of active forces (Q_i) and forces of inertia ($Q_i^{(\Phi)}$) . Taking into account the notation (5.24), Eq. (5.23) is reduced to the form

$$\sum_{i=1}^{s}(q_i + Q_i^{(\Phi)})\delta q_i = 0. \tag{5.25}$$

Hence, due to the independence of the generalized coordinates and the arbitrariness of the corresponding possible displacements, we obtain

$$Q_i + Q_i^{(\Phi)} = 0, \ (i = 2 = 1, 2, \ldots, s). \tag{5.26}$$

Conditions (5.26) express *the d'Alembert principle in generalized forces*. If the forces of inertia of the points of the system are equal to zero, then the equilibrium conditions of the system (5.54) are obtained from formula (5.26).

5.6 Differential Equations of Motion of a Mechanical System in Generalized Coordinates

Equations (5.26) obtained in the previous section can be directly used to solve many problems of system dynamics. However, the process of drawing up these equations can be greatly simplified if all the generalized inertial forces included in them are expressed in terms of the kinetic energy of the system. For this purpose, we transform the right-hand side of the second of formulas (5.24) to the form

$$Q_i^{(\Phi)} = -\sum_{k=1}^{N} m_k \frac{dv_k}{dt} \cdot \frac{\partial r_k}{\partial q_i} = -\sum m_k \left[\frac{d}{dt}\left(v_k \cdot \frac{\partial r_k}{\partial q_i}\right) - v_k \frac{d}{dt}\left(\frac{\partial r_k}{\partial q_i}\right)\right], \tag{5.27}$$

where v_k is the speed of the k-th point determined by the radius vector r_k. The validity of equality (5.27) can be verified by differentiating the product in the first parenthesis.

Considering further that

$$\frac{dr_k}{dt} = \dot{r}_k = v_k \text{ and } \frac{dq_i}{dt} = \dot{q}_i,$$

by direct verification, you can verify the following:

$$\frac{d}{dt}\left(\frac{\partial r_k}{\partial q_i}\right) = \frac{\partial}{\partial q_i}\left(\frac{dr_k}{dt}\right) = \frac{\partial v_k}{\partial q_i}; \tag{5.28}$$

$$\frac{\partial r_k}{\partial q_i} = \frac{\partial \dot{r}_k}{\partial \dot{q}_i} = \frac{\partial v_k}{\partial \dot{q}_i}. \tag{5.29}$$

Using properties (5.28) and (5.29), we represent equality (5.27) in the form

$$Q_i^{(\Phi)} = -\sum_{k=1}^{N} m_k \left[\frac{d}{dt} \left(\frac{1}{2} \cdot \frac{\partial v_k^2}{\partial \dot{q}_i} \right) - \frac{1}{2} \cdot \frac{\partial v_k^2}{\partial q_i} \right]$$

or

$$-Q_i^{(\Phi)} = \frac{d}{dt} \left[\frac{\partial}{\partial \dot{q}_i} \left(\sum_{k=1}^{N} \frac{m_k v_k^2}{2} \right) \right] - \frac{\partial}{\partial q_i} \left(\sum_{k=1}^{N} \frac{m_k v_k^2}{2} \right).$$

In accordance with definition (3.84), the sums included in the last formula represent the kinetic energy (T) of a system composed of N material points. Therefore, we can write

$$\frac{d}{dt} \left(\frac{\partial T}{\partial \dot{q}_i} \right) - \frac{\partial T}{\partial q_i} = Q_i, \quad (1 = 1, 2, \ldots, s). \tag{5.30}$$

The system s of differential equations (5.30) is called *Lagrange equations of the second kind.* Integrating this system and determining the integration constants from the initial conditions, we obtain s equations of motion of the mechanical system in generalized coordinates:

$$q_i = q_i(t), \quad (i = 1, 2, \ldots, s). \tag{5.31}$$

5.6.1 The Case of Conservative Forces

In this case, the potential energy Π of the system depends only on the generalized coordinates $q_1, q_2, \ldots, q_s$, but does not depend on the generalized velocities ($\partial \Pi / \partial \dot{q}_i = 0$). Therefore, the Lagrange equations of the second kind can be represented in the form

$$\frac{d}{dt} \left(\frac{\partial T}{\partial \dot{q}_i} \right) - \frac{\partial T}{\partial q_i} + \frac{\partial \Pi}{\partial q_i} = 0 \text{ or } \frac{d}{dt} \left(\frac{\partial (T - \Pi)}{\partial \dot{q}_i} \right) - \frac{\partial (T - \Pi)}{\partial q_i} = 0.$$

Function

$$L = T - \Pi \tag{5.32}$$

called *the kinetic potential* or *the Lagrange function of the mechanical system.* Taking into account definition (5.32), the Lagrange equations of the second kind for a conservative system take on a particularly elegant form

$$\frac{d}{dt} \left(\frac{\partial L}{\partial \dot{q}_i} \right) - \frac{\partial L}{\partial q_i} = 0, \quad (i = 1, 2, \ldots, s). \tag{5.33}$$

In conclusion, we make the following remarks:

1. The type and number of Lagrange equations of the second kind do not depend either on the number of bodies (or material points) that make up the system or on the nature of the motion of these bodies. The number of Eqs. (5.30) or (5.33) is determined only by the number of degrees of freedom of the system.
2. With ideal constraints, these equations make it possible to exclude in advance from consideration the reactions of constraints.
3. The state of a mechanical system, on which only conservative forces act, is determined by specifying only one Lagrange function. Knowing this function, one can compose the differential equations of motion of the system.

Example A solid cylinder of weight p_1 is wrapped with a thread thrown over block O and attached to a weight A with weight p_2 (Fig. 5.11). The load can slide along a horizontal plane with a coefficient of friction f. Find the acceleration of the load and the center C of the cylinder when the system moves. Neglect the block and thread masses.

D e c i s i o n. The system has two degrees of freedom. Let us take as the generalized coordinates the distance $q_1 = x$ of the center of mass of the load A from any fixed point D on the sliding plane, as well as the distance $q_2 = y$ from some knot E of the thread to the point of contact B of the thread and the cylinder.

Let us compose for a given system the Lagrange equations of the second kind. In accordance with formula (5.30), we have

$$\frac{d}{dt}\left(\frac{\partial T}{\partial \dot{x}}\right) - \frac{\partial T}{\partial x} = Q_1;$$
$$\frac{d}{dt}\left(\frac{\partial T}{\partial \dot{y}}\right) - \frac{\partial T}{\partial y} = Q_2.$$

Let us calculate the generalized forces Q_1 and Q_2. Let us first give the system a possible displacement δs_A, at which only the x coordinate changes: $\delta x > 0$, $\delta y = 0$. On this movement, the work is performed by the force of gravity p_1 and the force of friction $F_{fr} = f \cdot p_2$. The elementary work of these forces will be $\delta A_1 = (p_1 - f \cdot p_2)\delta x$. Hence, generalized strength $Q_1 = p_1 - fp_2$. Then let us inform

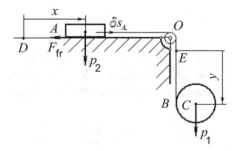

Fig. 5.11 For example, equations of motion in generalized coordinates

the system about the possible displacement $\delta y > 0$ at $x = const$. On this movement, only the cylinder gravity p_1 performs work $\delta A_2 = p_1 \cdot \delta y$; therefore, $Q_2 = p_1$.

In order to be able to use the written Lagrange equations, we will compose an expression for the kinetic energy (T) of the system. It is obvious that it is equal to the sum of the kinetic energies of the load A and the cylinder C: $T = T_A + T_C$. Body A performs translational motion and its kinetic energy is calculated by the formula

$$T_A = \frac{1}{2} \frac{p_2}{g} \dot{x}^2.$$

The cylinder makes a complex movement. The speed v_C of the cylinder axis is made up of the relative (with respect to the thread) speed, equal to $\dot{y}$, and the portable speed of the thread $\dot{x}$, and both components coincide in the direction (from top to bottom) for the selected reference directions. Therefore, $v_C = \dot{x} + \dot{y}$, and we calculate the kinetic energy of the cylinder using the formula (3.87)

$$T_C = \frac{1}{2} \cdot \frac{p_1}{g} v_C^2 + \frac{1}{2} J_C \omega^2.$$

The moment of inertia of the cylinder J_C about its C axis in accordance with the formula (3.63) will be

$$J_C = \frac{1}{2} \cdot \frac{p_1}{g \cdot r^2},$$

where r is the radius of the cylinder. The angular velocity of the cylinder ω can be expressed in terms of the relative velocity $\dot{y}$. Indeed, since the instantaneous center of velocities in the relative motion of the cylinder is at point B, then $\omega = \dot{y}/r$. Finally, we have the following expression for the kinetic energy:

$$T = T_A + T_C = \frac{p_2}{2g} \dot{x}^2 + \frac{p_1}{2g} \left[(\dot{x} + \dot{y})^2 + \frac{1}{2} \dot{y}^2 \right].$$

Further, by differentiating, we get

$$\frac{\partial T}{\partial x} = 0; \quad \frac{\partial T}{\partial \dot{x}} = \frac{p_2}{g} \dot{x} + \frac{p_1}{g} (\dot{x} + \dot{y});$$

$$\frac{\partial T}{\partial y} = 0; \quad \frac{\partial T}{\partial \dot{y}} = \frac{p_1}{g} \left(\dot{x} + \frac{3}{2} \dot{y} \right).$$

Substituting these expressions and the generalized forces Q_1, Q_2 into the Lagrange equations, we obtain the differential equations of motion of the system under study in the form

$$(p_1 + p_2)\ddot{x} + p_1 \ddot{y} = (p_1 - f p_2)g; \quad 2\ddot{x} + 3\ddot{y} - 2g.$$

Solving this system with respect to the desired accelerations $\ddot{x}$ and $\ddot{y}$, we find

$$\ddot{x} = \frac{p_1 - 3p_2 f}{p_1 + 3p_2} g; \quad \ddot{y} = \frac{2(1 + f)p_2}{p_1 + 3p_2} g.$$

The solution was found. In the absence of friction ($f = 0$), the accelerations will be

$$\ddot{x} = \frac{p_1}{p_1 + 3p_2} g; \quad \ddot{y} = \frac{2p_2}{p_1 + 3p_2} g. \tag{5.34}$$

Let's get these results using the Lagrange function. In the absence of friction, only conservative forces remain. If we assume that at $x = 0$, $y = 0$, the potential energy of the system is also zero ($\Pi = 0$), and then we can write

$$\Pi = -p_1(x + y).$$

In this case, according to definition (5.32), the kinetic potential has the form

$$L = T - \Pi = \frac{p_2}{2g}\dot{x}^2 + \frac{p_1}{2g}\left[(\dot{x} + \dot{y})^2 + \frac{1}{2}\dot{y}^2\right] + p_1(x + y).$$

Calculating the derivatives entering into Eqs. (5.33), for this system, we obtain

$$\frac{\partial L}{\partial x} = \frac{\partial L}{\partial y} = p_1; \quad \frac{\partial L}{\partial \dot{x}} = \frac{p_2}{g}\dot{x} + \frac{p_1}{g}(\dot{x} + \dot{y}); \quad \frac{\partial L}{\partial \dot{y}} = \frac{p_1}{g}(\dot{x} + \dot{y}) + \frac{p_1}{2g}\dot{y};$$

$$\frac{d}{dt}\left(\frac{\partial L}{\partial \dot{x}}\right) = \frac{p_2}{g}\ddot{x} + \frac{p_1}{g}(\ddot{x} + \ddot{y}); \quad \frac{d}{dt}\left(\frac{\partial L}{\partial \dot{y}}\right) = \frac{p_1}{g}(\ddot{x} + \ddot{y}) + \frac{p_1}{2g}\ddot{y}.$$

Substitution of these expressions into equations (53.7) gives

$$\frac{p_2}{g}\ddot{x} + \frac{p_1}{g}(\ddot{x} + \ddot{y}) - p_1 = 0; \quad \frac{p_1}{g}(\ddot{x} + \ddot{y}) + \frac{p_1}{2g}\ddot{y} - p_1 = 0.$$

Resolving this system with respect to accelerations $\ddot{x}$ and $\ddot{y}$, we arrive at the result (5.34).

Self-Test Questions

1. What is called a possible movement of a non-free mechanical system?
2. How are the possible and actual movements of the system interconnected?
3. What connections are called (a) stationary and (b) ideal?

4. Formulate the principle of possible displacements. Write down its formula expression.
5. Is it possible to apply the principle of virtual movement to systems with imperfect connections?
6. What connections are called holonomic?
7. What is called the number of degrees of freedom of a mechanical system?
8. What are the generalized coordinates of the system?
9. How many generalized coordinates does a non-free mechanical system have?
10. How many degrees of freedom does a steering wheel have?
11. What is called generalized power?
12. Write down the formula expressing the complete elementary work of all forces applied to the system in generalized coordinates.
13. How is the dimension of the generalized force determined?
14. How are generalized forces calculated in conservative systems?
15. What connections are called geometric?
16. Give a vector record of the principle of possible displacements.
17. Name the necessary and sufficient condition for the equilibrium of a mechanical system with ideal stationary geometric connections.
18. What property does the power function of a conservative system have in a state of equilibrium?
19. Write down one of the formulas expressing the general equation of the dynamics of a system with ideal constraints. What is the physical meaning of this equation?
20. What is called the generalized force of active forces applied to the system?
21. What is the generalized inertial force?
22. Formulate the d'Alembert principle in generalized forces.
23. Write down the system of Lagrange differential equations of the second kind.
24. How many Lagrange equations of the second kind can be made for a non-free mechanical system?
25. Does the number of Lagrange equations of the second kind for a mechanical system depend on the number of bodies included in the system?
26. What is called the kinetic potential of the system?
27. For which mechanical systems does the Lagrange function exist?

Control Tasks for the Section "Analytical Mechanics"

Task

Mechanical system of bodies (schemes of mechanisms for variants 1–10; see Fig. 5.12) moves under the influence of constant forces $\vec{P}$ and pairs of forces with moments M or only gravity (Yablonskii et al. 2000).

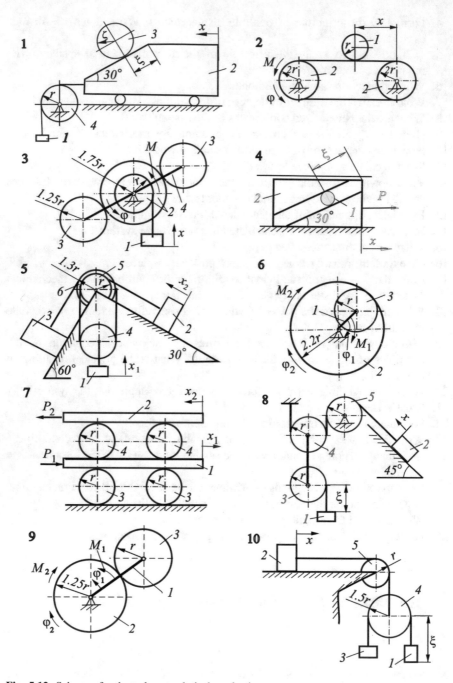

Fig. 5.12 Schemes for the task on analytical mechanics

Find the equations of motion of the system in the generalized coordinates q_1 and q_2 under given initial conditions. The necessary data are given in Table 5.1; there are also recommended generalized coordinates (x and φ are generalized coordinates for absolute motion and ξ for relative motion).

When solving the problem, the masses of threads should be ignored. Assume that the rolling of the wheels occurs without slipping. Rolling friction and resistance forces in bearings are not taken into account. Wheels for which the radii of inertia are not specified in the table are considered solid homogeneous disks. Drivers (cranks) are considered as thin homogeneous rods. Assume that in variants 6 and 9, the mechanism is located in a horizontal plane.

Additional data

1. Do not take into account the weight of the tape.
2. The moment M is applied to the driver.
3. Blocks 5 and 6 are mounted on a common axis freely, and their masses are the same.
4. The M_1 pair is attached to the driver.

Table 5.1 Initial data for options 1–10

Parameters	Variants									
	1	2	3	4	5	6	7	8	9	10
m_1	2m	m	m	m	m	m	3m	m	m	2m
m_2	6m	3m	3m	4m	2m	2m	3m	2m	2m	2m
m_3	m	–	2m	–	4m	3m	m	2m	3m	m
m_4	m	–	–	–	2m	–	m	2m	–	2m
m_5	–	–	–	–	2m	–	–	2m	–	m
i_{2_y}	–	–	$r\sqrt{2}$	–	–	$2r$	–	–	–	–
i_{3_y}	–	–	–	–	–	–	–	–	–	–
Forces P	–	–	–	–	–	–	$P_1; P_2$	–	–	–
Couples M	–	M	M	–	–	$M_1; M_2$	–	–	$M_1; M_2$	–
K_{fr}	–	–	–	0	f	–	–	f	–	f
Visc. res.K_{vr}	–	–	–	b	–	–	–	–	–	–
q_1	x	φ	φ	x	x_1	φ_1	x_1	x	φ_1	x
q_2	ξ	x	x	ξ	x_2	φ_2	x_2	ξ	φ_2	ξ
q_{10}	0	0	0	0	0	0	0	0	0	0
q_{20}	0	x_0	0	0	0	0	0	0	0	0
$\dot{q}_{10}$	0	0	0	$\dot{x}_0$	0	0	0	0	0	$\dot{x}_0$
$\dot{q}_{20}$	0	0	0	0	0	0	0	$\dot{\xi}_0$	0	0

Investigation of the Equations of Motion of a System with Two Degrees of Freedom by the Method of Lagrange Equations of the Second Kind (Example)

A mechanical system is given (Fig. 5.13). Masses of bodies: $m_1 = 3m$, $m_2 = 8m$, $m_3 = m_6 = m_6 = 2m$, $m_5 = 4m$, $m_7 = m$;[1] P is a constant force applied to the body 2; M is a constant moment applied to the wheel 6; b (Table 5.1) is the proportionality factor in the expression of the resistance body 5 $\vec{R} = -b\vec{v}_5$ $\vec{v}_5$—speed of the body 5), L—length of the filament 3; r is a radius of the wheels 4 and 6.

All wheels are considered solid homogeneous disks. The sliding friction of the body 2 is not taken into account.

Find the equations of motion of the system in the generalized coordinates $q_1 = x_1$, $q_2 = x_2$.

Initial conditions: $q_{10} = 0$ (the initial vertical distance from the lower end of thread 3 to its horizontal section is l_0), $q_{20} = 0$; $\dot{q}_{10} = 0$, $\dot{q}_{20} = 0$. In Fig. 5.13 the system is shown in the initial layout.

S o l u t i o n. To solve the problem, we apply the Lagrange equations of the second kind:

$$\frac{d}{dt}\frac{\partial T}{\partial \dot{x}_1} - \frac{\partial T}{\partial x_1} = -\frac{\partial \Pi}{\partial x_1} + Q_1; \qquad (5.35)$$

$$\frac{d}{dt}\frac{\partial T}{\partial \dot{x}_2} - \frac{\partial T}{\partial x_2} = -\frac{\partial \Pi}{\partial x_2} + Q_2; \qquad (5.36)$$

Here T is the kinetic energy of the system, Π is the potential energy, and Q_1, Q_2 are non-potential generalized forces.

For our example,

$$T = \sum_{i=1}^{7} T_i. \qquad (5.37)$$

Fig. 5.13 Mechanical system diagram

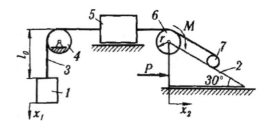

---

[1] Here, thread 3 is assumed to be weighty; this is a complication compared to the general condition of tasks 1–10. Thread slack is not taken into account.

We express the velocities of the centers of mass of the bodies of the system in terms of generalized velocities:

$$v_1 = v_5 = \dot{x}_1;$$
$$v_2 = v_6 = \dot{x}_2; \tag{5.38}$$
$$v_7^2 = (\dot{x}_1 + \dot{x}_2)^2/4 + \dot{x}_2^2 - (\dot{x}_1 + \dot{x}_2)\dot{x}_2 \cos\alpha,$$

or, keeping in mind that, $\alpha = 30°$

$$v_7^2 = 0.25[\dot{x}_1^2 + (5 - 2\sqrt{3})\dot{x}_2^2 - 2(\sqrt{3} - 1)\dot{x}_1\dot{x}_2].$$

Here it was taken into account that $(\dot{x}_1) + \dot{x}_2)/2$ is the velocity of the center of mass of body 7 relative to body 2, and $\dot{x}_2$ is its transport velocity (Fig. 5.14).

Angular velocities of bodies (see Figs. 5.12, 5.13, 5.14 and 5.15):

$$\omega_4 = \dot{x}_1/r;$$
$$\omega_6 = \omega_7 = (\dot{x}_1 + \dot{x}_2)/r. \tag{5.39}$$

Moments of inertia of the wheels relative to the central axles:

$$J_4 = J_6 = 2mr^2/2 = mr^2;$$
$$J_7 = (m/2)(r/2)^2 = mr^2/8.$$

Fig. 5.14 Body force system 7

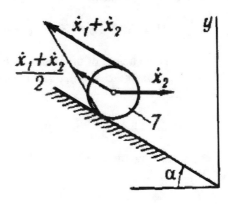

Fig. 5.15 The drive speed of wheel 7

The kinetic energy of the bodies 1, 2, and 4–7:

$$T_1 = m_1 v_1^2/2 = 3m\dot{x}_1^1/2;$$
$$T_2 = m_2 v_2^2/2 = 4m\dot{x}_2^2;$$
$$T_4 = J_4\omega_4^2/2 = m\dot{x}_1^2/2;$$
$$T_5 = m_5 v_5^2/2 = 2m\dot{x}_1^2;$$
$$T_6 = m_6 v_6^2/2 + J_6\omega_6^2/2 = 0.5m(\dot{x}_1^2 + 3\dot{x}_2^2 + 2\dot{x}_1\dot{x}_2);$$
$$T_7 = m_7 v_7^2/2 + J_7\omega_7^2/2 = \frac{m}{16}[3\dot{x}_1^2 + (11 - 4\sqrt{3})\dot{x}_2^2 - 2(2\sqrt{3} - 3)\dot{x}_1\dot{x}_2].$$

The kinetic energy of the thread 3, $T_3 = 0.5 \sum m_{3i} v_{3i}^2$, is found, given that the velocities of all its points are the same, i.e., $v_{3i} = v_3 = \dot{x}_1$ and $\sum m_{3i} = m$:

$$T_3 = m_3 v_3^2/2 = m\dot{x}_1^2.$$

Substituting the found values into the formula (5.37) gives

$$T = \frac{m}{16}[75\dot{x}_1^2 + (99 - 4\sqrt{3})\dot{x}_2^2 + 2(11 - 2\sqrt{3})\dot{x}_1\dot{x}_2]. \tag{5.40}$$

The potential energy of the system is found as the work of the gravity forces of solids 1 and 7 and threads 3 when they move from a given position with coordinates $(x_1; x_2)$ to some zero starting position, such as the one from which the generalized coordinates are counted:

$$\Pi = \Pi_1 + \Pi_7 + \Pi_3; \quad \Pi_1 = -m_1 g x_1; \quad \Pi_7 = m_7 g y_7, \tag{5.41}$$

where y_7 is the height of the center of the wheel 7 above the level where it was at the initial moment $(x_{10}; x_{20})$. Since the projection of the wheel center on the y axis (see Fig. 5.14)

$$v_{7y} = \dot{y}_7 = (\dot{x}_1 + \dot{x}_2) \cdot 0.5 \sin \alpha,$$

then, integrating this expression under zero initial conditions, we get

$$y_7 = (x_1 + x_2) \cdot 0.5 \sin \alpha.$$

Therefore,

$$\Pi_7 = m_7 g (x_1 + x_2) \cdot 0.5 \sin \alpha.$$

To determine Π_3, we note that the gravity of the thread 3 during its movement from position $a'b'$ in the position ab in which $x_1 = 0$ is equal to the work gravity of the plot threads bb' when it is moved in the position aa' (Fig. 5.16). Thus,

Fig. 5.16 To the potential
energy of the thread

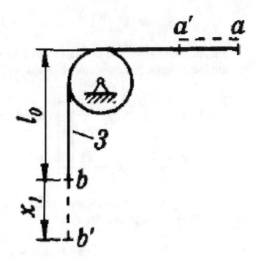

$$\Pi_3 = -m_3 x_1 g \left(\frac{x_1}{2} + l_0\right)/L.$$

Substituting these values in (5.41), we find

$$\Pi = g[-m_1 x_1 + m_7(x_1 + x_2) \cdot 0.5 \sin\alpha - m_3 x_1 (x_1/2 + l_0)/L],$$

or

$$\Pi = mg[-x_1^2/L - (2.75 + 2l_0/L)x_1 + x_2/4]. \tag{5.42}$$

The generalized forces Q_1 and Q_2 can be determined from expressions of the work of nonconservative forces on elementary displacements of the system corresponding to the variation of each generalized coordinate, or, equivalently, from expressions of the power N_1 and N_2 of nonconservative forces at possible velocities of the system corresponding to the increase of each generalized coordinate:

$$Q_1 = N_1/\dot{x}_1; \quad Q_2 = N_2/\dot{x}_2.$$

For our task,

$$N_1 = -(Rv_5 + M\omega_6)|_{(\dot{x}_2=0)}.$$

or, given (5.38) and (5.39),

$$Q_1 = -b\dot{x}_1 - M/r. \tag{5.43}$$

Similarly, under the initial condition $\dot{x}_2 = 0$, we get

$$Q_2 = P - M/r. \tag{5.44}$$

Substituting (5.40) and (5.42)–(5.44) into the expressions (5.35) and (5.36), we obtain the differential equations of motion of the system:

$$\frac{75}{8}m\ddot{x}_1 + \frac{11 - 2\sqrt{3}}{8}\ddot{x}_2 = 2mg\frac{x_1}{L} + mg\left(2.75 + \frac{2l_0}{L}\right) - b\dot{x}_1 - \frac{M}{r}; \tag{5.45}$$

$$\frac{99 - 4\sqrt{3}}{8}m\ddot{x}_2 + \frac{11 - 2\sqrt{3}}{8}m\ddot{x}_1 = -\frac{mg}{4} + P - \frac{M}{r}. \tag{5.46}$$

Expressing $\ddot{x}_2$ from (5.45) and substituting in (5.46), we obtain a second-order linear differential equation for x_1:

$$\ddot{x}_1 + 2n\dot{x}_1 - cx_1 = a. \tag{5.47}$$

Here $n = 0.0538b/m$; $c = 0.215g/L$; $a = g(0.298 + 0.215l_0/L) = 0.099M/(rm) - 0.0088P/m$.

The characteristic equation for (5.47) will be

$$z^2 + 2nz - c = 0.$$

Its roots are

$$z_{1,2} = -n \pm \sqrt{n^2 + c}.$$

The general integral of the differential equation (5.47) is

$$x_1 = \exp(-nt)\left(C_1 \exp(\sqrt{n^2 + c}t) + C_2 \exp(-\sqrt{n^2 + c}t)\right) - \frac{a}{c}. \tag{5.48}$$

Differentiating the Eq. (5.48) gives

$$\dot{x}_1 = \exp(-nt)[(-n + \sqrt{n^2 + c})C_1 \exp(\sqrt{n^2 + c}t) - \\ -(n + \sqrt{n^2 + c})C_2 \exp(-\sqrt{n^2 + c}t)]. \tag{5.49}$$

Using the initial conditions $t = 0 :\ x_1(0) = 0$; $\dot{x}_1(0) = 0$, from the Eqs. (5.48) and (5.49), we find

$$C_1 = \frac{a}{2c\sqrt{n^2 + c}}(n + \sqrt{n^2 + c});$$

$$C_2 = \frac{a}{2c\sqrt{n^2 + c}}(-n + \sqrt{n^2 + c}). \tag{5.50}$$

Thus, the formulas (5.48) and (5.50) define the law of change of the first generalized coordinate. To get the second equation of motion, we find from the formula (5.46)

$$\ddot{x}_2 = k - 0.0818\ddot{x}_1, \tag{5.51}$$

where

$$k = 0.0869 \left[\frac{1}{m} \left(P - \frac{M}{r} \right) - \frac{g}{4} \right].$$

Integrating the Eq. (5.51), we obtain

$$\dot{x}_2 = kt - 0.0818\dot{x}_1 + C_3; \tag{5.52}$$

$$x_2 = kt^2/2 - 0.0818x_1 + C_3t + C_4. \tag{5.53}$$

To determine the constants C_3 and C_4, we use all four initial conditions for generalized coordinates and their derivatives. As a result, we find the second generalized equation of motion of the system:

$$x_2 = \frac{kt^2}{2} - 0.01818 \left[e^{-nt} \left(C_1 e^{\sqrt{n^2+ct}} + C_2 e^{-\sqrt{n^2+ct}} \right) - \frac{a}{c} \right] + \dot{x}_{20}t. \tag{5.54}$$

References

N. Hand, J. Finch, 'Analytical mechanics' (1933). http://assets.cambridge.org/97805215/73276.

V. Molotnikov, *Mekhanika konstruktsiiyu Teoreticheskaya mekhanika. Soprotivlenie materialov [Construction Mechanics and Theoretical Mechanics. Mechanics of Materials]* (Lan' Publication, Sankt-Peterburg, 2012)

A. Yablonskii, S. Noreiko, S. Vol'fson, dr., *Sbornik zadanii dlya kursovykh rabot po teoreticheskoi mekhanike: Uchebnoe posobie dlya tekhnicheskikh vuzov, 5-e izd. [Collection of Tasks for Term Papers on Theoretical Mechanics: A Textbook for Technical Universities. – 5th edn.]* (Integral-Press, Moscow, 2000) Publ.

Chapter 6
Dynamics of Controlled Systems

Abstract Mechanical systems equipped with an adjustable power (control) motor are considered. The purpose of the control is the movement of the system, in which some extreme properties are fulfilled for a given period of time. A mathematical model of the system is constructed, for which the variational problem is formulated. The necessary information from the calculus of variations is given. As examples, the solution of the simplest problem about the optimal speed and about the flight of the drone to the maximum range is given.

Keywords Model · Phase vector · Control vector · Phase coordinates · Domain · Admissible control · Functional · Comparison curves · Minimum · Variation · The variation is total · Mayer problem · Maximum principle · First integral · Bellman principle · Traction control

6.1 Mathematical Model of the Controlled System

Consider a material point of mass $m = 1$, freely moving along a horizontal straight line and equipped with an engine that develops a traction or braking force u, in modulus not exceeding one, i.e., $|u| \leqslant 1$. Let us direct the x axis along the trajectory of the point and write, according to (3.9), the main differential equation of motion of the object under consideration:

$$\ddot{x} = u, \quad -1 \leqslant u \leqslant 1. \tag{6.1}$$

Let us transform the differential equation (6.1) into a system of two first-order differential equations. For this purpose, we denote

$$x_1 = v = \frac{dx}{dt}; \quad x_2 = \frac{dv}{dt} = \frac{d^2x}{dt^2} = \dot{x}_1.$$

where v is the speed of the point. In this case, Eq. (6.1) gives

$$\dot{x}_1 = x_2,$$
$$\dot{x}_2 = u. \tag{6.2}$$

Using a similar technique, the differential equations of motion of any mechanical system can be reduced to a system of first-order differential equations. Writing the differential equations of motion of the system in the form (6.2) was called *the Cauchy normal form*. Therefore, in the general case of the motion of a mechanical system, in what follows, we will consider an ordinary differential equation written in the normal Cauchy form in vector or coordinate form. The vector notation of the equation has the form

$$\dot{x} = X(x, u, t). \tag{6.3}$$

Here $x\{x_1, x_2, \ldots, x_n\}$ is the n-dimensional vector *of the state of the system*, which will also be called *the phase vector*, and its components x_1, x_2, $\ldots$, x_n, *the phase coordinates of the system*; $X(X_1, X_2, \ldots, X_n)$, vector of the generalized force applied to the system; $u\{u_1, u_2, \ldots, u_r\}$, r-dimensional *control vector*; and t, the time defined in the interval $\mathfrak{I} = \left[t_i, t_f\right]$ if $t_f < \infty$, or in the interval $\mathfrak{I} = \left[t_i, t_f\right)$ at $t_f = \infty$. In those cases when the generalized force X does not explicitly depend on the time t, we will assume $t_i = 0$, $t_f = T \leqslant \infty$.

Along with the vector notation of equation (6.3), we will also use its coordinate notation:

$$\dot{x}_k = X_k(x_1, x_2, \ldots, x_n; \ u_1, u_2, \ldots, u_r; \ t), \quad (k = 1, 2, \ldots, n). \tag{6.4}$$

The domain of definition of equation (6.3) or (6.4) is the set of points x, u, for which the projections of the generalized force X_k are determined for any $t \in \mathfrak{I}$. We denote this collection by $N \geqslant 0$ or in the form of equalities-inequalities:

$$N_\alpha(x, u, y) \geqslant 0, \quad (\alpha = 1, 2, \ldots, s). \tag{6.5}$$

Expressions (6.5) characterize the constraints imposed on the phase vector x and the control vector u. In the case of strict equality $(N = 0)$, the boundary of the domain N in coordinates x and u belongs to the domain of definition of equation (6.3). If $N > 0$, the domain remains open.

Consider Eq. (6.3) or (6.4) in more detail. Among all possible systems of these equations, we will consider only those for which the functions X_k and N_k in the domain $N \geqslant 0$ are continuous and continuously differentiable the required number of times. Functions X_k contain phase vector, control vector, and time as arguments, and the form of their dependence on these variables is known. Therefore, the phase vector x is determined by Eq. (6.3) for any choice of the vector u. The components of this vector can be the coordinates of the rudders, the components of the traction force, etc. Having properly arranged their behavior, we can impart certain properties to the movement of the system (Pontryagin 1985; Boltyanskii 1969; Wiener 1948).

Here it is necessary to immediately solve the question concerning the class U of functions from which the control u can be selected (Letov 1969). From the mathematical point of view, the widest possible class of functions is desirable, for example, the class D of piecewise-continuous functions with a finite number of discontinuity points and the domain of definition $N \geqslant 0$. However, from the point of view of the technical implementation of controls, it is clear in advance that some of these controls cannot be realized due to the inertia of the rudders, limited power of the steering drives, etc. Therefore, depending on the specific problem, we will consider U as a class D or a class of continuous functions C. Any function $u \in U$ and taking numerical values from the region $N \geqslant 0$ is called *admissible control.*

If the domain of definition of equations (6.3) is defined, the class of admissible controls and functions X_k, N_k is chosen, then for any $u \in U$ and any initial value $x_i \in N$, Eqs. (6.3) have a solution

$$x = x[x(t_i),\ u(t),\ t]. \tag{6.6}$$

In what follows, we will agree to call the system of differential equations (6.3) or (6.4) *a mathematical model of the controlled system,* if the following conditions are met:

(a) The area $N \geqslant 0$ of their definition is indicated.
(b) The interval $\Im$ of the change in time t is indicated.
(c) The class of admissible controls U is chosen.
(d) The functions X_k are smooth in all arguments x, u, and t at each point of the domain $N \geqslant 0$ and for $t \in \Im$.

To formulate the problem of the dynamics of a controlled system, it remains to supplement equations (6.3) with boundary conditions and a control goal.

As is known from the theory of ordinary differential equations, the conditions that obey the end values of the phase variables x at $t = t_i$ and $t = t_f$ are called boundary. In the dynamics of mechanical systems, three main types of boundary conditions are most often encountered at $t = t_i$ and (or) $t = t_f$:

(1) The values of the phase coordinates are fixed, which is equivalent to fixing the ends of the phase trajectory.
(2) Phase coordinates are free.
(3) The ends of the phase trajectory belong to some manifolds:

$$F_\rho(x, t) = 0; \quad \Phi_\rho(x, t) = 0.$$

In addition to these basic types of boundary conditions, various combinations of them can occur. But in any case, they express in mathematical form the fact that the system must reach the desired end state "f," starting from the initial state "i." In this case, it may turn out that the moments $t = t_i$ and $t = t_f$ themselves cannot be assigned in advance and they will be free to choose. Regardless of the type of boundary conditions, we will take for them the following symbolic notation:

$$(i,\ f) = 0. \tag{6.7}$$

The system can be transferred from the initial state "i" to the final state "f" in different ways, the set of which is determined by the controls $u \in U$. Among these methods, you can choose the one that provides the required quality of the controlled process. *The control quality indicator is characterized by a special mathematical quantity called functional.* The value of the functional that transfers the system from state "i" to state "f" determines the *payment* for achieving the control goal; therefore, it is also called the *payment function.*

A *variable* $I(y) = I[y(x)]$, *whose numerical value depends on the choice of one or more functions, is called a functional. The functional is,* for example, the integral

$$I(y) = \inf_{a}^{b} F(x, y, \dot{y})dx. \tag{6.8}$$

In accordance with the mapping (6.8), each function $y(x)$ from a certain class of functions is associated with the number I. Therefore, the functional $I(y)$ is given if:

(a) The class Y is indicated to which the function $y(x)$ belongs, considered on the interval $[a, b]$.
(b) A law is indicated according to which each function $y(x)$ from class Y is uniquely associated with a number $I(y)$, $(I(y) < \infty)$.

The functions $y(x) \in Y$ are called admissible curves or comparison curves for the functional $I[y(x)]$. The functional $I(y)$ is said to attain a minimum on a curve $y^0(x)$, if for any curve $y(x) \in Y$, the relation

$$I(y^0) \leqslant I(y). \tag{6.9}$$

The curve $y^0(x)$ that minimizes the functional $I(y)$ is called its *minimum.* Obviously, the functional $-I(y)$ will have a maximum at the minimum. Thus, we now have all the necessary means to formulate the problem of the dynamics of a controlled system.

A t a s k. It is required to find a control $u(t)$ for which the functional

$$I(u) = \Phi(x, u, t)|_i^f \tag{6.10}$$

attains a minimum on the admissible curves of the set $U \times X \times \mathfrak{I}$ of system (6.3) under the boundary conditions (6.7).

When compiling a functional in a specific control problem, it is necessary to ensure that two conditions are met:

(a) The class of admissible comparison curves should be clearly defined from the physical meaning of the problem.
(b) The record of the form of expression Φ in formula (6.10) should, as accurately as possible and in adequate mathematical symbols, reflect the payment for achieving the control goal.

6.2 Basic Information About Functionals and Function Spaces[1]

Problems that contain investigations of functionals to the maximum or minimum are called *variational*. The methods for solving these problems are very similar to the methods for finding the extremums of functions. To get a clear idea of these similarities and differences, the following is a summary of the theory of extremums of functions and in parallel introduces similar concepts for functionals.

In mathematical analysis, when studying the behavior of functions $f(x)$, geometric images in Euclidean space are widely used.

If x is an n-dimensional vector, then $|x| = +\sqrt{\sum_{k=1}^{n} x_k^2}$ is its norm (length). For all functions, the Euclidean space is uniform and is characterized by its own metric, which determines the distance between two points x_1 and x_2 by the formula

$$r(x_1, x_2) = |x_1 - x_2|. \tag{6.11}$$

Table 6.1 contains important definitions of the characteristics of functions and functionals. They will help to assimilate and forever remember both the similarities and differences between function and functionality.

When studying functionals, it is also convenient to use geometric images of the spaces in which the functionals are defined. A *functional* space is a space where each element—a point—corresponds to the definition of a function $y(x)$ that belongs to a certain class Y. Therefore, to define a function space, you must specify a function class. This means that it is not possible to define a single universal functional space of the Euclidean space type in analysis for the study of functionals. For example, the functional $\int_a^b F(x, y)dt$ is defined on the class C of continuous functions, and for the functional $\int_a^b F(x, y, \dot{y})dt$, the allowed curves are the class of smooth or piecewise-smooth functions.

---

[1] The reader familiar with the calculus of variations and the beginnings of functional analysis can skip this paragraph and the next.

Table 6.1 Features of functions and functionality

Functions	Functionals
1. A variable quantity $z = f(x)$ is called a *function* of a variable quantity x if each value of x from a certain area of its variation corresponds to a value of z, that is, the number x is associated with the number z according to the rule $f(x)$	1. A variable $v(y) = v[y(x)]$ is called a *functional* that depends on a function $y(x)$ from a certain class of functions Y if each function $y(x) \in Y$ corresponds to a value of v, i.e., to a function $y(x)$, there corresponds a number v according to the rule $v[y(x)]$. Functionals that depend on several functions and on functions of several variables are defined in a completely similar way
2. *The increment* Δx of the argument x of the function $f(x)$ is the difference between the two values variable: $\Delta x = x - x_1$, where x_1 is a point sufficiently close to x. If x is an independent variable, then $dx = \Delta x$	2. *The increment* or *variation* δy of the argument $y(x)$ of the functional $v[y(x)]$ is the difference between two sufficiently close functions $y(x) - y_1(x)$, belonging to the same class
3. A function $f(x)$ is called *continuous* if a small change in x corresponds to a small change in $f(x)$	3. A functional $v[y(x)]$ is called *continuous,* if a small change in $y(x)$ on the comparison curves corresponds to a small change in the value of the functional $v[y(x)]$. Comparison curves are distinguished by the degree of smallness of the change in $y(x)$. If $|y(x) - y_1(x)| < \varepsilon$, where ε is small, then the curves $y(x)$ and $y_1(x)$ are called close in the sense of closeness of order zero. If where $|y(x) - y_1(x)| < \varepsilon_1$ and $|y'(x) - y_1'(x)| < \varepsilon_2$ are ε_1 and ε_2—small, the curves are close in the sense of first-order proximity, etc.
4. The function $f(x)$ is continuous at $x = x_0$, if for any $\varepsilon > 0$ there is $\delta > 0$ such that $|f(x) - f(x_0)| < \varepsilon$ when $|x - x_0| < \delta$	4. The functional $v[y(x)]$ is *continuous* for $y = y_0(x)$ in the sense of k-th order proximity, if for any $\varepsilon > 0$ it is possible to choose such $\delta > 0$, that $|v[y(x)] - v[y_0(x)]| < \varepsilon$ for $$|y(x) - y_0(x)| < \delta;$$ $$|y'(x) - y_0'(x)| < \delta;$$ $$\dots\dots\dots\dots\dots$$ $$|y^{(k)}(x) - y_0^{(k)}(x)| < \delta$$
5. *A linear function* is a function $l(x)$, with the following properties: $$l(cx) = cl(x), \quad (c - \text{const});$$ $$l(x_1 + x_2) = l(x_1) + l(x_2)$$	5. *A linear functional* is a functional $L[cy(x)]$ that satisfies the following conditions: $$L[cy(x)] = cL[y(x)]. \quad (c - \text{const});$$ $$L[y_1(x) + y_2(x)] = L[y_1(x)] + L[y_2(x)]$$

(continued)

Table 6.1 (continued)

Functions	Functionals
6. If the function increment $\Delta f = f(x + \Delta x) - f(x)$ can be represented as $\Delta f = A(x) \cdot \Delta x + B(x, \Delta x) \cdot \Delta x$, and $A(x)$ does not depend on Δx, and $B(x, \Delta x) \to 0$ for $\Delta x \to 0$, then the function $f(x)$ is called differentiable. The linear part of the increment of a function with respect to Δx is called its differential, i.e., $df = f(x)dx$	6. If the increment of the functional is $\Delta v = v[y(x) + \delta y]$ can be represented as $22\Delta v = L[y(x), \delta y] + \beta(y(x), \delta y) \cdot \max|y|$, where $L[y(x), \delta y]$ is a linear functional with respect to δy and $\max|y|$ is the largest modulo value of the variation δy and $\beta(y(x), \delta y) \to 0$ together with $\max|\delta y| \to 0$, then the linear part of the increment of the functional relative to δy is called its variation and is denoted by $\delta v = L[y(x), \delta y]$
7. If the differentiable function $f(x)$ reaches a maximum or minimum at the inner point $x = x_0$ of the domain of its definition, then the differential of the function at this point vanishes, i.e., $df(x_0) = 0$	7. If a functional $v[y(x)]$ with a variation reaches a maximum or minimum on the curve $y = y_0(x)$, which belongs to the domain of the functional definition, then its variation on this curve is zero, i.e., $\delta v[y_0(x)] = 0$

However, with all the variety of functional spaces, a number of their common properties can be distinguished. Let a set R of elements x, y, z of arbitrary nature be given. *A collection R is said to form a linear space if the operations of addition and multiplication by a number are defined for any of its elements:*

A. $x + y = y + x$; $(x + y) + z = x + (y + z)$; $x + 0 = x$; $x + (-x) = 0$;
B. $x \cdot 1 - x$; $\alpha(\beta x) = (\alpha\beta)x$; $(\alpha + \beta)x = \alpha n x + \beta x$; $\alpha(x + y) = \alpha x + \alpha y$.

A linear space R is called *normalized* if each element of $x \in R$ corresponds to a nonnegative number $\|x\|$—the norm of the element x—such that

> *a)* $\|x\| = 0$ when $x = 0$;
> *b)* $\|\alpha x = |\alpha| \cdot \|x\|$, α — number;
> *c)* $\|x + y\| \leqslant \|x\| + \|y\|$.

For a linear normalized space, by analogy with the formula (6.1), the concept of the distance $r(x, y)$ between its two elements $x, y \in R$ is defined:

$$r(x, y) = \|x - y\|. \tag{6.12}$$

If the elements of a linear normalized space are a set of functions belonging to a certain class, then it becomes a function space. We give examples of such spaces.
I. Space ;C; of continuous functions in the interval $[a, b]$; its norm is expressed by the formula

$$\|y\| = \max|y(x), \ x \in [a, b]. \tag{6.13}$$

The distance between two points in space C is defined as

$$r(y_1, y_2) = \|y_1 - y_2\| = \max |y_1 - y_2|. \tag{6.14}$$

II. C_1 space of smooth functions with norm

$$\|y\| = \max |y(x) + \max |\dot{y}(x)|, \ x \in [a, \ b] \tag{6.15}$$

and the distance between two points

$$r(y_1, y_2)\|y_1 - y_2\| = \max |y_1 - y_2| + \max |\dot{y}_1 - \dot{y}_2|. \tag{6.16}$$

III. The space C_n of functions continuous together with their derivatives up to the n-th order inclusive. The norm of such a space is

$$\|y\| = \sum_{k=0}^{n} \max |y^{(k)}(x)|, \tag{6.17}$$

and the distance between its two points y_1 and y_2 is given by the formula

$$r(y_1, y_2) = \|y_1 - y_2\| = \sum_{k=0}^{n} \max |y_1^{(k)} - y_2^{(k)}|. \tag{6.18}$$

For normed linear function spaces, the concept of closeness of functions is often used in the sense of closeness of the points of the functional space representing them.

Let a number $\alpha > 0$ be given. By definition, *two elements* $y^K 0, y$ *of a given functional space have α-proximity in the interval $[a, b]$ if the distance between them satisfies the condition*

$$r(y^0, y) = \|y^0 - y\| \leqslant \alpha, \ x \in [a, \ b]. \tag{6.19}$$

where the norm is defined by the rule of a given functional space. In particular, if we are talking about the functional space of C continuous functions, then α-proximity means that the difference in the values of two functions modulo does not exceed α on each interval $[a, b]$. In the case of the space C_1 of smooth functions α, the proximity of two points (functions) includes restrictions not only on the difference in the values of the functions themselves but also on the difference in the values of their derivatives for each $x \in [a, b]$; i.e., in accordance with the formula (6.15) in this case

$$\|y^0 - y\| = \max |y^0 - y| + \max |\dot{y}^0 - \dot{y}| \leqslant \alpha.$$

6.3   Variations of Comparison Curves and Functionals

Consider a certain class Y of comparison curves of the variational problem formulated on p. 224. Let $y^0 \in Y$—be one of the comparison curves. Difference

$$\delta y = y - y^0 = \omega(t, \alpha), \quad (y \in Y, \ t \in \Im), \tag{6.20}$$

where α is a small parameter ($|\alpha| \leqslant \bar{\alpha}$, $\bar{\alpha} > 0$) and the function $\omega(t, \alpha)$ is such that $\omega(t, 0) = 0$ is called the *total variation* of the curve y^0. Among the total variations of comparison curves, *weak* and *strong* variations are distinguished.

The total variation of a curve is called weak if the following conditions are met: a) for a given α and any $t \in \Im$, there is a derivative $d\omega/dt$; b) $\lim\limits_{\alpha \to 0} \omega(t, \alpha) = 0$; $\lim\limits_{\alpha \to 0} \dot{\omega}(t, \alpha) = 0$.

In the case when the last of conditions (b) is not satisfied and the derivative $\bar{\omega}(t, \alpha)$ can take any values, the total variation is called *strong*.

Functional variations are used to identify their extrema.

In the case when the functional $v[y(x)]$ reaches an extremum on a curve $y = y_0(x)$ only with respect to curves $y = y(x)$ close to $y = y_0(x)$ in the sense of closeness of the first order, i.e., curves $y = y(x)$ and $y = y_0(x)$ are close not only in ordinates, but also in directions of tangents, the maximum or minimum is called *weak*.

If functional $v[y(x)]$ reaches on curve $y = y_0(x)$ a maximum or minimum with respect to all curves $y = y(x)$, close to $y = y_0(x)$ in the sense of closeness of the zero order, then the maximum or minimum of the functional is called *strong*.

Strong variation of comparison curves is used to identify a strong extremum of the functional. However, if a certain curve $y = y_0(x)$ does not stand comparison in the extreme test with strong variation, this does not mean that it will not stand comparison even with weak variation.

Conversely, if curve $y = y_0(x)$ compares well in the extreme test with strong variation, then it must all the more be comparable with weak variation. From this it follows that testing curve $y_0(x)$ for a weak extremum gives the necessary conditions for the existence of both a weak and a strong extremum of the given functional. Let $I(y)$ be some continuous functional and $y_0 \in Y$—a comparison curve. If $\delta y = y - y^0$ forms the total variation of the curve y^0, then the difference

$$\Delta I = I(y) - I(y^0) \tag{6.21}$$

is called the total variation of the functional at the point y^0.

Let's look at the expression $I(y)$ as an ordinary function of the independent variable y. If this function is differentiable, then it can be expanded in a Taylor series in a neighborhood of the point y^0 in increment $\alpha\eta$, where α is the same as in formula (6.20) and η is any admissible curve from the class Y under consideration. Expression kind

$$y = y^0 + \alpha \eta \tag{6.22}$$

is called the *Lagrangian form of variation* of an admissible curve. Performing the indicated decomposition, we obtain

$$I(y) = I(y^0) + \alpha \delta I + \frac{1}{2}\alpha * 2\delta * 2I + \ldots + O_m(\alpha), \tag{6.23}$$

where the remainder $O_m(\alpha)$ has the property

$$\lim_{\alpha \to 0} \frac{1}{\alpha^m} O_m(\alpha) = 0.$$

The expressions δI and $\frac{1}{2}\delta^2 I$, included in the expansion (6.23), are called the first and second variations of the functional $I(y)$, respectively. Taking into account the adopted definitions, Proposition 7 Table 6.1, written for functionals in the previous section, can be formulated as follows.

Theorem *In order for the curve y^0 to be an extremal, its first variation must vanish, that is, $\delta I = 0$.*

6.4   Statement of the Mayer Variational Problem

Consider a system whose motion is described by ordinary differential equations of the type (6.3) with generalized forces that do not explicitly depend on time (Kalman 1958). Such objects are called *stationary systems with lumped parameters*. Let us write out the equations of the mathematical model of the system under study:

$$\dot{x} = X(x, u) \tag{6.24}$$

or

$$\dot{x}_k = X_k(x_1, x_2, \ldots, x_n; \ u_1, u_2, \ldots, u_r).$$

Let, as before, Eqs. (6.24) are defined in an open or closed domain $N \geqslant 0$ for $t \in \mathfrak{I}$, the functions X_k are differentiable with respect to all their arguments, and the boundary conditions for the phase vector are symbolically represented in the form

$$(i, f) = 0, \tag{6.25}$$

Here, as before, the letter i denotes the conditions at the left, initial end of the admissible phase trajectory, and the letter f denotes the conditions at its right end.

Now let's decide on the optimizing functionality. We will assume that admissible controls belong to the class D of piecewise-continuous functions.

Consider some smooth scalar function:

$$G = G(x), \tag{6.26}$$

defined in the region $N \geqslant 0$ at $t \in \Im$. If by choosing some admissible control $u \in U$ the system is transferred from the initial state i to the final state f, then as a result the function G will receive an increment

$$\Delta G = G|_i^f, \tag{6.27}$$

equal to the difference between its values at the right and left ends of the permissible trajectory.

The variational problem (6.24)–(6.27), in which the optimized functional is represented by expression (6.27), is called the *Mayer problem*. Let us give its formulation.

A t a s k. Among the admissible curves u, x, find the one that minimizes functional (6.27).

For a geometric interpretation of the problem, consider the $(n + r + 1)$-dimensional space $x, u, \Delta G$ (Fig. 6.1). Points i and f represent the ends of the admissible phase trajectory. For different admissible controls u_1 and u_2, the functional ΔG receives different increments ΔG_1 and ΔG_2. The geometric meaning of the problem is to find the control u^0 for which the segment ΔG has the minimum length (Fig. 6.1).

The solution to the problem, if it exists, can be symbolically represented as

$$\begin{aligned} x &= x^0[t, (i, f)]; \\ u &= u^0[t, (i, f)]. \end{aligned} \tag{6.28}$$

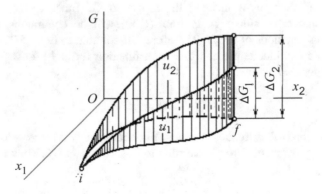

Fig. 6.1 Geometric interpretation of the problem of programming optimal trajectories

It depends on the boundary conditions, is expressed in parametric form, and is unique.

Solution (6.28) is called *programmed* or *brachistochronic* motion, and the problem of finding this kind of solutions is called *programming optimal trajectories*.

According to the terminology used in the mathematical theory of optimal processes, programming problems for optimal trajectories are classified as *nonclassical variational problems*. However, it can be shown that if the domain N of the definition of the problem is open, it becomes a classical extremal problem.

6.5   The Maximum Principle

We have already said (Sect. 6.1) that in real mechanical systems, the energy resources that can be used to control their behavior have limitations. Let us first of all consider the simplest types of nonclassical variational problems, in which the region $N \geqslant 0$ is limited only by the modulus of the control vector

$$\|u\| \leqslant \bar{u}, \tag{6.29}$$

where $\bar{u}$ is a given number.

Suppose that the right-hand sides X_k of Eqs. (6.24) are twice differentiable with respect to all arguments and define an auxiliary vector ψ as a solution to the equations:

$$\dot{\psi}_k = -\sum_{\alpha=1}^{r} \frac{\partial X_\alpha}{\partial x_k} \cdot \psi_\alpha, \ \ (k = 1, 2, \dots, n). \tag{6.30}$$

A vector ψ is called *completely nonzero*, if none of its components in the region $t \in \Im$ are identically zero.

Under the assumptions made with respect to the functions X_k, the partial derivatives under the sum sign in Eqs. (6.30) will be continuous, linear, and homogeneous functions of time t. Therefore, the solution to Eqs. (6.30) is a vector ψ, which is a continuous piecewise-smooth function for any $t \in \Im$. With this in mind, consider also the function

$$H(\psi, x, u) = \psi \cdot X, \tag{6.31}$$

composed as the scalar product of an auxiliary completely nonzero vector ψ and a vector of the generalized force X. Turning to Eqs. (6.24) and (6.30), we can obtain

$$\dot{x} = \frac{\partial H}{\partial \psi}; \ \dot{\psi} = -\frac{\partial H}{\partial x}. \tag{6.32}$$

Differential dependences (6.32) are called the *Euler-Lagrange equations*.

Function H plays an extremely important role in solving the problems under consideration. It allows you to find the optimal control by examining this function for maximum with respect to the variable u with the remaining variables fixed. Moreover, if the corresponding value of u is located inside the region $N \geqslant 0$, then this maximum is understood in the usual, classical sense:

$$\frac{\partial H}{\partial u} = 0; \quad \frac{\partial^2 H}{\partial u^2} < 0. \tag{6.33}$$

If u reaches the boundaries of the region $N \geqslant 0$, then the maximum $M(\psi, x)$ of the function H should be defined as the upper bound of the function H with respect to the variable u, i.e.,

$$M(\psi, x) = \max_{u \in N} H(\psi, x, u). \tag{6.34}$$

The legality of the described technique in solving optimization problems is justified by the following sentence, called the *maximum principle*.

Theorem *For the curve u^0, x^0 to provide a strong minimum to the functional in the Mayer problem, there must be a completely nonzero continuous vector $\psi\{\psi_1, \psi_2, \ldots, \psi_n\}$, defined by Eqs. (6.32), for which:*

(1) The function $H(\psi, x, u)$ reaches its maximum in the argument u
(2) The condition is satisfied:

$$[\delta G - H \delta t + \psi \cdot \delta x]|_i^f = 0. \tag{6.35}$$

Relation (6.35) is called the *transversality condition*. It reflects the requirements imposed on the behavior of the phase trajectory at its ends.

In the case when the functions G and H are linear in x and u, the conditions for the existence of a strong minimum of the functional G stipulated in Theorem 1 are also sufficient. This theorem in the form of a hypothesis was first stated by L.S. Pontryagin. The reader can find its proof in any of the sources on the mathematical theory of optimal processes indicated at the end of the book.

6.6 Erdmann-Weierstrass Conditions: First Integral

Choosing admissible controls u from the class D of piecewise-continuous functions, by virtue of equations (6.24), we find that the curves x belong to the class D_1 of piecewise-smooth functions. Consequently, the curves x will have a finite number of corner points $t_1^*, t_2^*, \ldots, t_s^*$ at each of which the derivatives on the right and on the left exist, but are discontinuous. Let us assume for simplicity that there is only one such point with coordinate t^*. It splits the interval $\Im$ into two intervals $[t_i, t^*]$ and $[t^*, t_f]$, in each of which u are continuous and x are smooth curves. Then, for each of these intervals, the conditions of Theorem 6.25 are satisfied. Therefore, if we want the curve (6.28) to be an extremal over the entire time interval $\Im$, it must be an extremal on any of the time intervals belonging to the region $\Im$. This position is sometimes called *Bellman principle*.

In the presence of one control discontinuity point t^*, the transversality condition (6.35) should be written in the form

$$[\delta G - H\delta t + \psi \delta x]_i^f = [\delta G - H\delta t + \psi \delta x]^f - [\delta G - H\delta t + \psi \delta x]_{t*}^+ +$$
$$+[\delta G - H\delta t + \psi \delta x]_-^{t^*} - [\delta G - H\delta t + \psi \delta x]_i = 0.$$

$$(6.36)$$

This notation reflects different values of the control u and functions $\dot{x}$ to the left $(-)$ and to the right $(+)$ of the break point. It is seen from expression (6.36) that, for the transversality condition to be satisfied only due to the boundary values at the ends of the phase trajectory, it is necessary and sufficient that the sum of the two inner brackets vanishes. For arbitrary variations of δx, the latter is possible only if the equalities

$$\psi^{(+)} = \psi^{(-)}; \quad H^{(+)} = H^{(-)} \tag{6.37}$$

at each discontinuity point of the control u. Equalities (6.37) expressing the conditions for the continuity of the auxiliary functions ψ and H on the extremals are called the *Erdmann-Weierstrass conditions*. Thus, the following has been proved.

Theorem *For the functional in the Mayer problem to reach an extremum in the class of piecewise-smooth functions x and piecewise-continuous controls u, all curves ψ, x, u satisfying the conditions of Theorem 6.25 must also satisfy the Erdman-Weierstrass conditions at each control discontinuity point.*

Let us now show that if the conditions of Theorem 6.25 are satisfied, then the variational Mayer problem has the first integral. Recall that in the theory of systems of ordinary differential equations of the first order, in which the argument is time, the first integral is a relation of the form

$$\Omega(x) = C,$$

where $\Omega(x)$ is any smooth function, the total time derivative of which, calculated by virtue of these equations, is zero, and C is an arbitrary constant.

Let us turn to Eqs. (6.24) of the variational problem for stationary systems and assume that the optimal control $u^0(\psi, x)$ is found. Consider a relation of the form

$$M(\psi, x) = H(\psi, x, u^0(\psi, x)) = C. \tag{6.38}$$

By definition, the function H is differentiable with respect to ψ, x everywhere in the open domain N, with the possible exception of the discontinuity points of the control u^0. Therefore, for any time interval that does not contain control break points, the following relation holds:

$$\frac{dH}{dt} = \frac{\partial H}{\partial \psi} \cdot \dot{\psi} + \frac{\partial H}{\partial x} \cdot \dot{x}.$$

By virtue of equations (6.32), the right-hand side of this relation vanishes. Therefore, for any point of the interval without discontinuity u, the ratio $H(\psi, x, u^0) = C$ remains constant. But by Theorem 6.26, the constant $\tilde{N}$ retains its value for the points of discontinuity of the control u^0. Thus, the relation remains constant for any $t \in \Im$; therefore, the following is proved.

Theorem *Aggregate equations (6.32) are the variational Mayer problems for stationary systems with bounded functions of the first integral* (6.38).

The outlined brief information from the theory of optimal processes makes it possible to move on to solving specific dynamics of controlled systems.

6.7   The Simplest Problem of Optimal Performance

In this and the next sections, we will consider methods for solving typical problems of programming optimal trajectories, which will illustrate the maximum principle.

In the problem of the dynamics of a controlled system, three main stages can be distinguished. At the first stage, the research engineer, who is well acquainted with the technical side of the problem, must accurately formulate the problem. Here it is extremely important to take into account the main features of the modeled system and discard the minor ones, maneuvering "... between the Swamp of overcomplication and the West of oversimplification" (R. Bellman).

After the equations of motion of the system are drawn up, the area of their definition is outlined, the boundary conditions are written down, and the control goal is formulated, the stage of the mathematical solution of the problem begins. Here it is already inadmissible to introduce additional hypotheses that change the formulation of the problem.

At the third stage, the result of the formal mathematical solution of the problem is comprehended. The resulting solution should correspond to experience and our understanding of the physical meaning of the found dependencies. If the study of the solution reveals a conflict with common sense or physical laws, then the error should be looked for first of all in the formulation of the problem itself.

6.7.1   Building a Mathematical Model

We will consider the problem of the rectilinear motion of a material point of unit
mass, equipped with an engine with thrust $|u| \leqslant 1$. It is required to dispose of
the engine thrust in such a way that the object, moving from an arbitrary initial
position at any initial speed, in the shortest possible time can be stopped at a
predetermined point. The described problem is known as the optimal performance
problem (Boltyanskii 1969).

We wrote down the differential equations of motion of a point in Cauchy normal
form in Sect. 6.1:

$$\dot{x}_1 = x_2; \ x_2 = u;$$
$$\left(\dot{x}_1 = \frac{dx_1}{dt}; \ \dot{x}_2 = \frac{dx_2}{dt} \right). \tag{6.39}$$

Equations (6.39) are defined in the domain $N \geqslant 0$ containing the constraints

$$-\infty < x_1 < \infty; \ -\infty < x_2 < \infty,$$

and on a finite time interval $\Im = [0, \ T]$, and we cannot set the value of T in advance.

6.7.2   Border Conditions

The boundary conditions must reflect the physical meaning of the problem. Let the
moving point at the moment $t = t_1$ have a coordinate $x_1 = x_{10}$ and have a speed
$x_2 = x_{20}$. Without loss of generality, for a stationary system, we can set $t_1 = 0$.
Then the conditions at the left end of the phase trajectory can be represented as

$$t = t_1 = 0 : \ x_1 = x_{10}, \ x_2 = x_{20}. \tag{6.40}$$

Align the object's stopping point with the origin. At the right end of the trajectory,
we have

$$t_f = T : \ x_1 = x_{1T} = 0, \ x_2 = x_{2T} = 0. \tag{6.41}$$

6.7.3   Functional

The purpose of the control is to minimize the time of movement of the object from
state i to state f. We put $G = t$. The admissible curves are piecewise-smooth
functions $x_1(t)$ and $x_2(t)$ a piecewise-continuous function $u(t)$. Functionality is
defined on them:

$$\Delta G = t|_i^f. \tag{6.42}$$

A t a s k. Among the admissible curves of the variational problem (6.39)–(6.42), it is required to find the one that minimizes functional (6.42). Physically, this corresponds to the goal set in the description of the problem: it is required to find such a program for controlling the engine thrust $u(t)$ for which the time of movement of a point from state i to state f is minimal.

Let's turn to solving the problem. We will build the solution based on the maximum principle, so we need additional relations.

6.7.4 Getting Additional Ratios

In accordance with what has been said in Sect. 6.6, we introduce into consideration a completely nonzero vector $\psi(\psi_1, \psi_2)$ and compose the function H as the scalar product of the vector ψ and the vector $X(X_1, X_2)$ whose components are the right-hand sides of the system of equations (6.39). We have

$$H = \psi \cdot X = \psi_1 x_2 + \psi_2 u, \tag{6.43}$$

where the components of the vector ψ satisfy the Euler-Lagrange equations. For system (6.39), based on formulas (6.32), we obtain

$$\dot{\psi}_1 = -\frac{\partial H}{\partial x_1} = 0; \quad \dot{\psi}_2 = -\frac{\partial H}{\partial x_2} = -\psi_1. \tag{6.44}$$

By Theorem 6.27, the equations of the problem being solved have the first integral

$$H = \psi \cdot X = \psi_1 x_2 + \psi_2 u = C, \quad (c - \text{const}). \tag{6.45}$$

Let us write down the transversality condition. By formula (6.35) with functional (6.42) and the function H given by expression (6.43), we have

$$[\delta t - (\psi + 1x_2 + \psi_2 u)\delta t + \delta x_1 + \delta x_2]_i^f = 0. \tag{6.46}$$

But, according to the boundary conditions (6.40) and (6.41), the ends of the trajectory are fixed, so the variations in the phase coordinates at both ends vanish. Moreover, by virtue of an arbitrary choice of the variation δt, from the transversality condition (6.46) and formula (6.45), we obtain

$$C = 1. \tag{6.47}$$

Thus, in the problem under consideration, the function $H = 1$ on the entire optimal trajectory, including the discontinuity points of the control u, if there are any.

6.7.5 Traction Control Program

According to expression (6.43), the function H is linear in all its arguments. Consequently, the conditions for the existence of a strong minimum of the functional specified in Theorem 6.25, for the problem being solved, are necessary and sufficient. Then, in accordance with formula (6.34), we obtain the following engine thrust control program:

$$u = \left| \begin{array}{l} 1 \text{ for } \psi_2 > 0; \\ -1 \text{ for } \psi_2 < 0. \end{array} \right. \tag{6.48}$$

So, the optimal engine thrust control program, if it exists, consists of a certain sequence of thrust and braking engagement by the limiting value $\bar{u} = 1$. The exact composition of this program can be established with a more detailed study.

6.7.6 Integration of Differential Equations of the Problem

Dividing the first equation of system (6.39) by the second, we obtain

$$\frac{dx_1}{dx_2} = \frac{x_2}{u}. \tag{6.49}$$

Since the program (6.48) gives a piecewise constant function for the control $u(t)$, the general solution of equation (6.49) has the form

$$x_1 = \left| \begin{array}{l} \frac{1}{2}x_2^2 + C_{11} \text{ for } u = 1; \\ -\frac{1}{2}x_2^2 + C_{12} \text{ for } u = -1, \end{array} \right. \tag{6.50}$$

where C_{11}, C_{12} are constants, different, generally speaking, on different pieces of phase trajectories.

Let us draw the phase plane Ox_1x_2 and draw on it the families of parabolas given by Eqs. (6.50) (Fig. 6.2). The family shown in position (a) is obtained for the control $u = 1$ for different values of the constant C_{11}. Phase points move along these parabolas from bottom to top, because $\dot{x}_2 = u = 1$, $\dot{x}_2 > 0$. Similarly, for $u = -1$, the parabolas of the family of curves shown in Fig. 5.2b are obtained from each other by shifting in the direction of the x_1 axis. The phase points move from top to bottom along the curves of this family, since here $\dot{x}_2 = u = -1$, $\dot{x}_2 < 0$.

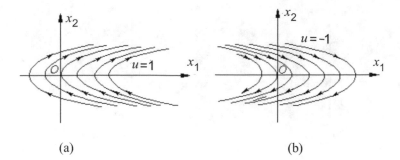

(a)   (b)

Fig. 6.2 Trajectories in the phase plane

Fig. 6.3 To the choice of the
optimal trajectory

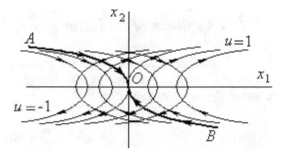

It follows from the Euler-Lagrange equations (6.44) that $\psi_2(t)$ is a linear function. But any linear function on a finite interval changes its sign at most once. It follows from this and expression (6.48) that the optimal control $u(t)$ in region $\Im = [t_i, \ T]$ has at most two intervals of constancy.

Keeping the last result in mind, we place both families of parabolas (6.50) in the same phase plane Ox_1x_2. In Fig. 6.3 is shown the entire family of phase trajectories. The arc OA highlighted here by a bold line belongs to parabola $x_1 = x_2^2/2$, obtained with control $u = -1$ and entering the origin of coordinates O. The arc OB is a branch of the parabola $x_1 = x_2^2/2$ that represents the phase trajectory for $u = 1$ and leads to the origin.

The engine thrust control program is now clear. If the starting point $x_0(x_{}10), x(20)$ is located in the phase plane above the line AOB, then in order to get to the origin, it must first move with the control $u = -1$ until it hits the arc OB at some moment of time t^*. At the moment t^*, the value of the control u must be switched to $u = +1$ and remain so until the moment it hits the origin. The phase trajectory, corresponding to such a sequence of controls, is shown in Fig. 6.4a.

If, at the initial moment $t = t_i = 0$, the phase point $x_0(x_{10}, x_{20})$ is located below the AOB line, then on the contrary, first you should turn on the control $u = +1$, which does not change until the moment the point hits the arc AO, and at the moment it hits this arc, the value u switches and becomes equal $u = -1$ to until the point stops at the origin. The corresponding phase trajectory is shown in Fig. 6.4b.

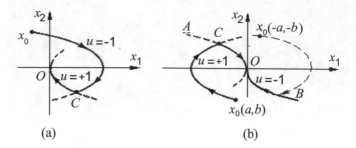

Fig. 6.4 The final stage of choosing the optimal trajectories

6.7.7 Completion of the Integration of Problem Equations

Let us first consider the case when the point x_0 lies above the curve AOB. For $0 \leqslant t \leqslant t^*$ $u(t) = -1$. Using the boundary condition (6.40), we determine the constant C_{12} in the solution (6.50):

$$x_{10} = -\frac{x_{20}^2}{2} + C_{12},$$

from where

$$C_{12} = x_{10} + \frac{x_{20}^2}{2}.$$

Then the equation of the first link of the optimal trajectory will be

$$x_1 = x_{10} - \frac{1}{2}(x_2^2 - x_{20}^2). \tag{6.51}$$

In the control switching point C shown in Fig. 6.4a, we find by joint solution of equation (6.51) and the equation of the parabola: OB $x_1 = x_2^2/2$. Denoting the coordinates of the point C through x_1^*, x_2^*, we get

$$x_{10} - \frac{1}{2}((x_2^*)^2 - x_{20}^2) = \frac{1}{2}(x_2^*)^2.$$

From here we find

$$x_2^* = \pm\sqrt{x_{10} + \frac{1}{2}x_{20}^2}.$$

For point C in this case, you should keep the negative sign, since C belongs to the arc OB and lies below the abscissa axis.

Thus, the coordinates of the control switch point will be

$$x_1^* = \frac{1}{2}\left(x_{10} + \frac{1}{2}x_{20}^2\right);$$

$$x_2^* = \sqrt{x_{10} + \frac{1}{2}x_{20}^2},$$

(6.52)

and the equation of the phase trajectory has the form

$$x_1 = \begin{vmatrix} x_{10} - \frac{1}{2}(x_2^2 = x_{20}^2) \text{ at } 0 \leqslant t \leqslant t^*; \\ \frac{1}{2}x_2^2 \text{ at } t^* \leqslant t \leqslant T. \end{vmatrix}$$

(6.53)

To find the moment $t*$ of switching control and the time T of the object's motion from the point x_0 to the origin, we obtain the parametric equations of the phase trajectory. Integrating equations (6.39) with boundary conditions (6.40) and (6.41), we obtain

On the first link of the phase trajectory ($u = -1,\ 0 \leqslant t \leqslant t*$):

$$x_(t) = -\frac{t^2}{2} + x_{20}t + x_{10};$$

$$x_2(t) = -t + x_{20};$$

(6.54)

On the second link ($u = +1,\ t^* \leqslant t \leqslant T$):

$$x_1(t) = \frac{t^2}{2} - tT + \frac{T^2}{2};$$

$$x_2(t) = t - T.$$

(6.55)

Полагая во второй из формул (6.54) $t = t^*$, $x_2(t^*) = x_2^*$, найдём момент переключения тяги:

$$t^* = x_{20} + \sqrt{x_{10} + \frac{1}{2}x_{20}^2}.$$

(6.56)

Moreover, the second of formulas (6.55) gives

$$T = x_{20} + 2\sqrt{x_{10} + \frac{1}{2}x_{20}^2}.$$

(6.57)

To fully solve the problem, it remains to integrate the Euler-Lagrange equations and show that the signs of the function $\psi_2(t)$ on each piece of the phase trajectory agree with the chosen controls.

Integrating equations (6.44), we find

$$\psi + 1 = \text{const everywhere with} 0 \leqslant t \leqslant T;$$
$$\psi + 2(t) = -\psi_1(t) + c_2, \quad (C_2 - \text{const}).$$
(6.58)

The constants ψ_1 and C_2 included in this solution are determined from the Erdmann-Weierstrass conditions (6.37). We get

$$\psi_1 = -\frac{1}{\sqrt{x_{10} + \frac{1}{2}x_{20}^2}}; \quad C_2 = -\frac{x_{20}}{\sqrt{x_{10} + \frac{1}{2}x_{20}^2}} - 1.$$
(6.59)

Thus, the function $\psi_2(t)$, on the sign of which the choice of control (6.48) depends, has the form

$$\psi_2(t) = \frac{t - t^*}{\sqrt{x_{10} + \frac{1}{2}x_{20}^2}} \text{ everywhere with } 0 \leqslant t \leqslant T.$$
(6.60)

It is obvious from formula (6.60) that for the case under consideration, when the initial point is located above the curve AOB, in the interval $0 \leqslant t < t^*$ the function $\psi_2(t) < 0$, and therefore $u = -1$. In the interval $t^* \leqslant t < T$ $\psi_2(t) > 0$ and in accordance with the program (6.48) $u = +1$.

We have fully considered the solution to the optimal performance problem for points x_0 located above the AOB line. The reader will easily understand that if the point x_0 with coordinates (a, b) lies below the curve AOB, then the point x_0' with coordinates $(-a, -b)$, symmetric to point x_0 relative to the origin, lies above the line AOB, (Fig. 6.4b). Therefore, there is no need to rebuild the solution to the problem for the starting points located below the AOB line. We will obtain the results we are interested in from the existing solutions by changing the signs of the initial parameters x_{10} and x_{20}. So, for example, the time of movement of an object to the origin will be

$$T(x_0) = \left| \begin{array}{l} x_{20} + 2\sqrt{x_{10} + \frac{1}{2}x_{20}^2}, \text{ if point } x_0 \text{ lies above the line } AOB; \\ \\ -x_{20} + 2\sqrt{-x_{10} + \frac{1}{2}x_{20}^2}, \text{ if point } x_0 \text{ lies below the line } AOB. \end{array} \right.$$
(6.61)

Note also that if the point x_0 lies on the curve AOB, then both formulas (6.61) give the same values to $T(x_0)$.

6.8 Flying the Drone to the Maximum Range

To master the techniques of modeling controlled systems, and methods of formulating and solving variational problems of dynamics, let us consider another example. An aircraft of mass m_0, equipped with a jet thrust engine $T = c\beta$, where β is the fuel consumption, and c is a constant, starts from the airfield. It is required to find such a fuel consumption program in which the device will fly the maximum distance (Letov 1969).

6.8.1 Building a Mathematical Model

Let us formulate a number of assumptions regarding the flight conditions. The purpose of these assumptions is a certain idealization of the process under study, which will simplify the equations of motion of the object. So, suppose the following: (a) The aircraft moves in the vertical xy plane. (b) The apparatus can be considered as a material point of variable mass $m(t)$. (c) The Earth is flat and its rotation during the flight of the aircraft can be neglected. (d) There is no atmosphere, and the acceleration g of gravity is constant. (e) The aircraft engine is inertialess. (f) Steering gears controlling the direction of the engine thrust force along the pitch (to the horizon) are also inertialess.

Let us denote by u and v, respectively, the horizontal and vertical components of the aircraft speed, and by ω—the angle made by the engine thrust vector with the horizon. The basic equation of dynamics and the accepted hypotheses make it possible to write the following differential equations of the mathematical model of the object:

$$
\begin{aligned}
\dot{x} &= u; \\
\dot{u} &= \frac{c\beta}{m} \cos; \\
\dot{v} &= \frac{c\beta}{m} \sin -g; \\
\dot{m} &= -\beta.
\end{aligned}
\tag{6.62}
$$

From the physical essence of the problem, it follows that the region $N \geqslant 0$ contains restrictions:

$$
0 \leqslant \beta \leqslant \bar{\beta}; \ y \geqslant 0,
\tag{6.63}
$$

where $\bar{\beta}$ is the maximum allowable fuel consumption. Equations (6.62) are defined on a finite time interval $\Im \in [0, T]$, and the flight duration T is unknown and cannot be assigned in advance.

6.8.2 Border Conditions

In accordance with the verbal description of the content of the task, the aircraft starts from the airfield. Therefore, choosing the origin xy at the starting point, at the left end (i), we will have

$$t = t_i = 0; \ x_i = y_i = u_i = v_i = 0, \ m_i = m_0. \tag{6.64}$$

Thus, at the left end of the phase trajectory, all variables are fixed. At the right end (f), only two variables can be fixed:

$$t = t_f = T; \ y_f = 0; \ m_f = m_T, \tag{6.65}$$

where m_T is the mass of the vehicle at the moment of landing. Consequently, the difference $m_0 - m_T$ determines the fuel supply on board the aircraft. Note that for a manned vehicle, in addition to conditions (6.65), it would be natural to require the vertical velocity component to vanish ($v_T = 0$), which would physically mean a soft landing. For an unmanned projectile aircraft, this limitation is unnecessary. So, the quantities T, v_T and u_T at the right end of the trajectory remain free.

6.8.3 Minimizing Functional

In accordance with the set control goal, we want to maximize the flight range x_T. This goal is met by minimizing the functional:

$$\Delta G = -x|_i^f = -x_t. \tag{6.66}$$

The functional is defined on admissible curves of the formulated variational problem: these are piecewise-smooth functions x, y, u, v, m and piecewise-continuous controls β, ω.

A t a s k. Among the admissible curves of the variational problem (6.62)–(6.66), it is required to find the one that minimizes functional (6.66).

6.8.4 Additional Ratios

Let us compose the function H as the scalar product of a completely nonzero vector ψ and a vector X, the components of which are the right-hand sides of Eqs. (6.62):

$$H = u\psi_1 + v\psi_2 - g\psi_4 + k_\beta\beta, \tag{6.67}$$

where indicated

$$k_\beta = \frac{c}{m} k_\omega - \psi_5; \quad k_\omega = \psi_3 \cos + \psi_4 \sin. \tag{6.68}$$

Calculating the corresponding derivatives, using formulas (6.32), we find the Euler-Lagrange equations of the problem under consideration:

$$
\begin{aligned}
\dot{\psi}_1 &= 0; \\
\dot{\psi}_2 &= 0; \\
\dot{\psi}_3 &= -\psi_1; \\
\dot{\psi}_4 &= -\psi_2; \\
\dot{\psi}_5 &= \frac{c\beta}{m^2} k_\omega.
\end{aligned}
\tag{6.69}
$$

By Theorem 6.27, the equations of the problem under study have the first integral:

$$H = u\psi_1 + v\psi_2 - g\psi_4 + k_\beta \beta = C, \quad (C - \text{const}). \tag{6.70}$$

Taking into account the boundary conditions (6.64)–(6.65), the transversality condition has the form

$$[(\psi_1 - 1)\delta x - C\delta t + \psi_2 \delta y + \psi_3 \delta u + \psi_4 \delta v + \psi_5 \delta m]_i^f = 0. \tag{6.71}$$

6.8.5 Necessary Conditions for a Strong and Weak Minimum

From what was said in Sect. 6.4, it follows that the condition for a strong minimum of the functional of the analyzed problem is reduced to the choice of controls β and ω, for which the function (6.70) attains its maximum in the variables β and ω.

The condition for a weak minimum according to formulas (6.33) has the form

$$
\begin{aligned}
\frac{\partial H}{\partial \omega} &= 0; \\
\frac{\partial^2 H}{\partial \omega^2}(\delta \omega)^2 &= -\frac{c\beta}{m}(\delta \omega)^2 \leqslant 0.
\end{aligned}
\tag{6.72}
$$

This implies that $k_\omega > 0$.

6.8.6 Preparing to Integrate the Aggregate Equations of the Problem

Five equations (6.62) of the mathematical model of the object supplemented the five equations (6.69) of Euler-Lagrange. In total, we have ten differential equations for the functions $x, y, u, v, m, \psi_1, \ldots, \psi_5$.

When integrating them, it will be necessary to determine ten constants C_n $(n = 1, \ldots, 10)$. The 11th unknown will be the duration of the flight T. These 11 unknowns must be determined from the boundary conditions and the transversality condition. Let us show that the total number of conditions is 11. Indeed, at the left end of the optimal trajectory, five boundary conditions (6.64) and two conditions (6.65) are given at its right end. Taking into account that the variations of the fixed variables at the ends of the trajectory vanish, the transversality condition (6.71) gives the equality

$$[(\psi_1 - 1)\delta x - C\delta t + \psi_3 \delta u + \psi_4 \delta v]_i^f = 0. \tag{6.73}$$

By virtue of the arbitrariness of variations $\delta x, \delta t, \delta u, \delta v$ at the right end of the extremal $(t = T)$, equality (6.73) gives

$$\psi_{1T} = 1; \quad C = 0; \quad \psi_{3T} = 0; \quad \psi_{4T} = 0' \tag{6.74}$$

Using the result (6.74), from the first four equations (6.69), we find

$$\psi_1 = C_1; \quad \psi_2 = C_2; \quad \psi_3 = C_3 - C_1 t; \quad \psi_4 = C_4 - C_2 t. \tag{6.75}$$

6.8.7 Traction Control Program

Conditions (6.72) for the maximum of H in the variable ω give

$$\frac{\partial H}{\partial \omega} = \frac{c\beta}{m}[\psi_4 \cos -\psi_3] \sin \omega = 0;$$
$$\frac{\partial^2 H}{\partial \omega^2} = -\frac{c\beta}{m} k_\omega < 0. \tag{6.76}$$

From here we find the thrust direction control program:

$$\tan \omega = \frac{\psi_4}{\psi_3}; \quad \cos \omega = \frac{\psi_3}{k_\omega}; \quad \sin \omega = \frac{\psi_4}{k_\omega}. \tag{6.77}$$

The fuel consumption program is obtained from the condition of minimum H with respect to the variable β. Since H is linear in the argument β, it reaches a maximum at the boundaries of the domain of β. Therefore, the fuel consumption control program has the form

$$\beta = \begin{vmatrix} \bar{\beta} \text{ for } k_\beta > 0; \\ o \text{ for } k_\beta < 0. \end{vmatrix} \tag{6.78}$$

From relations (6.74)–(6.75), we find

$$C_1 = 1; \ \ C_3 = T; C_4 = C_2 T; \ \ \psi_3 = T - t; \ \ \psi_4 = C_2(T - t), \tag{6.79}$$

and by the first of formulas (6.77), we obtain

$$\tan \omega = C_2. \tag{6.80}$$

Therefore, despite the presence of break points t^* in the fuel consumption program, the thrust direction remains constant throughout the interval $\Im$. This conclusion follows from the Erdmann-Weierstrass conditions and the independence of the first four equations (6.69) from β.

Let us now find out the switching points of the fuel consumption control. From formulas (6.68), (6.69), and (6.79), we have

$$\dot{k}_\beta = \frac{c}{m} \dot{k}_\omega$$

or

$$\dot{k}_\beta = -\frac{c}{m}\sqrt{1 + C_2^2}. \tag{6.81}$$

From expression (6.81), it follows that for any $0 \leqslant \beta \leqslant \bar{\beta} \, k_\beta$, there is a monotone function with at most one sign reversal point. Since the derivative of this function is negative, it is decreasing and can occupy one of the three locations shown in Fig. 6.5.

Fig. 6.5 To the choice of the optimal fuel consumption programs

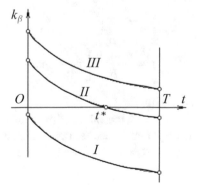

In case I, everywhere $k_\beta < 0$, $\beta \equiv 0$ no fuel is required at all and the flight will not take place. In case III over the entire interval $\Im\, k_\beta > 0$, $\beta = \bar{\beta}$ and the fuel supply $m_0 - m_T$ will not be enough. Therefore, the second case of the location of k_β corresponds to the optimal control.

The equation for the curve $k_\beta(t)$ is obtained by integrating expression (6.81):

$$
k_\beta = \left| \begin{array}{l} \dfrac{c}{\beta}\sqrt{1 + C_2^2}\, \ln \dfrac{m_0 - \bar{\beta}t}{m_0 - \bar{\beta}t^*} \text{ for } t \in [0, t^*]; \\[2ex] \dfrac{c}{m_T}\sqrt{1 + C_2^2}(t^* - t) \text{ for } t \in [t^*, T]. \end{array} \right.
\tag{6.82}
$$

Here one more, 12th in a row, constant t^* appeared, to be determined.

6.8.8 Completion of the Integration of the Differential Equations of the Problem

Integration of equations should be carried out in two stages. The first stage corresponds to the combustion of all fuel during the time $[0, t^*]$ at the maximum consumption $\bar{\beta}$. During the time t^*, the aircraft consumes the entire fuel supply, i.e.,

$$
m_0 - m_T = \bar{\beta}t^*.
\tag{6.83}
$$

At the second stage, during the time $[t^*, T]$, the apparatus continues to move with the engine off as a freely thrown body.

Using the initial conditions (6.64) at the left end of the extremal, we find the following integrals:

$$
\begin{aligned}
m &= m_0 - \bar{\beta}t, \\
u &= c \cos \ln \frac{m_0}{m_0 - \bar{\beta}t}, \\
v &= -gt + c \sin \omega \cdot \ln \frac{m_0}{m_0 - \bar{\beta}t}, \\
x &= \frac{c \cos \omega}{\bar{\beta}} \left[m_0 + (m_0 - \bar{\beta}t)\left(\ln\left(1 - \frac{\bar{\beta}t}{m_0}\right) - 1 \right) \right], \\
y &= \frac{c \sin \omega}{\bar{\beta}} \left[m_0 + (m_0 - \bar{\beta}t)\left(\ln\left(1 - \frac{\bar{\beta}t}{m_0}\right) - 1 \right) \right] - \frac{gt^2}{2}
\end{aligned}
\tag{6.84}
$$

for $t \in [0, t^*]$.

At $t = t^*$, the values of the phase coordinates determined by these formulas serve as the initial values for the continuation of the integration at the second stage of the flight at $\beta = 0$. Thus, we obtain the following integrals:

$$
\begin{aligned}
m &= m_T, \\
u &= u(t^*), \\
v &= v(t^*) - gt, \\
x &= x(t^*) + u(t^*) \cdot t, \\
y &= y(t^*) + v(t^*) \cdot t_{\frac{gt^2}{2}}
\end{aligned}
\tag{6.85}
$$

for $t \in [0, T - t^*]$.

Taking into account the boundary conditions (6.65) at the right end of the extremal, the switching condition $k_\beta = 0$, and the first integral (6.70), we obtain the following equalities:

$$
\begin{aligned}
y(t^*) + v(t^*)(T - t^*) - \frac{g(T - t^*)^2}{2} &= 0; \\
u(t^*) + C_2[v(t^*) - g(T - t^*)] &= 0.
\end{aligned}
\tag{6.86}
$$

Hence, we have

$$
\begin{aligned}
T &= t^* + \frac{1}{g}\left[v(t^*) + \sqrt{v^2(t^*) + 2gy(t^*)} \right]; \\
C_2 &= \frac{u(t^*)}{\sqrt{v^2(t^*) + 2gy(t^*)}}.
\end{aligned}
\tag{6.87}
$$

Now you can calculate the value of the functional ΔG and the flight range:

$$
x_t = x(t^*) + u(t^*) \cdot (T - t^*).
\tag{6.88}
$$

At this point, the solution to the problem can be considered complete.

Self-Test Questions

1. Give the vector and coordinate notation of differential equations in the Cauchy form.
2. What is a phase vector? Control vector?
3. What is called the domain of definition of a system of differential equations in the Cauchy form?
4. Name the three main types of boundary conditions.
5. What does "functional is set" mean?
6. What is called a comparison curve?
7. What is called the total variation of the comparison curve?
8. Give the definition of the weak and strong extremum of the functional.
9. What is total functionality variation?
10. Formulate the necessary condition for the extremum of the functional.

11. What mechanical system is called stationary?
12. Write down the kind of functional in the Mayer problem.
13. What is the essence of the problem of programming optimal trajectories?
14. Give the formulation of the maximum principle of L.S. Pontryagin.
15. Try to give a geometric interpretation of the transversality condition.
16. In what cases are the conditions for the existence of a strong minimum of a functional, required by the maximum principle, are not only necessary, but also sufficient?
17. What is the essence of the Bellman principle?
18. Does the number of Erdmann-Weierstrass conditions depend on the number of control break points?
19. What is called the first integral of a system of ordinary differential equations of the first order?
20. Can the variational Mayer problem have several first integrals?
21. Name the main stages of setting and solving the problem of dynamics of a controlled system.
22. What is the physical essence of the problem of optimal performance?
23. Outline the sequence of solving the Mayer problem.
24. Does the considered problem have a physical meaning if no restrictions are imposed on fuel consumption?
25. Where is hypothesis e) used when formulating the model?
26. Write down all the constants of integration and the conditions from which they are determined.
27. Prove that the thrust vector is perpendicular to the velocity vector at the landing point along the entire flight path.
28. To what class does the control function? What is the class of functions $u(t)$ in the problem under consideration.

References

V. Boltyanskii, *Matematicheskaya teoriya optimal'nogo upravleniya* [*Mathematical Theory of Optimal Control*] (Nauka Publ., Moscow, 1969)

R.E. Kalman, J. e Bertram, General synthesis procedure for computer control of single and multi-loop linear systems. Trans. Am. Inst. Electr. Eng. **77**(II), 602–609 (1958)

A. Letov, *Dinamika poleta i upravlenie* [*Flight Dynamics and Control*] (Nauka Publ., Moscow, 1969)

L. Pontryagin, *K 50-letiyu instituta: Tr. MIAN SSSR, 1985, tom 169, stranitsy 119–158* [Tr. MIAN USSR, 1985, vol. 169], (MIAN Publ., Moscow, 1985), chapter *Matematicheskaya teoriya optimal'nykh protsessov i differentsial'nye igry* [*Mathematical Theory of Optimal Processes and Differential Games*], pp. 119–158

N. Wiener, *Cybernetics or Control and Communication in the Animal and the Machine* (Wiley, New York, 1948)

Chapter 7
Stability of Mechanical Systems

Abstract The basic concepts of the theory of stability of the equilibrium of mechanical systems are determined. The classification of forces is given depending on the characteristics of the work they perform. The formulations of the most important stability theorems are given. For the main types of stability loss, methods for determining critical load are indicated. An example of studying the stability of a system under a non-conservative load is given.

Keywords Stability · Instability · Asymptotic stability · Conservative system · Non-conservative system · Dissipative forces · Lagrange theorem · Lyapunov-Chetaev theorem · First approximation · Critical load · Raus-Hurwitz criterion · Perturbed motion

7.1 Stability and Instability

Consider some mechanical system that is in equilibrium. Imagine that the points of this system were given infinitesimal speeds. The system will start a movement called *perturbed*. If with an unlimited increase in time, the deviations of the system points from the equilibrium position remain small for all perturbed movements, then the equilibrium of the system is called s t a b l e.

In the event that the deviations of the system from the equilibrium position decrease indefinitely, the equilibrium is called a s y m p t o t i c a l l y   s t a b l e. Otherwise, the equilibrium is called u n s t a b l e.

7.2 Work and Classification of Forces

Let some constant force be applied at an arbitrary point on the body. Let, further, the body make infinitely small movements. If a force moves with its point of application, then they say that the work of a given force is equal to the product of the magnitude of the force and the projection of its movement in the direction of the force. If the

© The Author(s), under exclusive license to Springer Nature Switzerland AG 2023   251
V. Molotnikov, A. Molotnikova, *Theoretical and Applied Mechanics*,
https://doi.org/10.1007/978-3-031-09312-8_7

displacements of the force and the points of its application do not coincide, the work should be defined as the product of the modulus of force and the corresponding projection of the displacement of the initial point of its application. With this in mind, we consider the work of a force applied to a completely rigid body subject to the following conditions:

(a) the force does not change the line of its action;
(b) the force is associated with the same point of application and when the body moves, it is always perpendicular to some line invariably connected with the body (*tracking force*).

If the body in question moves translationally (any straight line segment, rigidly connected with the moving body, remains parallel to its original position), then the work will be equal to the scalar product of force and displacement. If the body rotates around a certain center located on the line of action of the force, then the displacements of the points will be perpendicular to this line, and the work of the force is zero.

It is known (Molotnikov 2004; Molotnikov and Molotnikova 2021) that any plane motion of a solid can be represented as the sum of the translational motion with a displacement vector equal to the displacement vector of the body at an arbitrary given point, called *pole*, and of rotation around the specified pole.

We show that the work of force in the case of *a* or *b* depends on the order in which the body moves from one position to another. Let, for example, the body be a bar $ABCD$ (Fig. 7.1), subject to the action of the tracking force F. We move the body of $ABCD$ from position I to position II in two ways. In the first case, as a pole, we take some point O lying on the line of action of the force before moving the body (Fig. 7.1a).

Turn the rod counterclockwise by the angle $\pi/2$. The force F will occupy the position F', not doing any work. Then we give the body a vertical movement to position II. On this displacement, the work of the force F' is also zero. Thus, in the first method of moving the body from position I to position II, the total work is zero.

Fig. 7.1 To the unconservative nature of the tracking force

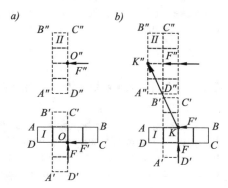

In the second way of moving, we select the point K (Fig. 7.1b) on the line of action of the force F as a pole and turn the rod around the selected pole by the angle $\pi/2$ counterclockwise. In this case, the force F will move to the position F', without doing any work. The body will occupy the position of $A'B'C'D'$ according to Fig. 7.1b. Now move the body vertically up so that the displacement vector of its pole is $\overrightarrow{KK'}$. Then move the body horizontally to the left by $\overrightarrow{K'K''}$. In the last move, the force F will do the work $A = F \cdot K'K''$ and take position II, the same as at the position a of Fig. 7.1.

Thus, it turns out that in the considered example, the work of force performed during the movement of the body depends on the method in which the movement is performed. Forces with the latter property are called n o n - c o n s e r v a t i v e. Forces whose work is determined only by the initial and final position of the system are called c o n s e r v a t i v e. By definition, the potential (potential energy) of such forces is the work done by them on the movements of the body from the final position to the initial. In this case, the initial position can be arbitrarily selected. It follows that the potential energy of the system is determined with an accuracy up to a constant.

In addition to conservative and non-conservative forces in real mechanical systems, there are always forces whose work during actual motion is negative (e.g., friction forces). Such forces are called d i s s i p a t i v e. They are a special case of non-conservative forces.

7.3 Stability with Conservative and Dissipative Forces

Lagrange Theorem[1] *A system subject to the action of only conservative and dissipative forces will be stable if in the equilibrium position the potential energy of conservative forces has a minimum.*

Proof The position of the system will be determined by the generalized coordinates $q_1, q_2, \ldots, q_n$, where n is the number of degrees of freedom of the system. We assume that in the equilibrium position $q_1 = q_2 = \ldots = q_n = 0$. The potential of the conservative forces $\Pi(q_1, q_2, \ldots, q_n)$ in the equilibrium position is also assumed to be zero, i.e., $\Pi(0, 0, \ldots, 0) = 0$.

Suppose that the potential of conservative forces is monotonously increasing, but this increase cannot be unlimited. After the potential reaches a certain value Π_1, it can decrease. There may be several such values of Π_1. The smaller of them is denoted by E and is called *potential barrier*. The range of generalized coordinates in the vicinity of the equilibrium under study, in which the potential of conservative forces does not reach the potential barrier E, is called the minimum region of

---

[1] The wording given here is borrowed from Leonov (1963) and is somewhat different from the generally accepted ones.

the function Π, or *potential well*. Under the above assumptions, the function Pi vanishes only at the point $(q_1, q_2, \ldots, q_n = 0)$, if the system does not go out of the potential well.

We derive the system from the considered equilibrium position by communicating to its elements such velocities at which the kinetic energy (T) does not reach the potential barrier E. From the kinetic energy theorem (Molotnikov 2004) in this case, it follows that the potential energy will always be less than E, i.e., the system cannot go beyond the potential barrier and will be in a potential well. With an unlimited decrease in the initial velocities of the perturbed motion, the potential energy will tend to zero, and the position of the system will differ infinitely little from the equilibrium or coincide with it, which proves the Lagrange theorem.

The Lagrange theorem defines only a sufficient, but, generally speaking, not a necessary condition for stability. In particular, in systems with dissipative forces such as Coulomb friction, the equilibrium can be stable in the absence of a minimum of potential energy.

7.4 Lyapunov-Chetaev Theorem

Suppose that there is a certain equilibrium position of a conservative system in which its potential energy is not minimal. The following statement is true.

Theorem *If the absence of a minimum of potential energy is determined by the lowest order terms that are in the expansion of the potential energy in a power series in generalized coordinates, then the equilibrium state under consideration is u n s t a b l e.*

The proof of this theorem is very complicated. In order not to bother the reader with mathematical calculations, here we give only an explanation of the theorem. Let, for example, the potential energy of the system have a decomposition

$$\Pi(q_1, q_2, \ldots, q_n) = \frac{1}{2} \sum_{i=1}^{n} a_i q_i^2 + \varepsilon(q_1, q_2, \ldots, q_n), \tag{7.1}$$

where $\varepsilon(q_1, q_2, \ldots, q_n)$ is a small value of the third or higher order, when the values of the generalized coordinates $(q_1, q_2, \ldots, q_n)$ are small.

For $a_i > 0 \, \forall \, i \in 1..n$, the function Π has a minimum at the point $(q_1..q_n = 0)$, and the equilibrium is stable. If at least one of the coefficients a_i is negative, then the equilibrium is unstable.[2] In the case when some of the coefficients a_i are positive, and the rest are equal to zero and the potential energy of the system does not turn to a minimum, the question of the stability or instability of the system remains open.

---

[2] This result belongs to Lyapunov.

However, if all the coefficients a_i are equal to zero and the expansion (7.1) actually begins with third-order small values, then the minimum potential energy is no longer possible, and the equilibrium is unstable.[3]

7.5 Instability in the First Approximation

Let the equations of perturbed motion near the considered equilibrium position $(q_1, q_2, \ldots, q_n = 0)$ contain only terms that are linear with respect to generalized coordinates, their velocities, and accelerations. Such equations are called *first approximation equations*. If they have only bounded solutions, then the considered equilibrium state is called *stable in the first approximation*.

Theorem *If only conservative forces act on the system, then its stability is guaranteed by stability as a first approximation.*

An interested reader will find a proof of the theorem in a course on the theory of oscillations (see, e.g., Il'in et al. (2003)). It is based on the use of the main (normal) coordinates, at which the kinetic and potential energy for the perturbed motion in the first approximation are represented in the form

$$T = \frac{1}{2} \sum_{i=1}^{n} m_i \dot{q}_i^2, \quad \Pi = \frac{1}{2} \sum_{i=1}^{n} a_i q_i^2. \tag{7.2}$$

The corresponding equations of perturbed motion in the case of the action of only conservative forces will have the form

$$m_i \ddot{q}_i + a_i q_i = 0 \ (m_i > 0). \tag{7.3}$$

Equations (7.3) can have bounded (periodic) solutions only for positive a_i. On the other hand, if all the coefficients are positive, then the potential energy (7.2) has a minimum. The latter means that the condition for the boundedness of the solution of equations (7.3) is a sufficient condition for stability under a conservative load.

7.6 Critical Load

Let all the forces applied at various points of the elastic system arise as a result of their monotonic growth from zero values. At sufficiently small loads, the deformation of the studied system is uniquely determined by the load. Therefore,

---

[3] The proof and generalization of the latter statement was given by Chetaev.

for simplicity, we can assume that the load arose by the growth of all forces in proportion to the same parameter (H). We call H the load parameter.

Let us denote by H_k such a value of the load parameter that for $H < H_k$, the equilibrium of the system is stable and for an infinitesimal excess of the value H_k, the equilibrium is either unstable or impossible. The load corresponding to H_k is called c r i t i c a l.

The critical load is determined either by a direct study of perturbed motion (dynamic method) or by a study of potential energy using the Lagrange and Lyapunov-Chetaev theorems (energy method) or by a static method involving the so-called Euler forces. The latter method is familiar to the reader from the course of resistance of materials.

The problem of determining the critical load is often replaced by a simpler problem of investigating first-approximation equations or conditions for the existence of extreme values of potential energy (in the presence of a force potential) or a state of indifferent equilibrium, etc. Despite the fact that this substitution is not always legal, these particular methods of determining the critical load are widely used in many computer programs for analyzing structures.

7.7   The Theorem on Stability by the First Approximation

In the case of a system with a finite (n) number of degrees of freedom, the first approximation equations of perturbed motion are a system of ordinary homogeneous differential equations with constant coefficients depending on the load parameter. Representing the generalized coordinates in this case as

$$q_i = A_i \exp \lambda t \quad (i = 1, 2, \ldots, n), \tag{7.4}$$

we obtain from the first approximation equations a known (characteristic) equation for determining the c h a r a c t e r i s t i c   i n d i c a t o r λ

$$p_0 \lambda^{2n} + p_1 \lambda^{2n-1} + \ldots + p_{2n} = 0 \quad (p_0 > 0), \tag{7.5}$$

where $p_0, \ldots, p_{2n}$ are coefficients that depend in a known way on the system and the value of the load parameter.

For these systems, Lyapunov proved the following theorems.

Theorem 1 *If the real parts of all the roots of equation (7.5) are negative, then the considered equilibrium state will be stable (asymptotically).*

Theorem 2 *If among the roots of equation (7.5) there is at least one with a positive real part, then the considered equilibrium state is unstable.*

Note The study of (non)stability by the first approximation, generally speaking, cannot give a solution to the problem of (non)stability only in the case when the real parts of the roots of the characteristic equation have zero and negative values.

7.8   The Raus-Hurwitz Criterion

In order for all the roots of equation (7.5) to have negative real parts, it is necessary and sufficient to perform the inequalities

$$\Delta_1 > 0, \ \Delta_2 > 0, \ \ldots, \ \Delta_{2n-1} > 0, \tag{7.6}$$

where

$$\Delta_1 = p_1, \ \ \Delta_2 = \begin{vmatrix} p_1 & p_0 \\ p_3 & p_2 \end{vmatrix}, \ \ \Delta_3 = \begin{vmatrix} p_1 & p_0 & 0 \\ p_3 & p_2 & p_1 \\ p_5 & p_4 & p_3 \end{vmatrix},$$

$$\Delta_{2n} = \begin{vmatrix} p_1 & p_0 & 0 & 0 & 0 & \ldots & 0 \\ p_3 & p_2 & p_1 & p_0 & 9 & 0 & \ldots & 0 \\ p_5 & p_4 & p_3 & p_2 & p_1 & p_0 & \cdots & 0 \\ \ldots & \ldots & & & \ldots & & & 0 \\ \ldots & \ldots & & & \ldots & & & 0 \\ p_{4n-1} & p_{4n-2} & & & \cdots & & & p_{2n} \end{vmatrix},$$

Moreover, p_i should be replaced with zero if $i > 2n$. For example, for a system with two degrees of freedom ($n = 2$), i.e., for the equation

$$p_0\lambda^4 + p_1\lambda^3 + p_2\lambda^2 + p_3\lambda + p_4 = 0$$

the determinant Δ_{2n} takes the form

$$\Delta_4 = \begin{vmatrix} p_1 & p_0 & 0 & 0 \\ p_3 & p_2 & p_1 & p_0 \\ 0 & p_4 & p_3 & p_2 \\ 0 & 0 & 0 & p_4 \end{vmatrix} = p_4 \begin{vmatrix} p_1 & p_0 & 0 \\ p_3 & p_2 & p_1 \\ 0 & p_4 & p_3 \end{vmatrix} = p_4 \cdot \Delta_3.$$

It follows from the above explanation that

$$\Delta_2 = p_{2n}\Delta_{2n-1}. \tag{7.7}$$

7.9 Main Types of Stability Loss

Let us consider a system that is asymptotically stable at a fairly low load. As the load parameter increases, the coefficients of the characteristic equation for p_i $(i = 0, 1, \ldots, 2n)$ and hence the values Δ_j $(j = 2, \ldots, 2n)$ usually change. We will consider these changes continuous. In this case, the loss of stability can occur either when at least one of the roots λ_i of equation (7.5), passing through zero, becomes positive or when two complex-conjugate roots with negative real parts turn into purely imaginary ones and then their real parts become positive. It is known (Chetaev 1965) that in the first case, the coefficient p_{2n} vanishes and in the second the value Δ_{2n-1}. It follows that if instability occurs with the growth of the load parameter, then under these conditions, the critical value of the parameter coincides with the smallest root of one of the two equations

$$p_{2n} = 0, \quad \Delta_{2n-1} = 0, \tag{7.8}$$

or, which is the same thing, with the smallest root of the equation $\Delta_{2n} = 0$.

Keeping in mind that for $p_{2n} = 0$, Eq. (7.5) has a zero root ($\lambda = 0$), which corresponds to arbitrary constant values of the generalized coordinates (7.4), we have in this case a state of indifferent equilibrium. If the load is critical, we will say that the elastic system loses *Euler stability*.

From the above, it follows that for $\Delta_{2n-1} = 0$, at least two characteristic numbers are purely imaginary ($\lambda_{1,2} = \pm\omega i$). Since the real parts of the remaining roots of equation (7.5) are negative, the system sets the periodic motion. This type of stability loss is called s e l f - o s c i l l a t i n g.

7.10 Methods for Determining Critical Load

The smallest load at which the total potential energy of the elastic system under the influence of forces loses a minimum in the state of equilibrium under study is called e n e r g y - c r i t i c a l. The determination of this load is the e n e r g y   m e t h o d task of studying the stability of elastic systems.

The value of the load parameter, at which infinitely close forms of equilibrium are possible, is called *critical in Euler's sense*, and the corresponding load is *Euler's*.[4]

The value of the load parameter, at an arbitrarily small excess of which there is no form of equilibrium infinitely close to the studied one, is called the *ultimate*, and the corresponding load is called the u l t i m a t e d. Determining Euler and limit loads is the task of a static method for studying the stability of elastic systems.

---

[4] This load is also sometimes called the bifurcation or branching load of equilibrium forms.

In the presence of non-conservative forces in an elastic system, the concept of energy load, generally speaking, loses its meaning. Below (p. 261) we show that the Euler load in such systems can be critical under certain conditions.

In some special cases, the Euler and energy-critical loads in conservative systems may differ. For example, when the Euler value of the load parameter $H = H_e$ does not change the sign of any of the coefficients $a_n(H)$ in the expression (7.2), the potential energy may not lose its minimum at $H = H_e$, i.e., the Euler load may be lower than the energy-critical one. The Euler load may not exist when the coefficient $a_n(H)$ changes the sign after going to infinity, and the energy-critical load exists.

It follows from the above, in particular, that the presence of the potential of forces acting on an elastic system is neither necessary nor sufficient for the existence of an Euler load or its coincidence with the critical load.

7.11 The Perturbed Motion of the Compressed Rod

Let the resistance force of the external environment be proportional to the speed. The equation of perturbed motion of a longitudinally bent rod (Fig. 7.2) is obtained by adding inertial forces to the acting forces. Assuming that the intensity (p) of the transverse load is equal to

$$p = -b\frac{\partial v}{\partial t} - m\frac{\partial^2 v}{\partial t^2}, \tag{7.9}$$

where b is a certain coefficient of friction and m is the linear mass of a unit of length of the rod, these values will be considered variables along the length of the rod.

The equation of the bending of a beam exposed to a given transverse load $p(z)$ and the compressive force H has the form (Leonov 1963)

$$D\frac{d^4v}{dz^4} + H\frac{d^2v}{dz^2} = p(z), \tag{7.10}$$

where for a rectangular-section rod with a height of h

$$D = \frac{Eh^2}{12(1 - v^2)};$$

Fig. 7.2 Stability in longitudinal and transverse bending

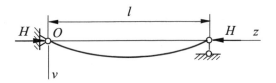

here E and v, respectively, are Young's modulus of elasticity and Poisson's ratio of the bar material.

If we substitute the expression pz into Eq. (7.10) by formula (7.9), and then replace the ordinary derivatives with partial ones, then we obtain

$$D\frac{\partial^4 v}{\partial z^4} + H\frac{\partial^2 v}{\partial z^2} + b(z)\frac{\partial v}{\partial z} + m(z)\frac{\partial^2 v}{\partial z^2} = 0. \tag{7.11}$$

We will look for a partial solution of this equation in the form

$$v = f(z)T(t). \tag{7.12}$$

In this case, Eq. (7.11) takes the form

$$(Df^{IV} + Hf^{II})T + [m(z)\ddot{T} + b(z)\dot{T}]f = 0 \tag{7.13}$$

or

$$\frac{Df^{IV}(z) + H^{II}f(z)}{m(z)f(z)} = -\frac{\ddot{T}(t) + \dfrac{b(z)}{m(z)}\dot{T}(t)}{T(t)}. \tag{7.14}$$

If $b(z)/m(z) = const$, i.e., if the coefficient of friction is proportional to the linear mass of the bar, then the right part of the latter equality will not depend on the variable z. Since here the left part does not depend on the variable t, it follows that the right and left parts of Eq. (7.14) must be equal to a constant, which we call e i g e n v a l u e λ, i.e.,

$$Df^{IV}(z) + Hf^{II}(z) - \lambda m f(z) = 0, \tag{7.15}$$

$$\ddot{T}(t) + 2c\dot{T}(t) + \lambda T(t) = 0, \tag{7.16}$$

where

$$c = \frac{1}{2}b(z)/m(z) \tag{7.17}$$

is assumed to be a constant called a r e d u c e d   c o e f f i c i e n t   o f   f r i c - t i o n.

In the case of the rod shown in Fig. 7.2, we have $v(0) = v(l) = 0$,

$$\frac{d^2 v}{dz^2}\Big|_{z=0} = \frac{d^2 v}{dz^2}\Big|_{z=l} = 0.$$

We will satisfy these conditions by assuming

$$f = \sin \frac{\pi n}{l} z \quad (n = 1, 2, \ldots, \infty),$$

and from Eq. (7.15) for $m = const$, we find $\lambda = \lambda_n$, where

$$\lambda_n = \frac{\pi^2 n^2}{ml^2} \left(n^2 D \frac{\pi^2}{l^2} - H \right) \quad (n = 1, 2, \ldots). \tag{7.18}$$

Note that

$$\lambda_{min} = \lambda_1 = \frac{\pi^2}{ml^2} \left(D \frac{\pi^2}{l^2} - H \right). \tag{7.19}$$

If all eigenvalues are positive ($\lambda_1 > 0$), then the solutions of equation (7.16) will fade over time at $c > 0$ (or they will be periodic in the absence of friction when $c = 0$). If λ_1 becomes negative, which is possible only when the compressive force exceeds the Euler force, then Eq. (7.16) for $\lambda = \lambda_1$ will have a solution that grows indefinitely over time according to the potential law. For the case under consideration, the Euler force is critical.

When $\lambda = 0$, Eq. (7.15) becomes independent of the nature of the mass distribution and passes into the static equation, from which the Euler forces are determined. Thus, the Euler load can be defined as the smallest load, at which the eigenvalue of the corresponding dynamic problem vanishes.

7.12   Stability Under Non-conservative Load (Example)

The study of equilibrium stability in the general case of non-conservative loads is a complex mathematical problem. Therefore, examples of this type of problem in publications are quite rare. This section only provides an illustration of the effect of non-conservative loads on the oscillation and stability of the simplest model of a cantilever rod compressed by spaced masses.

7.12.1   Equations of Perturbed Motion

We will consider an elastic rod of length l with two separated masses at the free end, compressed by the gravity G of the end masses and the tracking force H (Fig. 7.3a). Ignoring the mass of the rod, we consider here a system with two degrees of freedom.

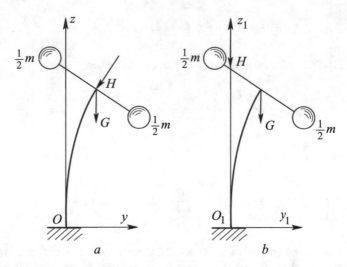

Fig. 7.3 The simplest non-conservative system

Let us make up the equations of small vibrations of the rod near the rectilinear form of equilibrium. For generalized coordinates, we take the deflection $v(l)$ and the angle of rotation $\varphi = \dfrac{dv}{dz}\Big|_{z=l}$ of the free end.

Let us assume that in the system under consideration, there are small friction forces proportional to the speeds. Under these conditions, small fluctuations of the system are described by the equations

$$m\frac{d^2v}{dt^2} = P_0 - h_1\frac{dv}{dt}, \quad I\frac{d^2\varphi}{dt^2} = M_0 - h_2\frac{d\varphi}{dt}, \tag{7.20}$$

where P_0 is the force, M_0 is the moment acting on the concentrated masses from the rod side, h_1 and h_2 are some small positive parameters, and $I = m\rho^2$ is the central moment of inertia of the end masses, the distance between which is equal to 2ρ.

The values P_0 and M_0 depend in a certain way on the movement and angle of rotation of the free end of the rod, namely (Leonov 1963, p. 66),

$$v = (P_0 - G\varphi)a_{11} + M_0 a_{12},$$
$$\varphi = (P_0 - G\varphi)a_{21} + M_0 a_{22}, \tag{7.21}$$

where

$$a_{11} = \frac{1}{k(G+H)}(kl\cos kl - \sin kl), \quad a_{12} = \frac{1}{G+H}(1 - \cos kl - kl\sin kl),$$

$$a_{21} = \frac{1}{G+H}(\cos kl - 1), \quad a_{22} = -\frac{1}{G+H}k\sin kl,$$

$$k = \sqrt{\frac{G+H}{D}}. $$

$$\tag{7.22}$$

Solving the system (7.22) with respect to P_0 and M_0, we get

$$P_0 = c_{11}v - c_{12}\varphi, \quad M_0 == -c_{21}v - c_{22}\varphi, \tag{7.23}$$

where

$$
\begin{aligned}
c_{11} &= \frac{G+H}{\Gamma}k\sin kl, \quad c_{12} = \frac{G+H}{\Gamma}(\cos kl - 1 + \eta\Gamma), \\
c_{21} &= \frac{G+H}{\Gamma}(\cos kl - 1), \quad c_{22} = \frac{G+H}{k\Gamma}(\sin kl - kl\cos kl);
\end{aligned}
\tag{7.24}
$$

$$\Gamma = 2 - 2\cos kl - kl\sin kl, \quad \frac{H}{G+H}. \tag{7.25}$$

Taking into account the results (7.23), the Eq. (7.20) of small vibrations of the system can be written as

$$
\begin{aligned}
m\frac{d^2v}{dt^2} + h_1\frac{dv}{dt} + c_{11}v + c_{12}\varphi &= 0, \\
I\frac{d^2\varphi}{dt^2} + h_2\frac{d\varphi}{dt} + c_{21}v + c_{22}\varphi &= 0.
\end{aligned}
\tag{7.26}
$$

We look for solutions to equations (7.26) in the form

$$v = A\exp\lambda t, \quad \varphi = B\exp\lambda t, \tag{7.27}$$

where A, B are constants and λ is a characteristic indicator.

Substituting the solutions (7.27) into Eq. (7.26) gives a system of homogeneous algebraic equations with respect to A and B. The condition for the existence of a nonzero solution of this system is that its determinant is equal to zero:

$$
\begin{vmatrix}
m\lambda_2 + b_1\lambda + c_{11} & c_{12} \\
c_{21} & l\lambda^2 + h_2\lambda + c_{22}
\end{vmatrix} = 0.
$$

By revealing this determinant, we obtain the characteristic equation

$$p_0\lambda^4 + p_1\lambda^3 + p_2\lambda^2 + p_3\lambda + p_4 = 0, \tag{7.28}$$

where

$$
\begin{aligned}
p_0 &= lm, \quad p_1 = h_2 m(\mu\rho^2 + 1), \\
p_2 &= m(\rho^2 c_{11} + c_{22}) + b_1 b_2, \quad p_3 = h_2(\mu c_{22} + c_{11}), \\
p_4 &= c_{11}c_{22} - c_{12}c_{21}, \quad \mu = h_1/h_2.
\end{aligned}
\tag{7.29}
$$

In this case, in the coefficient p_2, the product of small quantities b_1, b_2 will be ignored in the future (unless otherwise specified).

In conclusion of this point, we note the following. The oscillations of the system shown in Fig. 7.3b are described under similar assumptions by the following equations:

$$
\begin{aligned}
m \frac{d^2 v_1}{dt^2} + b_1 \frac{dv_1}{dt} + c_{11} v_1 + c_{21} \varphi_1 &= 0, \\
I \frac{d^2 \varphi_1}{dt^2} + b_2 \frac{d\varphi_1}{dt} + c_{12} v_1 + c_{22} \varphi &= 0,
\end{aligned}
\tag{7.30}
$$

Moreover, the coefficients c_{ij} $(i, j = 1, 2)$ are determined by formulas (7.24, 7.25). The systems of equations (7.26) and (7.30) differ from each other only in the arrangement of the coefficients c_{12} and c_{21}. These values are included in the stability conditions only as a product of $(c_{12} c_{21})$. It follows that the systems in Fig. 7.3a and b are equivalent with respect to the stability of their equilibrium.

7.12.2 Area of Valid Stability

Below, we will investigate the system of equations (7.26). To determine the critical load (see p. 256), it is enough to examine the coefficient p_4 and the value

$$
\Delta \mu \equiv p_3 (p_1 p_2 - p_0 p_3) - p_1^2 p_4.
\tag{7.31}
$$

It is easy to show that the function $\Gamma(kl)$ defined by formula (7.25) on the interval $(0, 2\pi)$ has no zeros. It follows that the functions (7.24), as well as $p_2(kl)$, $p_3(kl)$, $p_4(kl)$, and $\Delta\mu(kl)$, are continuous when $0 < kl < 2\pi$. Let us limit ourselves to the specified interval for now.

Using formulas (7.24) and (7.25), the coefficient p_4 can be converted as follows:

$$
p_4 = \frac{(G + H)^2}{\Gamma(\gamma)} [\eta + (1 - \eta) \cos \gamma] \ (\gamma = kl).
\tag{7.32}
$$

This shows that p_4 can change the sign for those values of γ that nullify either the expression in the square brackets or the denominator of the right side of the equality (7.32). The denominator $\Gamma(\gamma)$ of formula (7.32) changes the sign for $\gamma > 0$ for the first time when $\gamma_1 = 2\pi$. Equating the numerator to zero, we get

$$
\cos \gamma = -\frac{\eta}{1 - \eta}.
\tag{7.33}
$$

The equality (7.33) is possible if and only if

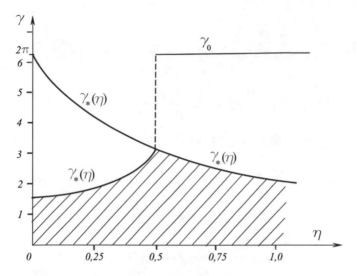

Fig. 7.4 Research on the stability of a non-conservative system

$$\left| -\frac{\eta}{1-\eta} \right| \leqslant 1,$$

i.e., for $\eta \leqslant \frac{1}{2}$. Setting the values $\eta \left(0 < \eta < \frac{1}{2} \right)$ from Eq. (7.33), we find the smallest values of $\gamma_0(\eta)$, which changes the sign of the expression in the square brackets of formula (7.32). The visualization of the calculation results is shown in Fig. 7.4.

The figure shows that the value p_4 for the values η that satisfy the condition $0 \leqslant \eta < \frac{1}{2}$ becomes negative when $\gamma > \gamma_0(\eta)$ and the inequality $\gamma_0(0) < \gamma_0(\beta) < \gamma_0 \left(\frac{1}{2} \right)$ is right; for the values $\eta \geqslant \frac{1}{2}$, the coefficient p_4 goes to negative values when γ exceeds 2π. Note that in the interval $\left(0, \frac{1}{2} \right)$ for η, the value p_4 passes from positive to negative values through zero (indifferent equilibrium), and p_4 reaches zero without changing the sign, at $\gamma = \pi$, when $\eta = \frac{1}{2}$; for $\eta \geqslant \frac{1}{2}$, the coefficient p_4 changes the sign, turning to infinity.

7.12.3   Investigation of the Value $\Delta\mu$, (Formula (7.31))

Using formulas (7.29), $\Delta\mu$ can be converted to the form

$$\Delta\mu = mh_2^2(\mu\rho^2 + 1)L(f, \rho^2, c_{ij}), \tag{7.34}$$

where

$$L(f, \rho^2, c_{ij}) = (\rho^2 c_{11} - c_{22}) f(\mu) + c_{12} c_{21}, \tag{7.35}$$

$$f(\mu) = \frac{\mu}{(\mu \rho^2 + 1)}^2. \tag{7.36}$$

Substituting in (7.35) formula (7.24), we get

$$L(f, \rho^2, c_{ij}) = \equiv \frac{(G+H)^2}{\Gamma^2} [f S^2 + (\cos \gamma - 1)(\cos \gamma - 1 - \eta \Gamma)], \tag{7.37}$$

where

$$S = \left[a^2 \gamma \sin \gamma - \frac{1}{\gamma} (\sin \gamma - \gamma \cos \gamma) \right] l, \quad \left(a = \frac{\rho}{l} \right). \tag{7.38}$$

From the identity (7.37), it can be seen that the sign of the function $L(f, \rho^2, c_{ij})$ or, what is the same, the value $\Delta \mu$ coincides with the sign of the expression

$$f S^2 + (\cos \gamma - 1)(\cos \gamma - 1 + \eta \Gamma). \tag{7.39}$$

Define the minimum positive values of $\gamma_*(\eta)$, where the value (7.39) or (7.31) goes from positive to negative values. Since in the expression (7.39), the first term is non-negative, and the second for small γ, as we will see below, is positive, the values we are interested in are $\gamma_*(\eta)$ we will have when the value $f S^2$ is the smallest. It is easy to make sure that the function (7.36) is always positive, and

$$f_{min} < f(\mu) \leqslant f_{max}, \tag{7.40}$$

where

$$f_{min} = \lim_{\mu \to 0} f(\mu) = 0; \quad f_{max} = f \Big|_{\mu = \frac{1}{\rho^2}} = \frac{1}{4 \rho^2}.$$

Since the relation (μ) of unknown friction coefficients can take any value from the range $(0, \infty)$, then the function $f(\mu)$ can be arbitrarily small. Since the value $S(\gamma)$ (7.38) is bounded at $\gamma > 0$, then in the expression (7.39), the value $f S^2$ can also be infinitesimal. Hence, $\gamma_*(\eta)$ are the smallest positive roots of the equation

$$(\cos \gamma - 1)[\cos \gamma - 1 + \eta \Gamma(\gamma)] = 0.$$

The first multiplier can only reach zero, but it does not change the sign. Equating the second multiplicand to zero and taking into account (7.25), we get

$$\cos \gamma - 1 + \eta(2 - 2\cos\gamma - \gamma\sin\gamma) = 0.$$

The latter equation is easily converted to the form

$$-2\sin\frac{\gamma}{2}\left[\sin\frac{\gamma}{2} + \eta\left(\gamma\cos\frac{\gamma}{2} - 2\sin\frac{\gamma}{2}\right)\right] = 0. \tag{7.41}$$

The smallest positive root of the equation $\sin\frac{\gamma}{2} = 0$ is $\gamma_1 = 2\pi$ and does not depend on η. Equating to zero the multiplier in (7.41) in the square brackets, we get

$$\sin\frac{\gamma}{2} + \eta\left(\gamma\cos\frac{\gamma}{2} - 2\sin\frac{\gamma}{2}\right) = 0. \tag{7.42}$$

Hence, for $\eta = \dfrac{1}{2}$, we find that $\gamma_*\left(\dfrac{1}{2}\right) = \pi$. Let $\eta \neq \dfrac{1}{2}$; then equation (7.42) can be converted:

$$\text{tg}\,\frac{\gamma}{2} = \frac{\dfrac{\gamma}{2}}{1 - \dfrac{1}{2\eta}}. \tag{7.43}$$

Setting η, one can find the corresponding value of $\gamma_0(\eta)$ from formula (7.43). These values are shown graphically in Fig. 7.4.

From the above arguments, it follows that the equilibrium of the system under consideration is always stable in the area that is covered with shading in Fig. 7.4. Let us call it the *reliable stability* area. In addition, one can say the following:

(1) when the parameter η is in the range $\left(0, \dfrac{1}{2}\right)$ or, equivalently, $H < G$ ($H \geqslant 0$, $G > 0$), the loss of stability occurs statically, and the critical load coincides with the Euler load. The critical parameter $\gamma_0(\eta)$ increases with an increase in the non-conservative component of the load, which indicates the stabilizing influence of H;

(2) for values η that satisfy the condition $\dfrac{1}{2} \leqslant \eta \leqslant 1$, in other words, $H \geqslant G$ ($H > 0$, $G \geqslant 0$), the loss of stability is self-oscillating; for $\eta > \dfrac{1}{2}$, the Euler load does not exist. The critical parameter $\gamma_*(\eta)$ decreases with an increase in the non-conservative component of the load. This means that the tracking force can have a destabilizing effect.

7.12.4 Investigation of the Effect of Friction

Now we will show that in this problem, taking into account small friction can sometimes have a significant impact on the stability of the system equilibrium.

Let us first assume that there is no friction ($b_1 = b_2 = 0$). Then instead of Eqs. (7.26), we have the system

$$m\frac{d^2v}{dt^2} + c_{11}v + c_{12}\varphi = 0,$$
$$I\frac{d^2\varphi}{dt^2} + c_{21}v + c_{22}\varphi = 0. \tag{7.44}$$

The characteristic equation of the system (7.44) has the form

$$p_0\omega^4 + p_2\omega^2 + p_4 = 0, \tag{7.45}$$

and the coefficients p_0, p_2, p_4 are still determined by formulas (7.29).

The zero solution of the system (7.44) is stable by the first approximation if and only if all the coefficients of equation (7.45) and its discriminant

$$\Delta_0 = p_2^2 - 4p_0p_4 \tag{7.46}$$

are positive because all the roots of equation (7.45) are purely imaginary.

Reasoning similarly, as in the presence of friction, it is not difficult to establish that a trivial solution of the system (7.44) can become unstable due to violation of one of the inequalities

$$p_4 > 0, \quad \Delta_0 > 0. \tag{7.47}$$

If the first one is violated, the system under study loses stability statically, passing through a state of indifferent equilibrium at $p_4 = 0$; if the second one is violated, the loss of stability is self-oscillating.

As shown above (see p. 264), in the presence of friction, the necessary and sufficient stability conditions for the zero solution of the system (7.26) had the form

$$p_4 > 0, \quad \Delta_\mu > 0. \tag{7.48}$$

Comparing the conditions (7.47) and (7.48), we see that in the case of s t a t i c stability loss, small friction does not affect the critical parameters.

Let the stability loss be self-oscillating $\left(\frac{1}{2} \leqslant \eta \leqslant 1\right)$. Compare the conditions $\Delta_0 > 0$ and $\Delta_\mu > 0$. Substituting the values of the coefficients (7.29) and (7.46), in the absence of friction, we will have the following stability condition:

$$\frac{1}{4\rho^2}(\rho^2c_{11} - c_{22})^2 + c_{12}c_{21} > 0. \tag{7.49}$$

In the presence of friction, the stability condition according to formula (7.35) has the form

$$(\rho^2 c_{11} - c_{22})^2 \cdot f(\mu) + c_{12} c_{21} > 0. \tag{7.50}$$

Since the left side of the latter inequality depends on the ratio of small friction coefficients μ, at $b_1 \to 0$ and $b_2 \to 0$, the condition (7.50) does not go, generally speaking, to the condition (7.49). In other words, the critical parameters determined with low friction are usually different from the critical parameters found without it.

The question arises: how significant can this difference be?

Obviously, the inequality (7.50) goes to (7.49) when $f = \dfrac{1}{4\rho^2}$. As already noted, this value is the maximum and is reached when

$$b_2 = b_1 \rho^2. \tag{7.51}$$

The expressions (7.37) and (7.38) show that the left side of the inequality (7.50) depends on the parameters γ, μ, a, and η. Denote the smallest positive values of the parameter γ, for which the condition (7.50) is violated, by $\gamma_\mu^0(a, \eta)$, and those of them that correspond to R_{max} will be denoted by $\gamma_*^0(a, \eta)$.

Since the function $f(\mu)$ is positive, it is obvious that at the given parameters a and η, the value of $\gamma_*^0(a, \eta)$ exceeds, as a rule, the corresponding value of $\gamma_\mu^0(a, \eta)$. Therefore, at $b_2 = b_1 \rho^2$, or at $\mu = \dfrac{1}{\rho^2}$, the stability loss usually occurs later than at other values of μ. Therefore, if the coefficients of small friction satisfy the relation (7.51), then friction will be called b e s t.

The minimum values of $\gamma_\mu^0(a, \eta)$, as previously defined, are $\gamma_*(\eta)$ and are reached at $f = 0$ when $\mu \to 0$ or $\mu \to \infty$. In this case, friction is called w o r s t.

Figure 7.5 shows graphs of the values γ_* and $\gamma_*^0(a)$ for the extreme values of the parameter $\eta = \dfrac{1}{2}$ and $\eta = 1$. Note that the values $\gamma_\mu^0\left(a, \dfrac{1}{2}\right)$ fill the area bounded by the straight line $\gamma_* = \pi$ and the curve $\gamma_*^0\left(a, \dfrac{1}{2}\right)$; the value area $\gamma_\mu^0(a, 1)$ in Fig. 7.5 is covered with hatching. It can be shown that for any value $\eta = \eta_0$ that belongs to the interval $\left(\dfrac{1}{2}, 1\right)$, curves $\gamma_\mu^0(a, \eta_0)$ fill some similar area.

From the above and Fig. 7.5, the following follows. Low friction in the system under consideration can cause destabilization; the critical load value obtained with any small difference from the best friction is usually less than the value obtained without taking into account friction. This discrepancy can be significant; for example, for $\eta = 1$ and small values of the parameter a, the worst-case friction reduces the critical load by almost 4 times.

Since the ratio μ of small friction coefficients is uncertain, the critical parameters have to be determined from the assumption that the friction is the worst. In this

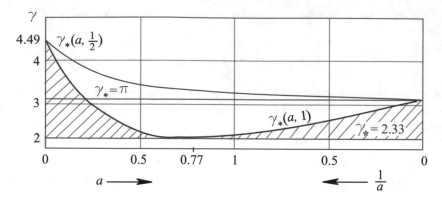

Fig. 7.5 Investigation of the effect of friction on stability

way, the area of reliable stability was determined above (see p. 264), where the equilibrium of the system is stable under arbitrary small friction.

7.12.5   The Influence of the Spacing of the End Masses

Let us show that the effect of friction on the value of the critical load depends significantly on the separation of the end masses, set by the parameter a. For this purpose, let us consider the case when $\eta = 1$. At the same time (see Fig. 7.5), the straight line $\gamma_* = 2.33$ touches the curve $\gamma_*^0(a, 1)$ at the point $a_0 = 0.77$. Obviously, for the values $a = 0.77$, the critical load is minimal (γ_0), and its value does not depend on the relation (μ) of low friction coefficients.

Let $\gamma_0 \neq 1$. It can be shown that when η is continuously reduced from 1 to $\frac{1}{2}$, the point $a_0(\eta)$ will move along the numeric axis from the value $a_0(1) = 0.77$ to $a_0\left(\frac{1}{2}\right) = \infty$, respectively. Each point $a_0(\eta_0)$ gives the minimum of the corresponding curve $\gamma_*^0(a, \eta_0)$ and has similar properties as $a_0(1) = 0.77$. At such points of the minimum, there is an equality

$$\rho^2 c_{11} = c_{22}. \tag{7.52}$$

In this case, the conditions (7.49) and (7.50) obviously coincide for any μ; the critical load is minimal, and there is no destabilization.

In the area of reliable stability (Fig. 7.4), the equilibrium of the system is stable for any values of the parameter a. Obviously, if the separation of the end masses is unfavorable ($a = a_0$), then the loss of stability can occur directly above the curve $\gamma_*(\eta)$.

The significant effect of low friction on critical parameters in self-oscillating loss of stability becomes apparent if we consider the general condition (7.31) in more detail.

In fact, let the loss of stability occur due to a violation of the condition (7.31), which can be represented as

$$p_1^2 \left[\frac{p_3}{p_1} p_2 - \left(\frac{p_3}{p_1}\right)^2 p_0 - p_4 \right] > 0. \tag{7.53}$$

All coefficients of the characteristic equation (7.28) in this case are positive.[5] Hence, the area of equilibrium stability is determined by the inequality

$$p_0 \left(\frac{p_2}{p_1}\right)^2 - p_2 \left(\frac{p_3}{p_1}\right) + p_4 < 0, \tag{7.54}$$

or

$$\frac{p_2 - \sqrt{\Delta}}{2p_0} < \frac{p_3}{p_1} < \frac{p_2 + \sqrt{\Delta}}{2p_0}, \tag{7.55}$$

where

$$\Delta = p_2^2 - 4p_0 p_4. \tag{7.56}$$

Note that the value of $\frac{p_3}{p_1}$ depends on the ratio (μ) of small friction coefficients. In the absence of friction, the characteristic equation goes into the following:

$$p_0 \lambda^4 + p_2^* \lambda^2 + p_4 = 0, \tag{7.57}$$

and also the equality is true:

$$p_2 = p_2^* + \varepsilon^2 \ (\varepsilon^2 = b_1 b_2). \tag{7.58}$$

Based on Eq. (7.57), we find the boundary of instability of the equilibrium by equating its discriminant Δ_0 to zero:

$$\Delta_0 = p_2^{*2} - 4p_0 p_4. \tag{7.59}$$

Comparing formulas (7.56) and (7.57), we get

$$\Delta = \Delta_0 + \varepsilon^2 (2p_2^* + \varepsilon^2). \tag{7.60}$$

---

[5] Here it is assumed that the product of small values of b_1 and b_2 is stored in the coefficient p_2.

Let the friction be so small that the value of ε^2 in equality (7.58) can be neglected. Then, with the accuracy of the values of the second order of smallness $p_2 = p_2^*$ and $\Delta = \Delta_0$, the following statements arise from the conditions (7.55):

1. Low friction, as a rule, causes destabilization and its value depends on the parameter μ.
2. Destabilization is absent when and only when the system parameters satisfy the condition

$$\frac{p_3}{p_1} = \frac{p_2}{2p_0}. \tag{7.61}$$

In fact, if the equality (7.61) is fulfilled, then, as can be seen from formula (7.55), stability occurs as long as the value (7.56) is positive, so that the boundaries of the stability ($\Delta = 0$) and instability ($\Delta_0 = 0$) regions in this case coincide. On the contrary, when there is no destabilization, the system parameters must obviously meet the condition (7.61).

Note that small friction can stabilize an unstable equilibrium, while taking into account the quantities of the second order of smallness (ε^2) is significant. Indeed, the presence of the second term in the right part of formula (7.60) may slightly expand the stability area (7.55). The latter becomes important in cases when the system without taking into account friction is unstable or is located on the boundary of the instability region ($\Delta_0 \leqslant 0$).

Self-Test Questions

1. What motion of a mechanical system is called disturbed?
2. What equilibrium of the system is called: (a) stable; (b) unstable; (c) asymptotically stable?
3. What property does the potential of conservative forces have in the equilibrium position?
4. What is called a potential barrier? A potential pit?
5. Formulate the Lyapunov-Chetaev theorem.
6. What are the first approximation equations?
7. Formulate Lyapunov's stability theorems in the first approximation.
8. What conclusion can be drawn about the stability of a mechanical system based on the Routh-Hurwitz criterion?

Stability of the Equilibrium State of a Conservative Mechanical System

Task (Yablonskii et al. 2000; Molotnikov and Molotnikova 2021)
A conservative mechanical system with one degree of freedom requires:

1. Determine the position of equilibrium, neglecting the masses of elastic elements;
2. Investigate the stability of the found equilibrium positions.

Mechanical system options are shown in Fig. 7.6, and the required ratios are given in Table 7.1. Select the φ angle as the generalized coordinate. In Fig. 7.6 show mechanical systems at some positive angle φ. In all variants, the wheels roll without slipping, and there is no friction in the joints. When solving the problem, consider all rods and disks to be homogeneous.

The table indicates G_1, G_2, body weights; ρ, weight per unit length of heavy thread, tape; L, thread length; c, coefficient of spring stiffness; f, spring deformation at $\varphi = 0$; γ, weight of the bar length unit; l_0, length of an undeformed spring; R, disk radius; b, l, constructional dimensions.

An Example of the Task
The conservative mechanical system (Fig. 7.8) consists of a homogeneous rod AB of length $2l$, bodies 1 and 2, a spring with a stiffness coefficient c, and a heavy thread BE of length L. The weight of the thread length unit is ρ. Let's take the angle φ as a generalized coordinate. When $\varphi = 0$, the spring is compressed by f. The weights of bodies 1 and 2 are G_1 and G_2, respectively. Ignore the sagging of the thread.
The system parameters meet the conditions:

$$L = 3l; \quad G_2 = cf - 7\rho l/4; \quad G_1 = (\rho + c)l.$$

Find the equilibrium positions of the system and examine them for stability.
D e c i s i o n. 1. *Determination of equilibrium positions of the system.* To study the equilibrium states of the system, we will compose the expression of its potential energy. In our case, the potential energy of the Π system is the sum of the potential energy of gravity Π_{gf} and the potential elastic force energy Π_{ef} of the spring:

$$\Pi = \Pi_{gf} + \Pi_{ef}. \tag{7.62}$$

For the zero level of the potential energy of gravity forces, we take the horizontal plane passing through the Ox axis (Fig. 7.8). Then, for each i-th weighty element for the chosen direction of the Oy axis, we can write

$$\Pi_{gf}(G_i) = G_i y_i,$$

where G_i is the weight of the i-th element and y_i is the ordinate of its center of gravity. For the system under consideration, we have

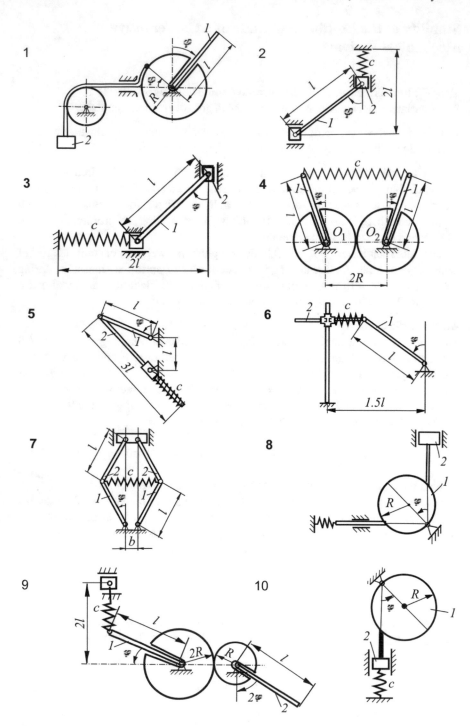

Fig. 7.6 Mechanical systems for the stability task

Table 7.1 Initial data for the stability problem

Option number (Figs. 7.6, 7.7)	Relationships between parameters	Notes
1	$lG_1 = 8RG_2$	Disregard the weight of the thread
2	$G_2 = 1.5G_1$; $l_0 = 2l$; $5cl = 12G_1$	
3	$G_2 = 1.5G_1$; $cl = 4G_1$; $l_0 = 2l$	
4	$16cl = 5G_1$; $l_0 = 2R$	
5	$G_2 = 3G_1$; $l_0 = 2l$; $cl = 8G_1$	
6	$G_2 = 2.5G_1$; $l_0 = 1.5l$; $cl = 12G_1$	
7	$G_2 = G_1$; $4cl = 5G_1$; $l_0 = b$	
8	$G_1 + 2G_2 = 2cR$; $f = 0$	
9	$G_1 = G_2 = cl$; $l_0 = l$	Friction neglected
10	$G_1 = 2G_2$; $cR = 6G_2$; $f = 2R$	

Fig. 7.7 For example, studying the stability of the system

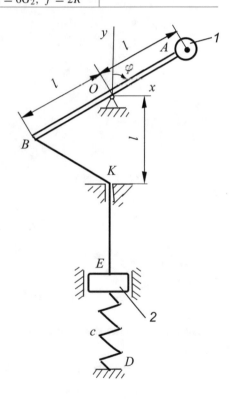

$$\Pi_{gf} = G_1 y_1 + G_2 y_2 + G_{BK} y_2 + G_{KE} y_4.$$

Here G_{BK} and G_{KE} are the weights of the sections BK and KE, respectively, and y_3, y_4 are the ordinates of their centers of gravity, and

$$G_{BK} = \rho \cdot BK; \quad G_{KE} = \rho \cdot KE = \rho(L - BK).$$

Fig. 7.8 The system of
forces in a rough diagram

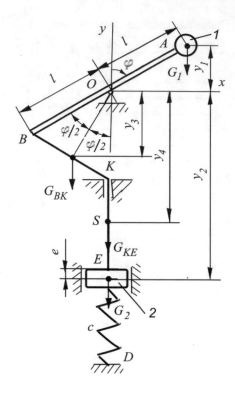

The length of BK is determined from the isosceles triangle OBK, in which $OB = OK = l$:

$$BK = 2l \left| \sin \frac{\varphi}{2} \right|.$$

Then we have

$$G_{BK} = 2\rho l \left| \sin \frac{\varphi}{2} \right|; \quad G_{KE} = \rho \left(L - 2l \left| \sin \frac{\varphi}{2} \right| \right).$$

Let's write out the values of ordinates y_i:

$$y_1 = l \cos \varphi = l \left(1 - 2 \sin^2 \frac{\varphi}{2} \right);$$
$$y_3 = -OK \cos \frac{\varphi}{2} = -l \cos^2 \frac{\varphi}{2} = -l \left(1 - \sin^2 \frac{\varphi}{2} \right);$$
$$y_4 = - \left\{ l + \frac{1}{2} \cdot KE \right\} = - \left\{ l + \frac{1}{2} \left(L - 2l \left| \sin \frac{\varphi}{2} \right| \right) \right\};$$
$$y_2 = - \{ l + KE + e \} = - \left\{ l + e + L - 2l \left| \sin \frac{\varphi}{2} \right| \right\}.$$

Substitution of the found relations into the expression for Π_{gf} gives

$$\Pi_{gf} = G_1 l \left(1 - 2\sin^2 \frac{\varphi}{2}\right) - G_2 \left[l + e + L - 2l \left|\sin\frac{\varphi}{2}\right|\right] -$$
$$-2\rho l^2 \left|\sin\frac{\varphi}{2}\right| \left(1 - \sin^2\frac{\varphi}{2}\right) - \rho \left(L - 2l \left|\sin\frac{\varphi}{2}\right|\right) \times$$
$$\times \left\{l + \frac{1}{2}\left(L - 2l \left|\sin\frac{\varphi}{2}\right|\right)\right\}.$$

The potential energy of the spring elastic force is determined by the equality

$$\Pi_{fe} = \frac{1}{2c\lambda^2},$$

where $\lambda ==$ is the spring deformation determined by the formula

$$\lambda = f + BK = f + 2l \left|\sin\frac{\varphi}{2}\right|;$$
$$f > 0, \text{ if the spring is stretched at} \varphi = 0,$$
$$f < 0, \text{ if at} \varphi = 0 \text{the spring is compressed},$$
$$\Pi_{fe} = \frac{c}{2}\left(f + 2l \left|\sin\frac{\varphi}{2}\right|\right)^2.$$

By adding the expressions for Π_{fg} and Π_{fe}, we obtain a formula for the potential energy of the system in the form

$$\Pi = \Pi\varphi = G_1 l \left(1 - \sin^2\frac{\varphi}{2}\right) - G_2 \left(l + e + L - 2l \left|\sin\frac{\varphi}{2}\right|\right) -$$
$$-2\rho l^2 \left|\sin\frac{\varphi}{2}\right| \left(1 - \sin^2\frac{\varphi}{2}\right) - \rho \left(L - 2l \left|\sin\frac{\varphi}{2}\right|\right) \times$$
$$\times \left[l + 0.5\left(L - 2l \left|\sin\frac{\varphi}{2}\right|\right)\right] + \frac{c}{2}\left(f + 2l \left|\sin\frac{\varphi}{2}\right|\right)^2.$$

The $\Pi(\varphi)$ function depends only on $\sin^2\frac{\varphi}{2}$ and $\left|\sin\frac{\varphi}{2}\right|$ and therefore is even. In addition, the functions $\sin^2\frac{\varphi}{2}$ and satisfy $\left|\sin\frac{\varphi}{2}\right|$

$$\sin^2\frac{\varphi}{2} = \sin^2\left(k\pi + \frac{\varphi}{2}\right); \quad \left|\sin\frac{\varphi}{2}\right| = \left|\sin\left(k\pi + \frac{\varphi}{2}\right)\right|$$

for any integer k. Therefore, the graph $\Pi(\varphi)$ is symmetric with respect to the verticals $\varphi = k\pi$. This allows the analysis of $\Pi(varphi)$ on the interval $[0; \pi]$ to be extended to any other interval.

Let us introduce the notation

$$x = \sin\frac{\varphi}{2}.$$

Then the formula for the potential energy of the system takes the form

$$\Pi = \Pi(x(\varphi)) = G_1 l(1 - 2x^2) - G_2(l + e + L - 2l|x|)-$$
$$-2\rho l^2 |x|(1 - x^2) - \rho(L - 2l|x|)[l + 0.5(L - 2l|x|)]+$$
$$+\frac{c}{2}(f + 2l|x|)^2. \tag{7.63}$$

The equilibrium positions are determined from the condition of equality to zero of the generalized force Q:

$$Q = -\frac{\partial \Pi}{\partial \varphi} = 0 \text{ or } \frac{\partial \Pi}{\partial \varphi} = \frac{\partial \Pi}{\partial x}\frac{\partial x}{\partial \varphi}. \tag{7.64}$$

The expression (7.63) contains a function $|x|$, the graph of which has a break at $x = 0$ ($\varphi = 0$), i.e., the derivative $\frac{\partial \Pi}{\partial x}$ at the point $x = 0$ is undefined. Therefore, we first determine the roots of the equation (7.64) for $x > 0$, and then we will investigate the position $x = 0$.

For the $0 < \varphi \leqslant \pi$ interval, the potential energy can be represented by the expression

$$\Pi(x(\varphi)) = 2l\left(\frac{a}{3}x^3 - bx^2 + dx\right) + const, \tag{7.65}$$

where $a = 3\rho l$; $b = G_1 + (\rho - c)l$; $d = G_2 + \rho L + cf$. Taking into account the relationships between the parameters specified in the condition, we can write

$$a = 3\rho l; \quad b = (\rho + c)l + (\rho - c)l = 2\rho l;$$
$$d = c|f| + \frac{1}{4}\rho l + \rho l - c|f| = \frac{5}{4}\rho l;$$

$$\Pi(x(\varphi)) = 2\rho l^2\left(x^3 - 2x^2 + 1.25x\right) + const. \tag{7.66}$$

The terms on the right-hand sides of the formulas (7.65)–(7.66) do not depend on x, so there is no need to write out their form in detail.

Performing differentiation, we get

$$\frac{\partial \Pi}{\partial x} = 2\rho l^2(3x^2 - 4x + 1, 23); \quad \frac{\partial x}{\partial \varphi} = 0.5\cos\left(\frac{\varphi}{2}\right).$$

Condition (7.64) gives two equations:

$$\frac{\partial x}{\partial \varphi} = 0 \text{ or } \cos\frac{\varphi}{2} = 0; \tag{7.67}$$

$$\frac{\partial \Pi}{\partial x} = 0. \text{ or } 3x^2 - 4x + 1.25 = 0. \tag{7.68}$$

In the interval $0 < \varphi \leqslant \pi$, Eq. (7.67) has a single root $\varphi_1 = \pi$ ($\varphi_1 = 180°$). Consider a quadratic equation (7.68). Its roots

$$x_1 = \frac{4 + \sqrt{16 - 4 \cdot 3 \cdot 1.25}}{6} = \frac{5}{6}; \ x_2 = \frac{4 - \sqrt{16 - 4 \cdot 3 \cdot 1.25}}{6} = \frac{1}{2}.$$

Since both roots do not exceed unity, the corresponding states of rest of the system are determined by the equalities:

$$\sin \frac{\varphi_2}{2} = x_1; \ \sin \frac{\varphi_3}{2} = x_2.$$

In the $0 < \varphi \leqslant \pi$ interval, we obtain two equilibrium positions:

$$\varphi_2 = 2 \arcsin x_1 = 2 \arcsin(5/6) = 112.89°,$$
$$\varphi_3 = 2 \arcsin x_2 = 2 \arcsin(1/2) = 60°,$$

Let us now investigate the position of the system at $\varphi = 0$ ($x = 0$).

It is easy to see (Fig. 7.8) that for $\varphi = 0$, all active forces are vertical and balanced by the vertical reaction of the hinged support U. Thus, in the interval $0 < \varphi \leqslant \pi$, the system has four equilibrium positions: $\varphi_1 = 180°$, $\varphi_2 = 112.89°$, $\varphi_3 = 60°$, and $\varphi_4 = 0°$.

The equilibrium positions of the system in the $\pi \leqslant \varphi \leqslant 2\pi$ interval are determined from the parity condition of the function $\Pi(\varphi)$ and are equal:

$$\varphi_5 = -\varphi_3 = -60° = 300°;$$
$$\varphi_6 = -\varphi_2 = -112.89° = 247.11°.$$

2. *Study of the stability of equilibrium positions.* Let us first investigate the stability of the equilibrium position from the interval $0 \leqslant \varphi \leqslant \pi$. For this, we define the form of the extrema of $\Pi(\varphi)$ at the equilibrium points. Let's calculate the second derivative of the function Π with respect to the argument φ:

$$\frac{\partial^2 \Pi}{\partial \varphi^2} = \frac{\partial^2 \Pi}{\partial x^2} \left(\frac{\partial x}{\partial \varphi}\right)^2 + \frac{\partial \Pi}{\partial x} \frac{\partial^2 x}{\partial \varphi^2}. \tag{7.69}$$

For $\varphi = \varphi_1 = \pi$, the first derivative of x with respect to φ becomes 0 (see (7.67); therefore from the formula (7.69), we have

$$\frac{\partial^2 \Pi}{\partial \varphi^2}\bigg|_{\varphi=\varphi_1} = \left[\frac{\partial \Pi}{\partial x}\left(\frac{\partial^2 x}{\partial \varphi^2}\right)\right]_{\varphi=\varphi_1}.$$

Taking into account that for $\varphi_1 = \pi$ $x = 1$, we obtain

$$\frac{\partial^2 \Pi}{\partial \varphi^2}\bigg|_\phi = 2\rho l^2 (3 - 4 + 1.25)(-0.25 \sin \frac{pi}{2}) = \frac{\rho l^2}{8} < 0.$$

Therefore, $\varphi = \varphi_1 = \pi$ is the maximum point of the function $\Pi(\varphi)$. For $\varphi = \varphi_2$ and $\varphi = \varphi_3$, the derivative $\dfrac{\partial \Pi}{\partial x} = 0$ (see 7.68), therefore

$$\frac{\partial^2 \Pi}{\partial \varphi^2}\bigg|_{\varphi_2,\varphi_3} = \left[\frac{\partial^2 \Pi}{\partial x^2}\left(\frac{\partial x}{\partial \varphi}\right)^2\right]_{\varphi_2,\varphi_3}. \tag{7.70}$$

From the formula (7.70) it can be seen that the sign of the second derivative at the equilibrium points φ_2 and φ_3 is determined by the sign of $\dfrac{\partial^2 \Pi}{\partial x^2}$.

It is easy to see that

$$\frac{\partial^2 \Pi}{\partial x^2} = 2\rho l^1(6x-4) = \begin{vmatrix} 1 > 0 \text{ for } \varphi = \varphi_2 \ (x = x_1 = 5/6) \text{ and} \\ -2\rho l^2 < 0 \text{ for } \varphi = \varphi_3 \ (x = x_2 = 0.5). \end{vmatrix}$$

Thus, at the point φ_2, the function $\Pi(\varphi)$ has a minimum, and at the point φ_3 it has a maximum. Figure 7.9 shows a graph of the behavior of the function $\Pi(\varphi)$. Therefore, by the Chetaev-Lyapunov theorem, the equilibrium in the position φ_2 is stable, and in the position φ_3, it is unstable. The stability of the system in other equilibrium positions is investigated similarly. The results are summarized in Table 7.2. Equilibrium positions are marked in Fig. 7.9.

Fig. 7.9 Extrema of the potential energy of the system

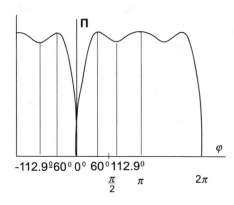

-112.9⁰ 60° 0° 60° 112.9⁰
$\dfrac{\pi}{2}$ π 2π

Table 7.2 (Un)stability of the equilibrium of the system

Equilibrium position	Stability of the equilibrium position	Equilibrium position	Stability of the equilibrium position
$\varphi_1 = 180°$	Unstable	$\varphi_4 = 0$	Steadily
$\varphi_2 = 112.89°$	Steadily	$\varphi_5 = -60°$	Unstable
$\varphi_3 = 60°$	Unstable	$\varphi_6 = -112.89°$	Steadily

References

N. Chetaev, *Ustoichivost' dvizheniya* [*Dynamic Stability*] (Nauka Publ., Moscow, 1965)

M. Il'in, K. Kolesnikov, S. Yu., *Teoriya kolebanii: ucheb. dlya vuzov* [*Theory of Vibrations: Textbook for Universities*] (MGTU im. N.Eh. Baumana Publ., Moscow, 2003)

M. Leonov, *Osnovy mekhaniki uprugogo tela* [*Fundamentals of Elastic Body Mechanics*] (Izd-vo AS Kirg. SSR., Frunze, 1963)

V. Molotnikov, *Osnovy teoreticheskoi mekhaniki* [*Fundamentals of Theoretical Mechanics*] (Rostov-on-Don, Fenix Publ., 2004)

V. Molotnikov, A. Molotnikova, *The Theory of Elasticity and Plasticity* (Springer Nature, Berlin, 2021)

A. Yablonskii, S. Noreiko, S. Vol'fson, dr., *Sbornik zadanii dlya kursovykh rabot po teoreticheskoi mekhanike: Uchebnoe posobie dlya tekhnicheskikh vuzov, 5-e izd.* [*Collection of Tasks for Term Papers on Theoretical Mechanics: A Textbook for Technical Universities. – 5th ed.*] (Integral-Press Publ., Moscow, 2000)

Part II
Elements of the Theory of Mechanisms and Machines

Chapter 8
Basic Concepts

Abstract This chapter provides a minimum of information that is necessary to study the basics of the course of mechanisms and machines. It does not include historical and literary-bibliographic reviews, as well as detailed literary references. Concepts and definitions necessary for studying the course are introduced. A generally accepted classification of machines is given. To study the structure of machines and mechanisms, the concept of a link is defined as a part (or a group of rigidly interconnected parts) participating in the transformation of motion. The development of the concept "link" is kinematic pairs and chains. A classification of kinematic pairs and kinematic chains is presented.

Keywords Power machines · Technological machines · Information machines · Detail · Link · Rack · Kinematic chain · Kinematic pair · Degree of freedom

8.1 Conditional Classification of Machines

Modern production is impossible without the use of machines and mechanisms. *A machine* is a device for converting energy and (or) movement and accumulation and processing of information.

A *mechanism* is a part of a machine in which a work process is carried out by means of mechanical movements. A mechanism is a system of interconnected bodies designed to transform one movement into another.

According to the purpose, the machines are conventionally divided into three groups.

A. *Power machines,* in which some kind of energy is converted into mechanical work or vice versa. Among these machines, there are machine motors (electric motors, thermal and nuclear engines) and machine converters (compressors, electric generators, etc.).

B. *Technological* or *working machines* designed to perform production processes to change the shape, properties, or position of objects of labor (metal-cutting machines, sewing, mining, and other machines).

C. *Information* or *control and control machines*, in which the input information is converted to control, regulate, and control technological processes.

Depending on the control method, there are manual control machines (at the built-in workplace or remotely), semi-automatic and automatic operation. A machine in which the transformation of energy, materials, or information occurs without the direct participation of a person is called an automatic machine. The set of automatic machines, interconnected by transport devices and intended for the execution of a specific technological process, forms an *automatic line*.

The main characteristics of machines are purpose and scope, control method, power and performance, efficiency, weight, dimensions, cost, etc. These characteristics are indicated in the passport of the machine.

8.2 Components of the Mechanism

All machines, mechanisms, devices, apparatus, and devices are made from parts. *A part* is a structural element made of a material of the same brand without the use of assembly operations (bolt, nut, gear, shaft, etc.). The set of parts connected by assembly operations (screwing, welding, riveting, etc.) and intended for joint work is called an *assembly unit*.

When studying the structure of mechanisms, the concept of "link" is used. All mechanisms are considered to have a common structural basis. They consist of *links*—bodies participating in the transformation of motion. *Link*—a part or a group of rigidly connected parts (solid link). Links can also be flexible (cables, belts, chains), liquid, or gaseous.

Among the links of the mechanism, it is always possible to single out a fixed link or a link, which can be conventionally taken for a fixed link. This link is called fixed member. The rack can be the bed of a metal-working machine, the cylinder block of an internal combustion engine, the hull of a ship, mechanism board, etc. Each movable part or group of parts that forms one rigid movable system of bodies is called a *movable link* of the mechanism. Movable links enter into connections with each other or with a fixed link so that there is always the possibility of movement of one link relative to the other.

The movable link, to which the movement from the engine is imparted, is called the *input* or *driving link*. Link committing the movement for which the mechanism is intended is called *output*. The rest of the moving links are *intermediate*.

8.3 Kinematic Pairs and Chains

The movable connection of two contacting links forms a kinematic pair.

The set of surfaces, lines, and points of possible contact of the links of a pair are called the *elements of a pair*. The system of links connected by kinematic

Fig. 8.1 Articulated
four-link

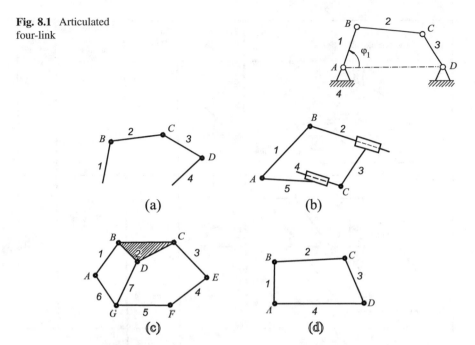

Fig. 8.2 Examples of kinematic chains

pairs forms a kinematic chain. The kinematic chain is the basis of any mechanism. Examples of various kinematic chains are shown in Fig. 8.1 and 8.2.

The positions of all links of the mechanism relative to the fixed link (rack) can be specified by generalized coordinates. The number of generalized coordinates corresponds to the number of degrees of freedom of the mechanism. Figure 8.1 shows a block diagram of a flat hinged four-link. Here $1 \ldots 4$ are links: A, B, C, D—kinematic pairs. The position of a mechanism with one degree of freedom is uniquely determined by specifying one coordinate (e.g., the angle φ_1 of crank rotation 1, which can be taken as a generalized coordinate). The link, the coordinates of which are chosen as the generalized coordinates of the mechanism, is called *the initial link* (Fig. 8.2).

Depending on the type of elements that make up the kinematic pair, *lower* and *higher* kinematic pairs are distinguished.

The lowest kinematic pairs, whose elements are surfaces, include translational, rotational, spherical, helical, and plane pairs (see Figs. 8.3, 8.4).

The highest are kinematic pairs, among the elements of which there are points and lines. In Fig. 8.3, these are the "ball on the plane" and the "cylinder on the plane." The advantage of lower pairs is their ability to transmit significant forces with less wear compared to higher pairs, while higher pairs can reproduce rather complex relative movements.

Kinematic pairs impose certain restrictions on the relative movements of the links that make up the pair. As we already know, in theoretical mechanics, such constraints are called constraints. It is known that a free body has six degrees of

Kinematic pairs schemes	Conditional image	The number of bonds S	Number of degrees of freedom W	Pair class
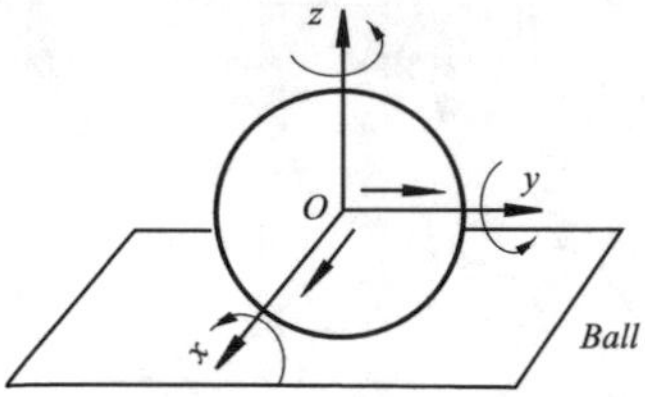 Ball on a plane	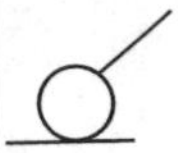	1	5	1
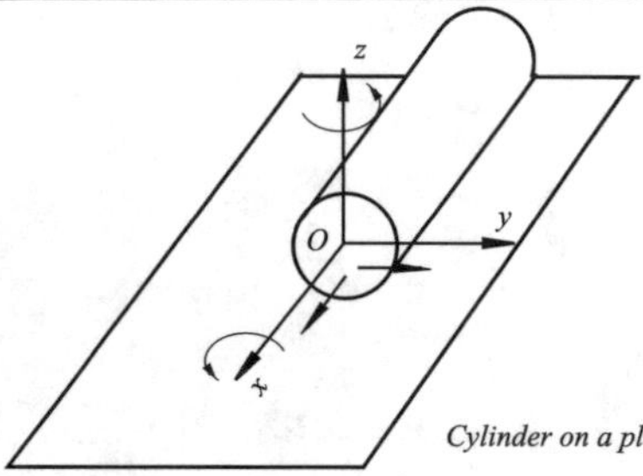 Cylinder on a plane	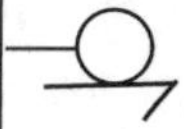	2	4	2
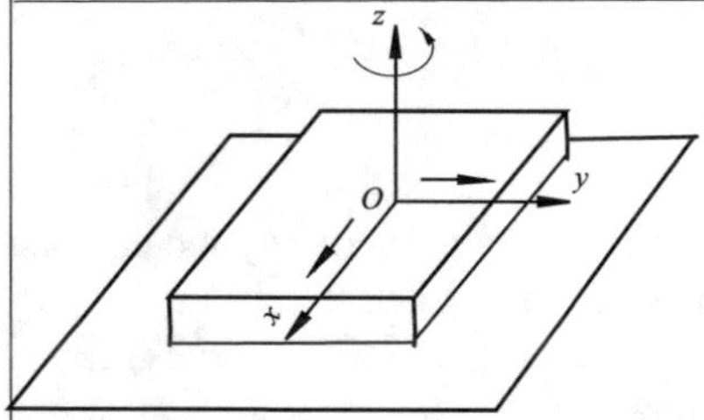 Prism on a plane	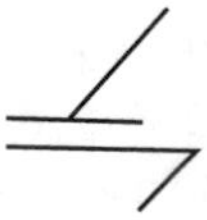	3	3	3
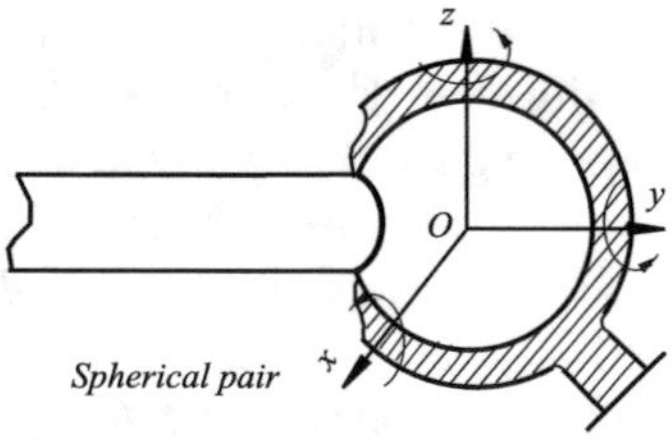 Spherical pair	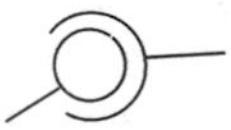	3	3	3

Fig. 8.3 Types of kinematic pairs

freedom. Therefore, the number of S connections limiting the relative movements of the links is determined by the equality

$$S = 6 - W,$$

where W is the number of degrees of freedom of the links included in the pair.

Kinematic pairs schemes	Conditional image	Number of links S	Number of degrees of freedom W	Pair class
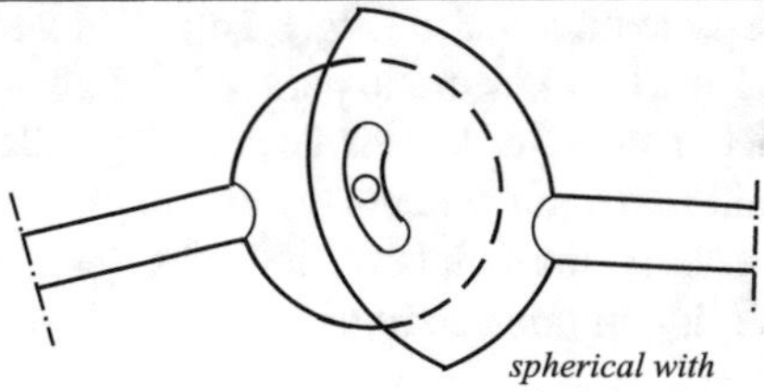 *spherical with finger*	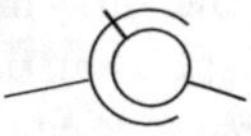	4	2	4
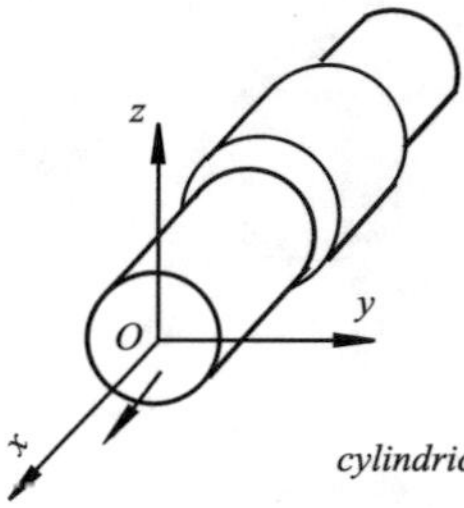 *cylindrical*		4	2	4
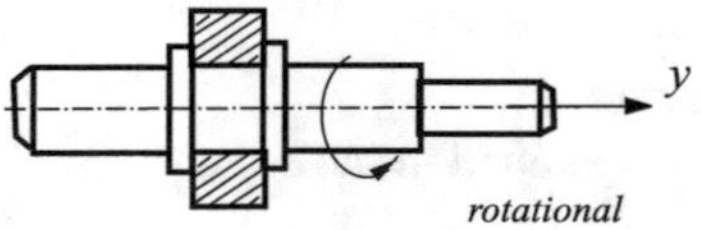 *rotational*		5	1	5
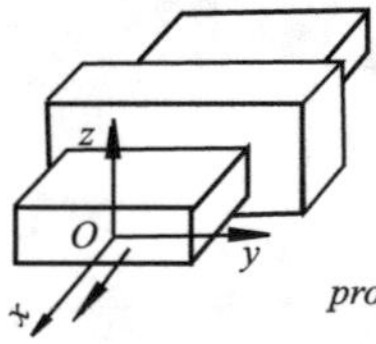 *progressive*		5	1	5
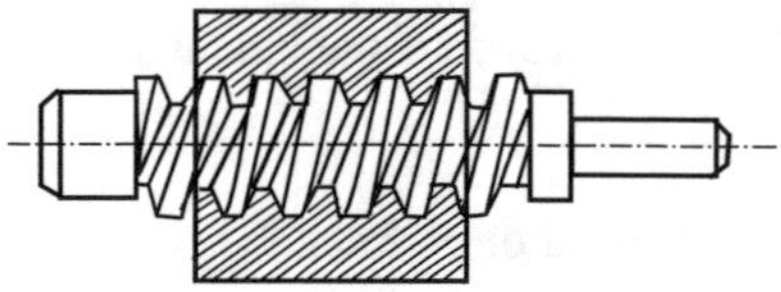 *helical*		5	1	5

Fig. 8.4 Types of kinematic pairs (continued)

According to both classical (Artobolevsky 1988) and modern researchers (Zhonghe et al. 2017), kinematic pairs are divided into *classes*. The class of a pair coincides with the number of its ties S.

Among the kinematic chains (Fig. 8.2), there are flat and spatial, simple and complex, and closed and non-closed. Simple chains include chains in which each link is included in at most two kinematic pairs (Fig. 8.2a,b,d): to complex, chains that have links included in three and more kinematic pairs (Fig. 8.2c); to closed, chains in which each link is included in at least two kinematic pairs (Fig. 8.2b–d); and to open, chains, which have links included in only one kinematic pair (Fig. 8.2a).

All movable links of a flat kinematic chain make movements parallel to one and the same fixed plane. In spatial kinematic chains, the points of the links describe spatial curves either move along flat curves lying in intersecting planes.

Self-Test Questions

1. What is a machine, a mechanism, what is their difference?
2. What is called a link of the machine (mechanism)?
3. How are machines classified by purpose?
4. Give examples of energy machines.
5. What are technological machines designed for?
6. What technological machines do you know?
7. What are information machines for?
8. How are machines classified by control method?
9. What are the characteristics of the machine specified in the e " e passport?
10. What link of the machine (mechanism) is called a rack?
11. What is the leading link?
12. What are the elements of a kinematic pair called?
13. Which link is called the initial one?
14. Define the kinematic pair.
15. What are the lowest and highest kinematic pairs?
16. How are kinematic pairs classified?
17. What is called the kinematic pair class?
18. What is called a kinematic chain?
19. Describe the movements made by the links of flat and spatial kinematic chains.
20. Simple and complex kinematic chains.
21. Closed and unclosed kinematic chains.
22. In which devices are open kinematic circuits most often found?

References

I. Artobolevsky, *Teoriya mekhanizmov i mashin* [*Theory of Mechanisms and Machines*] (Nauka Publ., Moscow, 1988)
Y. Zhonghe, L. Zhaohui, M. Smith, *Mechanisms and Machines Theory*, 1st edn. (Medtech Publisher, USA, 2017)

Chapter 9
Structural Analysis of Mechanisms

Abstract The task of the structural analysis of the mechanism is to determine the quantity moving links, the number of kinematic pairs included in its composition, as well as finding the number of degrees of mobility of each kinematic pair and the degree of mobility of the entire mechanism. The task of structural analysis also includes the sequential division of the mechanism into structural groups. This structural decomposition of the mechanism greatly simplifies its geometric, kinematic, and dynamic study, since structural groups, as a rule, are described by independent systems of corresponding equations of a small order. The connections imposed on the relative movements of the links can be determined using a structural diagram—a diagram of a mechanism, made taking into account the conventions of kinematic pairs and links. Some of the connections or mobility can be detected only on the kinematic diagram of the mechanism, where the geometric dimensions of the links are taken into account.

Keywords Degree of mobility · Kinematic chain · Chebyshev formula · Assur method · Assur group · Dyad · Group class · Mechanism class

9.1 The Degree of Mobility of the Kinematic Chain

Degree of mobility (W) kinematic of a chain is the number of degrees of freedom of the chain relative to one of the links. It can be determined by subtracting from the total number of degrees of freedom of all the moving links in the chain the number of connections imposed on the relative movement of links in kinematic pairs. Let denote by n the number of moving links of the spatial kinematic chain; P_i is the number of kinematic pairs of the i-th class $(i = 1, \ldots, 5)$. Then $6n$ is the total number of degrees of freedom of n chain links provided there are no links between them, and $i P_i$ is the total number of links imposed on the links of the mechanism by kinematic pairs o i-th class. With these designations, the degree of mobility of the kinematic chain will be

$$W = 6n - \sum_{i=1}^{5} i P_i,$$

or in expanded form

$$W = 6n - 5P_5 - 4P_4 - 3P_3 - 2P_2 - P_1. \tag{9.1}$$

The expression (9.1) is called the *Somov-Malyshev formula*. In the case of a plane kinematic chain, (9.1) implies

$$W = 3n - 2P_5 - P_4. \tag{9.2}$$

The expression (9.2) is called the *Chebyshev formula*.

Consider, for example, the crank-slide mechanism, the scheme of which is shown in Fig. 9.1a. Here, the leading link (slider 1) moves rectilinearly, and the generalized coordinate can be taken as the linear coordinate S, which determines the position of the slider. For this mechanism, as well as for the articulated four-ring (Fig. 8.1), we have $n = 3$, $P_5 = 4$, $P_4 = 0$. Then the degree of mobility of the kinematic chain according to the Chebyshev formula will be

$$W = 3 \cdot 3 - 2 \cdot 4 = 1.$$

The number of links available in the structure of the mechanism may include a certain number of q *redundant* (repeated) links that duplicate other links, not reducing the mobility of the mechanism, but turning it into a statically indeterminate system. An example of this type of device is shown in Fig. 9.1b dual parallelogram mechanism. Here, the link FG, introduced into the mechanism to increase its rigidity, ensures that the contour $OABC$ retains the shape of a parallelogram during operation. Redundant connections are also called *passivns*. The degree of mobility of a kinematic chain with redundant bonds is calculated by the formula

Let us consider, for example, a crank-slider mechanism, the diagram of which is shown in Fig. 9.1a. Here the leading link (slider 1) moves in a straight line, and the linear coordinate S, which determines the position of the slider, can be taken as

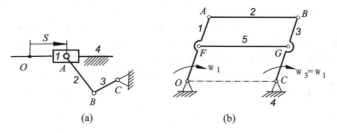

(a)   (b)

Fig. 9.1 The simplest mechanisms: (**a**) a crank-slider mechanism; (**b**) double parallelogram

the generalized coordinate. For this mechanism, as well as for the hinged four-link
link (Fig. 8.1), we have $n = 3$, $P_5 = 4$, $P_4 = 0$. Then the degree of mobility of the
kinematic chain according to the Chebyshev formula will be

$$W = 3 \cdot 3 - 2 \cdot 4 = 1.$$

The number of links available in the structure of the mechanism may include
a certain number of q *redundant* (repeated) links, which duplicate other links,
without reducing the mobility of the mechanism, but turning it into a statically
indefinable system. An example of such a device is the double parallelogram
mechanism shown in Fig. 9.1b. Here, the link FG, introduced into the mechanism
to increase its rigidity, during operation ensures that the contour $OABS$ of the
parallelogram shape is preserved.

Redundant connections are also referred to as *passive*. The degree of mobility
of the kinematic chain with redundant connections is calculated by the formula

$$W = 6n - 5P_5 - 4P_4 - 3P_3 - 2P_2 - P_1 + q.$$

9.2   Structural Classification of Mechanisms

The method of classification of plane mechanisms and the principles of their
construction were developed in 1914 by the Russian and Soviet scientist L.V. Assur
(1952). The Assur method makes it possible to determine the degree of complexity
of the mechanism (its class).) with the methods of its construction and research. In
relation to spatial mechanisms, the classification method was developed by acad. I.
I. Artobolevskij (1939).

According to Assur, the construction of the mechanism consists in the sequential
connection to the leading links and the rack of special kinematic chains, called
groups of Assur, without changing the degree of mobility of the mechanism as a
whole. The Assur group is a kinematic chain with a zero degree of mobility relative
to those links to which it is attached by its elements and which does not break up
into simpler kinematic chains with a zero degree of mobility.

Assuming that any kinematic pair can be reduced to the simplest rotational
kinematic pair, we denote, as before (see p. 291), by n the number of links in the
kinematic group and P_5 the number of rotational pairs in the group. By definition,
the Assur group has zero degree of mobility. Then from the Chebyshev formula (9.2)
for $3n - 2P_5 = 0$ for the Assur group should be

$$P_5 = \frac{3}{2}n. \tag{9.3}$$

Bearing in mind that n and P_5 are integers, the equality (9.3) corresponds to the
values $n = 2, 4, 6, \ldots$ и $P_5 = 3, 6, 9, \ldots$

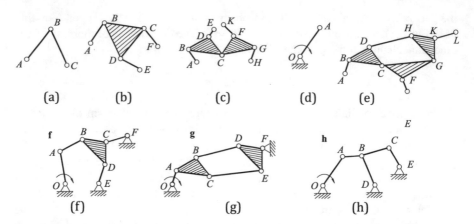

Fig. 9.2 Kinematic groups of mechanisms: (**a**) two-lead group (dyad); (**b**), (**c**), (**e**), (**f**), (**g**) central groups; (**d**) the original first order mechanism; (**h**) articulated six-link

This means that the set of kinematic Assur groups is defined by the combinations of the absolute number of links n and the number of kinematic pairs defined by the formula (9.3)

$$P_5 = \begin{vmatrix} 3 \text{ when } n = 2; \\ 6 \text{ when } n = 4; \\ 9 \text{ when } n = 6 \end{vmatrix}$$

etc.

Figure 9.2a,b shows groups with two, four, and six links, respectively. Here the links AB, CB (Fig. 9.2a); AB, ED, and FC (Fig. 9.2b); and AB, ED, KF, and HG (Fig. 9.2b) are called *leashes*. Rigid links represented by shaded triangles (Fig. 9.2b,c,d–m) are called *central*. The kinematic group depicted at the position a of the figure under consideration is called *a two-wire* or *dyad*. If one of the leashes of the dyad is turned into a rack (Fig. 9.2d), then the resulting chain is called (Assur) *is the initial mechanism of the first order.*

As part of the Assur group, internal and external kinematic pairs are distinguished. Internal pairs connect the link groups, and the outer pairs of the group join the rest of the kinematic chain. The *class of the structural group* is determined by the number of kinematic pairs included in the most complex closed loop, and the *order of the group* is determined by the number of external kinematic pairs. For example, the chain shown in Fig. 9.2(e) is an Assur group of class IV of the third order.

The dyad is a second-order group. It has two links and three rotational pairs. This combination of links and pairs is called *is the first type of class II group.* All subsequent types of class II group can be obtained by replacing individual rotational pairs with translational pairs. The second type is the one in which the translational pair is replaced by one of the extreme rotational pairs (Fig. 9.3a). The third view is

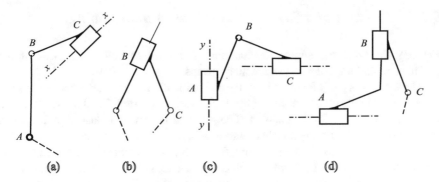

Fig. 9.3 (**a**) the second type of group of mechanisms of the second class; (**b**) a third kind of mechanisms of the second class; (**c**) the fourth type of group of mechanisms of the second class; (**d**) the fifth kind of mechanisms of the second class

shown in Fig. 9.3b. Here, the translational pair is replaced by the average rotational pair. The fourth view is shown in Fig. 9.3c. Here, the two extreme rotational pairs are replaced by two translational pairs. The fifth view is shown in Fig. 9.3d. Here, the translational pairs are replaced by the extreme and middle rotational pairs.

A four-stage structural groups with $n = 4$; $P_5 = 6$, can be trendovima class III of the third order (Fig. 9.2b) and the second four-bar order with a rolling four-sided circuit (Fig. 9.2c). A distinctive feature of the three-wire group is the presence of the internal base link, which is part of three kinematic pairs.

Each subsequent Assur group can be obtained by the so-called method of leash development, in which the leash is replaced by a group consisting of a central link with two leashes (Fig. 9.2b,c,d).

Any planar mechanism with rotational pairs can be represented as a connection of the original first-order mechanism with other kinematic groups. For example, joining the original dyad mechanism gives a four-link articulated $ABCD$ (Fig. 8.1). Attaching another dyad BCE to this four-link, we get a hinged six-link (Fig. ??h). The *class of the* mechanism is the highest class of the structural group in the mechanism.

Thus, we have defined all the necessary concepts to perform a *structural analysis of the mechanism,* which includes:

- defining the number of mechanism links, number, and class its kinematic pairs;
- determining the degree of mobility of the mechanism;
- splitting a mechanism into initial mechanisms and groups;
- defining the class and order of structural groups.

The result of the structural analysis is the determination of the class of the entire mechanism that corresponds to the highest the class of the Assur group that is part of the mechanism.

9.2.1 Example of a Structural Analysis of a Mechanism

Figure 9.4a shows the engine mechanism. You need to define the class of the mechanism and the order of the attached groups.

The 2 crank is part of a class V rotary pair with a 1rack. Next, the connecting rod 3 is included in the rotational pair of the V class with the crank 2 and in the rotational pair of the V class with the piston 4. The 4 piston is part of a class V translational pair with a cylinder rigidly attached with a rack of 1. With the connecting rod 3, the rotary pair of the class V includes a link 5, which in turn enters the rotary pair of the class V with a link 6. The 6 link is included in the class V rotational pair with the 1 rack and in the rotational pair class V with connecting rod7 compressor. Connecting rod 7 is included in the rotational pair class V with a compressor piston 8, which in turn is part of a forward pair of V class with a cylinder rigidly attached to the 1 rack. Consequently, the mechanism consists of eight rotational pairs of class V, two translational pairs of class V, and seven movable links.

Thus, we have $n = 7$ and $P_5 = 10$. Since there are no extra degrees of freedom and passive connections in the mechanism, the degree of freedom of the mechanism is determined by the Chebyshev formula

$$W = 3n - 2P_5 = 3 \cdot 7 - 2 \cdot 10 = 1,$$

i.e., the mechanism has one degree of freedom and, therefore, must have one initial link.

To determine the class of the mechanism and the order of the attached groups, we make a kinematic diagram of the mechanism under consideration (Fig. 9.4b). If the

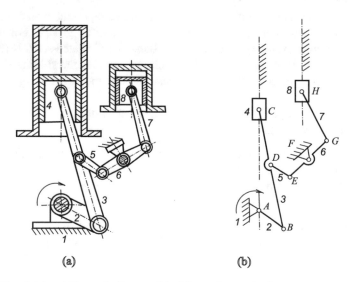

(a)   (b)

Fig. 9.4 Sketch (**a**) and kinematic diagram (**b**) of the engine mechanism

initial link is taken as the link 2 (crank AB of the engine), then the mechanism should be classified as a class II mechanism, since it is formed by three groups of class II, of which the first group is formed by links 3 and 4, included in two rotational pairs 2, 3 and 3, 4 and one translational pair 4, 1, and is a class II group of the second kind. The second group is formed by the links 5 and 6, which are part of three rotational pairs 3, 5; 5, 6; and 6, 1, and is a class II group of the first kind. Finally, the third group is formed by the links 7 and 8, included in two rotational pairs 6, 7 and 7, 8 and in one translational pair the pair 8, 1, and is a class II group of the second kind.

With the initial link 8 (compressor piston), the mechanism should be classified as a class III mechanism, since in this case the remaining links and the pairs in which they are included form two groups, of which one is a three-wire group of class III and the other is class II class. The first group is formed links 2, 3, 4, and 5 within five rotational pairs 2, 1; 2, 3; 3, 5; and 5, 6, and one sustained a couple of 4, 1, and the second group of class II of the first kind is formed by the links 6 and 7, which are included in three rotational pairs 6, 1; 6, 7; and 7, 8.

Finally, with the initial link 4 (engine piston), the mechanism must be assigned to class II mechanisms, because, just as in the first case, we get three groups of class II.

Self-Test Questions

1. What does the structural analysis of the mechanism include?
2. What is the method of development of the leash?
3. What determines the class of the structural group?
4. How to calculate the degree of mobility of the mechanism?
5. What is the physical meaning of the degree of mobility of the mechanism?
6. What are redundant links? Give examples of useful and harmful redundant links.
7. What is the principle of structural formation of mechanisms, according to L. V. Assur?
8. What is the Assur group, how are the Assur groups classified?
9. How is the class of the Assur group defined?
10. What is the degree of mobility of the Assur group?
11. What is the difference between the internal and external kinematic pair?
12. What are the criteria for classifying mechanisms in terms of their structure?

References

I. Artobolevskij, *Osnovy' edinoj klassifikacii mexanizmov* [*Fundamentals of the Unified Classification of Mechanisms*] (IMASh AN SSSR Publ., 1939)

L. Assur, *Issledovanie ploskix sterzhnevy'x mexanizmov s nizshimi parami* [*Investigation of Flat Rod Mechanisms with Lower Pairs*] (AN SSSR Publ., 1952)

Chapter 10
Kinematic Analysis of Mechanisms

Abstract Kinematic analysis is the study of the movement of the links of a mechanism without taking into account the forces that cause this movement. In the kinematic analysis, the following tasks are solved: determination of the positions of the links that they occupy during the operation of the mechanism, as well as the construction of trajectories of movement of individual points of the mechanism; determination of the speeds of characteristic points of the mechanism and determination of the angular speeds of its links; and determination of the acceleration of individual points of the mechanism and the angular acceleration of its links. When solving the problems of kinematic analysis, the following methods are used: graphic, graphic-analytical (method of plans of speeds and accelerations), and analytical.

Keywords Mechanism plan · Scale of the plan · Kinematic diagram · Speed diagram · Acceleration diagram · Speed plan · Acceleration plan · Analytical method · MathCad charts

10.1 Objectives and Methods of Kinematic Analysis

The study of the movement of mechanisms and machines is performed with the aim of (Artobolevsky 1988):

- determining the values of the forces acting on the links;
- establishing the correspondence of the trajectories of the working bodies of the machine to the technological processes for the execution of which it is intended;
- determining the geometric space required to accommodate the mechanism.

In the first stage of machine design, the geometries and materials of the links are usually unknown. Therefore, the study of the movement of the mechanism taking into account the acting forces is difficult. Due to these circumstances, one resorts to an approximate determination of the motion parameters—displacements, speeds, and accelerations of the links and their points—without taking into account the acting forces. Such studies are carried out by methods of kinematics of mechanisms.

To perform the kinematic analysis of the mechanism, its kinematic scheme, the dimensions of the links, as well as the dependence of the motion of the input links on time or other motion parameters must be specified. Kinematic studies of mechanisms are carried out by *analytical, graphical* , or *experimental* methods.

The analytical method of kinematic analysis is based on the use of the laws of kinematics of a point and a rigid body. In the conditions of modern availability of computer facilities and software products, this method is the most preferable. It provides high accuracy in determining the motion parameters and the ability to visualize the results both at any fixed point in time and direct display of the motion.

Graphical methods are based on direct geometric construction of trajectories of the most characteristic points of the plane mechanism. In this case, the drawing shows the actual configuration of the mechanism at different times and the actual shape of the trajectory of the point under study. Despite the seeming utility, graphical methods often lead to results of interest to the researcher much faster than other methods.

Experimental methods are based on instrumental measurements of various motion parameters on prototypes of mechanisms or on their models. A significant disadvantage of the methods is their high cost.

10.2   Building Plans for the Provisions of the Mechanism

The plan of positions or simply *the plan of the mechanism* is a graphical representation of the mutual arrangement of the links of the mechanism corresponding to a given moment in time. Position plans are built on a certain scale by the serif method in accordance with the mechanism structure formula. In this case, the linear dimensions of all links must be specified (Fig. 10.1).

In graphic constructions, the lengths of the links are shown in the drawing in a certain scale, characterized by the scale factor μ_l. For example, if on the plan a 10 mm segment represents a link 20 cm long, then $\mu_l = 0.2\text{m}/10\text{mm} = 0.02\,\text{m/mm}$. The construction uses the *serif method.* So, for example, the position of the point B of the link 2 for each position of the leading link 1 is defined as the point of

Fig. 10.1 Planning a mechanism using the serif method

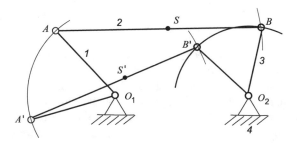

intersection of circular arcs centered at points A and O_2 and radii AB and O_2B, respectively.

After constructing several combined plans of the mechanism positions (Fig. 10.1), if necessary, you can graphically determine the trajectories of characteristic points of links that have complex movement, for example, the center of gravity S of the connecting rod AB (Fig. 10.1).

An example of constructing a plan for the positions of the planing machine mechanism is shown in Fig. 10.2. Here, the leading link is the crank OA, which performs a rotational movement. The circle represents the trajectory of the movable end of the crank. The *extreme positions* of the slider performing translational movement together with the rigidly connected backstage represent the points C_0 and C_4. To make the drawing more visible, the plan is constructed for six ($k = 6$) positions of the crank (in practice, $k = 12$ is usually accepted). The generalized coordinate S_C is calculated from the left extreme position C_0. The initial position of the crank is assumed to be OA_0, at which the slider C takes the leftmost position.

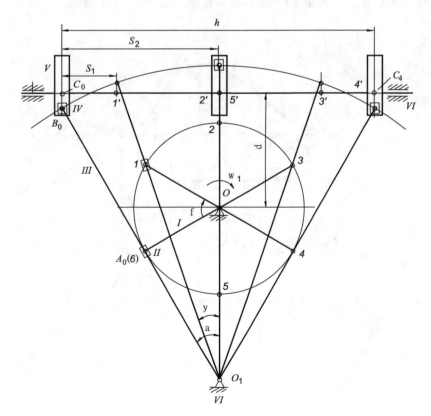

Fig. 10.2 Plan of positions of the planing machine mechanism

10.3   Kinematic Diagrams

Kinematic diagram is a graphical representation of changes in one of the kinematic parameters (displacement, speed, acceleration) of a point or link of the investigated mechanism as a function of time or displacement of the leading link.

10.3.1   Displacement Diagrams

Let it be required to construct a diagram of the displacements of the point C of the planer slider from its extreme left position C_0, assuming that the crank OA rotates uniformly at a frequency of n_1 rpm. Having a plan of the mechanism positions (Fig. 10.2) on a certain scale μ_l, we do so. We build the coordinate axes $S_C \sim t$ (or $S_C \sim \varphi$, $\varphi = \dfrac{\pi n_1}{30} t$) (see Fig. 10.3a), and on the abscissa, we plot the segment l, mm, representing the time T of one complete revolution of the crank on the scale μ_t.

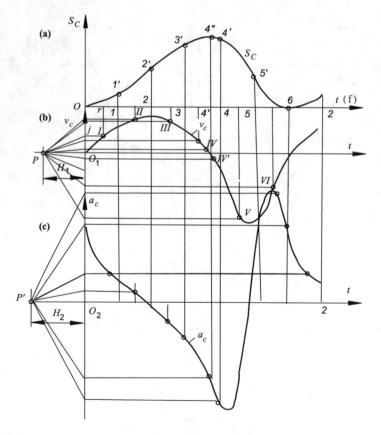

Fig. 10.3 Kinematic diagrams of the point C of the working link

Bearing in mind that

$$T = \frac{60}{n_1} = \mu_l t, \text{ c,}$$

we find

$$\mu_t = \frac{60}{n_1 l}, \text{ c/мм.} \tag{10.1}$$

Divide the segment l into k equal parts (in Fig. 10.3 $k = 6$) and at the corresponding points 1, 2, ... put in the direction of the ordinate segments S_1 S_2, ..., taken from the plan of the mechanism positions (Fig. 10.2). Connecting the ends of these segments with a curved curve, we obtain a diagram of the displacements of the point $S_C(t)$. In the considered case, the scale μ_S of the displacement diagram of the point C will coincide with the scale μ_l of the position plan. If, for some reason, the lengths of the segments S_1, S_2, ... have to be changed m times, then the scale along the ordinate axis of the displacement diagram will be

$$\mu_S = m\mu_l. \tag{10.2}$$

10.3.2 Plotting Velocity and Acceleration Diagrams

The velocity diagram $V_C(t)$ of the point C is constructed in the following sequence:

1. Under the displacement diagram $S_C(t)$, we build the coordinate system O_1, t, v_c (Fig. 10.3b), and on the continuation of the $O_1 t$ axis to the left, we put off the segment $O_1 P = H_1$ mm.
2. From the point P parallel to the chord $O1'$ draw the ray to the intersection with the axis $O_1 v_c$ at the point j (Fig. 10.3a). Draw a horizontal line from the point j and a vertical (dashed) line from the point r in the middle of $O1$. The ordinate of the point I of the intersection of these straight lines is proportional to the average velocity C on the arc $O1'$.
3. Carrying out the constructions of item 2 for chords $1'2'$, $2'3'$, ..., we construct points II, III, ... Connecting points I, II, III, ... a smooth curve, we get a diagram of the velocity $v_c \sim t$. The scale μ_v of this diagram will be

$$\mu_v = \frac{\mu_S}{\mu_t H_1}, \tag{10.3}$$

where the parameters μ_S and μ_t are defined by the formulas (10.2) and (10.1), respectively, and the value of H_1 is equal to the length of the segment $O_1 P$ in millimeters.

Having the diagram $v_c \sim t$, by differentiating it we construct the diagram of accelerations $a_c \sim t$ of the point C (Fig. 10.3c). Scale this diagram, by analogy with the formula (10.3), will be

$$\mu_a = \frac{\mu_v}{\mu_t H_2},\tag{10.4}$$

with $H_2 = P'O_2$.

A more detailed description of the method of graphical differentiation in the construction of kinematic diagrams, the reader will find, for example, in § 22 of the textbook I. I. Artobolevsky (Artobolevsky 1988) or in § 4 chap. VI of the manual A. S. Korenyako (and others) (Korenyako 1970). More than one generation of engineers in many countries of the world studied these wonderful books.

10.4 Speed and Acceleration Plans

The method of plans is one of the most illustrative. Linear velocities and accelerations of individual points and angular velocities and accelerations of links are subject to determination. In this case, the vector equations for the velocities and accelerations of the points of the links performing complex motion are preliminarily compiled.

10.4.1 Building a Speed Plan

We will construct a plan of speeds using the example of a planer mechanism (Fig. 10.2) with the crank in position 1, when it has turned from the left extreme position by an angle $\pi/3$ (Molotnikov 2017). The mechanism in this position is shown in Fig. 10.4a.

For definiteness, the following values of the initial data are taken: length of the link 1 (crank) $l_1 = 0.12$ m; its angular velocity $\omega_1 = 6$ c^{-1}; link 1 that rotates uniformly (angular acceleration $\varepsilon_1 = 0$); length of the swing stage (link 3) $l_3 = 0.44$ m; center distance $OO_1 = 0.24$ m; and $\delta = 0.16$ m.

Linear velocity v_{a_1} of point a of link 1 is calculated by the formula

$$v_{a_1} = \omega_1 \cdot l_1 = 6 \cdot 0, 12 = 0, 72 \text{ m} \cdot \text{c}^{-1}.$$

Choose an arbitrary point P (Fig. 10.4b), called the pole of the plan of velocities, and construct the vector $\boldsymbol{v}_{a_1}$ starting at point P directed perpendicular to Oa in the direction of the crank rotation. The end of the vector is marked with the point a_1. The speed of the point a_2, which belongs to the link 2, coincides with v_{a_1}; therefore, in the plan of speeds, a_2 is added after a_1, separated by commas.

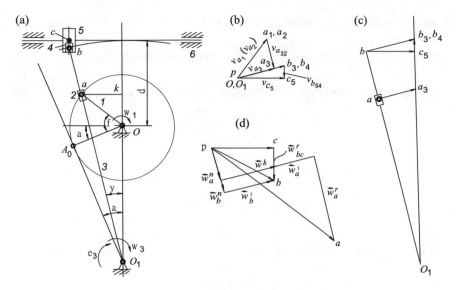

Fig. 10.4 Building plans for speeds and accelerations

The speed of the point a_3, belonging to the third link, can be found by graphically solving the vector equation

$$\boldsymbol{v}_{a_3} = \boldsymbol{v}_{a_1} + \boldsymbol{v}_{a_{32}}, \tag{10.5}$$

where $\boldsymbol{v}_{a_3} \perp O_1 a$ and $\boldsymbol{v}_{a_{32}} \parallel O_1 a$ (see Fig. 10.4b). Performing the construction, we get

$$v_{a_3} = 0,544 \text{ m/s}; \quad v_{a_{32}} = 0,472 \text{ m/s}.$$

To determine the velocities of the points b and c, you can use the graphical construction shown in Fig. 10.4c. After the construction indicated here, the upper three vectors are transferred to the plan of velocities (Fig. 10.4b).

An analytical method can be used instead of a graphical solution. Indeed, from the triangle $O_1 ak$ ($ak \perp OO_1$) find

$$\text{tg } \psi_1 = \frac{l_1 \sin \dfrac{\pi}{3}}{OO_1 + l_1 \cos \dfrac{\pi}{3}}.$$

Making calculations, we find

$$\psi_1 = 0,334.$$

Bearing in mind that, according to the initial data, $\alpha = \pi/6$, $\varphi_1 = \pi/3$, calculate

$$O_1 a = l_1 \cos\left(\frac{\pi}{3} - \psi_1\right) + OO_1 \cos\psi_1 = 0,317 \text{ m.}$$

From the similarity of the triangles $O_1 O a_3$ and $O_1 b b_3$ (see Fig. 10.4c), we have

$$\frac{v_{b_3}}{v_{a_3}} = \frac{l_3}{O_1 a}.$$

Calculation gives

$$v_{b_3} = v_{b_4} = 0,755 \text{ м/с.}$$

Projecting now the vector equality

$$\mathbf{v}_{c_5} = \mathbf{v}_{b_4} + \mathbf{v}_{b_{54}}$$

on the direction of movement of the slider 5 and the rocker stone 4, we find

$$v_{c_5} = v_{b_4} \cos\psi_1 = 0,732 \text{ м/с}; \quad v_{b_{54}} = v_{b_4} \sin\psi_1 = 0,254 \text{ м/с.}$$

Using the results of the last calculations and the scale of the velocity plan μ_v, we finish building the plan.

In conclusion, let us calculate the angular velocity of the link 3:

$$\omega_3 = \frac{v_{b_3}}{l_3} = \frac{0,755}{0,44} = 1,716 \text{ c}^{-1}.$$

10.4.2   Building an Acceleration Plan

Acceleration of point a of link 1 under condition $\varepsilon_1 = 0$ will be

$$w_{a_1} = \omega_1^2 l_1 = 6^2 \cdot 0,12 = 4,32 \text{ м} \cdot \text{c}^{-2}.$$

The vector of this acceleration is directed parallel to aO from a to O. Given a certain scale μ_w, we postpone this vector from an arbitrary pole π (Fig. 10.4d) of the acceleration plan. We got the point a on the plan.

The acceleration of the point a_3 of the link 3 is determined from the vector equations of the complex motion of the point a_3:

$$w_{a_2} = w_{a_3}^n + w_{a_3}^\tau + w_{a_{23}}^r + w_{a_{23}}^k,$$

$$w_{a_3} = w_{a_2}.$$

(10.6)

The system (10.6) will be solved graphically. To do this, first we calculate the normal acceleration w_a^n of the point a_3 of the link 3 and the Coriolis acceleration w^k of the point a_2 of the slider 2, for which the rotation of the stage 3 is a transferable motion, and the sliding along the stage is relative movement. We have

$$w_a^n = \omega_3^2 \cdot O_1 a = 1.7162 \cdot 0.317 = 0.933 \text{ m/s}^2,$$

$$w^k = 2\omega_3 v_{a_{32}} = 2 \cdot 1.716 \cdot 0.472 = 1.62 \text{ m/s}^2.$$

Now you can start solving the system (10.6). From the pole π (Fig. 10.4d) on the scale μ_w, we construct the vector $w_{a_3}^n$, by the end of which add the vector $w_{a_{23}}^k$. Further, from the point a of the plan, draw a straight line parallel to $O_1 a$. The point of intersection of this straight line with the line of the vector $w_{a_{23}}^k$ defines the required quantities, namely, the vector of tangential acceleration $w_{a_3}^\tau$ points a of link 3 and vector $w_{a_{23}}^r$ of relative acceleration of point a of link 2 when it moves relative to the point of the same name of the link 3.

It remains to construct the accelerations of the points b and c. To do this, calculate the normal w_b^n acceleration of the point b of the link 3:

$$w_b^n = \omega_3^2 \cdot l_3 = 1.716^2 \cdot 0.44 = 1.296 \text{ m} \cdot \text{s}^{-2}.$$

Next, we determine the angular acceleration of the link 3 by the formula

$$\varepsilon_3 = w_a^\tau / O_1 a,$$

and w_a^τ is determined by multiplying the length of the segment representing it (Fig. 10.4d) by the scale μ_w. In our case $w_a^\tau = 1.056 \text{ m} \cdot \text{c}^{-2}$ is obtained. Then

$$\varepsilon_3 = 1.056/0.317 = 3.31 \text{ c}^{-2} \text{ и } w_b^\tau = \varepsilon_3 \cdot l_3 = 3.31 \cdot 0.44 = 1.466 \text{ м} \cdot \text{c}^{-2}.$$

From the known components w_b^n and w_b^τ in Fig. 10.4c, the acceleration vector of the point b of the link 3 is constructed.

We find the acceleration of the point c by graphically solving the vector equation:

$$w_c = w_b + w_{bc}^r.$$

The solution to this equation is given by the point c lying at the intersection of $bc \perp ak$ with the horizontal ray outgoing from the pole π. Then the segment bc represents on the scale mu_w the vector of relative acceleration w_{bc}^r, and the segment πc is the vector of absolute acceleration of any point of the link 5.

10.5   Analytical Study of the Kinematics of Mechanisms

Graphical methods do not always have the accuracy that is necessary in some specific problems of mechanism analysis. In these cases, it is preferable to use analytical methods, with the help of which the study of the kinematics of mechanisms can be done with any degree of accuracy. In addition, analytical dependencies allow us to identify the relationship of the kinematic parameters of the mechanism with its metric parameters, i.e., the size of the links. The role of analytical methods of kinematic analysis of mechanisms has especially increased in recent years, due to the fact that, having analytical expressions linking the main kinematic and structural parameters of the mechanism, it is possible to obtain all the necessary results with the help of a computer.

Analytical methods for the kinematic study of mechanisms are based on the application of analytical geometry, tensor-matrix operations, the theory of functions of a complex variable, etc. However, in many cases, the analytical solution of problems of kinematics of mechanisms is possible by elementary mathematical means.

Let us consider the analytical method using the example of the study of the six-link planer mechanism. Let it be required to construct kinematic diagrams of some point c of the link 5. As we did in the previous paragraph, we find from the triangle $O_1 ak$ (Fig. 10.4a) the value of the angle ψ as a function of the generalized coordinate φ, which determines the position of the input link 1. In Fig. 10.4a the angle $\varphi \in (0; 2\pi)$ and is counted from the initial position $O A_0$ of the link 1 in the direction of its rotation. We have

$$\psi(\varphi) = \operatorname{arctg} \left(\frac{l_1 \cos(\varphi - \alpha)}{l_6 + l_1 \sin(\varphi - \alpha)} \right).$$

Then the position of the point c of the output link 5 is determined by the formula

$$S(\varphi) = \frac{h}{2} - l_3 \sin[\psi(\varphi)], \tag{10.7}$$

where h is the range of displacements of the 5 slider (Fig. 10.2).

Dependency (10.7) is called *displacement function* or *transfer function of order zero*. *First order transfer function* $S'(\varphi)$ or *analog speed* is the first derivative of the function of displacement along the coordinate φ input link. Differentiating the function (10.7), we get

$$S'(\varphi) = \frac{d S(\varphi)}{d\varphi} = \frac{l_1 l_3 [l_1 + l_6 \sin(\varphi - \alpha)]}{L} \cos[\psi(\varphi)], \tag{10.8}$$

where indicated

$$L = l_1^2 + l_6^2 + 2 l_1 l_6 \sin(\varphi - \alpha). \tag{10.9}$$

Fig. 10.5 Kinematic
MathCad Charts

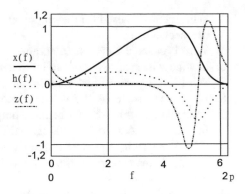

Differentiating the formula (10.8), we obtain a *transfer function of the second order* or *analogue of acceleration* of the 5 link

$$S''(\varphi) = \frac{dS'(\varphi)}{d\varphi} = l_1^2 l_3 \frac{[l_1 + l_6 \sin(\varphi - \alpha)]^2}{L^2} \sin[\psi(\varphi)] +$$

$$+ l_1 l_3 l_6 (l_1 - l_6) \frac{[l_1 - l_6 + 2l_6 \sin(\varphi - \alpha)] \cos(\varphi - \alpha)}{L^2} \cos[\psi(\varphi)], \qquad (10.10)$$

where L is defined by the formula (10.9).

In Fig. 10.5, using the MathCad application, normalized kinematic diagrams are built for the point c of the output link of the planer mechanism. In this case, the same geometric dimensions and angular velocity of the leading link are used as in the previous paragraph (see p. 304). Rationing was carried out for the purpose of rational use of the drawing field. It is supposed:

$$\xi(\varphi) = \frac{S(\varphi)}{h}, \quad \upsilon(\varphi) = \frac{S'(\varphi)}{\omega_1 l_1}, \quad \eta(\varphi) = \frac{S''(\varphi)}{\omega_1^2 l_1 / 5}. \qquad (10.11)$$

Note that the normalized transfer functions (10.11) are dimensionless quantities.

In cases where the mechanism has several input links, the transfer functions will depend on many variables, for which, as a rule, generalized coordinates are taken. In this situation, the transfer functions of the first and subsequent orders for individual generalized coordinates will contain partial derivatives.

Questions for Self-Test

1. Name the goals of the kinematic analysis of the mechanism.
2. What do you need to have in order to perform a kinematic analysis of the mechanism?
3. List the methods of kinematic analysis of mechanisms.

4. What method is used for the graphical construction of plans for the positions of the mechanism?
5. What is called a kinematic diagram?
6. List the sequence of actions to build a speed plan.
7. What is the speed and acceleration of the end of the crank?
8. How is the scale of the velocity and acceleration plans determined in the graphical construction of kinematic diagrams?
9. In what case does the Coriolis acceleration of the material point occur?
10. How to determine the direction of Coriolis acceleration?
11. Formulate the essence of the analytical method for studying the kinematics of the mechanism.
12. What is the advantage of the analytical method of kinematic investigation of mechanisms?
13. What is called the analog of speed (acceleration)?
14. What is the zero-order transfer function?
15. Why is the normalization of kinematic diagrams performed?

References

I. Artobolevsky, *Teoriya mekhanizmov i mashin* [*Theory of Mechanisms and Machines*] (Nauka Publ., Moscow, 1988)

A. Korenyako, *Kursovoe proektirovanie po teorii mexanizmov i mashin* [*Course Presentation on the Theory of Mechanisms and Machines*] (High School Publ., Kiev, 1970)

V. Molotnikov, *Texnicheskaya mexanika* [*Technical Mechanics*] (Lan' Publ. Sant-Peterburg, 2017)

Chapter 11
Dynamic Analysis of Mechanisms

Abstract This chapter examines the forces acting on the links of mechanisms. When designing links and elements of kinematic pairs of mechanisms and machines, it is necessary to solve the problem of ensuring the necessary strength, rigidity, and durability. To do this, you need to know the power load on the links and kinematic pairs. The tasks and methods of studying the forces acting on the links of the mechanism are presented in the section of the theory called *kinetostatics*.

Keywords Classification of forces · Kinetostatics · Kinctostatic problems · Method of forces · Decomposition of forces · Zhukovsky's lever · Pole · Balance · Balancing force · Assur group

11.1 Classification of Forces

There are the following forces acting on the mechanism:

- driving forces F_d or moments M_d, accelerating the movement of input links and doing positive work;
- forces F_s or moments M_s of resistance, slowing down the movement of the input links and doing negative work; they can be forces of useful resistance, for the sake of overcoming which a mechanism was created, and forces of harmful resistance (e.g., friction forces);
- reactions in kinematic pairs F_{ij} arising in the link supports and being internal forces for the mechanism as a whole and external for each individual link;
- gravity of G links;
- inertial forces F_i or moments of inertial forces M_i.

Fig. 11.1 Active and reactive force

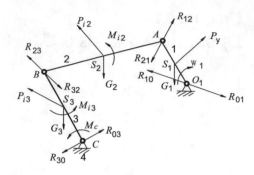

11.2   Kinetostatics Problems

Kinetostatics is designed to solve two main problems:

- determination of reactions in kinematic pairs of mechanisms for the purpose of their use in the subsequent construction of links and elements of kinematic pairs, calculation of friction forces, etc.;
- determination of the balancing force P_y or the balancing moment M_y on the leading link.

To solve these problems, it is necessary to have:

(a) the kinematic scheme of the mechanism and the geometric dimensions of its links;
(b) the masses and moments of inertia of the links;
(c) external forces acting in machines (in technological machines, forces of useful resistance; in motor machines, driving forces or moments).

Let's explain this with an example. In Fig. 11.1 a diagram of a four-link mechanism is presented, to the link 3 of which the executive body is attached. The moment of resistance M_c, which is the payload, acts on this link against the direction of its movement. The movement of the mechanism occurs as a result of the action of a driving force on the driving crank 1, which is simultaneously the balancing force P_y, which rotates the crank in the direction of clockwise movement with an angular velocity ω_1. The k-th link of the mechanism is affected by the force of its weight G_k ($k = 1, 2, 3$), inertial forces P_{ik}, moments of inertial forces M_{ik}, and reactions in kinematic pairs $R_{kn} = R_{nk}$ ($n = 1, 2, 3$).

As mentioned above, as a result of the kinetostatic analysis, the reactions $R_{kn} = R_{nk}$ and the balancing force P_y (or the balancing moment) should be determined. Power calculation is performed according to Assur groups, starting with the farthest from the leading link of the group. The input (leading) link is calculated last.

11.3 Methods of Force Calculation of Mechanisms

The following methods are most often used in the power calculation:

- force plans method;
- force decomposition method;
- analytical method;
- method of "hard lever" N.E. Zhukovsky;
- experimental method.

11.3.1 Force Plans Method

Consider the method of plans of forces for various groups of Assur II class. The diagrams of these groups are shown in Figs. 9.2a and 9.3. Let's start, for example, with the simplest group of the II class—the group of the fifth kind (Fig. 9.3d).

The group diagram is shown in Fig. 11.2a. On the right, a plan of forces is plotted on a certain scale. Its construction should be started by summing the known vectors (P_3, P_{i3}, G_3) (Fig. 11.2b). Then, from the beginning of the first and the end of the last of these vectors, lines are drawn parallel to the directions of the vectors R_{03} and R_{12} .... The point of intersection of the specified lines completes the construction of the plan of forces.

The peculiarity of the study of the two-flood dyad (Fig. 9.2a) is that the reactions in the end rotational pairs R_{12} and R_{43} (Fig. 11.3a) are not known either in size or direction. Therefore, we do this. Let us expand the indicated forces into normal (R_{12}^n, R_{43}^n) and tangential $(R_{12}^\tau, R_{43}^\tau)$ (Fig. 11.3a). The tangential components can be determined by equating to zero the sum of the moments about the point B of all forces acting on each of the links of the dyad. Find

Fig. 11.2 Force analysis of the Assur group II class of the fifth type

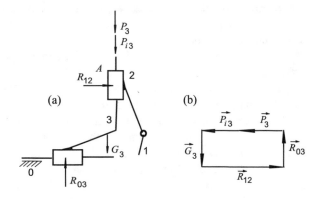

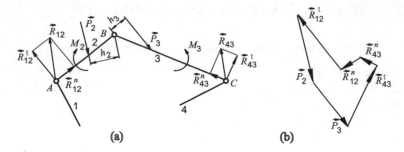

Fig. 11.3 Force analysis of the Assur double-water group

Fig. 11.4 Power analysis of
the leading link

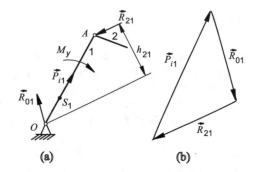

link 2: $\Sigma M_B^{(2)} = 0$; $R_{12}^\tau \cdot \overline{AB} + M_2 - P_2 \cdot h_2 = 0$;

$$R_{12}^\tau = \frac{P_2 \cdot h_2 - M_2}{\overline{AB}};$$

link 3: $\Sigma M_B^{(3)} = 0$; $R_{43}^\tau \cdot \overline{CB} + M_3 - P_3 \cdot h_3 = 0$,

$$R_{43}^\tau = \frac{P + 3 \cdot h_3 - M_3}{\overline{BC}},$$

where P_2, P_3 are the given forces applied to the links 2 and 3, respectively.

Given the R_{12}^τ and R_{43}^τ found, a force plan can be constructed (Fig. 11.3b), from which the components R_{12}^n and R_{43}^n.

After performing the force analysis of all kinematic groups of the Assur, the force calculation of the leading link is performed (Fig. 11.4a). At the point A, the reaction R_{21} is applied from the side of the discarded connecting rod 2, which is equal to and opposite to the previously found reaction R_{12} (Fig. 11.3).

At the center of mass S_1 of the crank, an inertial force P_{i1} is applied. Let m_1 be the mass of the link 1. When the crank rotates uniformly, the inertial force is $P_{i1} = m_1 \cdot \omega_1^2 \cdot \overline{OS_1}$ and is directed to the point A.

At the point O of the crank, the reaction R_{01} acts on the side of the rack, which must be determined. In addition, the crank must be applied to the so-called balancing moment M_y, acting on it from the side of the machine-engine and driving this mechanism. Instead of a balancing moment, you can apply a balancing force P_y by setting the point of application and the direction of the force on the connecting rod to be arbitrary.

Fig. 11.5 To the method of decomposition of forces

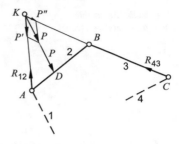

In the case of the balancing moment, its value is determined from the equilibrium equation of the crank in the form of the sum of moments equal to zero with respect to the point O:

$$R_{21}h_{21} - M_y = 0; \Rightarrow M_y = R_{21} \cdot h_{21}.$$

To find the reaction R_{01}, a plan of forces is constructed (Fig. 11.4b).

11.3.2   Decomposition of Forces

It is known Molotnikov (2017) that if the body is in equilibrium under the influence of three forces, then the lines of action of these forces intersect at one point. Let one of the three forces be a given external force P applied to the link AB at a certain point D (Fig. 11.5), and the other two forces are the reactions in the kinematic pairs of this link.

In the absence of active forces on the link BC, the reaction R_{43} in the hinge C (and the same reaction in the hinge B) is directed along BC. Thus, the point K of intersection of the lines of action of the forces P and R_{43} is simultaneously the point of intersection of the above three forces. This means that the reaction in the kinematic pair A is directed along the line AK.

Therefore, placing the beginning of the vector P at the point K and decomposing it into components in the directions KA and KB, we find the required reactions in the rotational pairs of the link 2:

$$R_{12} = -P' \text{ and } R_{32} = -P''.$$

Note that using the above-described method of decomposition of forces in each kinematic pair, a separate component of the reaction from each of the specified external forces is obtained. At the last stage of the calculation, it is necessary to find the resultant of these components. Therefore, it is advisable to apply this method with a small amount of external forces; otherwise it will require a large number of expansions and the subsequent calculation of the total reactions in each pair, which will be very cumbersome.

Fig. 11.6 Toward an
analytical study of forces

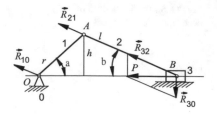

11.3.3 Analytical Method

The above methods for determining reactions are, in essence, graphoanalytical. However, the graphical solution of vector equations in both the force plans method and the force decomposition method can be replaced by analytical calculations. Let's show it.

It is shown above (p. 313) that in the method of plans of forces, the tangential components of reactions in rotational pairs of Assur groups are analytically determined. Here you can continue analytical calculations, calculating the normal components from the equilibrium equations of the entire Assur group as a whole in the form of moments of all forces relative to the points C and A (see Fig. 11.3). In the calculated reactions in extreme rotational pairs, the components of the reaction in the average rotational pair B are analytically determined from the equilibrium equations for any of the links of the dyad.

We will show the application of the analytical method by the example of a crank-slider mechanism (Fig. 11.6). The position of the mechanism is determined by the angle α of the crank rotation 1 with the length r. A connecting rod 2 of length l transfers the movement from the crank to the slider 3. An external force P is applied at the point B of the slide. Expanding the force P into components (Fig. 11.6), we get

$$R_{32} = \frac{P}{\cos \beta}, \quad R_{30} = P \operatorname{tg} \beta \quad \left(\sin \beta = \frac{r}{l} \sin \alpha\right).$$

The pressure in the kinematic pairs A and O is determined by the force $\boldsymbol{R}_{32}$.

11.3.4 Determination of Balancing Force

Determination of the Balancing Force Based on the Principle of Possible Displacements

In the dynamics of systems, the d'Alembert principle already known to us from theoretical mechanics *is fundamental: if inertial forces are applied to a moving system, then it can be considered as motionless, in equilibrium.*

We also remind (Molotnikov 2017) that the principle of possible displacements states if the system of forces is in equilibrium, then the sum of the elementary work of these forces on possible displacements of the points of their application is equal to zero.

If we consider all elementary work for an infinitely small period of time, then this principle can be formulated differently: if the system of forces is in equilibrium, then the sum of the instantaneous powers of these forces is zero:

$$\Sigma N(P_i) + \Sigma N M_j + N(P_y) = 0, \tag{11.1}$$

where

$$N(P_i) = P_i V_i \cos \delta_i, \quad N(M_j) = M_j \omega_j,$$

where P_i is the force applied at the i th point, V_i is the linear velocity of this point, and $\delta_i = \widehat{(P_i, V_i)}$; M_j is the moment applied to the j-th link at its angular velocity ω_j.

Note that under the sign of the sum in the first term of the formula (11.1) are the instantaneous powers of external forces and in the second term the instantaneous powers of external moments, and the third term is the moment of the balancing force: $N(P_y) = P_y V_y \cos(\widehat{P_y, V_y})$.

Substituting the formulas for calculating the powers into the Eq. (11.1), we obtain

$$\Sigma P_i V_i \cos \delta_i + \Sigma M_j \omega_j + P_y V_y \cos(\widehat{P_y, V_y}) = 0. \tag{11.2}$$

For given external forces and moments, using the Eq. (11.2), it is easy to determine the balancing force P_y. If the driving crank is acted upon not by the balancing force, but by the balancing moment, then instead of the third term in the formulas (11.1) and (11.2), there will be $M_y \omega_1$. Since the angular velocity of the leading link is known, (11.2) is easily solved relative to M_y.

Determination of the Balancing Force Using the "rigid lever" N.E. Zhukovsky
Let some point i of some link of the mechanism move with linear velocity V_i (Fig. 11.7). Let us agree to denote by $\overline{V}_i$ the length of the vector V_i, depicted on the plan of velocities at a certain scale μ_v. Suppose also that at the point i, a given force P_i is applied to the link at an angle $delta_i$ to the velocity vector.

Let us rotate the velocity vector $\overrightarrow{V}_i$ by the angle $\pi/2$ in any direction and match the end of the rotated vector $\overrightarrow{V}_i^{\Pi}$ with the beginning $\overrightarrow{V}_i$ (see Fig. 11.7). Let us then lower the perpendicular from the beginning Π of the rotated velocity vector to the line of action of the force. The length of this perpendicular will be $\overline{h}_i = \overline{V}_i^{\Pi} \cos \delta_i$ or $\overline{h}_i = \frac{V_i}{\mu_v} \cos \delta_i$. Considering the segment $\overline{h}_i$ as a shoulder of the force P_i relative to the center Π, we arrive at the equality

$$P_i \cdot \frac{V_i}{\mu_v} \cdot \cos \delta_i = P_i \cdot \overline{h}_i \quad \text{or} \quad \frac{N_i}{\mu_v} = M_i.$$

Fig. 11.7 To the concept of
Zhukovsky's lever

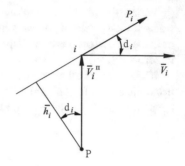

This means that the instantaneous power of a force can be represented as the moment of this force relative to the beginning of the velocity vector of the point of application of the force rotated by 90° and the magnitude of the moment is μ_v times less than the magnitude of the power of this force.

Performing such an operation for all external forces acting on the mechanism, we come (N.E. Zhukovsky) to the following.

Theorem *If the vectors of all forces applied to different points of the links and balanced on the mechanism are transferred parallel to themselves to the same points of the speed plan rotated by 90°, taking the figure of the plan for a rigid lever, then the sum of the moments of all the indicated forces relative to the pole of the plan will be zero.*

Thus, the equilibrium condition of the speed plan as a conditional rigid lever will be written in the form

$$\Sigma M_\Pi(P_i) + M_\Pi^y = 0. \qquad (11.3)$$

The velocity plan rotated by 90° and hinged at its pole is called the *Zhukovsky lever*. The formula (11.3) allows elementary constructions to determine the balancing force (or balancing moment). We will consider the above on the example of the crank-slide mechanism shown in Fig. 11.8a. The links 2 and 3 are affected by the external specified forces P_2 and P_3. You need to find the balancing force P_y ($P_y \perp OA$).

To solve this problem, do the following. 1) Draw the velocity diagram and rotate it by 90 degrees opposite to the rotation of the link OA (Fig. 11.8b). Such a construction is called a Zhukovsky lever. 2) Let us move the force vectors parallel to ourselves to the corresponding points of the plan. 3) Write equilibrium equation equating the sum of moments of all forces with respect to pole P to zero:

$$P_3 h_3 + P_2 h_2 - P_y h_y = 0.$$

From here we find

$$P_y = \frac{1}{h_y}(P_3 h_3 + P_2 h_2).$$

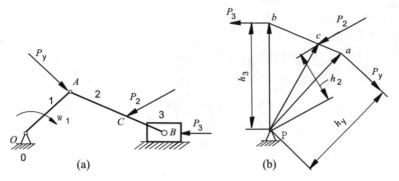

Fig. 11.8 For example, using the Zhukovsky lever

Note 1 If there are concentrated moments among the external loads, then it is advisable to represent them in the form of pairs of forces, the shoulders of which are equal to the lengths of the corresponding links.

Note 2 Interested readers are referred to fundamental sources (Martin 1982; Artobolevsky 1988), as well as to manuals for computer applications Solid Works and Universal Mechanism.

Self-Test Questions

1. Name the two main tasks of kinetostatics.
2. What data should be known to solve the problems of kinetostatics?
3. List the forces acting on the links of the mechanism.
4. What is the of d'Alembert?
5. What is the essence of the method of plans of forces in kinetostatics?
6. What parameters of forces are known and what are unknown in kinematic pairs?
7. How are the tangential components of reactions determined?
8. What is the peculiarity of the power calculation of the leading link of the mechanism?
9. What is balancing moment (balancing force)? From what condition is he (she) determined?
10. Formulate N.Ye. Zhukovsky.

References

I. Artobolevsky, *Teoriya mekhanizmov i mashin* [*Theory of Mechanisms and Machines*] (Nauka Publ., Moscow, 1988)

G. Martin, *Kinematics and Dynamics of Machines* (McGraw-Hill IBC Publ., 1982)

V. Molotnikov, *Texnicheskaya mexanika* [Technical Mechanics] (Lan' Publ., Sant-Peterburg, 2017)

Part III
Strength of Materials

Chapter 12
Initial Concepts and Definitions

Abstract In contrast to theoretical mechanics, the resistance of materials considers problems where the properties of deformable bodies are the most important, and the laws of motion of a body as a rigid whole not only recede into the background, but in some cases are simply insignificant. The resistance of materials is intended to create practically acceptable simple methods for calculating the typical, most common structural elements. This chapter discusses the classification of structural elements from the standpoint of common calculation methods. The concepts of internal forces, deformations, and displacements are introduced. The simplest experiments for determining the basic mechanical characteristics of materials are described.

Keywords Strength · Stability · Rigidity · Hardness · Calculation scheme · The Saint-Venant principle · Full · Normal · Tangential stress · Internal efforts · Relative extension · Poisson's ratio · Young's modulus · Hooke's law

12.1 Objectives of the Course on Strength of Materials

As you know, theoretical mechanics deals with a material point and an absolutely solid body. For those phenomena where the deformations of the body can be ignored, the conclusions of theoretical mechanics are accurate and quite sufficient. However, for the construction of the mechanics of a real solid body, the laws of theoretical mechanics are not enough, and they have to be supplemented with physical laws or hypotheses about the interaction between the points that make up the system. The branch of mechanics that deals with the behavior of real solids under load is called the resistance of materials (Timoshenko 1947). The practical goal that the resistance of materials sets itself is the calculation of structural elements for strength, stability, and rigidity.

The *strength* refers to the ability of a structure to withstand specified loads without destroying its elements. If at the same time the system does not lose the ability to maintain to the desired extent the known law of its motion or the specified limits of its deformation, then it is called *stable*. Otherwise, the system is called

Fig. 12.1 Calculation
diagrams of structural
elements: (**a**), timber; (**b**),
shell; (**c**), plate; (**d**), massive

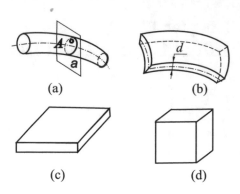

unstable. By *destruction* , we will understand the process of dismemberment of the
body into parts, caused by overcoming the internal forces of connection between its
parts.

The property of the system to maintain the specified limits of deformation is
called its *rigidity.* The question of quantifying the permissible limits of deformation
is solved in each individual case separately. So, the deflection of the bridge span
under the influence of the weight of the moving train, which is several centimeters, is
considered permissible. On the other hand, the deflection of the lathe caliper guide,
which is a fraction of a millimeter, is completely unacceptable, since it excludes the
possibility of any precise processing of parts.

The objects of study in the strength of materials are not abstract bodies, but
parts of machines and engineering structures made of real structural materials.
The geometric shape of structural elements is usually quite complex. Accurate
accounting of all the geometric features of parts is impossible, and often impractical,
as it leads to complex calculations. Therefore, in engineering calculations, a
simplification is introduced into the geometry of the part, leading it to one or another
scheme. Depending on the geometric shape, the structural element is considered as
bar, shell, plate , or *massive body* (Fig. 12.1).

A *bar* is a structural element that has one dimension (length) significantly larger
than the other two dimensions (Fig. 12.1a). The *cross-section* of a beam passing
through some point A (Fig. 12.1a) is the cross-section of the beam with the plane
α containing this point and having a minimum area. The geometric location of the
points that coincide with the centers of gravity of the cross-sections of the beam is
called its *axis.* Bars are the main object of research in the resistance of materials.

A structural element that has one dimension (thickness) significantly smaller than
the other two dimensions is called a *shell* (Fig. 12.1b). The surface equidistant from
the inner and outer surfaces of the shell is called the *median surface.*

The shell whose median surface is a plane is called the *plate* (Fig. 12.1c). The
thickness of the δ plate or shell can be either constant or variable.

A part whose dimensions differ little from each other in all directions is treated
as a *array* or *massive body.* For example, shown in Fig. 12.1d, the foundation
stone should be treated as an array.

12.2 External Forces: The Saint-Venant Principle

The interaction forces of solid body particles form a system of internal forces. Part of them, acting according to Newton's law of gravity, is of interest only in the case of large astronomical bodies. Such forces between individual parts of engineering structures are negligible, and we will not take them into account. The remaining part of the internal forces of a solid for its two elementary particles is essential, if only the particles in question are separated from each other by no more than the distance between the molecules. Therefore, mentally separating one part of the body from another, we will imagine the interaction of these parts by forces distributed over the surface of the cross-section.

For any of the severed parts of the body, the system of internal forces replaces the action of the discarded part with the one under consideration. Thus, the method of imaginary sections allows, firstly, to detect internal forces and, secondly, translates internal forces into the category of external loads applied to each of the cut-off parts of the body. According to the principle of the relationship of action and reaction, the specified loads applied to each of the sides of the "cross-section" are equal in magnitude and opposite in direction.

Let us explain what has been said using graphic images (Fig. 12.2) (Molotnikov 2017, 2012). Let an arbitrary solid be loaded with the forces P_1, P_2, $\ldots$, P_k. Some of these forces (or even all of them) can be reactive. The set of external forces is denoted by the symbol (P_e). Mentally dissect the body by some surface (e.g., by the plane Q (Fig. 12.2a)) into two parts-the right (R) and the left (L). Let's separate these parts (Fig. 12.2b). The action of each part on the other is replaced by internal forces applied to each side of the cut (see Fig. 12.2b) and are continuously distributed over the cross-section surface in a complex way. The set of these forces in the cross section is denoted by (P_i). The internal forces must be such that the conditions of balance for the right and left parts of the body separately are satisfied. Symbolically, this can be written as

$$(P_e)_R + (P_i) = 0, \tag{12.1}$$

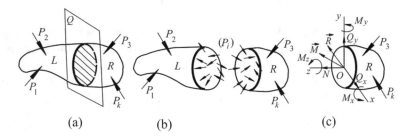

(a)   (b)   (c)

Fig. 12.2 To the section method

and also

$$(P_e)_L + (P_i) = 0, \tag{12.2}$$

where the symbols $(P_e)_N$ and $(P_e)_L$ are understood as a set of external forces applied to the right and left parts, respectively. Thus, the system of internal forces (P_i) is completely determined from the condition of equilibrium of the right or left cut-off part of the body.

Consider, for example, the right part (N). Let us use the rules of statics and bring the system of internal forces to the center of gravity of the section (point O, Fig. 12.2 *in*). As a result, we get the main vector R and the main moment M. Enter the Cartesian coordinates, directing the z axis along the external normal and the x and y axes in the cross-sectional plane (Fig. 12.2*in*). The projections of the main vector R on the axis x, y, z are denoted by N, Q_x, Q_y, and the projections of the main moment M are denoted by M_x, M_y, M_z. These components of internal forces are called *internal forces* (or internal force factors) in the cross-section. The component of N is called the longitudinal or normal force in the cross-section. The components Q_x and Q_y are called transverse or shearing forces. The M_z component the main moment is called the torque, and the components M_x, M_y are called the bending moments.

The introduced designations allow passing from symbolic notations (12.1) and (12.2) of equilibrium conditions to equilibrium equations in coordinate form (1.31). For the considered right-hand side of the body, these equations have the form

$$
\begin{aligned}
&1)\ \Sigma X_i = 0: \quad Q_x + \Sigma \mathrm{np}_x (P_e)_\Pi = 0; \\
&2)\ \Sigma Y_i = 0: \quad Q_y + \Sigma \mathrm{np}_y (P_e)_\Pi = 0; \\
&3)\ \Sigma Z_i = 0: \quad N + \Sigma \mathrm{np}_z (P_e)_\Pi = 0; \\
&4)\ \Sigma M_{xi} = 0: \quad \Sigma M_x (P_e)_\Pi = 0; \\
&5)\ \Sigma M_{yi} = 0: \quad \Sigma M_y (P_e)_\Pi = 0; \\
&6)\ \Sigma M_{zi} = 0: \quad \Sigma M_z (P_e)_\Pi = 0,
\end{aligned}
\tag{12.3}
$$

where the symbol $\Sigma \mathrm{pr}_j (P_e)_R$ denotes the sum of the projections on the j axis of all external forces applied to the right side of the body and through $\Sigma M_j (P_e)_R$ is the sum of the moments of these forces relative to the j-th axis ($j = x,\ y,\ z$).

Note that each of the Eqs. (12.3) includes only one internal force, which greatly simplifies the problem of their formal definition from the equilibrium equations.

As already mentioned, equations of the type (12.3) can also be written for the left part of the body. It is clear that the result of determining the internal forces in the cross-section does not depend on which part of the body—right or left—is considered.

12.3   The Simplest Stress States

The measure of internal forces at each point of a deformed body is characterized by a physical quantity called stress. Let us draw some section through the point A of the body (Fig. 12.3).

Let's select in the vicinity of the point A the area ΔF. The resulting internal forces acting on the area ΔF will be denoted by ΔR. The total stress at point A over the area ΔF of a given section is a vector, the value of which is

$$p = \lim_{\Delta F \to 0} \frac{\Delta R}{\Delta F}. \tag{12.4}$$

Stress has a force/area dimension; in the system SI, it is N/m^2 (pascal, Pa); in the technical system of units MKGSS (LFT), it is kgf/cm^2. The total stress p can be decomposed into two components, one of which is directed along the normal (z) to the section plane (Fig. 12.3) and the second is located in this plane. The component of the total stress normal to the section is called normal stress. Let us agree to denote it by σ (Fig. 12.3). The second component (τ), lying in the section plane, is called the tangential stress at the point A along the area ΔF.

Let us choose in the section plane arbitrary orthogonal axes x and y and expand the tangent dressing τ along the axes x, y, denoting them by τ_x and τ_y, and the subscript coincides with the designation of the axis parallel to which the shear stress vector is directed.

Summing up the stresses in the cross-section of the bar in the appropriate way, we obtain the formulas for the relationship between stresses and internal forces:

$$N = \int_F \sigma dF, \quad Q_x = \int_F \tau_x dF, \quad Q_y = \int_F \tau_y dF,$$

$$M_x = \int_F \sigma y dF, \quad M_y = \int_F \sigma x dF, \quad M_z = \int_F (\tau_y x - \tau_x y) dF. \tag{12.5}$$

Fig. 12.3 To the concept of stress

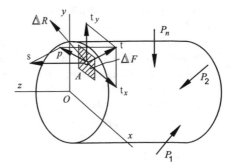

Fig. 12.4 Uniaxial tension
and compression

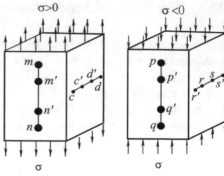

Fig. 12.5 Pure shift

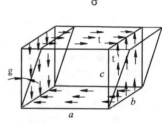

If a different section is drawn through the point A, then the orientation of the area
ΔF will change and the stress p at the point A will, generally speaking, be different.
The aggregate of stresses across a set of areas passing through a given point of the
body forms a stress state at the point. In the general case, it is determined by six
components that form the so-called second-rank tensor. We will not develop the
general theory of the stress state here, since to solve the most important engineering
problems, it will be enough for us to start from the simplest types of stress state. Let
us imagine a prismatic body, to the bases of which are applied uniformly distributed
normal stresses of intensity σ per unit area (Fig. 12.4). The normal stress directed
along the external normal to the site of its action is considered positive and is called
stretching. And, conversely, the normal voltage directed along the internal normal
will be considered negative and called compressive. The stress state depicted in
Fig. 12.4 is called uniaxial stretching (left) and compression (right).

Another simplest type of stress state is the so-called net shift. This state is
experienced by a rectangular parallelepiped with dimensions a, b, c, shown in
Fig. 12.5.

Let the upper face be affected by uniformly distributed tangent forces parallel
to the edge a, the intensity of which is equal to τ per unit area. In order for the
main vector of external forces to be zero, oppositely directed forces of the same
intensity must be applied to the lower face. The forces on the horizontal faces have
the resultant $\tau a b$ and form a pair with the shoulder c. The moment of this pair is
equal to $\tau a b c$. In order to balance this moment, we apply evenly distributed vertical
tangential forces of intensity τ' to the face bc and the same forces directed opposite
to τ' to the opposite bc face. The system of these forces is paired with the moment

$\tau'bca$. There will be an equilibrium only if the specified moments are equal. From here we get

$$\tau' = \tau. \tag{12.6}$$

Equality (12.6) expresses the law of pairing of tangential stresses: *components of tangential stresses on two mutually orthogonal areas, perpendicular to their intersection lines, are equal to each other.*

12.4 Deformations and Displacements

Let's turn again to the prisms shown in Fig. 12.4. If we choose a segment mn of some straight line parallel to the generator, the length of which before the deformation was l, then after the deformation the points m and n will take new positions m' and n'. The length of the segment $m'n'$ is denoted by l'. The quantity

$$\Delta l = l' - l \tag{12.7}$$

is called *absolute strain (elongation).* In the case of uniaxial compression, on the contrary, the length of the fibers (pq), parallel to the generatrix, decreases and the absolute deformation will be negative. Quantity

$$\varepsilon = \frac{\Delta l}{l} \tag{12.8}$$

is called the relative elongation or relative linear deformation in a given direction. If the material is homogeneous, then the value of ε does not depend on the length of the segment mn (or pq), i.e., does not depend on the measurement base.

For real materials, under uniaxial tension or compression, the transverse dimensions of the body also change (see Fig. 12.4). If the length of the segment cd perpendicular to the generator was before the deformation l_1, and after the deformation it became l'_1 (the segment $c'd'$), then the relative transverse deformation will be

$$\varepsilon' = \frac{l'_1 - l_1}{l_1} = \frac{\Delta l_1}{l_1}. \tag{12.9}$$

Experiential research shows that the values ε and ε' are always of opposite sign. In addition, if the stress σ does not exceed a certain value constant for a given material, then the relative transverse deformation is proportional to the linear deformation in the direction of tension (compression), i.e.,

$$\varepsilon' = -\mu\varepsilon. \tag{12.10}$$

Fig. 12.6 Linear and angular
movements

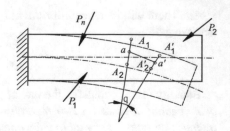

The material constant μ entering the formula (12.10) is called Poisson's ratio. For
real materials, the values this ratio are in the range $0 < \mu \leqslant 0.5$.

The deformation of a parallelepiped under pure shear has a different character
(Fig. 12.5). Experiments show that the lengths of the edges of the parallelepiped do
not change in this case and initially the right angles between the edges a and c are
distorted. Such deformation is called *shear deformation* or simply *shear.* The
measure of the shift is the value of the angle γ by which the original right angle
between the edges a and c has changed.

In the general case, at each point of a deformed body, there can be linear defor-
mations ε_x, ε_y, ε_z in three orthogonal directions (x, y, z) and shifts γ_{xy}, γ_{yz}, γ_{zx}
in three coordinate planes. The set of these six quantities characterizes *deformed
state* at a point of the body.

The deformation of the body causes its points to move to new positions, and
the infinitesimal segments connecting each pair of points are rotated by a certain
angle. Consider, for example, the timber shown in Fig. 12.6. In the absence of loads,
we note the position of a certain point a of the beam and an arbitrary infinitesimal
segment $A_1 A_2$ containing the specified point. Now load the beam with the system
of forces P_1, P_2, $\ldots$, P_n. Under the influence of loads, the beam is deformed and
occupies a certain position, shown in Fig. 12.6 with dashed lines. In this case, the
point a will move to the position a', and the segment $A_1 A_2$ will take a new position
$A_1' A_2'$. The length of the segment aa' is called *linear by moving* (or shifting) the
point a and the angle *vartheta* by turning or *by angular moving at that point.*

12.5   Material Testing: Hooke's Law

For engineering calculations, it is necessary to know a number of characteristics
of the behavior of the material under load. These characteristics are determined
experimentally by testing standard samples of material on special testing machines.
For ductile materials such as structural steels, aluminum, and some other alloys,
the most characteristic properties are determined from tensile tests and for brittle
materials such as cast iron, stone, glass, etc., from compression tests.

The appearance of such a sample for tensile tests before (dashed lines) and after
(solid lines) testing is shown in Fig. 12.7.

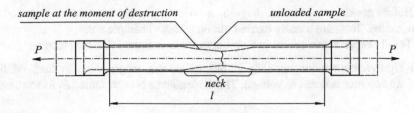

Fig. 12.7 Tensile test piece

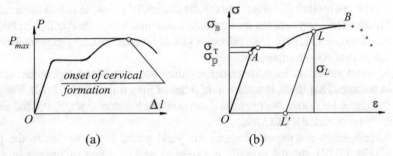

Fig. 12.8 Experimental study of uniaxial tension

Figure 12.8a shows an example dependency diagram between the value of the resultant (P) of the tensile forces structural steel sample and lengthening (Δl) of the sample. This schedule is automatically drawn on a diagrammatic apparatus when performing experiments on tensile on almost all existing tensile testing machines.

In such tests, a uniform deformation of the sample can be observed everywhere at a sufficient distance from its ends. In Fig. 12.8a, the dashed line marks the part of the diagram that corresponds to the inhomogeneous deformation of the sample due to the formation of a local narrowing (neck), shown in Fig. 12.7.

Recall (p. 329) that the relative longitudinal strain in the direction of the tensile force is called the value

$$\varepsilon_1 = \frac{\Delta l}{l},$$

and l is the working length of the sample (Fig. 12.7). *Transverse relative strain* in the direction perpendicular to the sample axis is denoted by ε_2.

The normal voltage is the force per unit area F of the normal cross-section of the undeformed rod, i.e.,

$$\sigma_1 = \frac{P}{F}. \tag{12.11}$$

Graphical representation of the relationship between σ_1 and ε_1 (Fig. 12.8b) is called *diagram stretching*.

Before necking, the stretch diagram is shape-independent cross-section and its dimensions. Tests are usually carried out on standard sample sizes.

Tensile experiments reveal in a first approximation the following results.

(A) Up to some tension σ_{pr} (Fig. 12.8b), called the *proportionality limit,* deformation proportional to voltage. This dependence is represented as follows way:

$$\varepsilon_1 = \frac{\sigma_1}{E}, \quad \varepsilon_2 = -\mu\varepsilon_1 = -\mu\frac{\sigma_1}{E}. \tag{12.12}$$

The coefficient E, which has the dimension of stress, is called the modulus Young's elasticity, and the dimensionless parameter μ is the coefficient Poisson (see formula (12.10)). Dependencies (12.12) mathematically express Hooke's law in tension—compression.

(B) At some stress, an increase in deformation occurs without significant change in voltage. This stress is called *yield point* of material σ_t (Fig. 12.12). They say that under tension σ_t the material flows, and the horizontal section of the stretch diagram is called a pad yield.

(C) After loading the sample beyond the yield point, for example, to the point L (Fig. 12.12), the relationship between ε_1 and σ_1 when unloading in a first approximation, you can represent the straight line LL'. Before shaping the neck (local constriction), this straight line is almost parallel to AO.

(D) The highest stress σ_{vr}, corresponding to point B on the diagram, is called *ultimate strength* or *ultimate resistance.* At this moment, the destruction of the sample occurs in the zone of the formed neck (Fig. 12.7).

In mechanical engineering technology, the plastic properties of materials are characterized by the value

$$\delta = \frac{l_p - l_0}{l_0} \cdot 100\%, \tag{12.13}$$

where l_0 is the length of the working part of the specimen before testing and l_p is the length of this part after rupture. This value is called *residual elongation.* Another characteristic of plastic properties is *residual relative constriction:*

$$\phi = \frac{F_0 - F_{neck}}{F_0} \cdot 100\%, \tag{12.14}$$

Moreover, F_0 is the cross-sectional area of the sample before testing, and F_r is the cross-sectional area at the site of the sample rupture.

Not all plastic materials have a yield point on the tension diagram. If there is no yield point, then a conditional value is taken for the yield strength of the material-the stress at which the residual deformation is 0.002 or 0.2%. For the conditional yield strength, the notation $\sigma_{0.2}$ is accepted. For plastic bodies, the tension and compression diagrams are almost the same.

Brittle materials (cast iron, concrete, etc.) have different diagrams under tension and compression. For them, the tests determine the tensile strength σ_{bp}^{p} and the compressive strength σ_{bp}^{c} and $\sigma_{bp}^{c} \gg \sigma_{bp}^{p}$.

Testing samples of materials for tension (compression), although it gives an objective assessment of their mechanical properties, but for production conditions are often non-operational and sometimes impracticable. Therefore, in many cases, to judge the strength and wear resistance of the material of the part, designers and technologists use a comparative assessment of the properties of the material using a hardness test. The hardness of a material is understood as its ability to resist the mechanical penetration of another, more solid body into it. Brinell and Rockwell hardness tests are the most widely used. In the first case, a steel ball with a diameter of 10 mm is pressed into the surface of the part under study and in the second a diamond cone. The hardness is judged by the measurement of the resulting print. The Brinell hardness is usually denoted *HB*, and the Rockwell hardness is *HRC*. The change in the hardness of the material is achieved by using heat treatment or other methods.

Self-Test Questions

1. Name the subject and purpose of studying the science of the resistance of materials.
2. What is meant by the strength of structural elements?
3. List the objects of study in the resistance of materials.
4. Give the formulation of the Saint-Venant principle.
5. Formulate the essence of the cross-section method.
6. Define the total, normal, and tangential stresses at a point belonging to a certain section of the body.
7. The expression of internal forces through tension.
8. Give the formulation of the law of pairness of tangential stresses.
9. What is called an absolute deformation in a certain direction?
10. Name the units of relative strain measurement.
11. What units are used to measure the shear strain?
12. Name the types of displacements of the points of the deformable body.
13. List the main mechanical characteristics of materials determined in experiments on tension and compression.

References

V. Molotnikov, *Mekhanika konstruktsii* [*Mechanics of Structures*] (Lan' Publ., SPb., Moscow, Krasnodar, 2012)

V. Molotnikov, *Texnicheskaya mexanika* [*Technical Mechanics*] (Lan' Publ., Sant-Peterburg, 2017)

S. Timoshenko, *Strength of Materials. Part I* (Macmillan and Company Publ., 1947)

Chapter 13
Calculation of Parts in Tension-Compression

Abstract Tension (compression) is understood as a type of loading in which only longitudinal forces arise in the cross-sections of the beam, and all other force factors (transverse forces, torsional and bending moments) are equal to zero. This is the simplest and most common type of deformation. It usually occurs when an external load acts along the longitudinal axis of the beam. The tension-compression beam is called the *bar* . The locus of the centers of gravity of the cross-sections of the bar is called its axis. This chapter deals with the tasks of determining internal forces, stresses, and strains in a bar. The problems of strength verification and design of tensioned (compressed) rods are also solved. Studied are also rod systems, the forces in the elements of which are not determined by static methods.

Keywords Timber · Kernel · The axis of the rod · Longitudinal forces · Diagrams · Stress · Deformation · Displacement of sections · Potential energy · Permissible Stress · Strength calculation · Stiffness calculation · Static indeterminacy

13.1 Internal Tensile-Compressive Forces

Stretching (compression) is a type of deformation of a bar in which only one internal force is different from zero in the cross-sections—the longitudinal force. A bar working in tension (compression) is called an *bar.* The design diagram of a bar is used to design bolts, studs, truss elements, and many others (Timoshenko 1947; Molotnikov 2017).

The longitudinal force N in any cross-section of the rod is determined using the cross-section method. The longitudinal force directed from the cross-section is considered positive, and the force directed to the cross-section is considered negative.

In other words, the positive longitudinal force causes stretching in this section and is directed along the external normal to the section. The negative longitudinal force causes, on the contrary, compression of the material in the considered section and coincides in the direction with the internal normal.

© The Author(s), under exclusive license to Springer Nature Switzerland AG 2023   335
V. Molotnikov, A. Molotnikova, *Theoretical and Applied Mechanics*,
https://doi.org/10.1007/978-3-031-09312-8_13

Fig. 13.1 Central extension-compression of the rod: (**a**)—loading diagram; (**b**)—longitudinal force diagram

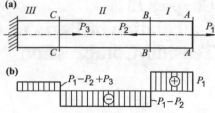

Diagram N, (indicate the dimension of the force)

Consider, for example, the rod shown in Fig. 13.1a. To determine the longitudinal force in its cross-sections, we consistently apply the cross-section method. Suppose we want to know the longitudinal force in an arbitrary normal section of the section I located between $A - A$ and $B - B$. Discarding the left part of the rod, we are convinced that for the balance of the right, the remaining part in the section we are interested in, we must apply a longitudinal force N^I, equal in modulus to P_1 and directed along the external normal to the section. In the same way, the longitudinal force in an arbitrary section of the section II between $B - B$ and $C - C$ is determined: the right part of the rod is stretched by the force P_1 and compressed by the force P_2. Hence, the longitudinal force on the section II will be $N^{II} = P_1 - P_2$. Similarly, $N^{III} = P_1 - P_2 + P_3$.

To analyze the work of a bar, it is convenient to have a graph of changes in internal forces along the length of the bar. Such plots are called *plots*. In Fig. 13.1b, a diagram of longitudinal forces is built. Plots are plotted at a certain scale. The axis of the plot is parallel to the axis of the bar, and the axis of values is perpendicular to it. The values of the investigated quantity are plotted from the axis of the diagram, taking into account the signs.[1]

13.2   Stresses

Under tension (compression), each element of the body volume in places sufficiently remote from the anchors and points of application of external forces is deformed in exactly the same way. The ends of the prisms representing such elements in Fig. 12.4 will move apart or closer together, but will remain flat and parallel to each other. We will formulate this fact as follows.

Sections that are flat and perpendicular to the beam axis before deformation remain flat and perpendicular to the axis after deformation.

---

[1] Sometimes the diagrams are covered with hatching with lines perpendicular to the axis of the diagram. Apparently, this reflects the utilitarian use of a caliper, with the help of which the ordinate of the plot was transferred to a scale to determine the value of the parameter.

Fig. 13.2 Stresses on
inclined platforms

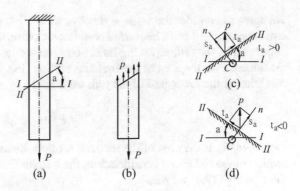

(a) (b) (c) (d)

The reduced position is called the (Bernoulli) *hypothesis of flat cross-sections*.[2] On the basis of this hypothesis and the Saint-Venant principle, the distribution of normal stresses in the cross-sections of the rod (outside the zones of influence of anchors, points of application of forces, etc.) can be considered uniform.

Then, based on the formula (12.11), the normal stress in the cross-section of the rod under conditions of uniaxial (central) tension-compression will be

$$\sigma = \frac{N}{F}. \tag{13.1}$$

Let us now consider the stresses in the inclined sections of the rod. Let's draw a cross-section $I - I$ (Fig. 13.2a) and at an angle α to it an inclined section $II - II$. The angle α is assumed to be positive if, to align with the cross-section $II - II$, the cross-section $I - I$ must be rotated counterclockwise. Otherwise, $\alpha < 0$. Discard the upper part of the rod, replacing its action with the lower part by stresses p, evenly distributed over the area of the inclined section (Fig. 13.2b). From the equilibrium condition of the lower part of the rod, we get

$$P = pF_\alpha, \tag{13.2}$$

Moreover, F_α is the area of the inclined section. Bearing in mind that in any cross-section of the rod under consideration, the longitudinal force N is equal to P, from the equality (13.4), we find

$$p = \frac{P}{F_\alpha} = \frac{N}{F/\cos\alpha} = \frac{N}{F}\cos\alpha = \sigma\cos\alpha, \tag{13.3}$$

where σ is the normal stress in the cross-section.

Let us now consider the stresses in the inclined sections of the bar. Let's draw a cross-section $I - I$ (Fig. 13.2a) and an inclined section $II - II$ at an angle α to it.

---

[2] A generalization of this hypothesis in relation to plates and shells is given by Kirchhoff and Love.

We agree to consider the angle α positive if, to align with the section $II - II$, the cross-section $I - I$ must be rotated counterclockwise. Otherwise, $\alpha < 0$. Discard the upper part of the rod, replacing its action with the lower part by stresses p, evenly distributed over the area of the inclined section (Fig. 13.2b). From the equilibrium condition of the lower part of the rod, we obtain

$$P = pF_\alpha, \tag{13.4}$$

moreover, F_α is the area of the inclined section. Bearing in mind that in any cross-section of the rod under consideration, the longitudinal force N is equal to P, from the equality (13.4) we find

$$p = \frac{P}{F_\alpha} = \frac{N}{F/\cos\alpha} = \frac{N}{F}\cos\alpha = \sigma\cos\alpha, \tag{13.5}$$

where σ is the normal stress in the cross-section.

Decompose the total voltage p into two components: normal voltage σ_α (see Fig. 13.2c,d) and the tangential stress τ_α lying in the plane of the inclined section $II - II$. Let's focus on the rule of signs for stresses. For normal voltages, this rule has already been introduced earlier (p. 328): the normal stress is positive if it is tensile ($N > 0$) and negative if it is compressive ($N < 0$). The tangent stress τ_α will be considered positive if the vector representing it tends to rotate the body relative to any point (C, Fig. 13.2b,d) of the internal normal in the direction of clockwise movement (see Fig. 13.2c). Otherwise, $\tau_\alpha < 0$ (see Fig. 13.2c).

After performing the specified expansion of the voltage p, we get

$$\sigma_\alpha = p\cos\alpha = \sigma\cos^2\alpha,$$
$$\tau_\alpha = p\sin\alpha = \sigma\sin\alpha\cos\alpha = \frac{\sigma}{2}\sin 2\alpha. \tag{13.6}$$

We note some properties of the stress state under tension-compression, which follow from the last formulas.

1. *The sum of the normal stresses on two mutually orthogonal cross-sections is constant and equal to the normal stress in the cross-section.*

    In fact, let $\alpha' = \alpha + \frac{\pi}{2}$. Then based on the first of the formulas (13.6)

$$\sigma_{\alpha'} = \sigma\cos^2\left(\alpha + \frac{\pi}{2}\right) = \sigma\sin^2\alpha,$$

$$\sigma_\alpha + \sigma_{\alpha'} = \sigma\cos^2\alpha + \sigma\sin^2\alpha = \sigma.$$

2. *Tangential stresses are maximal on the pads inclined to the rod axis at an angle 45°.*

    To prove the validity of this proposition, we investigate the extremum function τ_α, given by the second of the formulas (13.6). Denoting by α_0 the value of α at which the derivative of this function vanishes, we get

$$\frac{d\tau_\alpha}{d\alpha}\bigg|_{\alpha=\alpha_0} = \frac{\sigma}{2} \cdot 2\cos 2\alpha_0 = 0;$$

from here

$$\alpha_0 = \pm\frac{\pi}{4} + k\pi, \quad (k = 0, 1, \ldots). \tag{13.7}$$

In this way

$$\tau\big|_{\alpha=\alpha_0} = \tau_{\max} = \frac{\sigma}{2}. \tag{13.8}$$

3. *On areas perpendicular to the rod axis, normal stresses are maximum, and shear stresses are zero. There are no stresses on sites parallel to the axis.*

It is easy to check the validity of the last sentences using formulas (13.6).

13.3 Determination of Deformations and Displacements

In the overwhelming majority of cases, structural elements work under stresses, when the material remains elastic and Hooke's law is valid (12.12). Situations in which the appearance of inelastic deformations is inevitable, possibly or even desirable, occur most often in metal forming technology and require special consideration. We will deal mainly with elastic bodies. For an elastic bar, using Hooke's law (12.12) and the formula (12.8), we get

$$\varepsilon = \frac{N}{EF}. \tag{13.9}$$

The product EF is called the *stiffness* of the bar under tension-compression. Further, from the definition (12.8) and the formula (13.9) it follows

$$\Delta l = \frac{Nl}{EF} \text{ or } \Delta l = \frac{\sigma l}{E}. \tag{13.10}$$

In the case when the longitudinal force and (or) the stiffness of the bar changes along its length, the formula (13.10) should be applied to the element of infinitesimal length dl. Then the absolute deformation of the bar can be found by integrating

$$\Delta l = \int_0^l \frac{N\,dl}{EF}. \tag{13.11}$$

If on each i-th section of the rod the integrand in the formula (13.11) remains constant, integration can be replaced by summation

$$\Delta l = \sum_{i=1}^{n} \frac{N_i l_i}{E_i F_i},\qquad (13.12)$$

where n is the number of rod sections.

The formulas (13.10)–(13.12) allow you to define the displacement of any cross-section caused by the deformation of the bar. Obviously, this displacement is numerically equal to the absolute deformation of the part of the rod enclosed between the stationary (fixed) and the considered sections.

13.4   Potential Deformation Energy of the Body

The force applied to the deformable body does the work (A). If this work is entirely spent on imparting velocities to the particles of the body (kinetic energy T) and on the accumulation of completely reversible potential energy (U), then, as you know, such a mechanical system is called (Molotnikov 2017) conservative. The energy balance equation for such a system will be

$$A = T + U. \qquad (13.13)$$

Under static loading, the velocities acquired by the body particles due to its deformation are negligible. Therefore, we can assume that the work of external forces completely transforms into the potential energy of elastic deformation:

To calculate the potential energy of elastic deformation under tension (compression), consider the bar shown in Fig. 13.3. Since during deformation the force P does not remain constant, but increases from zero to the final value P, the process can be represented as a sequence of infinitely small increments of elongation $d(\Delta l)$ at a fixed load $\bar{P}$, Fig. 13.3:

$$A = T + U.$$

Fig. 13.3 Calculating the energy of elastic deformation

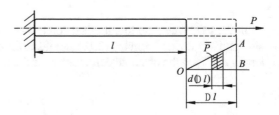

The work of the force $\bar{P}$ on the path $d(\Delta l)$ is graphically depicted by the area of the shaded quadrangle in the figure, i.e., $dA = \bar{P}d(\Delta l)$. Summing up the elementary work along the entire path Δl, we find that the total work under loading up to the value of the force P is numerically equal to the area of the triangle OAB:

$$A = U = \frac{1}{2}P\Delta l.$$

Expressing the absolute deformation Δl by the formula (13.10), we get

$$U = \frac{P^2l}{2EF}. \qquad (13.14)$$

In the case when the normal force N and (or) the stiffness of the rod EF change along the length l, one should consider the deformations of the elements dl. For an elementary section by the formula (13.14), we have

$$dU = \frac{N^2dl}{2EF},$$

and for the entire rod, the potential energy of deformation will be

$$U = \int_0^l \frac{N^2dl}{2EF}. \qquad (13.15)$$

The formula (13.14) can be represented differently. Bearing in mind that $P = \sigma F$, and the product Fl is the volume V of the rod material, we get

$$U = \frac{\sigma^2 Fl}{2E} \text{ или } U = \frac{\sigma^2 V}{2E}. \qquad (13.16)$$

Dividing both sides of the formula (13.16) by the volume V, we obtain the amount of potential energy of elastic deformation per unit volume of the material. This value is called *specific potential energy of deformation*. Denoting it by u, we can write

$$u = \frac{\sigma^2}{2E}. \qquad (13.17)$$

Energy representations are widely used in calculating displacements in elastic systems. This issue will be discussed later.

13.5 Strength Calculations

The structure is considered strong if the operating stresses in any of its elements do not exceed the limit values. For plastic materials, the limit is the stress corresponding to the yield strength σ_t. Despite the fact that the occurrence of fluidity in any element does not yet cause its destruction, the structure may lose its serviceability due to the appearance of unacceptably large deformations. For brittle materials, the limiting stresses are those corresponding to the time resistance under tension (σ_{vr}^p) or compression (σ_{vr}^c), since reaching such stresses means destroying the element.

But the magnitude of the loads, the geometric dimensions of the parts, as well as the values of the limit stresses cannot be set absolutely accurately. Therefore, to ensure the guaranteed strength, it is necessary that the greatest stresses obtained as a result of the design calculation (design stresses) do not exceed a certain value less than the limit stress. This value is called *allowed voltage* $[\sigma]$ and is determined by dividing the limit voltage $\sigma_{\lim}$ by the so-called coefficient of the safety margin (n), i.e.,

$$[\sigma] = \frac{\sigma_{\lim}}{n}. \tag{13.18}$$

As mentioned above, for plastic materials, the ultimate stress is the yield point: $\sigma_{\lim} = \sigma_t$. The safety factor for such materials will be denoted by n_t. This value is called the safety factor in relation to the yield strength. With this in mind, the permissible stress when calculating parts made of plastic materials will be

$$[\sigma] = \frac{\sigma_t}{n_t}. \tag{13.19}$$

The values of the coefficient n_t in general mechanical engineering take from 1.5 to 2.0. Taking into account the notation (13.19) *strength condition for plastic materials* can be written as

$$\sigma \leqslant [\sigma], \tag{13.20}$$

Moreover, σ is the highest calculated voltage in absolute value.

For brittle materials, a distinction is made between the permissible tensile stress $[\sigma]_s$ and the compressive stress $[\sigma]_c$. They are defined by the formulas

$$[\sigma]_s = \frac{\sigma_{vr}^s}{n_s}, \quad [\sigma]_c = \frac{\sigma_{Bp}^c}{n_s}, \tag{13.21}$$

where n_s is *standard safety factor of strength in relation to ultimate strength*. As a rule, for the safety factor of brittle materials, take on higher values than for plastic bodies ($n_s \in [2; 5]$). Strength conditions for brittle materials are written as

$$\sigma_s \leqslant [\sigma]_s, \quad \sigma_c \leqslant [\sigma]_c, \tag{13.22}$$

where σ_s, $and\sigma_c$ are the highest absolute design stresses in the stretched and compressed element, respectively.

The strength analysis based on the conditions (13.20) or (13.22) is called the *allowable stress analysis*. The maximum load at which these conditions are met is called the allowable load. Strength conditions (13.20) and (13.22) allow solving three types of problems.

1. Verification of stresses by comparing the highest absolute value of the calculated stress in the most loaded element with the permissible value. Such a calculation is sometimes called *verification*.
2. Selection of the cross-section of the bar. Bearing in mind that $\sigma = |N|/F$, the required area (F_{ra}) of the cross-section of the bar is determined from the strength condition:

$$F_{ra} \geqslant \frac{|N|}{[\sigma]}. \tag{13.23}$$

Calculation by condition (13.23) is called *projection*.
3. Determination of the *carrying capacity* of the structural element. For example, let a steel bar be stretched by forces P. Then by the formula (13.20), we have

$$\sigma = \frac{P}{F} \leqslant [\sigma].$$

From here, the largest possible load value is determined:

$$P_{доп} \leqslant [\sigma]F. \tag{13.24}$$

The load determined by the formula (13.24) while holding the equal sign in it is called *the permissible load when calculating the permissible stresses*.

13.6 Statically Indeterminate Tasks

Consider the core system shown in Fig. 13.4. The system consists of three rods 1, 2, and 3 of the same stiffness EF, converging at the node A. A load P is applied to the node along rod 3. It is required to find the forces in the rods N_1, N_2, N_3.

If we try to find the forces in the rods from the equilibrium equations of the node A freed from constraints, then we immediately find those two equilibrium equations:

$$\sum X = 0: \quad N_1 \sin\alpha + N_2 \sin\beta = 0;$$
$$\sum Y = 0: \quad P - N_3 - N_2 \cos\beta - N_1 \cos\alpha = 0. \tag{13.25}$$

Fig. 13.4 Statically
indeterminate system

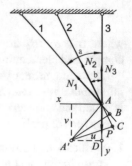

This will not be enough to find the three unknown efforts. In this situation, it is said that the equilibrium equations contain «extra» unknown.

The number of extra unknown forces determines *the degree of static indeterminacy of the system*. The definition of all unknown forces is called the *disclosure of static indeterminacy*.

The discovery of the static indeterminacy of the system becomes possible after drawing up as many additional equations as there are extra unknowns in the equilibrium equations. Additional equations are obtained from the consideration of the deformed state of the system, mathematically expressing the features of its geometric connections. Such relations that complement the equilibrium equations are called *compatibility of deformations*.

The system under consideration is once statically indeterminate, and to reveal the static indeterminacy, it is necessary to make one equation of the compatibility of deformations. To get the missing equation, consider the deformed state of the structure.

To do this, we give the node A (Fig. 13.4) an arbitrary displacement AA', and from the point A', we drop the perpendiculars $A'B$, $A'C$, and $A'D$ on the centerlines of the members. The indicated perpendiculars replace circular arcs along which the ends of the deformed bars should be moved to connect them at the point A'. Such a replacement is admissible on the basis of the *hypothesis of small deformations*. Segments AB, AC, and AD are equal to bar extensions Δl_1, Δl_2 and Δl_3, respectively.

Denoting the projections of AA' on the axis x and y by u and v, we project the polyline ADA' onto the directions of the rods. We get

$$\Delta l_3 = v,$$

$$\Delta l_1 = u \sin\alpha + v \cos\alpha,$$

$$\Delta l_2 = v \cos\beta + u \sin\beta.$$

Eliminating u and v from these ratios, we obtain the dependence between the deformations of the rods:

$$\Delta l_3 \sin(\beta - \alpha) = \Delta l_1 \sin\beta - \Delta l_2 \sin\alpha. \tag{13.26}$$

Fig. 13.5 To the calculation of the permissible loads

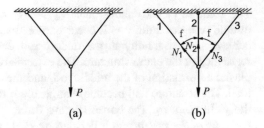

(a)   (b)

This is the deformation compatibility equation. Replacing deformations in it through efforts according to Hooke's law (13.10), we obtain the equation of compatibility in efforts:

$$2N_3 \sin(\beta - \alpha) \cos \alpha \cos \beta = N_1 \sin 2\beta - N_2 \sin 2\alpha. \qquad (13.27)$$

13.7 Calculation of Statically Indeterminate Systems by Permissible Loads

Consider a statically definable system of two identical plastic rods (Fig. 13.24a), loaded at the node by the force P. When the yield point is reached, the point of application of the force will receive significant displacement, even if only one rod is flowing. In general, in any statically definable system, fluidity in at least one element means the loss of connection and makes the system changeable.

Statically indeterminate systems behave differently. Let us determine the forces in the elements of a system of three identical ideally plastic rods (Fig. 13.5b). The equilibrium equations of a node freed from bonds will be

$$N_1 = N_3,$$

$$N_2 - P + (N_1 + N_3) \cos \varphi = 0. \qquad (13.28)$$

The equation of compatibility of deformations for the considered system will be

$$\Delta l_1 = \Delta l_2 \cos \varphi$$

or

$$N_1 = N_2 \cos \varphi. \qquad (13.29)$$

Joint solution of equations (13.28) and (13.29) gives

$$N_1 = N_3 = P \frac{\cos^2 \varphi}{1 + 2 \cos^3 \varphi}, \quad N_2 = P \frac{1}{1 + 2 \cos^3 \varphi}. \qquad (13.30)$$

The solution (13.30) shows that $N_2 > N_2$ and as the force P increases in the middle rod, the fluidity will occur earlier than in the extreme rods. However, the achievement of fluidity in the middle rod does not mean the exhaustion of the resistance of the entire structure. The extreme rods, while remaining elastic, prevent plastic deformation of the middle rod, and the system is able to carry an increasing load. The structure will become changeable if the stresses in the extreme rods reach the yield point σ_t. The corresponding force P in this state is called the *carrying capacity of the system* and is denoted by P_t. Let's define it from the condition

$$N_1 = N_2 = N_3 = \sigma_T F, \tag{13.31}$$

where F are the cross-sectional areas of the rods. Under the condition (13.31), from the equilibrium equations, we find

$$P_T = \sigma_T F (1 + 2 \cos \varphi). \tag{13.32}$$

Taking the same safety factor n for all rods, we obtain the following expressions for the permissible load:

(a) when calculating the allowable stresses:

$$\left(N_2 = \frac{\sigma_t}{n} F, \ N_1 = N_3 < \frac{\sigma_t}{n} F \right);$$

$$P_{av} \leqslant \frac{\sigma_t}{n} F (1 + 2 \cos^3 \varphi); \tag{13.33}$$

(b) when calculating the permissible loads:

$$\left(N_1 = N_2 = N_3 = \frac{\sigma_T}{n} F \right);$$

$$P_{av}^t \leqslant \frac{\sigma_t}{n} F (1 + 2 \cos \varphi). \tag{13.34}$$

Comparing the results of (13.33) and (13.34), we come to the conclusion that with the second method of calculation, the permissible load is greater than with the first (for $\varphi = 30°$ by 19%). Therefore, the calculation of statically indeterminate systems made of plastic materials for permissible loads is more preferable than the calculation for permissible stresses. The onset of fluidity in all elements of a system made of ideally plastic material is called its *limit state,* and the calculation based on permissible loads is often called *limit state calculation.*

It is taken into account that the lengths of rods 1 and 2 are expressed in terms of the length of the third rod in an elementary way. The joint solution of the equations (13.25) and (13.27) now allows us to find the forces in the rods.

Thus, the general plan for solving a statically indeterminate problem is as follows.

1. We make the equilibrium equations.
2. Considering the possible (allowed by the connections) displacements of the points of the system, we make equations that connect the deformations of individual elements (the equations of the compatibility of deformations).
3. We express in the equations of the compatibility of deformations the values of deformations in terms of forces (or stresses) according to Hooke's law.
4. We solve together the equations of equilibrium and the equations of compatibility of deformations in forces.

Self-Test Questions

1. What kind of loading of a bar is called stretching (compression)?
2. What is called the stiffness of the rod in tension (compression)?
3. Formulate the hypothesis of flat sections.
4. Internal forces during tension (compression) and the method of their determination.
5. How are the normal stresses in the cross-sections of the rod determined?
6. What is the difference between tension and internal effort?
7. Write down Hooke's law for uniaxial tension (compression).
8. Formulate the properties of the stress state under tension (compression).
9. How are deformations and displacements under tension and compression determined?
10. How are the movements of the nodes of the rod systems determined?
11. How to calculate the potential energy of elastic deformation of the rod under tension-compression?
12. What is the specific potential energy of elastic deformation?
13. How is the permissible stress determined in the case of plastic and brittle material?
14. Write down the strength conditions for ductile and brittle materials under tension (compression).
15. In which cases do the displacements of a centrally compressed rod not satisfy Hooke's law?
16. What is the essence of the calculation of structures by load-bearing capacity? What are the advantages of this calculation method?
17. To which structures can the load-bearing capacity calculation be applied?

References

V. Molotnikov, *Texnicheskaya mexanika* [*Technical Mechanics*] (Lan' Publ., Sant-Peterburg, 2017)
S. Timoshenko, *Strength of Materials. Part I* (Macmillan and Company Publ., 1947)

Chapter 14
Tense State: Strength Theories

Abstract Based on the assumption of the solidity of the body, it can be assumed that the internal forces are continuously distributed inside the body. The set of stresses acting on the areas drawn through the point under study constitutes the stress state at the point under consideration. Here, the elements of the theory of the stress state are considered, which are necessary for strength calculations for an arbitrary stress state at a point in the body. For such calculations in the strength of materials, one or another hypothesis is introduced about the predominant influence of a certain stress function on the strength of the material. These hypotheses are called strength theories.

Keywords Continuity hypothesis · Tense state · Main stresses · Generalized Hooke's law · Strength theory · Working theory of strength

14.1 Material Strength Hypotheses

It is impossible to reflect all the variety of properties of real materials in the theory of engineering calculation; therefore, a number of simplifying assumptions have to be made. These assumptions are called the (Timoshenko 1947; Rabotnov 1962) hypotheses.

Let us list these hypotheses.

1. All bodies are considered to be solid, homogeneous, and isotropic.
2. The body is considered to be absolutely elastic, i.e., after eliminating the causes of deformation, it completely restores its original shape and size.
3. The deformations of the structural material at each point are directly proportional to the stresses at this point (Hooke's law). Hooke's law is valid only at voltages not exceeding the proportionality limit.
4. In most cases, the deformations of structural elements are so small that one can ignore their influence on the mutual arrangement of loads and on the distance from the loads to any points of the structure.

5. The result of the impact on the structure of the system of loads is equal to the sum of the results of the impact of each load separately (the principle of independence of the action of forces).[1]
6. A cross-section that is flat before deformation remains flat after deformation (the Bernoulli flat section hypothesis)

14.2   Types of Stress

In Sect. 12.3, we introduced the concept of stress at a point in the material. From the definition of this value, it follows that the voltage at a point depends not only on the loads but also on the orientation of the area of its action.

In the general case, among this infinite set of areas, there are three mutually perpendicular areas on which there are no shear stresses. Such pads are called *principal,* and the normal stresses on them are called *principal stresses.* For principal stresses, the designations σ_1, σ_2, σ_3 are accepted, and the indices (1, 2, 3) are arranged so that the relations $\sigma_1 \geqslant \sigma_2 \geqslant \sigma_3$ are fulfilled.

In the case where only one main voltage is different from zero, the stress state is called *uniaxial or linear* (Fig. 14.1a,b). This stress state occurs in uniaxial tension-compression, and we were introduced to it in detail in the previous chapter.

Fig. 14.1 Types of stress

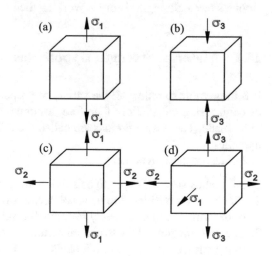

---

[1] The principle of the independence of the action of forces does not apply to the work of external and internal forces and to potential energy.

If one and only one of the three main stresses is zero, then such a stress state is called *flat* or *biaxial* (Fig. 14.1c). In the conditions of a flat stressed state, many machine parts work: thin plates, receiver walls, bent bars, etc.

The stress state at the point at which all three main stresses are not equal to zero is called *volumetric or triaxial* (Fig. 14.1d). The volumetric stress state is characteristic of the points of the material located in the vicinity of the contact zones of the contacting bodies (balls and bearing clips, shafts and bushings, etc.).

Note that in those special cases where all three or at least two main voltages are equal, the main sites become not three, but infinitely many.

14.3 Generalized Hooke's Law

As already mentioned (p. 330), homogeneous deformation is characterized by three relative elongations ($\varepsilon_x, \varepsilon_y, \varepsilon_z$) in the direction of the coordinate axes and three shear deformations of elements parallel to the coordinate axes ($\gamma_{xy}, \gamma_{yz}, \gamma_{zx}$). Using the formula (12.12) and the superposition principle, we obtain the following results: in an elastic body, deformations are associated with stresses *by the generalized law of Hooke*

$$\varepsilon_x = \frac{1}{E}[\sigma_x - \mu(\sigma_y + \sigma_z)],$$

$$\varepsilon_y = \frac{1}{E}[\sigma_y - \mu(\sigma_z + \sigma_x)],$$

$$\varepsilon_z = \frac{1}{E}[\sigma_z - \mu(\sigma_x + \sigma_y)]; \tag{14.1}$$

$$\gamma_{xy} = \frac{\tau_{xy}}{G}, \quad \gamma_{yz} = \frac{\tau_{yz}}{G}, \quad \gamma_{zx} = \frac{\tau_{zx}}{G}, \tag{14.2}$$

where G is the *shear modulus* associated (Molotnikov and Molotnikova 2015) with Young's modulus E and Poisson's ratio μ by envy

$$G = \frac{E}{2(1 + \mu)}. \tag{14.3}$$

The ratios (14.2) express Hooke's law at shear.

Formulas (14.1) and (14.2) are applicable for certain restrictions imposed on the values of stresses $\sigma_x, \ldots, \tau_{zx}$. In engineering structures, the specified restrictions often met by meeting the requirements dictated by conditions strength and (or) rigidity.

14.4   Plane Stress-Strain State

14.4.1   General Provisions

Let the relative elongation in the direction of one of the axes is equal to zero
(e.g., $\varepsilon_z = 0$). In this case, the movements of all points of the bodies during its
deformation occur parallel to the plane $z = 0$. Such a deformation is called *flat*.
   Assuming $\varepsilon_z = 0$, the last of the formulas (14.1) gives

$$\sigma_z = \mu(\sigma_x + \sigma_y),\tag{14.4}$$

and the first two of these formulas can be written as

$$\varepsilon_x = \frac{1}{E^*}(\sigma_x - \mu^*\sigma_y); \quad \varepsilon_y = \frac{1}{E^*}(\sigma_y - \mu^*\sigma_x),\tag{14.5}$$

where

$$E^* = \frac{E}{1 - \mu^2}, \quad \mu^* = \frac{\mu}{1 - \mu}.\tag{14.6}$$

   Keeping in mind that

$$\frac{E^*}{1 + \mu^*} = \frac{E}{1 + \mu},\tag{14.7}$$

we come to the following conclusion:
*relationship between relative elongations and normal stresses at plane strain ($\varepsilon_z =$
0) can be obtained from the corresponding plane stress state dependencies ($\sigma_z = 0$)
by replacing E and μ with E^* and μ^* using the formulas* (14.6).
   With the indicated replacement, the dependence between the shear stresses and
shifts due to the relations (14.7) and (14.3) will not change.

14.4.2   Analysis of the Plane Stress State

We cut out an infinitesimal parallelepiped with edges parallel to the axes x, y, z in
the vicinity of the point of the body. Let the stresses be zero on the areas parallel to
the xOy plane. This means that the areas normal to the Oz axis are the main ones.
Due to this condition and the law of tangential stress pairness, the components of
the tangential stress in the direction of the Oz axis at sites parallel to this axis must
be equal to zero. Such a general case of a plane stress state is shown in Fig. 14.2a.
   Let the stress state components σ_x, σ_y, τ be given. Let's set the task: to determine
the stresses on the site, the normal to which is with the axis Ox a certain angle
α. This angle is assumed to be positive if, in order to align with the one shown

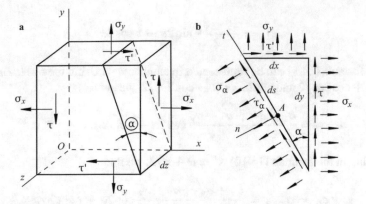

Fig. 14.2 To the analysis of the plane stress state

in Fig. 14.2a, inclined cross-section pad with normal voltage σ_x should be rotated counterclockwise.

Let us denote the normal stress on the inclined platform by σ_α and the tangent stress by τ_α. Figure 14.2b shows a frontal projection of a triangular prism cut off from the parallelepiped by an inclined plane. The lengths of the edges of the prism are denoted by dx, dy, ds, and dz (see Fig. 14.2). Through the point A, lying in the middle of the segment ds, draw the axis n directed along the normal to the inclined section, as well as the axis t perpendicular to it. Let us write down the conditions for the equilibrium of the prism, equating to zero the sum of the moments of all forces relative to the point A and the sum of the projections of all forces on the n and t axes:

$$\sum M_A = 0: \ \tau dydz\frac{dx}{2} - \tau'dxdz\frac{dy}{2} = 0;$$

$$\sum np_n F_i = 0: \ \sigma_\alpha dsdz - \sigma_y dxdz \sin\alpha - \tau'dxdz \cos\alpha -$$

$$-\tau dydz \sin\alpha - \sigma_x dydz \cos\alpha = 0;$$

$$\sum np_t F_i = 0: \ \tau_\alpha dsdz + \sigma_y dxdz \cos\alpha - \tau'dxdz \sin\alpha +$$

$$+\tau dydz \cos\alpha - \sigma_x dydz \sin\alpha = 0.$$

From the first equation, we obtain

$$\tau = \tau',$$

which expresses the already familiar (p. 329) law of pairing shear stresses. The other two equations give

$$\sigma_\alpha = \sigma_x \cos^2\alpha + \sigma_y \sin^2\sigma + \tau \sin 2\alpha, \tag{14.8}$$

$$\tau_\alpha = \frac{\sigma_x - \sigma_y}{2} \sin 2\alpha - \tau \cos 2\alpha. \tag{14.9}$$

The formula (14.8) can be represented in another way. Using the $\cos$ $identities^2$ $\alpha = (1 + \cos 2\alpha)/2$ and $\sin 2\alpha = (1 - \cos 2\alpha)/2$, we get

$$\sigma_\alpha = \frac{\sigma_x + \sigma_y}{2} + \frac{\sigma_x - \sigma_y}{2} \cos 2\alpha + \tau \sin 2\alpha. \tag{14.10}$$

Assuming in the formula (14.10) $\alpha' = \alpha + \pi/2$, we find

$$\sigma_{\alpha'} = \sigma_{\alpha + \frac{\pi}{2}} = \frac{\sigma_x + \sigma_y}{2} - \frac{\sigma_x - \sigma_y}{2} \cos 2\alpha - \tau \sin 2\alpha.$$

Summing the left and right sides of the last two formulas, we have

$$\sigma_\alpha + \sigma_{\alpha + \frac{\pi}{2}} = \sigma_x + \sigma_y. \tag{14.11}$$

Hence, as in the case of tension-compression, the *sum of the normal stresses on two mutually orthogonal sites is a constant value.*

We investigate at what value α the normal voltage σ_α has an extremum. Let's denote this value of the angle α by α_0. Equating to zero the derivative of the function (14.8) at $\alpha = \alpha_0$, we get

$$\text{tg} 2\alpha_0 = \frac{2\tau}{\sigma_x - \sigma_y}. \tag{14.12}$$

The formula (14.12) defines two mutually orthogonal sites where the normal stresses are σ_{α_0} and $\sigma_{\alpha_0 + \frac{\pi}{2}}$ extreme. It is easy to make sure that the tangential stresses at these sites are zero. This means that these sites are the main ones.

To calculate the principal stresses, we use the identities:

$$\cos 2\alpha_0 = \pm \frac{1}{\sqrt{1 + \text{tg}^2 2\alpha_0}}, \quad \sin 2\alpha_0 = \pm \frac{\text{tg} 2\alpha_0}{\sqrt{1 + \text{tg}^2 2\alpha_0}},$$

or, taking into account (14.12),

$$\cos 2\alpha_0 = \pm \frac{\sigma_x - \sigma_y}{\sqrt{(\sigma_x - \sigma_y)^2 + 4\tau^2}}, \quad \sin 2\alpha_0 = \pm \frac{2\tau}{\sqrt{(\sigma_x - \sigma_y)^2 + 4\tau^2}}.$$

Substituting these values into the formula (14.10) for $\alpha = \alpha_0$, we find the main stresses:

$$\sigma_{\substack{\max \\ \min}} = \frac{\sigma_x + \sigma_y}{2} \pm \frac{1}{2} \sqrt{(\sigma_x - \sigma_y)^2 + 4\tau^2}. \tag{14.13}$$

Thus, an arbitrary plane stress state is reduced to tension-compression in two mutually perpendicular directions. The main area over which the maximum stress σ_{max} acts is obtained by turning by an angle α_0 a site with a large normal stress in the algebraic sense in the direction in which the tangent stress vector on this site tends to rotate the element relative to any point on the internal normal.

We now examine the tangent stress τ_α at the extremum. Denoting by α_1 the value of α that delivers the extremum of the function (14.9), we get

$$\frac{d\tau_\alpha}{d\alpha}\bigg|_{\alpha=\alpha_1} = (\sigma_x - \sigma_y)\cos 2\alpha_1 + 2\tau \sin 2\alpha_1 = 0,$$

from here we find

$$\text{tg}2\alpha_1 = -\frac{\sigma_x - \sigma_y}{2\tau}. \tag{14.14}$$

The formula (14.14) gives two angle values α_1 that differ by $\pi/2$. Sites with extreme tangential stresses are called *shear sites*. On one of them, the maximum tangential stress is (τ_{max}), and on the other the minimum (τ_{min}), and by virtue of the law of pairness $\tau_{max} = -\tau_{min}$.

Comparing the formulas (14.14) and (14.12), we find

$$\alpha_1 = \alpha_0 + \frac{\pi}{4}. \tag{14.15}$$

Hence, the *shift pads are inclined to the main pads at an angle of* 45°. To calculate the extreme tangential stresses, put in the formula (14.9) $\sigma_x = \sigma_{max}$, $\sigma_y = \sigma_{min}$, $\alpha = \pm\pi/4$. We get

$$\tau_{\substack{max\\min}} = \pm\frac{\sigma_{max} - \sigma_{min}}{2}. \tag{14.16}$$

Replacing the main stresses in the formula (14.16) with their expressions in (14.13) gives the expression of extreme tangential stresses in terms of non-main stresses:

$$\tau_{\substack{max\\min}} = \pm\frac{1}{2}\sqrt{(\sigma_x - \sigma_y)^2 + 4\tau^2}. \tag{14.17}$$

14.5 Strength Theories

14.5.1 Question Statement

The most common task of engineering calculation is to estimate the strength of a part based on a known stress state. For simple types of stress states, such as uniaxial tension-compression or pure shear, this problem is solved simply. In fact, with these types of stress states, it is possible to determine experimentally for each material the amount of stress corresponding to the appearance of unacceptably large deformations or destruction. Such stresses are called *limit* or *dangerous*. For plastic materials, the ultimate stress is most often taken as the yield strength and for brittle materials the ultimate strength. As we have already said (see Sect. 13.5), the permissible stresses are set for dangerous stresses, which provide a given margin of safety. If at no point in the part does the current voltage exceed the permissible voltage, then the strength of the part is guaranteed.

However, most structural elements operate under complex stress conditions, where all three or at least two main stresses are nonzero. In these cases, the dangerous state can occur at different values of the relations $(\sigma_1/\sigma_2,\ \sigma_2/\sigma_3)$. The question arises: how can the set of stresses at the point of the material be compared with its mechanical properties in order to obtain a reliable conclusion about the strength?

This problem has occupied the minds of engineers since the time of Galileo. To judge the strength under a complex stress state, in principle, it is necessary to put an infinitely large number of experiments on the destruction of material samples under various combinations of main stresses. In practice, this is not feasible. Therefore, to establish the criterion of strength, researchers are forced to go the other way. One or another hypothesis about the predominant influence on the strength of a factor is introduced. These hypotheses are called *strength theories*.

14.5.2 Working Strength Theories

Of the many variants of the strength theory, the so-called third and fourth strength theories are widely used in modern strength calculations of products made of plastic materials that are equally resistant to tension and compression.

Is the Third strength theory proposed by Cod and Darwin in the late eighteenth century, based on the assumption that there is a limit value of the maximum tangential stress $\tau_{\lim}$, the excess of which causes the fluidity of the material. The yield condition is written as

$$\tau_{\max} = \tau_{\lim}. \tag{14.18}$$

To determine $\tau_{\lim}$, you can use the uniaxial tension experiment. In this case, based on the formula (13.8), we have

$$\tau_{\lim} = \frac{1}{2}\sigma_{\lim}.$$

Then, taking into account the expression (14.16), the limit state condition (14.18) is reduced to the form

$$\sigma_1 - \sigma_3 = \sigma_{\text{lim}}, \tag{14.19}$$

and the strength condition providing a given safety margin n will be

$$\sigma_1 - \sigma_3 \leqslant [\sigma] \quad \left([\sigma] = \frac{\sigma_{\text{lim}}}{n}\right). \tag{14.20}$$

The value on the left side of the limit equilibrium condition or strength condition according to one theory or another is called *equivalent stress*. We will denote it by the symbol σ_{equiv} indicating the number of the corresponding strength theory in the superscript. In the case of the third theory of strength, we have

$$\sigma_{\text{equiv}}^{III} = \sigma_1 - \sigma_3. \tag{14.21}$$

In the case of a plane stress state, when

$$\sigma_1 = \sigma_{\max}, \quad \sigma_2 = \sigma_{\min}, \quad \sigma_3 = 0,$$

the formula (14.21) takes the form

$$\sigma_{\text{equiv}}^{III} = \sqrt{(\sigma_x - \sigma_y)^2 + 4\tau^2}. \tag{14.22}$$

Moreover, in particular, if $\sigma_y = 0$, $\sigma_x = \sigma$, the last formula takes the form

$$\sigma_{\text{equiv}}^{III} = \sqrt{\sigma^2 + 4\tau^2}. \tag{14.23}$$

The fourth (energy) theory of strength was proposed at the beginning of the twentieth century. Huber and Mises. According to this theory, it is assumed that all stress states with the same specific potential energy of shape change are equally dangerous.

Generalizing the definition (13.17) of the specific potential energy of elastic deformation for the case of a triaxial stress state using Hooke's law (14.2), we can obtain

$$u = \frac{1}{2E}[\sigma_1^2 + \sigma_2^2 + \sigma_3^2 - 2\mu(\sigma_1\sigma_2 + \sigma_2\sigma_3 + \sigma_3\sigma_1)]. \tag{14.24}$$

Assuming in the formula (14.24) $\sigma_1 = \sigma_2 = \sigma_3 = \frac{1}{3}(\sigma_1 + \sigma_2 + \sigma_3)$, we get *the specific potential energy of the volume change:*

$$u_{\text{v}} = \frac{1 - 2\mu}{6E}(\sigma_1 + \sigma_2 + \sigma_3)^2. \tag{14.25}$$

The difference between the total specific potential energy (14.24) of elastic deformation and the specific elastic potential energy of volume change (14.25) determines the *specific potential energy of shape change:*

$$u_f = u - u_v = \frac{1+\mu}{6E}[(\sigma_1 - \sigma_2)^2 + (\sigma_2 - \sigma_3)^2 + (\sigma_3 - \sigma_1)^2]. \tag{14.26}$$

Using the expression (14.26), the equivalent stress according to the fourth strength theory is written as

$$\sigma_{equiv}^{IV} = \frac{1}{\sqrt{2}}\sqrt{(\sigma_1 - \sigma_2)^2 + (\sigma_2 - \sigma_3)^2 + (\sigma_3 - \sigma_1)^2}. \tag{14.27}$$

In a flat stress state ($\sigma_1 = \sigma_{max}$, $\sigma_2 = \sigma_{min}$, $\sigma_3 = 0$), the formula (14.27) takes the form

$$\sigma_{equiv}^{IV} = \frac{1}{\sqrt{2}}\sqrt{(\sigma_1 - \sigma_2)^2 + \sigma_1^2 + \sigma_2^2}, \tag{14.28}$$

where

$$\sigma_{1,2} = \frac{\sigma_x + \sigma_y}{2} \pm \frac{1}{2}\sqrt{(\sigma_x - \sigma_y)^2 + 4\tau^2}.$$

In particular, for $\sigma_y = 0$, $\sigma_x = \sigma$ the formula (14.28) takes the form

$$\sigma_{equiv}^{IV} = \sqrt{\sigma^2 + 3\tau^2}. \tag{14.29}$$

Calculations based on the third and fourth strength theories give similar results. Therefore, in the strength calculations of parts made of plastic materials, you can use any of these theories.

Fifth strength theory is suggested by Mora. The strength condition and the equivalent stress according to the fifth strength theory have the form:

$$\sigma_1 - \frac{\sigma_p}{\sigma_c}\sigma_3 \leqslant [\sigma]_p,$$

$$\sigma_{equiv}^{V} = \sigma_1 - \frac{\sigma_p}{\sigma_c}\sigma_3, \tag{14.30}$$

where σ_p, σ_c are the tensile and compressive strength limits of the material, respectively, and $[\sigma]_p$ is the allowable tensile stress.

A comparison of the dependencies (14.30) with the formulas (14.20) and (14.19) leads to the conclusion that for plastic materials with equal limits of tensile and compressive strength, Mohr's theory passes into the third theory of strength. Therefore, the fifth theory of strength can be successfully applied in the calculation of parts made of both plastic and brittle materials.

Despite its apparent universality, Mohr's theory suffers from the disadvantage that the strength condition is independent of the average main stress (σ_2). Experiments show that, in fact, the dependence of the strength condition on σ_2 is detected tactically for all materials.

Self-Test Questions

1. What is called the stress state at a point in the material?
2. Define the main sites and main stresses.
3. List the types of stress state at the point of the body.
4. Give an example of a tense state at a point of the body, when any platform is the main one.
5. Write down the formula for determining the extreme stresses in the plane stress state.
6. Calculate the normal stresses on the shear pads at the plane stress state.
7. Write down the mathematical expression of the generalized Hooke's law.
8. What is the limit voltage or dangerous voltage?
9. What is the equivalent voltage?
10. What is the essence of the Cod strength theory?
11. Formulate the fourth (energy) theory of strength.
12. Write down the expressions of the equivalent stress according to the third and fourth theories of strength.
13. To the calculation of which materials are the third and fourth theories of strength applicable?
14. Write down the equivalent stress and the Mora strength condition.
15. What are the disadvantages of Mohr's strength theory?

References

V. Molotnikov, A. Molotnikova, *Mexanika deformacij* [*Deformation Mechanics*] (RFNO Publ., Rostov-on-Don, 2015)

Y. Rabotnov, *Soprotivlenie materialov* [*Strength of Materials*] (Gosfiztexmzdat Publ., Moscow, 1962)

S. Timoshenko, *Strength of Materials. Part I* (Macmillan and Company Publ., 1947)

Chapter 15
Shear and Torsion

Abstract When shear in the cross-sections of the bar, only the shearing force is not zero. Shear deformation can be observed, for example, when cutting bars with scissors. There is a net shift, in which only tangential stresses act on the faces of the element. Torsion is a type of deformation in which only the torque acts in the cross-sections, and the other force factors are absent. Rotating and torsion bars are called shafts. Many machine parts are subject to shear or torsion. Therefore, this separate chapter is devoted to the methods of their calculation for strength and rigidity.

Keywords Slice · Shear · Torsion · Shafts · Stiffness · Polar moment of inertia · Polar moment of resistance · Load capacity · Linear twist angle · Strength condition · Stiffness condition

15.1 Shear

The phenomenon of shear can be observed, for example, when cutting a strip with scissors (Fig. 15.1) (Molotnikov 2006, 2017). The shift takes place when the shoulder h of the pair of forces Q is small.

Otherwise, shear is accompanied by bending. The deformation of a parallelepiped with a section of $1 - 2 - 3 - 4$ (Fig. 15.1) is different from the pure shear shown in Fig. 12.5, and, by virtue of the shear stress pairing law, the distribution of shear stresses on platforms $1 - 2$ and $3 - 4$ will not be uniform.

However, for small δ, one can conditionally accept

$$\tau = \frac{Q}{F},\qquad(15.1)$$

where F is the area of the slice.

The shear strength condition is written as

$$\tau \leqslant [\tau],\qquad(15.2)$$

Fig. 15.1 To conditional
calculations at shift

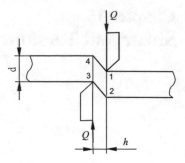

where $[\tau]$ is the allowable shear stress. In general engineering, take $[\tau] \cong (0.55 \ldots 0, 65)[\sigma]$.

The calculation of shear strength according to the formulas (15.2) and (15.1) is subjected to rivets, bolts, dowels, flank welds, and other structural elements.

15.2   Torsion of the Circular Shaft

Torsion is a type of loading in which only the torque occurs in the cross-sections of the beam and the remaining internal forces are zero. Twisted bars are often called *shafts*.

If the shaft transmits power N in kW at n rpm, then the torsional torque in N · m is determined by the formula

$$M = 9555\frac{N}{n}$$ (15.3)

or by the formula

$$M = 7026\frac{N'}{n},$$ (15.4)

where N'—horse power.

The torque in an arbitrary cross-section of a bar is determined by the section method as an algebraic sum of external torsion moments applied to any of the cut-off parts

$$M_k = \sum_{i=1}^{n} M_i + \int_0^x m(x)dx,$$ (15.5)

where M_i is the concentrated external moments and $m(x)$ is the intensity of the distributed torsion moment in the section with the coordinate x.

Fig. 15.2 Torsion of a
thin-walled tube

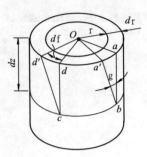

When constructing the theory of torsion of a round beam, we will take the
following assumptions:

1. In the cross-sections of the beam, only tangential stresses occur.
2. The beam is deformed by turning the cross-sections like hard disks.

Numerous experimental studies have shown that the actual nature of the torsion
deformation of circular shafts confirms the accepted hypotheses. From assumption
1 and the law of pairness of tangential stresses, it follows that in the axial sections
of the shaft, there are tangential stresses parallel to the generators.

We select a tube from the shaft with two coaxial surfaces of the radius-whiskers
ρ and $\rho + d\rho$ (Fig. 15.2). For an infinitesimal tube thickness, the distribution
of tangential stresses in its cross-section can be considered uniform. Two cross-
sections of the tube located at a distance of dz (Fig. 15.2), due to twisting, will
rotate relative to each other at an angle of $d\varphi$. In this case, the element $abcd$ will
undergo a shift, so that the generators ab and cd will rotate by a certain angle γ.
Calculating the arc length dd' from the triangles dcd' and Odd', we get

$$dd' = \rho d\varphi = \gamma dz.$$

From here

$$\gamma = \rho \frac{d\varphi}{dz}. \tag{15.6}$$

Let's introduce the notation

$$\Theta = \frac{d\varphi}{dz}. \tag{15.7}$$

This value gives the relative angle of rotation of two sections that are separated
from each other by a distance equal to one. We will call it the *linear twist angle.*
It follows from hypothesis 2 that for all coaxial tubes, the twist Θ in a given cross-
section will be the same. Using Hooke's law at shift (14.2), from the formula (15.6),
we find

$$\tau = G\Theta\rho. \tag{15.8}$$

Fig. 15.3 Static equivalent of
torsional stresses

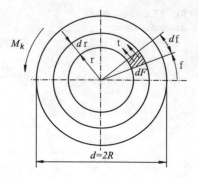

From the resulting formula, it follows that the tangential stresses in the cross-
section of the round shaft change in proportion to the distance from its axis.

Let us now consider the static side of the problem. Figure 15.3 shows a cross-
section of a shaft of radius R and an infinitely thin tube extracted from it. The
tube element dF, which is spaced from the axis at a distance ρ, is affected by the
tangential stress τ. The elementary force τdF gives a torque relative to the shaft
axis:

$$dM_k = \rho \tau dF.$$

By integrating over the cross-sectional area, we obtain the torque in the cross-
section:

$$M_k = \int_F \tau \rho dF = G\Theta \int_F \rho^2 dF. \tag{15.9}$$

The area integral F, which is included in the resulting formula, is called *of the
polar moment of inertia* and is denoted by

$$J_p = \int_F \rho^2 dF. \tag{15.10}$$

This integral for a circular cross-section is expressed in quadratures. In fact,
noting that in polar coordinates $dF = \rho d\rho d\varphi$ and the integration extends over
the entire area of the circle, we get

$$J_p = \int_0^{2\pi} d\varphi \int_0^R \rho^3 d\rho = \frac{\pi R^4}{2},$$

or through the diameter

$$J_p = \frac{\pi d^4}{32} \cong 0,1d^4. \tag{15.11}$$

Taking into account the notation (15.10), we rewrite the formula (15.9) in the form

$$\Theta = \frac{M_k}{GJ_p}. \tag{15.12}$$

The product of GJ_p is called *the torsional rigidity of the bar*. Knowing the linear angle of the twist Θ, it is easy to determine the angular displacement φ of sections that are separated from each other at a distance l. According to the definition (15.7) and the formula (15.12)

$$d\varphi = \frac{M_k dz}{GJ_p},$$

where from

$$\varphi = \int_0^l \frac{M_k dz}{GJ_p}. \tag{15.13}$$

If the torque and shaft stiffness are constant over the entire length of l, then

$$\varphi = \frac{M_k l}{GJ_p}. \tag{15.14}$$

Let us now return to the expression (15.8). Substituting into it the linear twist angle according to (15.12), we find

$$\tau = \frac{M_k \rho}{J_p}.$$

From the last result, it follows that the value of the greatest tangential stress will be

$$\tau_{max} = \frac{M_k \rho_{max}}{J_p}.$$

Denote

$$W_p = \frac{J_p}{\rho_{max}}. \tag{15.15}$$

This value is called *polar moment resistances*. Using the formula (15.11), for a circular shaft, it is not difficult to calculate that $W_p = 0.2d^3$.

Taking into account the notation (15.15), the formula for determining the maximum tangential stress takes the form

$$\tau_{max} = \frac{M_k}{W_p}. \tag{15.16}$$

The formulas (15.16) and (15.14) are the main calculation formulas for torsion of a round beam.

In conclusion, we note that the accepted hypotheses and the entire conclusion are also valid for tubular shafts. The only difference is that the geometric characteristics of J_p and W_p for the tubular shaft will be different. Denoting the inner diameter of the pipe by d_1, and the outer diameter by d, we get from the definitions (15.10) and (15.15)

$$J_p = \frac{\pi(d^4 - d_1^4)}{32}, \quad W_p = \frac{\pi d^3}{16}\left(1 - \frac{d_1^4}{d^4}\right). \tag{15.17}$$

15.3 Calculation of Shafts of Circular Cross-Section for Strength and Torsional Rigidity

By subjecting a thin-walled tube to torsion, we can experimentally establish the relationship between the tangential stress τ and the shear γ. A graphical representation of this relationship for low-carbon steel is shown in Fig. 15.4. Comparing it with the stretching diagram (Fig. 12.8), we find that these diagrams are similar to each other. As in the stretching diagram, here you can note the yield strength at the shear τ_t. Experiments show that for most structural materials, the shear yield strength is $(0.55\ldots0.6)\sigma_t$.

For the permissible shear stress for plastic materials, the value is taken as

$$[\tau] = \frac{\tau_T}{n_T}, \tag{15.18}$$

where n_t is the coefficient of the safety margin with respect to the yield strength.

For brittle materials, the permissible tangential stress can be approximated as the ultimate tensile strength:

$$[\tau] \cong [\sigma]_s. \tag{15.19}$$

Now you can write down the torsional strength condition:

$$\tau_{max} \leqslant [\tau], \tag{15.20}$$

Fig. 15.4 Tube torsion stress diagram

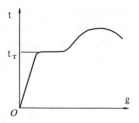

Moreover, τ_{max} is the greatest tangential stress in the dangerous section of the beam. For a round cross-section beam, the strength condition (15.20) takes the form

$$\frac{M_k}{0,2d^3} \leqslant [\tau], \tag{15.21}$$

where M_k is the torque in the dangerous cross-section.

For a project calculation, the condition (15.21) is usually represented by as

$$d \geqslant \sqrt[3]{\frac{M_k}{0.2[\tau]}}. \tag{15.22}$$

When determining the load capacity (permissible torque), the formula (15.21) gives

$$M_k^{\text{acc}} \leqslant 0.2[\tau]d^3 \text{ or } M_k^{\text{acc}} \leqslant \frac{\pi}{16}[\tau]d^3. \tag{15.23}$$

For real structures, the strength condition (15.20) is necessary, but not always sufficient. Often, even if this condition is met, unacceptably large twists can occur, which lead to the appearance of torsional vibrations of the shafts that are dangerous for the strength. Therefore, in addition to calculating the strength, you also need to calculate the stiffness. The torsional stiffness condition has the form

$$\Theta_{max} \leqslant [\Theta], \tag{15.24}$$

where Θ_{max} is the largest linear twist angle and $[\Theta]$ is its allowed value. In general mechanical engineering and in non-responsible calculations, most often take $[\Theta] = (2\ldots4) \cdot 10^{-5}$ рад/см.

Self-Test Questions

1. Write down the shear strength condition.
2. Name the scope of the cross-section calculations.
3. Write down the formulas for calculating the torsional torque when setting the transmitted power in kilowatts and in horsepower.
4. What assumptions are made when constructing the theory of torsion of a round cross-section beam?
5. By what law do the tangential stresses change during torsion of the circular shaft?
6. Write down the formulas for calculating the polar moments of inertia and the polar moments of resistance for a round and tubular shaft.

7. Give the formulas for calculating the tangential stresses and torsion angles when the circular shaft is twisted.
8. What is the relationship between the yield strength in tension-compression and torsion for plastic materials?
9. What is the value taken for the permissible tangential stress when torsion of a shaft made of brittle material?
10. Write down the strength condition and the torsional stiffness condition.

References

V. Molotnikov, *Kurs soprotivleniya materialov[The Course of Strength of Materials]*, (Lan' Publication, SPb., Moscow, 2006)
V. Molotnikov, *Tekhnicheskaya mekhanika [Technical Mechanics]* (Lan' Publication, Sankt-Peterburg, 2017)

Chapter 16
Bending

Abstract Bending is a type of deformation in which the axes of straight bars are bent or the curvature of the axes of curved bars changes. Bending occurs when bending moments act in the cross-sections of the beam. A straight bend of a beam occurs when, if the bending moment in a given cross-section of the beam acts in a plane passing through one of the main central axes of inertia of this section, the bend is called straight. Otherwise, they talk about an oblique bend. This chapter discusses methods for determining internal forces, stresses, and deformations in direct bending. The differential equation of the curved axis of the beam and the conditions of its strength and rigidity are also derived.

Keywords Bending · Straight bend · Transverse force · Bending moment · Deflection · Strength under permissible stresses · Zhuravsky formula · Strength under tangential stresses · Application of strength theories · Flexural displacements

16.1 The Concept of Bending Deformation

Consider a bar that is under the action of loads perpendicular to its axis. This type of loading of the beam is called *bending*. Bars that work mainly on bending are called *beams*. When elastic bodies are deformed, the action of a system of forces can be considered as the sum of the effects of applying these forces singly. This position is called the *superposition principle*.

Based on the superposition principle, for a qualitative consideration of the bend, we assume that a single force P is applied at the end of the bar, and the other end is pinched immobile (Fig. 16.1). Let the length of the bar be l, and the cross-section be a figure whose two dimensions b and h are of the same order. It is obvious that the destruction of the beam can occur in a cross-section close to the sealing. It is said that this most loaded cross-section is *dangerous*. Let's find out what stresses occur in the dangerous cross-section.

First of all, we note that in order to balance the moment of force P relative to the axis located in the cross-section plane, it is necessary to consider the normal

© The Author(s), under exclusive license to Springer Nature Switzerland AG 2023   369
V. Molotnikov, A. Molotnikova, *Theoretical and Applied Mechanics*,
https://doi.org/10.1007/978-3-031-09312-8_16

Fig. 16.1 To the concept of
bending

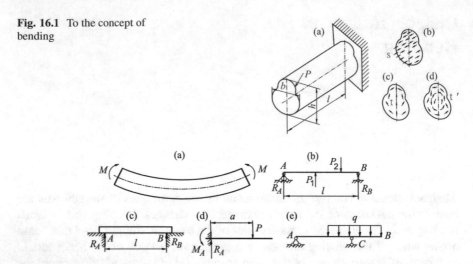

Fig. 16.2 Types of beams

stresses σ in the cross-section (Fig. 16.1b). Since there are no forces along the axis
of the beam, there will be normal stresses of various signs in the cross-section: in
the upper zone they are stretching, and in the lower zone they are compressing.

Next, the force P creates a scissor effect, i.e., it tends to cut the beam. To
balance the force P in the cross-section, you need to apply tangential stresses τ
directed upward (not necessarily parallel to the force P, Fig. 16.1c). The law of
their distribution is unknown. Let's call them *tangential bending stresses*.

Finally, the force P causes, generally speaking, torsion of the bar. Torsional
stresses, if they exist, are denoted by τ'; they are shown in Fig. 16.1 r. If the cross-
section has a plane of symmetry, and the force P lies in this plane, then, obviously,
torsional stresses will not occur. For any profile, there is a point called *the center of
the bend,* such that if the line of action of the force P passes through this point, then
torsion does not occur.

For the time being, we will assume that the force P passes through the center
of the bend, and the stresses τ' are excluded from consideration. Moreover, we
will here consider beams whose profile has at least one axis of symmetry and all
perpendicular axes of the load are located in the plane of symmetry. This case of
loading the beam is called *flat bending.*

The beam can have different types of anchors or have no supports at all. So,
Fig. 16.2a shows a bar in equilibrium under the action of two pairs of opposite
directions applied to its ends. It is not difficult to make sure that there are no
transverse forces in any section of this bar. Such a case of loading a bar is called
a *clean bend.*

Figure 16.2b shows a beam with freely supported ends. The support points A
and B are hinges, so that the ends of the beam can turn freely when bending.
The distance l between the supports is called *span.* Special devices that provide

Fig. 16.3 The pattern of bending deformation

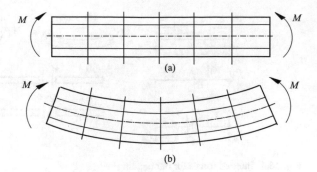

free rotation of the ends and horizontal movement of the support B are used only in beams of large spans (bridges, long shafts, etc.). For beams of small span, the supports are usually arranged as shown in Fig. 16.2c. With this support, the friction forces between the support surfaces and the beam will prevent the horizontal movement of its ends. These forces may have some significance in the case of flexible bars. For a rigid beam whose deflections are small compared to the span, these forces can be ignored, and the reactions R_A and R_B can be calculated (Fig. 16.2c), as for a freely supported beam.

Figure 16.2d shows the *console*. The end of A of such a beam is sealed and can neither turn nor move in any direction when bending. The second end of the *In* console is completely free, and when the beam is deformed, it can receive linear and angular movements in any direction.

Figure 16.2d shows a diagram of a double-hinged beam with an overhanging end. Such beams are called *cantilever beams*. They have a pivot-fixed support at the end of A and a movable support C. The end of the beam B is free and can move in any direction when deformed.

To identify the features of the bending deformation, imagine that parallel lines are drawn on the side surface of the beam in the longitudinal and transverse directions (Fig. 16.3a), forming a grid of rectangles. As a result of bending, these rectangles turn into curved quadrilaterals (Fig. 16.3b); their upper bases are shortened, and their lower ones are lengthened. The sides of the rectangles do not experience noticeable curvature, but become non-parallel.

Experiments show that not only the sides of the rectangles remain straight, but also any cross-section of the beam, initially flat, remains flat after bending and normal to the longitudinal fibers. Thus, when bending, the *hypothesis of flat sections (the Bernoulli hypothesis) is accepted.*

16.2 Internal Bending Forces

Consider the double-jointed beam shown in Fig. 16.4a. After freeing the beam from the bonds and determining the reactions, we will get the calculation scheme shown in Fig. 16.4b. Mentally cut the beam with the plane $n - n$, perpendicular to the axis

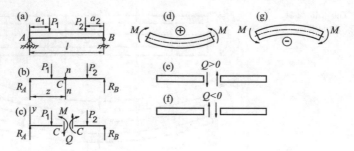

Fig. 16.4 Internal forces for flat bending

of the beam. To balance each of the cut-off parts of the beam in the cross-section, it is necessary to apply bending moments M and transverse forces Q (Fig. 16.4c). Let us assume that the bending moment in the cross-section is positive if it deforms a horizontally positioned beam with a downward bulge. Figure 16.4d; otherwise, the bending moment is negative (Fig. 16.4d). The transverse (cutting) force in the cross-section is considered positive if it rotates each of the cut-off parts of the beam clockwise, as shown in Fig. 16.4e. The negative transverse force has the opposite direction (Fig. 16.4f).

To determine the internal forces M and Q, we make the equilibrium equations of any of the cut-off parts of the beam:

– for the left part

$$\sum Y = 0 : \quad R_A - P_1 - Q = 0; \quad \Rightarrow \quad Q = R_A - P_1 = \sum_{\text{lev}} Y;$$

$$\sum M_C = 0 : \quad -R_A z + P_1(z - a_1) + M = 0; \quad \Rightarrow$$

$$M = R_A z - P_1(z - a_1) = -\sum_{\text{лев}} M_C(P_i), \tag{16.1}$$

where $\sum\limits_{\text{lev}} Y$ denotes the sum of projections on the y axis (Fig. 16.4c) of all external (active and reactive) forces located to the left of the section $n - n$, in which internal forces are defined; $\sum\limits_{\text{lev}} M_C(P_i)$ respectively denotes the sum of the moments of all the left forces relative to the point C;

– for the right part

$$\sum Y = 0 : \quad R_B - P_2 + Q = 0; \quad \Rightarrow \quad Q = P_2 - R_B = -\sum_{\text{прав}} Y;$$

$$\sum M_C = 0 : \quad R_B(l - z) = P_2 l - z - a_2 - M = 0; \quad \Rightarrow$$

$$M = R_B(l - z) - P_2(l - z - a_2) = \sum_{\text{прав}} M_C(P_i). \tag{16.2}$$

Thus, the following statements are true:

(1) The transverse force in any section of the beam is numerically equal to the algebraic sum of the projections on the y axis of all external forces located to the left of the section or with the sign « minus» the sum of the projections on the y axis of all external forces acting to the right of the section.
(2) The bending moment in the cross-section of the beam is numerically equal to the algebraic sum of the moments (calculated relative to the center of gravity C of the cross-section) of all external forces acting on one side of the cross-section.

16.2.1 *Example of Defining Internal Effort*

For the beam shown in Fig. 16.5a, determine the internal forces in any section and plot them.

   S o l u t i o n.

(1) We release the beam from the bonds (Fig. 16.5b) and determine the reference reactions:

$$\sum M_A = 0: \ -q \cdot 2a \cdot a + R_A \cdot 2a - P \cdot (2a + a) = 0; \ \Rightarrow \ R_B = \frac{5}{2}da;$$

$$\sum M_B = 0: \ -R_A \cdot 2a + q \cdot 2a \cdot a - Pa = 0; \ \Rightarrow \ R_A = \frac{1}{2}qa.$$

To check the correctness of the obtained results, we will make the third equation of equilibrium. Projecting all the forces on the vertical axis Y, we get

$$\sum Y = 0: \ R_A - q \cdot 2a + R_B - P = 0,$$

Fig. 16.5 Internal effort diagrams

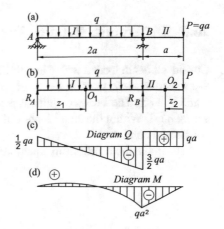

i.e.

$$\frac{1}{2}qa - 2qa + \frac{5}{2}qa - qa = 0.$$

From this, it can be seen that this equation is satisfied.

(2) Divide the beam into sections I and II. On the section I, we draw an arbitrary cross-section at a distance z_1 from the left support. Using the formulas (16.2) for the internal forces in the area under consideration, we can write

$$Q^I(z_1) = \sum_{\text{лев}} Y = R_A - qz_1; \ M^I(z_1) = R_A z_1 - qz_1 \cdot \frac{z_1}{2} \ (0 \leqslant z_1 \leqslant 2a).$$

Here and further on, the number of the section of the beam is indicated in Roman numerals in the upper indexes. Then, in the second section, we draw an arbitrary cross-section at a distance of z_2 from the right end of the beam. Using the formulas (16.2), we get

$$Q^{II}(z_2) = -\sum_{\text{rig}} Y = -(-P) = qa;$$

$$M^{II}(z_2) = \sum_{\text{rig}} M_{O_2} = -Pz_2 = -qaz_2;$$

$$(0 \leqslant z_2 \leqslant a).$$

According to the formulas found in Fig. 16.5c, d, plots of transverse forces and bending moments are constructed. The main purpose of the plots is to determine dangerous cross-sections in which the internal forces are either extreme or their combination gives an extreme equivalent stress according to a suitable strength theory.

16.3 Differential Bending Dependencies

Consider a beam loaded with a length-distributed load of intensity $q(z)$ (Fig. 16.6a). We select an element from the beam with two cross-sections with the coordinates z and $z + dz$. The selected element is shown in Fig. 16.6b together with the loads acting on it. Within the small length of the element dz, the load intensity q can be considered constant.

Let's make the equilibrium equations of the element:

Fig. 16.6 Differential dependencies in bending

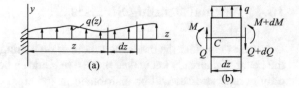

$$\sum Y = 0: Q + qdz - (Q + dQ) = 0; \Rightarrow \frac{dQ}{dz} = q;$$

$$\sum M_C = 0: -M + qdz \cdot \frac{dz}{2} - (Q + dQ)dz + M + dM = 0. \tag{16.3}$$

Ignoring in the second equation the summand $qdz \cdot \dfrac{dz}{2}$ as infinitesimal of higher order, we obtain after simple transformations

$$\frac{dM}{dz} = Q. \tag{16.4}$$

Differentiating the last equality (16.4) by the variable z, taking into account the formula (16.3), we find

$$\frac{d^2M}{dz^2} = q. \tag{16.5}$$

The dependencies (16.3)–(16.5) can be used to control the correctness of plotting internal forces in beams. In particular, the following properties of plots follow from them:

1. In those areas where there is no distributed load ($q = 0$), the plot Q is rectangular, and the plot M is rectilinear (see Fig. 16.5c, d (II)).
2. In areas where a uniformly distributed load q is applied, the plot Q is bounded by an inclined straight line, and the plot M is bounded by a square parabola. In this case, the line bounding the plot Q is tilted in the direction of the action q, and the parabola representing the plot M is turned by a bulge toward q (see Fig. 16.5c, d (I)).
3. In the section where the concentrated force is applied, the plot Q has a jump by the amount of this force, and the plot M has a fracture (corner point) (see the section B, Fig. 16.5).
4. The bending moment M reaches the extremum in those sections where $Q = 0$.
5. In the section where a concentrated moment is applied, the plot M has a jump by the value of this moment.

16.4 Normal Bending Stresses

Let us first consider the case when there is no cutting force in the cross-sections of the beam. We already know (see p. 370) that such a bend is called pure. In this case, only normal stresses will be distributed in the cross-section (Fig. 16.7a). Selecting an elementary area dF in the vicinity of an arbitrary cross-section point, which is removed from the x axis by a distance y, we calculate the elementary moment of the forces acting on dF, relative to the x axis:

$$dM_x = \sigma y dF.$$

Summing up such moments over the entire cross-section, we find

$$M_x = \int_{(F)} \sigma y dF. \qquad (16.6)$$

To find the law of distribution of normal stresses over the cross-section of a beam, consider its deformation. Figure 16.7b shows a section of a beam that is in pure bending conditions with moments M. Let us distinguish between two cross-sections $I - I$ and $II - II$ an element of length dz and note the fibers ab, cd, and ef parallel to the axis of the beam. Due to the deformation, the beam bends (Fig. 16.7c), so that the sections $I - I$ and $II - II$ will rotate relative to each other by an angle $d\vartheta$. In this case, the fibers located near the convex contour of the beam are lengthened, and on the side of the concave contour, they are shortened. Logic suggests that there is a *neutral layer of material whose* fibers do not change their length. The line of

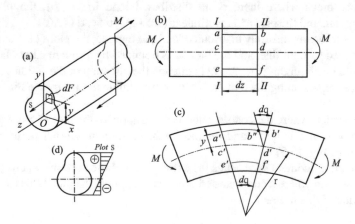

Fig. 16.7 Normal bending stresses

intersection of the neutral layer with the cross-sectional plane of the beam is called the *neutral line* or *neutral axis*.

Suppose that cd is a fiber of the neutral layer. ab and ef fibers after the deformation, they will go to $a'b'$ and $e'f'$, respectively. Let us draw from the point d' (Fig. 16.7c) a straight line $d'b''$ parallel to $a'c'$. Then the segment $b'b''$ represents the absolute elongation of the fiber ab due to the deformation of the beam. The relative elongation of this fiber will be

$$\varepsilon = \frac{a'b' - ab}{ab} = \frac{b'b''}{cd} = \frac{y d\vartheta}{\rho d\vartheta} = \frac{y}{\rho}. \tag{16.7}$$

In this formula, the radius of curvature of the fibers of the neutral layer is denoted by ρ (Fig. 16.7c).

Assume that the fibers of the beam do not press on each other, and the normal stresses in the direction perpendicular to the axis of the beam are zero. Under this assumption, the fibers parallel to the beam axis experience uniaxial tension or compression. Then by Hooke's law:

$$\sigma = \varepsilon E = \frac{yE}{\rho}, \tag{16.8}$$

where E is Young's modulus. Thus, *the normal bending stresses are proportional to the distance from the neutral layer.* In Fig. 16.7d a plot of normal stresses is constructed, graphically representing the law (16.8).

Consider the balance of the beam section shown in Fig. 16.7i. Projecting all the forces on its axis, we get

$$\sum Z = 0: \int\limits_{(F)} \sigma dF = 0.$$

Substituting the value of the normal voltage according to the formula (16.8), from here we find

$$\frac{E}{\rho} \int\limits_{(F)} y dF = 0.$$

Remembering (see Molotnikov 2004) that the number determined by the integral included in this formula is the static moment S_x of the cross-section relative to the x axis, we get

$$S_x = 0. \tag{16.9}$$

This means that the neutral axis x is central, i.e., it passes through the center of gravity of the section.

The second condition for the balance of the beam element under consideration is

$$\sum M_y = 0 : \Rightarrow \int\limits_{(F)} \sigma x dF = 0; \quad \frac{E}{\rho} \int\limits_{(F)} xy dF = 0.$$

The integral included in the last result is the centrifugal moment of inertia J_{xy} with respect to the axes x, y coinciding with the neutral axis and the force line. The reversal of this centrifugal moment to zero together with the result (16.9) means that the neutral axis and the force line are the main central axes of the cross-section of the beam.

The third condition of equilibrium gives

$$\sum M_x = 0 : \Rightarrow \int\limits_{(F)} \sigma y dF = M; \quad M = \frac{E}{\rho} \int\limits_{(F)} y^2 df.$$

The integral in the last formula is the moment of inertia J_x of the section relative to the x axis. Therefore

$$\frac{1}{\rho} = \frac{M}{E J_x}. \tag{16.10}$$

The product of the Young's modulus $E J_x$ at the moment of inertia of the section is called the *stiffness* of the beam when bending relative to the x axis. Substituting the result (16.10) into the formula (16.8), we obtain the following formula for normal stresses in pure bending:

$$\sigma = \frac{M y}{J_x}. \tag{16.11}$$

From the formula (16.11), it follows that the greatest modulo normal stresses occur in the fibers that are most distant from the neutral axis, i.e.,

$$\sigma_{max} = \frac{M}{W_x}, \tag{16.12}$$

where indicated

$$W_x = \frac{J_x}{|y_{max}|}. \tag{16.13}$$

The value W_x, defined by the formula (16.13), is called the *moment of resistance* of the cross-section relative to the x axis.

Experimental verification showed that the formulas (16.11)–(16.12) are also applicable for transverse bending, when, in addition to the bending moments, the transverse forces are different from zero.

16.5   Tangential Bending Stresses

Having established the law of distribution of normal stresses, we will now deal with the tangents bending stresses. Their appearance is caused by the presence of transverse forces. To determine the tangential stresses, first consider a beam of rectangular cross-section $b \times h$ (Fig. 16.8). Mentally, we will cut out an element from the beam in the form of a rectangular parallelepiped bounded by two cross-sections $I - I$ and $II - II$, separated from each other at a distance of dz, as well as a horizontal plane separated from the neutral layer at a distance of y.

The right side of the selected parallelepiped is affected by normal stresses distributed according to the law (16.11)

$$\sigma = \frac{My}{J_x},$$ (16.14)

where M is the bending moment in the cross-section $II - II$, as well as the tangential stresses, the law of distribution of which over the cross-section is still unknown. Denote by τ the value of the tangential stress on the site $II - II$ at the points adjacent to the lower right edge of the parallelepiped.

Based on the law of pairness of tangential stresses, the same tangential stresses act on the lower face of the selected parallelepiped. In the section $I - I$, removed from the right side by a distance dz, the bending moment will receive an increment dM and will be equal to $M+dM$. Then the normal voltages according to the formula (16.11) will be

$$\sigma + d\sigma = \frac{(M + dM)y}{J_x}.$$ (16.15)

The tangential stresses near the lower left edge of the parallelepiped on its left side are equal in magnitude to τ and are directed as shown in Fig. 16.8.

Consider the equilibrium of the selected parallelepiped. Equating to zero the sum of the projections on the z axis of all the forces applied to it, we have

$$\sum Z = 0: \int_{(F_1)} \sigma \, dF_1 + \tau b \, dz - \int_{(F_1)} (\sigma + d\sigma) dF_1 = 0,$$

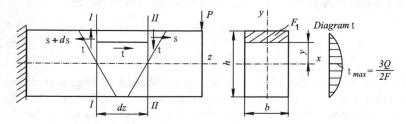

Fig. 16.8 To the conclusion of the Zhuravsky formula

where F_1 is the area of the side face of the parallelepiped, shaded in Fig. 16.8. Substituting into this equation the values of normal stresses according to the formulas (16.14) and (16.15), we can write

$$\frac{M}{J_x} \int\limits_{(F_1)} y\,dF_1 - \frac{M+dM}{J_x} \int\limits_{(F_1)} y\,dF_1 + \tau b\,dz = 0.$$

The integral included in this expression is the static moment relative to the x axis of the area F_1 cut off by the horizontal platform with the voltage τ. Denoting it by S_x^{ots}, i.e.,

$$S_x^{\text{ots}} = \int\limits_{(F_1)} y\,dF_1, \qquad (16.16)$$

we reduce the equilibrium equation to the form

$$\frac{S_x^{\text{ots}}}{J_x} \cdot \frac{dM}{dz} = \tau b.$$

But in Sect. 16.3 we showed that the derivative of the bending moment in the z coordinate is equal to the transverse force (see the formula (16.4)). Therefore, we finally have

$$\tau = \frac{Q S_x^{\text{ots}}}{J_x b}. \qquad (16.17)$$

This formula was obtained by the Russian engineer D.I. Zhuravsky and bears his name. Using the Zhuravsky formula, we investigate the distribution of tangential stresses over the beam cross-section. To do this, we calculate the static moment S_x^{ots} as the product of the area F_1 and the distance from its center of gravity to the axis x:

$$S_x^{\text{ots}} = \frac{b}{2}\left(\frac{h^2}{4} - y^2\right).$$

Considering also that the moment of inertia of the rectangle relative to the central axis x is equal to $J_x = bh^3/12$ (see, e.g., Molotnikov 2006), we get

$$\tau = \frac{6Q}{bh^3}\left(\frac{h^2}{4} - y^2\right). \qquad (16.18)$$

In the right position, Fig. 16.8 a plot of the distribution of tangential stresses is constructed using the formula (16.18). As follows from the formula (16.18), the tangential stresses in the cross-section change according to the law of the square parabola. The maximum tangential stresses occur at the level of the neutral layer of the beam and are equal to

$$\tau_{max} = \frac{3}{2} \cdot \frac{Q}{F}. \tag{16.19}$$

Here, Q/F is the shear force divided by the cross-sectional area, i.e., the average shear stress. Therefore, when a rectangular beam is cross-bent, the maximum tangential stress is one and a half times greater than the average shear stress.

Remark The Zhuravsky formula is obtained for a bar of rectangular cross-section. Experimental testing has shown that the approximate formula is also correct when bending bars of a different profile. It should only be remembered that the formula (16.18) gives the component of the tangential stress parallel to the line of force.

16.6 Strength Calculation Based on Permissible Stresses

As it was established above (p. 376), the normal bending stresses are determined by the formula

$$\sigma = \frac{My}{J_x}.$$

Keeping in mind that at $M > 0$, the upper fibers of the beam are stretched and the lower fibers are compressed, the positive direction of the y axis must be selected from top to bottom. After this remark, we will proceed to the strength calculations.

When calculating the strength of a beam according to the permissible stresses, we must require that at no point in the cross-section does the greatest tensile stress exceed the permissible tensile stress $[\sigma]_p$ and the greatest compressive stress modulo does not exceed the permissible compressive stress $[\sigma]_c$.

Let us denote the distances from the neutral axis to the extreme points in the stretched and compressed zones by h_p and h_c (Fig. 16.9). Therefore, the strength conditions for the permissible stresses will be

$$\frac{|M| h_c}{J_x} \leqslant [\sigma]_c; \quad \frac{|M| h_p}{J_x} \leqslant [\sigma]_p. \tag{16.20}$$

Fig. 16.9 To calculate the bending strength

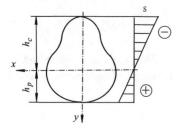

The most advantageous dimensions of the cross-section of the beam are obtained under the condition of equal reserves of tensile and compressive strength. This condition is satisfied by assuming

$$\frac{h_c}{h_p} = \frac{[\sigma]_c}{[\sigma]_p}. \tag{16.21}$$

If the material of the beam is equally resistant to tension and compression, then the strength calculation is carried out according to the highest modulus of stress, i.e., the calculated stress is taken as the

$$|\sigma|_{max} = \frac{|M_{max}| \cdot |y_{max}|}{J_x}, \tag{16.22}$$

Moreover, $|M_{max}|$ is the largest bending moment in absolute value.

The ratio $J_x/|y_{max}|$ is called the *moment of resistance* to bending and is denoted by W_x, i.e.,

$$W_x = \frac{J_x}{|y_{max}|}. \tag{16.23}$$

For the most important figures, the moments of resistance are given by the formulas:

• circle of diameter d;
• pipe with outer diameter d and inner diameter d_1

$$W_x = \frac{\pi d^3}{32} \left(1 - \frac{d_1^4}{d^4} \right); \tag{16.24}$$

• rectangle with base b and height h

$$W_x = \frac{bh^2}{6}.$$

Taking into account the notation (16.23) and the formula (16.22), the strength condition of a plastic material beam is written as

$$|\sigma|_{max} = \frac{|M_{max}|}{W_x} \leqslant [\sigma]. \tag{16.25}$$

Note that in a stretched or compressed rod, the condition $|\sigma_{max}| = [\sigma]$ is achieved simultaneously in all fibers of the dangerous cross-section, and when bending, it occurs only in the fibers of the dangerous cross-section that are farthest from the neutral layer. Therefore, for a more complete use of the strength properties of the beam material, sometimes instead of the main permissible stress $[\sigma]$, the formula (16.25) is substituted with the permissible bending stress $[\sigma]_b$, the value of which is greater than $[\sigma]$.

The dependences expressed by the formulas (16.20) and (16.25) are the main conditions for bending strength. Their implementation is necessary, but not always sufficient to ensure the strength of the beam. In the presence of large transverse forces, a violation of the bending strength can occur due to the cut. Therefore, in addition to checking the strength of normal stresses (conditions (16.20) or (16.25)), in these cases, it is also necessary to check the strength of tangential stresses. This check is done according to the formula

$$\tau_{max} = \frac{|Q_{max}|S_{max}}{J_x b} \leqslant [\tau], \tag{16.26}$$

where $|Q_{max}|$ is the largest cutting force in modulus;

S_{max} is the maximum static moment relative to the neutral axis of the cross-sectional area that is cut off by the platform with the maximum tangential stress; $[\tau]$ is permissible shear stress.

The condition (16.26) is written for the section of the beam in which the greatest modulo transverse force acts and the condition (16.26) is checked in those fibers of this section where the maximum tangential stresses occur.

The performance of the strength conditions for normal (16.25) and tangential (16.26) stresses separately is necessary, but not always sufficient to ensure the strength of the beam. The strength will be guaranteed when the beam is fully calculated, when both normal and tangential stresses are taken into account at the same time. As a rule, thin-walled and composite beams are subjected to a full calculation in the presence of significant transverse forces. When the calculation is complete, the following steps are performed.

1. Check the strength or select the cross-section of the beam based on the conditions (16.20), and the normal stress test is performed for the cross-section with the maximum bending moment in modulus.
2. According to the condition (16.26), check whether the cross-section width is sufficient (the wall thickness of the cross-section of the I-beam type). This condition is written for the points of the neutral cross-section layer with the maximum cutting force in modulus.
3. Test the strength according to a suitable strength theory with simultaneous consideration of both normal and tangential stresses. For example, for plastic materials according to the fourth theory of strength, we have

$$\sigma_{eq}^{IV} = \sqrt{\sigma^2 + 3\tau^2} \leqslant [\sigma].$$

This condition is checked in each dangerous section. Dangerous are those sections in which the modules of the bending moment and the cutting force are simultaneously large, although, perhaps, not maximum. There may be several such cross-sections, and it is easiest to outline them by constructing and analyzing plots of internal forces. In each dangerous section, the equivalent stress strength condition is checked at the points where both normal and tangential stresses are simultaneously large, although not maximum. For example, for an I-beam, the points to be checked belong to the wall in the area where it adjoins the shelf.

16.7   Bending Movements

When calculating the beam, the engineer may be interested not only in the stresses arising from the acting loads but also in the values of deflections and sometimes the angles of rotation of the cross-sections of the beam. The fact is that in many structures, in addition to strength, limited movement is required. They say that the structure should be rigid. In addition, the knowledge of displacements is used in the calculation of statically indeterminate structures.

16.7.1   Differential Equation of the Curved Beam Axis

Let the arc AB (Fig. 16.10) represent the axis of the beam after bending. This curve is called the *curved axis of the beam.*

To derive the differential equation of the curve AB, we introduce the coordinate axes xy, shown in the figure, and use the formula (16.10), which relates the curvature $1/\rho$ of the curved axis of the beam to the bending moment

$$\frac{1}{\rho} = \frac{M}{EJ},$$

Fig. 16.10   Curved beam axis

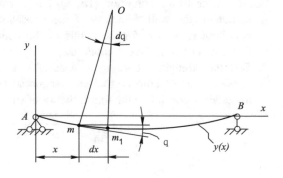

where EJ is the stiffness of the beam when bending relative to the main central axis z.[1]

The equation of the curve AB in the xy axes is written as $y = y(x)$, and we consider two adjacent points m and m_1 located on the curve at a distance of ds from each other. The angle formed by the tangent to AB at the point m with the abscissa axis is denoted by ϑ. As you know, the tangent of this angle is equal to the first derivative of the function $y(x)$ at the point m, i.e.,

$$\tan \vartheta = \frac{dy}{dx}.$$

Draw the normals to the curve $y(x)$ at the points m and m_1. The point O of the intersection of these normals at $ds \to 0$ defines the center of curvature of the curved axis of the beam at the point m. Denoting the angle between the normals by $d\vartheta$, we can write

$$ds = \rho d\vartheta \; \text{или} \; \frac{1}{\rho} = \frac{d\vartheta}{ds} = \frac{d\vartheta}{dx} \cdot \frac{dx}{ds}.$$

Whereas,

$$\vartheta = \operatorname{arctg} \left(\frac{dy}{dx} \right)$$

and

$$\frac{dx}{ds} = \cos \vartheta = \frac{1}{\sqrt{1 + \operatorname{tg}^2 \vartheta}} = \frac{1}{\left[1 + \left(\dfrac{dy}{dx} \right)^2 \right]^{1/2}},$$

after calculating the derivative of ϑ by x, we get

$$\frac{d^2 y}{dx^2} \left[1 + \left(\frac{dy}{dx} \right)^2 \right]^{-3/2} = \frac{M}{EJ}. \tag{16.27}$$

Equation (16.27) is the exact differential equation of the curved axis of a beam. To determine the deflections $y(x)$ and the angles of rotation $\vartheta(x)$ of the cross-sections of the beam, it must be integrated in each particular case. It is possible to perform such integration even for the simplest beams only by numerical methods.

---

[1] The bending occurs in the plane xy; here and below the indices z are omitted, which should not cause confusion.

For rigid beams, the deflections of which are very small, you can put

$$dx \approx ds, \quad \vartheta \approx \mathrm{tg}\vartheta = \frac{dy}{dx}.$$

Neglecting in Eq. (16.27) the square of the first derivative as infinitesimally small compared to one, we obtain an approximate differential equation of the curved axis of the beam:

$$\frac{d^2 y}{dx^2} = \frac{M}{EJ}. \tag{16.28}$$

Studies show (see, e.g., Rabotnov 1962) that replacing the exact differential equation (16.27) with an approximate one (16.28) gives an error in determining deflections of no more than 5%, even in the case of large deflections reaching one third of the beam length. Therefore, in engineering calculations, the approximate equation (16.28) is almost always and everywhere used.

16.7.2 Example of Determining Movements During Bending

Let us consider a double-jointed beam loaded with a force P in the middle of the span (Fig. 16.11a). In this case, the bending moment in the cross-sections of the beam can be written in the form of formulas

$$M(x) = \begin{vmatrix} \dfrac{P}{2} x & \text{при } 0 \leqslant x \leqslant \dfrac{l}{2}; \\ \dfrac{P}{2}(l-x) & \text{при } \dfrac{l}{2} \leqslant x \leqslant l. \end{vmatrix} \tag{16.29}$$

Substituting the function (16.29) into the approximate bending equation (16.28), we obtain after integration

$$EJ\vartheta(x) = \begin{vmatrix} \dfrac{Px^2}{4} + C_1 & \text{when } 0 \leqslant x \leqslant \dfrac{l}{2}; \\ \dfrac{P}{2}\left(lx - \dfrac{x^2}{2}\right) + C_2 & \text{when } \dfrac{l}{2} \leqslant x \leqslant l; \end{vmatrix} \tag{16.30}$$

$$EJy(x) = \begin{vmatrix} \dfrac{Px^3}{12} + C_1 x + C_3 & \text{when } 0 \leqslant x \leqslant \dfrac{l}{2}; \\ \dfrac{P}{2}\left(\dfrac{lx^2}{2} - \dfrac{x^3}{6}\right) + C_2 x + C_4 & \text{при } \dfrac{l}{2} \leqslant x \leqslant l. \end{vmatrix} \tag{16.31}$$

Fig. 16.11 Movements of
the axis of the double-hinged
beam

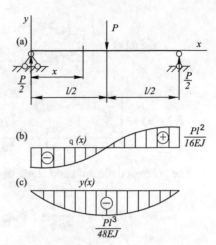

To determine the integration constants, we use the conditions $y(0) = 0$; $y(l) = 0$, as well as the continuity of the displacements of $vartheta(x)$ and $y(x)$ at the boundary of the plots, i.e.,

$$\vartheta\left(\frac{l}{2} - 0\right) = \vartheta\left(\frac{l}{2} + 0\right); \quad y\left(\frac{l}{2} - 0\right) = y\left(\frac{l}{2} + 0\right).$$

We obtain the following system of linear equations with respect to the constants $C_1, dotsc, C_4$:

$$(1) \ 0 = C_3;$$

$$(2) \ 0 = \frac{P}{2}\left(\frac{l^3}{2} - \frac{l^3}{6}\right) + C + 2l + C_4;$$

$$(3) \ \frac{Pl^2}{16} + C_1 = \frac{P}{2}\left(\frac{l^2}{2} - \frac{l^2}{8}\right) + C_2;$$

$$(4) \ \frac{Pl^3}{96} + C_1\frac{l}{2} + C_3 = \frac{P}{2}\left(\frac{l^3}{8} - \frac{l^3}{48}\right) + C_2\frac{l}{2} + C_4.$$

(16.32)

As a result of solving the system (16.32), we find

$$C_1 = -\frac{Pl^2}{16}; \quad C_2 = -\frac{3Pl^2}{16}; \quad C_3 = 0; \quad C_4 = \frac{Pl^3}{48}.$$

Substituting these constants into the general solutions (16.30) and (16.31), we finally have

$$EJ\vartheta(x) = \begin{vmatrix} \dfrac{P}{16}(4x^2 - l^2) & \text{при } 0 \leqslant x \leqslant l/2; \\ \dfrac{Px}{2}\left(l - \dfrac{x}{3}\right) - \dfrac{3Pl^2}{16} & \text{при } l/2 \leqslant x \leqslant l; \end{vmatrix} \tag{16.33}$$

$$EJy(x) = \begin{vmatrix} \dfrac{Px}{48}(4x^2 - 3l^2) & \text{при } 0 \leqslant x \leqslant l/2; \\ \dfrac{Px^2}{4}\left(l - \dfrac{x}{3}\right) - \dfrac{3Pl^2x}{16} + \dfrac{Pl^3}{48} & \text{при } l/2 \leqslant x \leqslant l. \end{vmatrix} \tag{16.34}$$

The greatest deflection and angle of rotation will be

$$y_{max} = y(l/2) = -\frac{Pl^3}{48EJ}; \quad \vartheta_{max} = \vartheta(l) = -\vartheta(0) = \frac{Pl^2}{16EJ}. \tag{16.35}$$

By formulas (16.33)–(16.35) in Fig. 16.11b, c constructed are displacement plots.

As you can see from the above example, determining the movements during bending is quite time-consuming work. Therefore, we note that there are other, less-expensive methods for calculating displacements. However, most often the design engineer uses the data from the reference book (see, e.g., Chernavskij and Reshchikov 1976, p. 210–213). In addition, many computer applications have been developed that successfully perform this work.

Self-Test Questions

1. Formulate the principle of superposition.
2. What is the loading of the beam called a flat bend?
3. Give the formulation of the hypothesis of flat sections.
4. What internal forces occur in the cross-section of the beam in a flat bend?
5. How are the transverse forces and bending moments in the cross-sections of the beam determined?
6. Why are internal effort plots constructed?
7. What are the differential relations between the internal forces that occur during bending?
8. What is the neutral axis in the beam section? How many such axes are there?
9. What is called the force line of the beam section?
10. Write down the formula for determining the normal bending stresses.
11. What causes the appearance of tangential stresses during bending?
12. Write down the Zhuravsky formula.
13. Name the points at which the tangential stresses are maximal. At what points are they equal to zero?
14. Name the main stages of calculating the bending strength.
15. Movements during bending: differential dependencies.

16. Write down the approximate differential equation of the curved axis of the beam.
17. From what conditions are the integration constants of the bending equation determined?

References

S. Chernavskij, V. Reshchikov, *Spravochnik metallista [Metalist's Handbook]* (Mashinostroenie Publication, Moscow, 1976)

V. Molotnikov, *Osnovy teoreticheskoi mekhaniki [Fundamentals of Theoretical Mechanics]*, (Fenix Publication, Rostov-on-Don, 2004)

V. Molotnikov, *Kurs soprotivleniya materialov [The Course of Strength of Materials]*, Lan' Publication, SPb., Moscow, 2006)

Y. Rabotnov, *Soprotivlenie materialov [Strength of Materials]* (Gosfiztexizdat Publication, Moscow, 1962)

Chapter 17
Combined Strength

Abstract Above, we considered the so-called simple types of beam deformation (axial tension or compression, shear, torsion, flat bending), in which only one internal force occurs in the cross-sections (in the case of transverse bending, two). In practice, most elements of structures and machines are subjected to the action of forces that simultaneously cause not one of these deformations, but two or more. Various combinations of simple deformations are called *complex resistance*. In this chapter, only those types of complex resistance that are most commonly found in engineering arc considered.

Keywords Simple loading · Complex resistance · Oblique bending · Off-center tension-compression · Pole · Cross-section core · Torsional bending · Springs · Contact stresses

17.1 On Complex Resistance Problems

In the previous sections, we investigated three types of simple loading of a bar, namely: central (uniaxial) tension (compression), torsion, and flat bending of a straight bar. In the first two cases, only one internal force differs from zero in the cross-sections of the beam and in the third two, of which one (the cutting force) is taken into account relatively rarely in the calculations.

In engineering practice, there are also more complex cases when several internal forces act simultaneously in the cross-sections of the beam, each of which has a significant impact on the strength. In such cases, it *is said that the beam experiences a complex resistance.*

For complex resistance, strength and stiffness calculations are based on the superposition principle. Most tasks are usually solved in this order.

1. External loads, including reactive loads, are projected on two main planes passing through the axis of the beam and the main axes of inertia of the section.
2. Plot the internal forces in the main planes.

3. Based on the diagrams of internal forces, outline possible dangerous sections of the beam.
4. For the planned cross-sections, the law of stress distribution over the cross-section is found out, and the positions of the points at which the maximum of the equivalent voltage is possible are determined.
5. At each designated point, on the basis of a suitable strength theory, equivalent stresses are calculated, comparing which establish the most dangerous point.
6. According to the value of the equivalent stress at the most dangerous point, the question of the strength of the beam is solved.
7. Calculate the displacements from each internal force and then the total displacements, and, if necessary, check the stiffness condition.

Later in this chapter, the most common types of complex timber resistance are considered.

17.2   Oblique Bend

An oblique bend is called such a case of bending a bar in which the plane of action of the bending moment does not lie in any of the main planes. For example, let a rectangular console be loaded with a force P, which is the angle α with the main axis Oy (Fig. 17.1).

Decompose the force P into components along the main central axes x and y:

$$P_x = P \sin\alpha; \quad P_y = P \cos\alpha.$$

For the selected directions of the coordinate axes, the bending moments in the section $I - I$, which is removed from the free end of the beam by a distance of z, will be

$$M_x = -P_y z = -Pz \cos\alpha; \quad M_y = -P_y z = -Pz \sin\alpha. \tag{17.1}$$

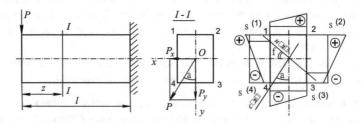

Fig. 17.1 Oblique bend

Thus, in each cross-section, two moments act simultaneously, bending the beam in the main planes. To calculate the normal stresses in the cross-sections of the beam, you need to use the superposition principle. In the case under consideration, this means that the resulting normal stress in the cross-section is found by summing the normal stresses from bending in each main plane. Using the formula (16.11), we find

$$\sigma = \frac{M_x y}{J_x} + \frac{M_y x}{J_y}. \tag{17.2}$$

At the angular points of the cross-section (Fig. 17.1), the coordinate modules x and y are extreme, and for the stresses at these points, the formula (17.2) can be written as

$$\sigma = \pm \frac{M_x}{W_x} \pm \frac{M_y}{W_y}, \tag{17.3}$$

where W_x, W_y are the moments of resistance of the beam section. Substituting into this expression the values of the bending moments according to the formulas (17.1), for normal stresses at angular points 1, 2, 3, and 4 (Fig. 17.1), we find

$$\sigma^{(1)} = \frac{P_y z}{W_x} - \frac{P_x z}{W_y}; \quad \sigma^{(2)} = \frac{P_y z}{W_x} + \frac{P_x z}{W_y};$$

$$\sigma^{(3)} = -\frac{P_y z}{W_x} + \frac{P_x z}{W_y}; \quad \sigma^{(4)} = -\frac{P_y z}{W_x} - \frac{P_x z}{W_y}. \tag{17.4}$$

Analyzing the obtained formulas, we notice that the greatest tensile stress occurs at the point 2 and the greatest compressive stress occurs at the point 4. In the right position, Fig. 17.1 plots of normal stresses along the sides of the rectangle are constructed from the stress values at the corner points.

The formulas (17.4) give the stress values at the points that have the greatest distances from both main axes. However, you can give examples of sections that do not have such points. In this case, it is impossible to name the most stressed points only by the known position of the main axes, and you should use the formula (17.2). Assuming in (17.2) $\sigma = 0$, we find the equation of the neutral line (n. l. in Fig. 17.1):

$$\sigma = \frac{M_x y}{J_x} + \frac{M_y x}{J_y} = 0.$$

From here we get

$$y = -\frac{J_x}{J_y} \cdot \frac{M_y}{M_x}. \tag{17.5}$$

We can always choose the directions of the x and y axes so that the bending moments have the same signs. Therefore, we will assume that this condition is met. Then the angular coefficient of the straight line (17.8) is negative and the neutral line passes through the second and fourth quadrants. Denoting by *varphi* the smallest angle between the neutral line and the x axis (Fig. 17.1), we have

$$\operatorname{tg}\varphi = \frac{J_x}{J_y} \cdot \frac{M_y}{M_x}. \tag{17.6}$$

Substituting the expressions (17.1) into this formula gives

$$\operatorname{tg}\varphi = \frac{J_x}{J_y}\operatorname{tg}\alpha, \tag{17.7}$$

where the angle α determines the position of the force line (see Fig. 17.1).

$$\sigma = \frac{M_x y}{J_x} + \frac{M_y x}{J_y} = 0.$$

From here we get

$$y = -\frac{J_x}{J_y} \cdot \frac{M_y}{M_x}. \tag{17.8}$$

We can always choose the directions of the x and y axes so that the bending moments have the same signs. Therefore, we will assume that this condition is met. Then the angular coefficient of the straight line (17.8) is negative and the neutral line passes through the second and fourth quadrants. Denoting by φ the smallest angle between the neutral line and the x axis (Fig. 17.1), we have

$$\operatorname{tg}\varphi = \frac{J_x}{J_y} \cdot \frac{M_y}{M_x}. \tag{17.9}$$

Substituting the expressions (17.1) into this formula gives

$$\operatorname{tg}\varphi = \frac{J_x}{J_y}\operatorname{tg}\alpha, \tag{17.10}$$

where the angle α determines the position of the force line (see Fig. 17.1).

From the result obtained, it follows that in the case of an oblique bend, the neutral line is generally not perpendicular to the force line. Orthogonality occurs only in the special case when $J_x = J_y$. But for such sections, any central axis is the main one, and the oblique bend turns into a flat one.

Let's go back to normal voltages. With a known position of the neutral line, we can now determine the points farthest from it. Substituting the coordinates of these points in the formula (17.2) gives the values of the greatest tensile and compressive stresses, which solves the problem of strength calculation of the beam.

If we are interested in displacements, then, as already mentioned in point Sect. 17.1, we need to determine the displacements in each of the main planes and then use the superposition principle to determine the total stresses.

17.3 Off-center Tension (Compression)

Let a massive prismatic column be loaded with a force P parallel to its axis (Fig. 17.2). Denote by x and y the main central axes of the cross-section.

The projection of the point of application of the force P on the plane xOy is called the *pole*, and the coordinates poles $C(x_p; y_p)$ are called *eccentricities*.

The internal forces in an arbitrary section of the column will be

$$N = P; \quad M_x = Py_p; \quad M_y = Px_p,$$

where N is the longitudinal force and M_x, M_y are the bending moments relative to the x and y axes, respectively. Therefore, in the case under consideration, there is a superimposition of uniaxial tension-compression on the oblique bend. Applying the superposition principle,[1] we find the normal stresses in the cross-sections of the column:

$$\sigma = \frac{N}{F} + \frac{M_x y}{J_x} + \frac{M_y x}{J_y}, \tag{17.11}$$

and F is the cross-sectional area of the column. Substituting here the expressions of internal efforts, we get

Fig. 17.2 Off-center stretching

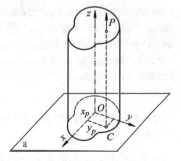

---

[1] Under the compressive load P, the superposition principle is applicable under the condition of small curvatures of the column caused by its deformation.

$$\sigma = \frac{P}{F}\left(1 + \frac{xx_p}{i_y^2} + \frac{yy_p}{i_x^2}\right), \tag{17.12}$$

where $i_x^2 = J_x/F$, $i_y^2 = J_y/F$. The values i_x, i_y are called *inertia radii* of the cross-section relative to the axes x and y, respectively.

The formula (17.12) is symmetric with respect to the coordinates x, y and x_p, y_p, which means that the following sentence is valid.

Theorem 1 *The normal stress at an arbitrary point A caused by a force parallel to the bar axis whose line of action passes through point B is equal to the stress at point B from the same force whose line of action passes through point A.*

Assuming in the formula (17.12) $\sigma = 0$, we get the equation of the neutral line:

$$\frac{xx_p}{i_y^2} + \frac{yy_p}{i_x^2} + 1 = 0. \tag{17.13}$$

To quickly construct the neutral axis, we write the formula (17.13) in the form of the equation of a straight line in segments

$$\frac{x}{a_x} + \frac{y}{a_y} = 1, \tag{17.14}$$

where a_x, a_y are the segments cut off by the straight line (17.14) on the coordinate axes. Comparing equations (17.13) and (17.14), we find

$$a_x = -\frac{i_y^2}{x_p}; \quad a_y = -\frac{i_x^2}{y_p}. \tag{17.15}$$

Thus, to build a neutral line, you need to calculate the segments a_x, a_y by the given pole coordinates x_p, y_p using the formulas (17.15) and postpone them (taking into account the signs) on the axes x, y from the origin. The straight line connecting the ends of these segments will be the neutral axis. Extreme normal stresses occur at the points farthest from the neutral axis. If the neutral axis intersects the cross-section, then there will be stresses of the opposite sign on different sides of it. For further investigation of the stress distribution in the cross-section, we prove the following.

Theorem 2 *When moving the pole in a straight line , the neutral axis rotates near a fixed point.*

Proof Let the neutral axis take the position $I - I$ at the pole K (Fig. 17.3). In this case, from the formulas (17.15), the coordinates of the pole K will be

$$x_K = -\frac{i_y^2}{a_x}; \quad y_K = -\frac{i_x^2}{a_y}. \tag{17.16}$$

Fig. 17.3 On the theorem on
the kernel of the section

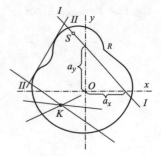

This means that if K is a pole, then at any point in the S line $I - I$, the voltage is zero. But according to Theorem 1, if, on the contrary, we take the point S as a pole, then at the point K the voltage will be zero. Since S is an arbitrary point of the line $I - I$, moving it along this line, we get a bundle of lines passing through the point K (Fig. 17.3), which we needed to prove.

In case of off-center loading of bars made of brittle material, compressive stresses are sought to take place at all points of the cross-section. This can be achieved by applying the force P inside some area of the end section of the beam. The geometric location of the poles, for which the stresses at all points of the section have the same sign, is called the *core of the section*. It is clear that the kernel of the cross-section must contain the center of gravity inside itself, since the central application of the force P obviously gives a stress of one sign in the cross-section.

The method of constructing the cross-section core follows from its definition and Theorem 2. If the pole is located on the contour of the cross-section core, then the neutral line must touch the cross-section contour of the beam. The angular point R (Fig. 17.3) of the section contour corresponds to the rectilinear segment of the boundary of the section core and, conversely, to the rectilinear section of the envelope $II - II$ corresponds to the corner point on the contour of the cross-section core. Thus, to build the core of the cross-section, you need to run a neutral line around the cross-section, not allowing it to intersect. In this case, the pole will draw the contours of the cross-section core. Hence, it is obvious that if the cross-section of the beam has the shape of a polygon, then the core of the cross-section will also be a polygon.

Let us explain this by using the example of constructing a cross-section kernel for a rectangular profile $b \times h$. Align the neutral line with the lower side of the section contour (Fig. 17.4). The segments cut off by the straight line $I - I$ on the coordinate axes will be

$$a_x = \infty, \quad a_y = -\frac{h}{2}.$$

Fig. 17.4 Rectangular core

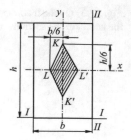

Next, we calculate

$$F = bh; \quad J_x = \frac{bh^3}{12}; \quad i_x^2 = \frac{J_x}{F} = \frac{h^2}{12}.$$

Swapping b and h, we similarly find

$$i_y^2 = \frac{b^2}{12}.$$

Using the formulas (17.15), we determine the coordinates of the pole (K) at the position of the neutral line $I - I$:

$$x_K = -\frac{i_y^2}{a_x} = 0; \quad y_K = -\frac{i_x^2}{a_y} = -\frac{h^2}{12 \cdot (-h/2)} = \frac{h}{6}.$$

Using the found coordinates, we construct the point K (Fig. 17.4) of the contour of the cross-section core.

Next, we combine the neutral line $II - II$ with the other side of the beam section contour (see Fig. 17.4) and find

$$a_x = \frac{b}{2}; \quad a_y = \infty; \quad x_L = -\frac{b^2 \cdot 2}{12 \cdot b} = -\frac{b}{6}; \quad y_l = 0,$$

Using the coordinates x_L, y_L, we construct the point L of the contour of the cross-section core.

The other two points K' and L' are constructed by the symmetry property. Now, on the basis of Theorem 2, connecting the constructed points with segments, we get the core of the section in the form of a rhombus, shaded in Fig. 17.4.

17.4   Bending with Torsion of Round Shafts

Consider a circular beam loaded with a force P normal to the axis and a twisting moment m (Fig. 17.5a). By bending plot and torques (Fig. 17.5b), it is easy to establish a dangerous cross-section, which in this example coincides with the sealing

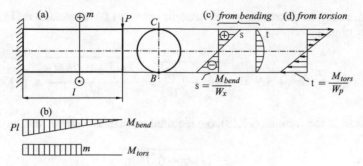

Fig. 17.5 Bending with torsion of the round shaft

site. Given the nature of the stress distribution over the cross-section (Fig. 17.5c, d), we can conclude that the points B and C are dangerous, but not the points located at the ends of the horizontal diameter of the shaft.

Elementary calculations show that in most cases, the shear stresses in a solid bar caused by the action of the transverse force are small in comparison with the tangential stresses from torsion, and they can be ignored. For this reason, only normal bending stresses and tangential torsional stresses are taken into account. In the case where there is a non-planar bend, plot the bending moments in two orthogonal planes, and in the dangerous section, calculate the total bending moment by the formula

$$M_f = \sqrt{M_x^2 + M_y^2},\tag{17.17}$$

where M_x, M_y are the bending moments relative to the orthogonal central axes x and y in the dangerous cross-section. There are, however, cases where the dangerous cross-section cannot be determined from the diagrams of internal forces. In this situation, it is necessary to check the strength in several sections, where an unfavorable combination of internal forces may occur. The dangerous section will be the section at the dangerous points of which the greatest equivalent voltage occurs.

The strength test in each dangerous section of the shaft is carried out according to one of the strength theories. When using the third theory of strength according to the formula (14.23), we have

$$\sigma_{equ}^{III} = \sqrt{\sigma^2 + 4\tau^2} = \sqrt{\frac{M_f^2}{W_x^2} + 4\frac{M_t^2}{W_p^2}} = \frac{\sqrt{M_f^2 + M_t^2}}{W_x} \leqslant [\sigma].$$

Here it is taken into account that the polar moment of the resistance of the circle is twice as large as the axial moment W_x. From the last formula for the project calculation, we get

$$W_x \geqslant \frac{\sqrt{M_f^2 + M_t^2}}{[\sigma]}.\tag{17.18}$$

When using the fourth theory of strength based on the dependence (14.29), we similarly obtain

$$\sigma_{equ}^{IV} = \sqrt{\sigma^2 + 3\tau^2} = \sqrt{\frac{M_f^2}{W_x^2} + 3\frac{M_t^2}{W_p^2}} = \frac{\sqrt{M_f^2 + 0,75M_t^2}}{W_x} \leqslant [\sigma],$$

and instead of the formula (17.18), the required resistance moment will be

$$W_x \geqslant \frac{\sqrt{M_f^2 + 0,75M_t^2}}{[\sigma]}. \tag{17.19}$$

For cast-iron shafts and shafts made of other brittle materials, the Mohr strength theory should be applied. Using the condition (14.30), we get

$$\sigma_{equ}^V = \sigma_1 - m\sigma_3 \leqslant [\sigma_p],$$

where $[\sigma_p]$ is the allowable tensile stress and m is the ratio of the tensile and compressive strength, i.e., $m = \sigma_p/\sigma_c$.

By the formula (14.13), we find for the considered stress state

$$\sigma_{1,3} = \sigma_{\substack{max \\ min}} = \frac{\sigma}{2} \pm \frac{1}{2}\sqrt{\sigma^2 + 4\tau^2}.$$

Then the strength condition takes the form

$$\sigma_{equ}^V = \frac{1-m}{2}\sigma + \frac{1+m}{2}\sqrt{\sigma^2 + 4\tau^2} \leqslant [\sigma]_p.$$

Expressing now the stresses in terms of moments and the corresponding moments of resistance, we get

$$\sigma_{экв}^V = \frac{1-m}{2} \cdot \frac{M_и}{W_x} + \frac{1+m}{2W_x}\sqrt{M_и^2 + M_к^2} \leqslant [\sigma]_p \tag{17.20}$$

or for the design calculation

$$W_x \geqslant \frac{M_f(1-m) + (1+m)\sqrt{M_f^2 + M_t^2}}{2[\sigma]_p}. \tag{17.21}$$

Example From the strength calculation, determine the diameter of the shaft shown in Fig. 17.6, with the following data:

- pulley diameters: $D_1 = 0.2$ m, $D_2 = 0.6$ m;
- weight of pulleys: $G_1 = 2$ kN, $G_2 = 4$ kN;

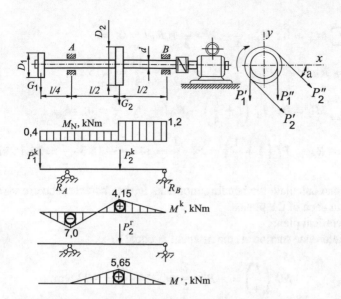

Fig. 17.6 For example, the calculation of the shaft when bending with torsion

- belt tension: $P_1' = 8$ kN, $P_1'' = 4$ kN, $P_2' = 10$ kN, $P_2'' = 6$ kN;
- the distance between the shaft supports $l = 2$ m; take the angle α to be $45°$.

The permissible stress of the shaft material is $[\sigma] = 100$ MPa.

S o l u t i o n. We bring the belt tension forces to the shaft axis and find the torsional torques at the places where the pulleys are installed:

$$M_t^{(1)} = (P_1' - P_1'') \cdot \frac{D_1}{2} = (8 - 4) \cdot \frac{0,2}{2} = 0.4 \text{ kNm};$$

$$M_\kappa^{(2)} = (P_2' - P_2'') \cdot \frac{D_2}{2} = (10 - 6) \cdot \frac{0,6}{2} = 1.2 \text{ kNm}.$$

Using the calculated values, we plot the torques M_t. Next, we project all the active forces on the horizontal and vertical planes. We get

$$P_1^v = G_1 + P_1' + P_1'' = 2 + 8 + 4 = 14 \text{ kN};$$
$$P_2^v = G_2 + (P_2' + P_2'') \sin \alpha = 4 + (10 + 6) \cdot 0.707 = 15.3 \text{ kN};$$
$$P_2^h = (P_2' - P_2'') \cos \alpha = (10 + 6) \cdot 0.707 = 11.3 \text{ kN}.$$

Here and further on in this example, the indices «v» and «h» indicate the values related to the vertical and horizontal planes, respectively.

We determine the vertical components of the reactions of the supports A and B:

$$\sum M_A^v = 0: \ P_1^v \cdot \frac{l}{4} - P_2^B \cdot \frac{l}{2} + R_B l = 0; \ \Rightarrow$$

$$\Rightarrow R_B = \frac{1}{2} \cdot P_2^v - \frac{1}{4} \cdot P_1^v = \frac{15.3}{2} - \frac{14}{4} = 4.15 \text{ kN};$$

$$\sum M_B^v = 0: \ P_1^v \left(l + \frac{l}{4} \right) - R_A l + P_2^v \cdot \frac{l}{2} = 0; \ \Rightarrow$$

$$\Rightarrow R_A = P_1^v \left(1 + \frac{1}{4} \right) + P_2^v \cdot \frac{1}{2} = 14 \cdot \frac{5}{4} + 15.3 \cdot \frac{1}{2} = 25.15 \text{ kN}.$$

Now you can calculate the bending moments in the characteristic cross-sections of the beam in each of the planes.

In the vertical plane:

in the reference section A, the moment is equal to

$$M_f^v \left(\frac{l}{4} \right) = -P_1^v \cdot \frac{l}{4} = -14 \cdot \frac{2}{4} = -7 \text{ kNm};$$

in the middle of the span we have

$$M_f^v \left(\frac{3}{4}l \right) = R_B \cdot \frac{l}{2} = 4,15 \cdot \frac{2}{2} = 4.15 \text{ kNm}.$$

In the horizontal plane, the moment in the middle of the span will be

$$M_f^h \left(\frac{3}{4}l \right) = \frac{P_2^h}{2} \cdot \frac{l}{2} = \frac{11,3}{2} \cdot \frac{2}{2} = 5.65 \text{ kNm}.$$

Using the calculated values, we plot the bending moments in the vertical and horizontal planes (Fig. 17.6). Analyzing the constructed plots, we come to the conclusion that either the reference section A or the section in the middle of the span can be dangerous. To uniquely determine the dangerous cross-section, calculate the total bending moment in the middle of the span:

$$M_f = \sqrt{4.15^2 + 5.65^2} \approx 7.01 \text{ kNm}.$$

Thus, the bending moments in the reference section A and in the middle of the span are almost the same, but to the right of the middle section, the torque is three times greater than the torque in the section A. Therefore, the average cross-section on the shaft span is dangerous.

Let's use the third theory of strength. Using the formula (17.18), we find

$$W_x \geqslant \frac{\sqrt{M_f^2 + M_t^2}}{[\sigma]} = \frac{\sqrt{7.01^2 + 1,2^2}}{100 \cdot 10^3} \frac{\text{(kNm)}}{\text{(кПа)}} = 7.11 \cdot 10^{-5} \text{ m}^3 = 71.1 \text{ cm}^3.$$

But $W_x \approx 0, 1d^3$. Then

$$d \geqslant \sqrt[3]{\frac{71.1}{0.1}} \approx 8.93 \text{ cm.}$$

Thus, it is possible to accept a shaft with a diameter of 90 mm. Problem solved.

17.5 Calculation of Cylindrical Coil Springs

Consider the helical with a small step, for which the angle α (Fig. 17.7a) between the tangent to the axis of the coil and the axis of the spring is not less than 75°. For simplicity, we will also assume that the rod has a circular cross-section. Mentally dissect the coil with a plane passing through the axis of the spring. Due to the smallness of the step, the cross-section of the turn with this plane, shaded in Fig. 17.7a, can be considered transverse. We bring the axial force P to the center of the coil section (Fig. 17.7b), adding the moment of force M_k relative to the new center.

Now we have a coil twisted by the moment M_k with the simultaneous action of the cutting force P. The tangential stresses from each of these forces will be denoted by the indices «k» and «p», respectively. Assuming that the distribution of the shear stresses from the force P is uniform, we obtain

$$\tau_p = \frac{P}{F} = \frac{4P}{\pi d^2}, \tag{17.22}$$

where F denotes the cross-sectional area of the coil and d denotes its diameter. The distribution of these stresses along the diameter CD is shown in Fig. 17.7*in*.

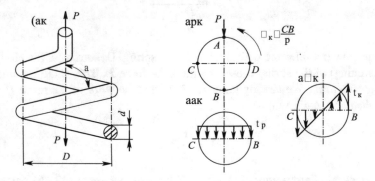

Fig. 17.7 To the calculation of helical springs

Tangential torsional stresses are distributed over the cross-section in proportion to the distance from the axis of the coil. The highest voltage occurs on the loop and, according to the formula (15.16), will be

$$\tau_\kappa = \frac{M_\kappa}{W_p} = \frac{PD \cdot 16}{2\pi d^3} = \frac{8PD}{\pi d^3},$$ (17.23)

where D is the average diameter of the spring (Fig.   refpruginaa). The plot of the distribution of tangential stresses along the diameter CD from the torsion of the coil is shown in Fig. 17.7r.

Analyzing the given diagrams of tangential stresses, we see that the most loaded points are C, located on the inner surfaces of the turns. The tangential stresses at these points are co-directional, so the total tangential stress will be equal to the sum of the stresses (17.22) and (17.23), i.e.,

$$\tau_{max} = \tau_p + \tau_\kappa = \frac{8PD}{\pi d^3}\left(1 + \frac{d}{2D}\right).$$ (17.24)

In real springs, usually the second term is small compared to the unit, and it can be neglected. In this case, the calculated tangential stress coincides with the maximum torsional stress (17.23). Clarification of this formula is sometimes carried out by introducing an amendment k, assuming

$$\tau_\kappa = \frac{8PD}{\pi d^3}k,$$ (17.25)

where the coefficient k is expressed in terms of the ratio of the diameters by the formula

$$k = \frac{\dfrac{D}{d} + 0,25}{\dfrac{D}{d} - 1}.$$

Let us now consider the deformation of the spring. Denote by λ the elongation (or shortening) of the spring caused by the axial force P. This value is sometimes called *spring draft*. We already know (p. 341) that the work of the force P on the elastic displacement λ will be

$$A = \frac{1}{2}P\lambda.$$ (17.26)

Under elastic deformation, the work of the external force (17.26) passes into the work of internal forces, numerically equal to the potential energy (U) of the deformation. Ignoring the influence of the cutting force, we can calculate this work as half the product of the moment M_k and the twist angle $varphi$. Using the formula (15.14), we get

$$U = \frac{1}{2}M_\kappa\varphi = \frac{4P^2D^3n}{Gd^4},\qquad(17.27)$$

where the calculation takes into account that πD is the length of one turn, the polar moment of inertia of the cross-section is $\pi d^4/32$, and n means the number of turns of the spring.

Equating the work A to the value of the potential energy U, we get the draft:

$$\lambda = \frac{8PD^3n}{Gd^4}.\qquad(17.28)$$

The amount of force P required for a single elongation of a spring is called its *stiffness*. Assuming in the formula (17.28) $\lambda = 1$ and denoting the spring stiffness by c, we find

$$c = \frac{Gd^4}{8D^3n}.\qquad(17.29)$$

As can be seen from the formula (17.29), the dimension of the stiffness is equal to the dimension of the force divided by the length (N/m, kgf/cm, etc.).

Thus, we have considered four cases of complex resistance that are often encountered in engineering: oblique bending, off-center tension-compression, bending with torsion, and cutting of a bar accompanied by torsion. Another case is practically important, when a spatial stress state occurs in a certain volume of the body. And although it does not formally refer to complex resistance, it is appropriate to consider it in this chapter.

17.6 Contact Stresses

Stresses and deformations arising from mutual pressing index Stresses! Contact of contacting bodies are called *contact*. In this case, the transfer of pressure from one body to another occurs over very small areas. Figure 17.8 on the left shows a wheel on the support head. The contacting bodies are compressed by the forces P. Studies show that the material near the contact area is under conditions of all-round volumetric compression, as shown in the figure to the right.

Contact stresses are local in nature and rapidly decrease as you move away from the contact site. The exact solution of the contact problem for elastic bodies was first given by Hertz. The solution is given under the assumption that the contacting surfaces are perfectly smooth, so that there are no friction forces in the contact area. In addition, it is considered that the size of the contact area is small compared to

Fig. 17.8 Stresses in the
contact zone

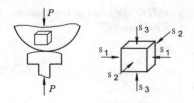

Table 17.1 Значения коэффициента α в формуле (17.32)

A/B	1,0	0,9	0,8	0,7	0,6	0,5	0,4	0,3	0,2	0,1	0,01
α	0,388	0,4	0,42	0,44	0,468	0,49	0,536	0,6	0,716	0,97	2,271

the minimum radius of curvature of the contacting surfaces at their initial point of
contact. The solution is obtained by the methods of the theory of elasticity, and we
use the ready-made results here.

In general, the contact area is the inner area of the ellipse:

$$Ax^2 + By^2 = C, \tag{17.30}$$

where C is the absolute convergence of the contacting bodies due to local deformation; the coefficients A and B depend on the main radii of curvature of the surfaces
at the point of contact.

The compressive stresses at the points of the contact area are distributed
according to the law of the ellipsoid. The greatest compressive stress occurs in the
center of the ellipse (17.30) and is determined by the formula

$$\sigma_{max} = \frac{3}{2} \frac{P}{\pi ab}, \tag{17.31}$$

where a and b are the semi-axes of the elliptical contact area. The product πab is
the area of an elliptical platform. Consequently, the maximum compressive stress is
one and a half times higher than the average pressure at the contact site.

If the semi-axes a and b are expressed in terms of the equation parameters
(17.30), then the formula (17.31) for the maximum compression stress takes the
form

$$\sigma_{max} = \alpha \sqrt[3]{4PA^2E^2}, \tag{17.32}$$

where α is a coefficient depending on the ratio A/B, and E is the elastic modulus
of the material of the contacting bodies. The values of the coefficient α for some
relations A/B are given in Table 17.1.

Table 17.2 shows the values of the coefficients A, B and the maximum compression stress σ_{max} for the most common cases of contact between two bodies.

By direct testing, we can make sure that if the values of the main stresses are
expressed in terms of the greatest compressive stress in the center of the contact

Table 17.2 To the calculation of contact stresses

Contact diagram	Parameters Body	A	B	σ_{max}
	Two balls of radii R_1 and R_2	$\dfrac{R_1 + R_2}{2R_1 R_2}$	$\dfrac{R_1 + R_2}{2R_1 R_2}$	$0,388\ \sqrt[3]{\dfrac{PE^2(R_1 + R_2)^2}{R_1^2 R_2^2}}$
	A ball of radius R_1 in a sphere of radius R_2	$\dfrac{R_2 - R_1}{2R_1 R_2}$	$\dfrac{R_2 - R_1}{2R_1 R_2}$	$0,388\ \sqrt[3]{\dfrac{PE^2(R_2 - R_1)^2}{R_1^2 R_2^2}}$
	A ball of radius R_1 in a trough of radius R_2	$\dfrac{1}{2R_1} - \dfrac{1}{2R_2}$	$\dfrac{1}{2R_2}$	$\alpha\ \sqrt[3]{\dfrac{PE^2(R_2 - R_1)^2}{R_1^2 R_2^2}}$
R_1 R_2 R_3 Ball Bea-ring	Ball Bea-ring	$\dfrac{1}{2R_1} - \dfrac{1}{2R_2}$	$\dfrac{1}{2R_1} + \dfrac{1}{2R_1}$	$\alpha\ \sqrt[3]{\dfrac{PE^2(R_2 - R_1)^2}{R_1^2 R_2^2}}$

area and the equivalent stress is calculated according to the fourth theory of strength, then, regardless of the shape of the contacting bodies, the approximate equality is fulfilled: $\sigma_{equ}^{IV} \approx 0,6\sigma_{max}$.

Then the strength condition can be written as

$$\sigma_{equ}^{IV} = 0,6\sigma_{max} \leqslant [\sigma]$$

or

$$\sigma_{max} \leqslant \frac{[\sigma]}{0,6} = [\sigma]_{con}. \qquad (17.33)$$

The value $[\sigma]_{con}$ represents the *permissible contact voltage*. From the formula (17.33), we notice that this value is almost twice the main permissible voltage.

The question of calculating contact stresses is discussed in detail in the courses of elasticity theory and in the special literature (see, for example, Molotnikov and Molotnikova 2015, 2021). Valuable research in this area was carried out by the Soviet scientist I. Staerman, who gave (Shtaerman 1949) a fundamentally new solution to the contact problem. The question of calculating contact stresses is discussed in detail in the courses of elasticity (Molotnikov and Molotnikova 2021) theory and in the special literature (see, for example, Molotnikov and Molotnikova 2015,). Valuable research in this area was carried out by the Soviet scientist I.Ya.Staerman, who gave (Shtaerman 1949) a fundamentally new solution to the contact problem.

Self-Test Questions

1. Describe the sequence of actions when solving complex resistance problems.
2. Give examples of elements of real structures operating under conditions of complex resistance.
3. What kind of loading is called oblique bending?
4. Write down the formula for determining the normal stresses in an oblique bend.
5. How to find the position of the neutral line in an oblique bend?
6. What is called off-center compression? In what elements of technical devices does it occur?
7. Give a formula for determining the normal stresses under off-center compression.
8. Write the calculation formulas for bending with torsion of round shafts according to the third and fourth strength theories.
9. At which points of a cylindrical helical spring do the greatest tangential stresses occur? How to calculate the draft and stiffness of such a spring?
10. What are the contact voltages?

References

V. Molotnikov, A. Molotnikova, *Mexanika deformacij [Deformation Mechanics]* (RFNO Publication, Rostov-on-Don, 2015)

V. Molotnikov, A. Molotnikova, *The Theory of Elasticity and Plasticity*, (Springer Nature, Berlin, 2021)

I. Shtaerman, *Kontaktnaya zadacha teorii uprugosti [Contact Problem of Elasticity Theory]* (Gostekhizdat Publication, Moscow, 1949)

Chapter 18
General Theorems of Mechanics

Abstract The chapter presents the main theorems of solid mechanics used in solving problems of technical mechanics; among them are Clapeyron's theorem on the work of external forces in linear elastic systems; the principle of reciprocity of work (Betty); the principle of reciprocity of displacements (Maxwell); Castigliano's theorem; integral of displacements; and graphical-analytical method for its calculation (Vereshchagin). On the basis of the theorems formulated, the so-called method of forces for calculating statically indeterminate systems.

Keywords Theorems of mechanics · Clapeyron's theorem · Maxwell's principle · Maxwell's principle · Castigliano's theorem · Vereshchagin's method · Mohr's integral · Method of forces

18.1 Work of External and Internal Forces

Let some system of material points be in equilibrium under the action of a given load. By virtue of the equilibrium condition, the resultant of all forces applied to each point of the system is zero. Let us give the points of the system infinitely small displacements, keeping the forces acting between them unchanged. In this case, the work of the given forces applied to any point of the system will be equal to zero (Molotnikov and Molotnikova 2021). It follows from this that for the entire system of material points, the work of the given forces in equilibrium is equal to zero. This means that the sum of the work of external and internal forces is equal to zero, i.e.,

$$A + W = 0, \tag{18.1}$$

where A is the work of external forces and W is the work of internal forces.

In the case of an absolutely rigid body, the movements occur without deformation, the distances between the particles do not change, and the internal forces do not perform any work. In this case, the principle of possible displacements follows from the formula (18.1).

Let us represent the work of external forces as the sum of the products of the parameters (generalized forces P_i) that determine the load, by some coefficients that depend on the type and magnitude of the displacements. These coefficients are called generalized displacements (u_i).

Similarly, the work of internal forces at infinitesimal displacements of a loaded system can be represented as

$$W = -\Sigma P_i u_i. \tag{18.2}$$

The work (18.2) is called virtual. It differs from the actual work of the forces P_i, produced during the loading process.

18.2   Linear Systems

Consider an elastic system loaded with a system of forces $P_1, \ldots, P_n$. Let us assume that in the process of loading, the inertia forces resulting from the displacements of the points of the system caused by the deformation are negligible. This loading is called *static*. If the value of the displacements of the points of the system increases by λ times under the action of the forces $\lambda P_1, \lambda P_2, \ldots, \lambda P_n$, then the elastic system is called *linear*.

Denote by u_i the elastic displacement corresponding to the force P_i from the load $P_1, P_2, \ldots, P_n$. Then the moves under load $\lambda P_1, \lambda P_2, \ldots, \lambda P_n$ will be $\lambda u_1, \lambda u_2, \ldots, \lambda u_n$. We give the parameter λ an infinitesimal increment of $d\lambda$ and calculate the work (dA) of the last forces on the actual movements of $u_1 d\lambda, u_2 d\lambda, \ldots, u_n d\lambda$:

$$dA = P_1 u_1 \lambda d\lambda + P_2 u_2 \lambda d\lambda + \cdots + P_n u_n \lambda d\lambda.$$

Summing up the work when λ changes from zero to one, we get

$$A = (P - 1u_1 + P_2 u_2 + \cdots + P_n u_n) \int_0^1 \lambda d\lambda,$$

or

$$A = \frac{1}{2}(P_1 u_1 + P_2 u_2 + \cdots + P_n u_n). \tag{18.3}$$

Thus, the following is proved.

The Theorem *The work of statically applied external forces on displacements caused by deformations of a linear elastic system is equal to the half-sum of the products of the final values of each force by the values of the corresponding displacements.*

The theorem was first formulated by Clapeyron (1852).

The work of the internal forces of an elastic body, taken with the opposite sign, is called the *potential energy of elastic deformations*. According to the formula (18.1), numerically it is equal to the work of external forces. For this reason, the potential energy of elastic deformation as well as the work of external forces will be denoted by A.

18.3 The Principle of Reciprocity of Work

Consider two different states of a linear elastic system loaded with two different loads. We agree to denote the loads, as well as the internal forces and movements of the system in these two states, by the indices 1 and 2. Let's imagine that in the beginning, the load 1 gradually began to act on the unloaded system. From load 1, the system received some deformations, and the load itself performed work, which is denoted by A_{11}. We will fix load 1 and gradually load the system with load 2. This load will cause additional deformation of the system. The work of load 2 on the displacements caused by additional deformation after the application of load 2 is called A_{22}.

However, with additional deformation, not only load 2 will perform the work, but also load 1, since the points of application of the forces of system 1 will receive additional displacements with additional deformation. The load 1 on these movements remains constant and will perform the work, which we denote by A_{12}. As a result, the work of external forces will be expressed as the sum of $A_{11} + A_{22} + A_{12}$. In these terms, the first index indicates which load does the work, and the second what forces are caused by the movement of this load.

Now let's change the loading order. First, we apply load 2 to the unloaded system, which will perform some work A_{22} on the movements caused by it. We fix the load 2 and gradually apply the load 1. As in the first case of loading, the load 1 will cause additional deformation of the system and perform the work A_{11} on the movements caused by it. At this time, load 2, while remaining constant, will perform additional work A_{21} on the movements caused by the application of load 1. As a result, the work of external forces under the second method of loading will be expressed as the sum of $A_{22} + A_{11} + A_{21}$.

The total work of the external forces in each of these loading cases is equal to the work of the internal forces taken with the opposite sign. This work is determined only by the final state of the system and does not depend on the loading sequence. Since the final states of the system in the two considered loading cases are the same, it follows that

$$A_{12} = A_{21}. \qquad (18.4)$$

Thus, the following *principle of reciprocity of works is proved (Batty)*.

Theorem 1 *The work of the forces of the first system on the displacements caused by the second by the system is equal to the work of the forces of the second system on the movements caused by the first system of forces.*

We substitute in the result (18.4) $A_{12} = P_1 \delta_{12} P_2$, where P_1 is the generalized force of the first system and δ_{12} is the generalized movement of the force P_1 in the direction of its action caused by the force $P_2 = 1$. Similarly, imagine $A_{21} = P_2 \delta_{21} P_1$, where δ_{21} is the generalized displacement of the force P_2 in the direction of its action caused by the force $P_1 = 1$. Then the condition (18.4) implies

$$\delta_{12} = \delta_{21}. \tag{18.5}$$

The equality (18.5) expresses (Maxwell) *is the principle of reciprocity of displacements,* to which we give the following formulation.

Theorem 2 *The movement of the point of application of the first force in the direction of its action caused by the second unit force is equal to the movement of the point of application of the second force in the direction of its action caused by the first unit force.*

18.4 Castigliano's Theorem

Consider an elastic system loaded with an arbitrary system of forces $P_1,\ P_2,\ \ldots,\ P_n$ and fixed so that its displacements as a rigid whole are excluded (Fig. 18.1). Due to the deformation, an arbitrary point A will take a new position A'. The segment $\overline{AA'}$ is called the complete displacement of the point A.

Take an arbitrary axis l passing through the point A and project the point A' onto it. As a result, we get the point A'' on the l axis. The segment $\overline{AA''}$ is called *by moving the point A in the direction l.* Thus, if, for example, $\overline{BB'}$ is the total displacement of the force point P_1, then $\delta_1 = \overline{BB''}$ is the displacement of the force point P_1 in the direction of its action.

Fig. 18.1 On Castigliano's theorem

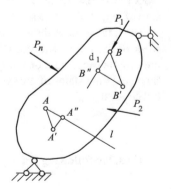

The potential energy of the deformation of the system by the forces P_1, P_2, ..., P_n is denoted by U.

Let's give one of the forces, for example, the force P_n, an infinitesimal increment of dP_n. In this case, the potential energy will also receive an infinitesimal increment and become equal to

$$U + dU = U + \frac{\partial U}{\partial P_n} dP_n. \tag{18.6}$$

Let's change the order of application of forces. First, we will apply only the dP_n force. Due to the deformation of the system, the point of application of this force in the direction of its action will receive some movement, which we denote by $d\delta_n$. The work of the dP_n force on the specified displacement based on the formula (18.3) will be $dP_n d\delta_n/2$. Now we apply the entire system of external forces P_1, P_2, ..., P_n. In the absence of the force dP_n, the potential energy of the system would be U. But since there is a force dP_n, this force will perform additional work on moving δ_n the point of application of the force P_n in its direction, and the movement of δ_n is caused by the entire system of external forces at $dP_n = const$. Therefore, the additional work of the dP_n force will be equal to the product of $dP_n \delta_n$. As a result, the potential energy of elastic deformation of the system under the second method of loading is expressed as the sum of

$$\frac{1}{2} dP_n d\delta_n + U + dP_n \delta_n. \tag{18.7}$$

Since the final state of the system in the first and second loading methods is the same, we can equate the sum (18.7) on the right-hand side of the formula (18.6). We get

$$U + \frac{\partial U}{\partial P_n} dP_n = \frac{1}{2} dP_n d\delta_n + U + dP_n \delta_n.$$

Neglecting the first term on the right-hand side of this equality as an infinitely small quantity of higher order, we finally obtain

$$\delta_n = \frac{\partial U}{\partial P_n}. \tag{18.8}$$

Thus, the following has been proved (Castigliano).

Theorem *Partial derivative of the potential energy of the system in force is equal to the displacement of the point of application of the force in the direction of its action.*

18.5   Potential Deformation Energy of a Bar

Mentally we will cut out an element of length dz from a bar with two cross-sections and fix its left end section (Fig. 18.2). We already know (p. 326) that in the general case, six internal forces occur in the cross-section of the beam: the longitudinal force N; the transverse forces Q_x, Q_y; the torque M_k, (M_z); as well as the bending moments M_x, M_y relative to the main central axes x and y.

With respect to the selected element, these six forces can be considered as external loads, so to calculate its potential energy, you can use the formula (18.3). You only need to find the u_i moves that correspond to each of the efforts.

For the longitudinal force N, this displacement is denoted by u_N. Obviously, this displacement is equal to the lengthening of the bar element with length dz, and by the formula (13.10), we have

$$u_N = \frac{N dz}{EF},$$

where E is the modulus of elasticity and F is the cross-sectional area of the bar. Then, under the action of only one force N, the potential energy of the selected element in accordance with the formula (18.3) will be

$$dU(N) = \frac{1}{2}N \cdot u_N = \frac{N^2 dz}{2EF}.$$

The potential energy of the beam from the force N we find by integrating $dU(N)$ the length of timber for l

$$U(N) = \int_l \frac{N^2 dz}{2EF},$$

which coincides with the formula (13.15).

Fig. 18.2 Internal forces in
the general case of beam
deformation

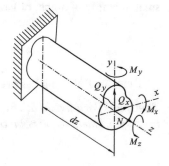

Calculating similarly the displacements and potential energy from the remaining five internal forces and summing up[1] then the components of the potential energy from all internal forces, we find the total energy of the deformation of the beam

$$U = \int_l \frac{N^2 dz}{2EF} + \int_l \frac{M_\kappa dz}{2GJ_p} + \int_l \frac{M_x^2 dz}{2EJ_x} + \int_l \frac{M_y^2 dz}{2EJ_y} + \int_l \eta_x \frac{Q_x^2 dz}{2GF} + \int_l \eta_y \frac{Q_y^2 dz}{2GF},$$

(18.9)

where η_x, η_y are parameters that depend in a known way on the shape of the cross-section of the beam (see, e.g., Molotnikov 2006, p. 219).

18.6 Integral of Displacements

Let some linear elastic system be given, for example, a spatial bar with a polyline axis (Fig. 18.3). The system carries the specified load P_1, P_2, ..., P_n. The internal forces in the cross-sections of the beam from this load will be marked with the subscript «p», so that, for example, N_p, M_{xp} mean, respectively, the longitudinal force and the bending moment relative to the x axis caused by the forces P_1, P_2, ..., P_n.

Let's set the task: find the move (δ_A) some point A of the system in a given direction, for example, its linear movement in the vertical direction.

To solve this problem, we cannot directly use Castigliano's theorem, since there is no force applied at the point A in the direction of the desired displacement. This difficulty can be circumvented by the following technique.

We apply at the point A in the direction of interest to us some force Φ, which we will call *a fictitious force*. Internal efforts caused by this force, let us agree to mark the index « Φ ». Let's represent them in the form

$$N_\Phi = N_1 \Phi, \quad M_{k\Phi} = M_{k1}\Phi, \quad \ldots, \quad Q_{y\Phi} = Q_{y1}\Phi,$$

where N_1, M_{k1}, ..., Q_{y1} are internal forces in the beam from the force $\Phi = 1$. Let's call these efforts *singular*. Based on the superposition principle in a system loaded with the forces P_1, P_2, ..., P_n and the fictitious force Φ, the internal forces

Fig. 18.3 To Mohr's formula

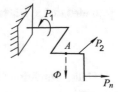

---

[1] The summation operation is legal here, since each internal force performs work only on the corresponding displacement.

will be

$$N_p + N_1 \Phi, \ M_{\kappa p} + M_{\kappa 1} \Phi, \ \ldots, \ Q_{yp} + Q_{y1} \Phi. \tag{18.10}$$

According to the formula (18.9), these forces correspond to the potential deformation energy of the bar:

$$U = \int_l \frac{(N_p + N_1 \Phi)^2 dz}{2EF} + \int_l \frac{(M_{\kappa p} + M_{\kappa 1} \Phi)^2 dz}{2GJ_p} + \cdots + \int_l eta_y \frac{(Q_{yp} + Q_{y1} \Phi)^2 dz}{2GF}. \tag{18.11}$$

To calculate the movement of the point A in the direction of interest, we now apply the Castigliano's theorem (18.8). To do this, we differentiate the potential energy (18.11) by the fictitious force Φ, and then we put $\Phi = 0$. As a result, we will find

$$\delta_A = \frac{\partial U}{\partial \Phi} \Big|_{\Phi=0} = \int_l \frac{N_p N_1 dz}{EF} + \int_l \frac{M_{\kappa p} M_{\kappa 1} dz}{GJ_p} + \int_l \frac{M_{xp} M_{x1} dz}{EJ_x} +$$

$$+ \int_l \frac{M_{yp} M_{y1} dz}{EJ_y} + \int_l \eta_x \frac{Q_{xp} Q_{x1} dz}{GF} + \int_l \eta_y \frac{Q_{yp} Q_{y1} dz}{GF}. \tag{18.12}$$

The resulting formula is called the *integral of displacement* or *the Mohr integral*. It finds extremely important application in the design of beams and bar systems.

18.7   Vereshchagin's Method

For rectilinear bars, a graphical-analytical method can be applied to the calculation of Mohr's integrals, which is called the *rule for multiplying diagrams* or *Vereshchagin's method*. Let's consider the essence of the method.

Let it be required to calculate the integral of the product of two functions on a segment of length l

$$I = \int_0^l f_1(z) f_2(z) dz \tag{18.13}$$

provided that at least one of these functions is linear. For definiteness, put

$$f_2(z) = a + bz \ \ (a, \ b-const). \tag{18.14}$$

Under condition (18.14), the integral I can be represented as the sum of two integrals:

$$I = a \int_0^l f_1(z) dz + b \int_0^l z f_1(z) dz. \tag{18.15}$$

Fig. 18.4 To substantiate
Vereshchagin's method

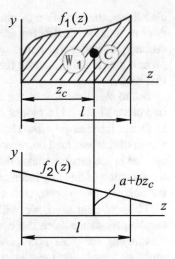

In the formula (18.15), the first integral is numerically equal to the area bounded by the line $y = f_1(z)$ or, in other words, equal to the area of the plot of the function $f_1(z)$, highlighted in Fig. 18.4 shading. Denoting this area by Ω_1, we can write

$$\int_0^l f_1(z)dz = \Omega_1. \tag{18.16}$$

The second integral in the formula (18.15) is the static moment of a figure with area Ω_1 relative to the y axis, and it can be calculated as the product of the area Ω_1 by the distance z_c (Fig. 18.4) from the center of gravity of this area C to the y axis, i.e.,

Substituting the expressions (18.16)–(18.17) into the formula (18.15), we get

$$I = \Omega_1 a + \Omega_1 b z_c = \Omega_1 (a + b z_c). \tag{18.17}$$

But $(a + b z_c) = f_2(z_c)$ is the ordinate of the linear function under the center of gravity of the area Ω_1 (Fig. 18.4). Therefore, the formula (18.17) can be rewritten as

$$\int_0^l f_1(z) f_2(z)dz = \Omega_1 f_2(z_c). \tag{18.18}$$

Thus, according to Vereshchagin's method, the integral on the segment from zero to l of the product of two functions, one of which is linear, is equal to the product of the area bounded by the first (in general, nonlinear) function and the value of the second (linear) function under the center of gravity of this area.

We see that in the considered graphoanalytic method, the integration operation is replaced by the multiplication of plots. In the case when the function $f_1(z)$ is also linear, the operation of multiplying plots has the property of commutativity. It does not matter whether the area of the first plot is multiplied by the ordinate of the second plot under the center of gravity of the first, or, conversely, the area of the second plot is multiplied by the ordinate under its center of gravity taken from the first plot.

So, to calculate any of the integrals of the formula (18.12), you need to build two plots. The first is a plot of the internal force from the system of specified loads. Let's call it the *cargo plot*. The second plot is the plot of the internal force from the action of the unit load corresponding to the desired displacement. The unit load is applied in the section whose displacement we want to determine, and it must act in the direction of this displacement. The plot from the action of a single load is called the *unit plot*. For a beam with a straight axis, the unit plot will always be straight.

To avoid confusion, we will agree to build both cargo and single diagrams of bending moments from the compressed fibers of the beam. Note also that when multiplying the plots , the product is assigned a positive sign if both plots to be multiplied are located on the same side relative to the axis of the beam. Otherwise, the product is taken with a negative sign.

18.7.1   Example of Determining Movements in Vereshchagin

Task. Determine the vertical, horizontal, and angular displacement of the end section B of a broken bar (Fig. 18.5a).

S o l u t i o n. Let's use the formula (18.12) and Vereshchagin's method. In this case, we are dealing with a flat bend of a bar with a polyline axis. Therefore, of all the internal forces, we will take into account only one thing, namely, the bending moment.

We plot the bending moments from a given load, i.e., the load plot M_p, as well as the unit plots M_1, M_2, and M_3 caused by the unit loads $\overline{P}_1 = 1$, $\overline{P}_2 = 1$, and $\overline{M} = 1$. These plots are constructed in Fig. 18.5b–d.

Let's start calculating the movements. The vertical movement of δ_b (Fig. 18.5e) is calculated by multiplying the cargo plot M_p by the unit plot M_1. At the same time, since the unit plot along the vertical bar is zero, we multiply the plots only

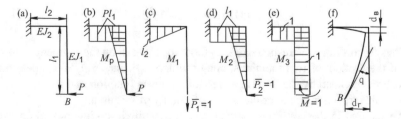

Fig. 18.5 For example, calculating displacements along Vereshchagin

on the horizontal section of the structure. Here, both plots being multiplied are linear, so it doesn't matter which area is multiplied by the ordinate of the other plot under the center of gravity of that area. Take, for example, the area of the plot M_1, which is $l_2^2/2$, and multiply it by the ordinate Pl_1 of the plot M_p, which is constant everywhere along the horizontal section of the broken bar. Thus, we have

$$\delta_c = \frac{1}{EJ_2} \cdot \frac{l_2^2}{2} \cdot Pl_1 = \frac{Pl_1l_2^2}{2EJ_2}.$$

Next, we calculate the horizontal movement δ_d of the section B. To do this, we multiply the unit plot M_2 and the cargo plot M_p. We have

$$\delta_d = \frac{1}{EJ_1} \cdot \frac{1}{2}Pl_1 \cdot l_1 \cdot \frac{2}{3}l_1 + \frac{1}{EJ_2} \cdot Pl_1 \cdot l_2 \cdot l_1 = \frac{Pl_1^3}{3EJ_1} + \frac{Pl_1^2l_2}{EJ_2}.$$

Finally, we define the angular displacement ϑ of the section B. To do this, multiply the plots M_3 and M_p. Finding it

$$\vartheta = \frac{1}{EJ_1} \cdot \frac{1}{2} \cdot Pl_1 \cdot l_1 + \frac{1}{EJ_2} \cdot \frac{1}{2} \cdot Pl_1 \cdot l_2 = \frac{P}{2E}\left(\frac{l_1^2}{J_1} + \frac{2l_1l_2}{J_2}\right).$$

In Fig. 18.5f a diagram of the deformed state of the system is given, and the displacements of the section B are indicated.

18.8 Force Method

We have already said (see p. 344) that to calculate a statically indeterminate system, one converts it into a statically definable one by discarding redundant connections and replacing their action with unknown forces. Recall that the design scheme of the structure transformed in this way is called the main system. When moving to the main system, we get as many unknown forces $X_1, X_2, \ldots, X_k$ as many times the system is statically undefined. To determine them, it is necessary to create k equations for the compatibility of deformations. Since the unknown forces are taken in the «extra» connections, the considered method of calculating statically indeterminate systems is called the *method of forces*. The calculation sequence is traced on the example of the frame, the scheme of which is shown in Fig. 18.6a.

The system has five links. Hence, the degree of its static indeterminability is equal to two. When choosing the main system, we need to discard two « extra» links. This can be done in many ways. For example, we drop the support A (Fig. 18.6b) and compensate for its action with unknown forces X_1 and X_2.

After selecting the main system, we must write down the equations of joint deformations. Physically, they should reflect the fact that the movements in the

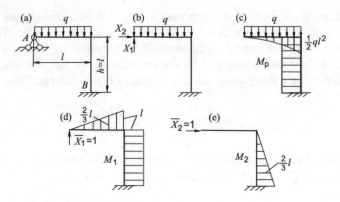

Fig. 18.6 To the method of forces

direction of each unknown force X_1 and X_2 are zero. Let's write these conditions in the following symbolic form:

$$\Delta_{X_1}(X_1,\, X_2,\, P) = 0; \quad \Delta_{X_2}(X_1,\, X_2,\, P) = 0, \tag{18.19}$$

where it is indicated in parentheses that the movements are caused by forces X_1, X_2 and external loads, the totality of which is conditionally denoted by P.

By virtue of the principle of independence of the action of forces, we can represent these displacements as the sum of the displacements from each action, i.e., the sum of the displacements from each action.

$$\Delta_{X_1}(X_1,\, X_2,\, P) = \Delta_{11} + \Delta_{12} + \Delta_{1P} = 0;$$
$$\Delta_{X_2}(X_1,\, X_2,\, P) = \Delta_{21} + \Delta_{22} + \Delta_{2P} = 0, \tag{18.20}$$

where the first index indicates in the direction of what force the movement is made and the second what force it is caused by.

Next, imagine

$$\Delta_{11} = X_1\delta_{11}, \quad \Delta_{12} = X_2\delta_{12},$$
$$\Delta_{21} = X_1\delta_{21}, \quad \Delta_{22} = X_2\delta_{22}. \tag{18.21}$$

Here, as before (see Sect. 18.3), δ_{ij} is the movement of the point of application of the force X_i in the direction of its action, caused by the unit force $\overline{X}_j = 1$.

Taking into account the notation (18.21), the equations (18.20) take the form:

$$\delta_{11}X_1 + \delta_{12}X_2 + \Delta_{1P} = 0,$$
$$\delta_{21}X_1 + \delta_{22}X_2 + \Delta_{2P} = 0. \tag{18.22}$$

Obviously, in the case of n times of a statically indeterminate system, we get n of such equations:

$$\delta_{11}X_1 + \delta_{12}X_2 + \ldots + \delta_{1n}X_n + \Delta_{1P} = 0,$$
$$\delta_{21}X_1 + \delta_{22}X_2 + \ldots + \delta_{2n}X_n + \Delta_{2P} = 0,$$
$$\ldots\ldots\ldots\ldots\ldots\ldots\ldots\ldots\ldots\ldots\ldots\ldots\ldots\ldots\ldots\ldots\ldots, \quad (18.23)$$
$$\delta_{n1}X_1 + \delta_{n2}X_2 + \ldots + \delta_{nn}X_n + \Delta_{nP} = 0.$$

The equations of joint deformations, written in a once-and-for-all established form (18.23), are called *canonical equations of the force method.*

Coefficients δ_{11}, δ_{12}, $\ldots$, δ_{nn} and free terms Δ_{1P}, $\ldots$, Δ_{nP} systems of canonical equations can be calculated using the displacement integral and for rectilinear structural elements using Vereshchagin's method. In this case, most often only the bending moments are taken into account, since the contribution from the longitudinal and transverse forces is negligible. Note also that, based on the reciprocity of displacement theorem, $\delta_{ij} = \delta_{ji}$.

18.8.1   *The Simplest Example Is*

Consider the beam shown in Fig. 18.7. The beam has four bonds, and to determine their reactions, we have three equations of equilibrium of the plane system of forces. Hence, the system is once statically indeterminate.

To reveal the static indeterminacy, we choose the main system, discarding the «extra» connection, for example, a hinged-movable support. The action of the dropped connection is replaced by an unknown force X_1. The selected primary system is shown in Fig. 18.7b. Note that the selected main system is not the only possible one.

From the system of canonical equations (18.23), in our case, only one thing remains:

$$\delta_{11}X_1 + \Delta_{1P} = 0. \quad (18.24)$$

We will search for the coefficients δ_{11} and Δ_{1P} using Vereshchagin's method. To do this (see Fig. 18.7d), plots of bending moments are constructed: load – M_p, from the action of a given external load q and unit—$\overline{M}_1$, from the action of a unit force $\overline{X}_1 = 1$.

Fig. 18.7 For example,
Sect. 18.8.1

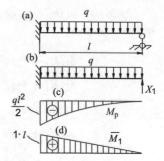

Keeping in mind that the area Ω_p of the plot M_p is equal to $1/3$ of the product of the legs, and its center of gravity is removed from the vertex by $3/4$ of the height, we calculate

$$\Omega_p = \frac{1}{3} \cdot l \cdot \frac{ql^2}{2} = \frac{1}{6}ql^3; \quad x_c = \frac{3}{4}l;$$

$$\overline{M}_1(x_c) = \frac{3}{4}l; \quad \delta_{11} = \frac{1}{EJ} \cdot \frac{1}{2}l^2 \cdot \frac{2}{3}l = \frac{l^3}{3EJ};$$

$$\Delta_{1P} = -\frac{1}{EJ}\Omega_p x_c = -\frac{1}{EJ} \cdot \frac{1}{6}ql^3 \cdot \frac{3}{4}l = -\frac{ql^4}{8EJ}.$$

Here, the product of the cargo and unit plots is taken with a negative sign, since the multiplied plots are located on different sides of the beam axis (see Fig. 18.7).

Substituting the found coefficients into the Eq. (18.24) and solving it with respect to X_1, we get $X_1 = 3ql/8$.

The static indeterminacy of the system is revealed.

Self-Test Questions

1. What is called the total displacement of a point of a deformable solid?
2. What is called moving a point in a given direction?
3. Give an example of a nonlinear elastic system.
4. How is the potential energy of elastic deformation of the beam calculated in the general case of its loading?
5. Formulate a theorem on the reciprocity of works.
6. Formulate the principle of reciprocity of movements.
7. Give a statement of Castigliano's theorem and write down its mathematical expression.
8. What is a fictitious force? A single force?
9. Write down the displacement integral (Mora) and explain the meaning of the integrands.

10. What is Vereshchagin's method of calculating the Mohr integral?
11. Write down the canonical equations of the force method and explain their physical meaning.
12. For which structural elements can the coefficients of the canonical equations be calculated by Vereshchagin's method?

References

V. Molotnikov, *Kurs soprotivleniya materialov [The Course of Strength of Materials]*, (Lan' Publication, SPb., Moscow, 2006)

V. Molotnikov, A. Molotnikova, *The Theory of Elasticity and Plasticity*, (Springer Nature, Berlin, 2021)

Chapter 19
Taking into Account the Forces of Inertia

Abstract Under static action, the loads increase from zero to the final value so slowly that the accelerations of the body particles during the deformation process can be ignored. Therefore, in statics, it is assumed that external and internal forces are mutually balanced. During operation, the design deals with a dynamic load that changes its value or position quite quickly. Dynamic loads cause large accelerations of the body particles during deformation, and therefore the problems of dynamics, in addition to external and internal forces, take into account the forces of inertia. This chapter discusses methods for calculating the forces of inertia. On the example of a rotating bar with a broken axis, the strength calculation is presented, taking into account the forces of inertia. The fundamentals of the theory of vibrations of structural elements are considered. An example of the calculation of stresses under vibrations of forced vibrations of a mechanical system is given.

Keywords d'Alembert principle · Coefficient of reduction · Mass · Growth rate of fluctuations · Dynamic coefficient · Own oscillations · Forced fluctuations · Embrittlement · Rotation frequency · Resonance

19.1 Dynamic Loads and Inertial Forces

The load applied to the system is called dynamic, if in the calculation the forces of inertia cannot be neglected, arising in the system or its elements. Strength calculations for dynamic loads are complicated for two reasons.

Firstly, to determine the internal forces and stresses, it is necessary to use the general principles and laws of mechanics (the d'Alembert principle, the laws of conservation, etc.). Secondly, under dynamic loads, the mothers discover some new (Molotnikov and Molotnikova 2011; Rabotnov 1962) properties (e.g., embrittlement) (Molotnikov and Molotnikova 2015).

Fig. 19.1 For example 1

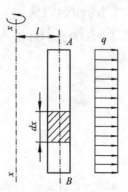

Elementary inertial force dP acting on an elementary volume dV, defined as the product of the mass dm of this volume by its acceleration, will be

$$dP = a \cdot dm = \frac{\gamma}{g} \cdot a \cdot dV, \qquad (19.1)$$

where γ is the specific gravity of the body material and g is the acceleration of gravity.

The forces of inertia are directed in the direction opposite to the direction of acceleration. We give examples of calculating these forces.

Example 1 A homogeneous rod AB (Fig. 19.1) rotates with a constant angular velocity ω around the axis $x - x$, spaced from the axis of the rod at a distance l. Find the intensity of the inertial forces. The cross-sectional area F and the specific gravity γ of the rod material are assumed to be given.

S o l u t i o n . With uniform rotational motion acceleration, a is equal to the centripetal acceleration of a_n, i.e.,

$$a = a_n = \omega^2 \cdot l.$$

According to (19.1), the elementary force of inertia in the case under consideration will be

$$dP = \frac{\gamma}{g} \cdot \omega^2 \cdot l \cdot dV = \frac{\gamma}{g} \cdot \omega^2 \cdot l \cdot F \cdot dx.$$

From here we find the intensity of the inertial forces, which is understood as the value of the inertial force per unit length of the rod:

$$q = \frac{dP}{dx} = \frac{\gamma F}{g} \cdot \omega^2 \cdot l. \qquad (19.2)$$

The plot of the inertial forces is shown on the right in Fig. 19.1.

Fig. 19.2 For example 2

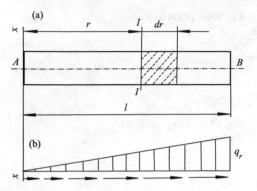

Example 2 A homogeneous rod AB of length l with a cross-sectional area F (Fig. 19.2a) rotates with a constant angular velocity ω around the $x - x$ axis perpendicular to the rod axis. Find the intensity of the inertial forces q_r.

S o l u t i o n . The shaded element (Fig. 19.2a) is affected by the elementary force of inertia:

$$dP = \frac{\gamma}{g}\omega^2 r \cdot F dr.$$

Then, according to the formula (19.2), we obtain

$$q_r = \frac{dP}{dr} = \frac{\gamma F}{g}\omega^2 \cdot r. \tag{19.3}$$

Next, we calculate for $r = 0$ $q_r(0) = 0$; when $r = l$

$$q_r(l) = \frac{\gamma F \omega^2 l}{g} = q.$$

According to the calculated values in Fig. 19.2 *used* to plot q_r. Here the arrows show the direction of the inertial forces.

The general method of calculating the dynamic load is based on the principle of d'Alembert, the essence of which is as follows: if the forces of inertia are added to the external forces acting on the body, then the moving body can be considered as being in a state of equilibrium.

19.2 Calculation of a Rotating Beam

Consider the following problem. The roller and a rigidly connected polyline rod of the same cross-section rotates with a constant angular velocity ω around the axis AB (Fig. 19.3). The following is required:

Fig. 19.3 Beam diagram

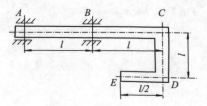

Fig. 19.4 Design scheme of
the beam

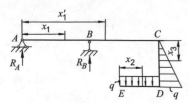

(1) plot the bending moments from the inertial forces arising on the vertical (CD) and horizontal (DE) sections of the broken rod; the inertial forces of the shaft itself can be ignored.

(2) find the permissible number of revolutions of the roller per minute at the permissible voltage $[\sigma] = 1000\,\text{kgf/cm}^2$ and $\gamma = 7.8\,\text{g/cm}^3$. When calculating, take $l = 10\,\text{cm}$; roller diameter $d = 20\,\text{mm}$; $g = 980\,\text{g/cm}^3$. from the forces of inertia arising on the vertical (CD) and horizontal (DE) sections of the broken rod; the inertial forces of the shaft itself can be ignored.

(3) find the permissible number of revolutions of the roller per minute at the permissible voltage $[\sigma] = 1000\,\text{kgf/cm}^2$ and $\gamma = 7.8\,\text{g/cm}^3$. When calculating, take: $l = 10\,\text{cm}$; roller diameter $d = 20\,\text{mm}$; $g = 980\,\text{g/cm}^3$.

Solution.

1. Calculate the inertial forces. Their intensity q along the section ED according to the formula (19.2) will be

$$q = \frac{\gamma \cdot F}{g}\omega^2 l,$$

and the intensity along the CD section is given by the formula (19.3).

The inertial forces found are directed opposite to the acceleration, i.e., away from the axis of rotation. The diagrams of these forces are shown in Fig. 19.4.

2. To plot the bending moments, we define the reactions in the supports A and B. We compose two equilibrium equations in the form of the sum of the moments of all forces equal to zero with respect to the points A and B (Fig. 19.4):

$$\sum M_A = 0, \quad R_B l - \frac{1}{2}ql \cdot 2l - q \cdot \frac{l}{2} \cdot \left(l + \frac{3}{4}l\right) = 0;$$

$$\sum M_B = 0, \quad -R_A l - \frac{1}{2}q \cdot l \cdot l - q \cdot \frac{l}{2} \cdot \frac{3}{4} \cdot l = 0.$$

Fig. 19.5 Plot of bending
moments

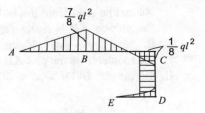

Here, the resultant of the inertial forces along the section CD is calculated
as the area of a triangular plot with a base q and a height l; when determining
the reactions, the resultant of this load can be considered applied at the point C.
The resultant of the inertial forces of the section DE is equal to the area of the
load plot, which is a rectangle with sides $l/2$ and q. The point of application of
this resultant lies in the middle of the segment DE; therefore, the shoulder of the
resultant relative to the point A will be $(l + 3l/4)$, and relative to the point B the
shoulder of this force is $3l/4$.

From the equilibrium equations, we find the reactions R_A and R_B:

$$R_A = -\frac{ql}{2} - \frac{3}{8}ql = -\frac{7}{8}ql,$$

$$R_B = ql + \frac{7}{8}ql = \frac{15}{8}ql.$$

Let's make a check. To do this, we write the third equation of the balance of
the beam in the form of equal to zero the sum of the projections of all forces on
the axis perpendicular to AB:

$$\sum Y = 0 : R_A + R_B - \frac{1}{2}ql - q \cdot \frac{l}{2} = 0;$$

$$-\frac{7}{8}ql + \frac{15}{8}ql - \frac{1}{2}ql = 0.$$

Calculating

$$\frac{ql}{8} \cdot (-7 + 15 - 4 - 4) = 0.$$

Therefore, the reactions are determined correctly.
3. Plot the bending moments for each section of the broken beam (Fig. 19.5).

I. On the plot AB $(0 \leqslant x_1 \leqslant l)$ (Fig. 19.4),

$$M^I(x_1) = R_A \cdot x_1 = -\frac{7}{8}qlx_1.$$

As can be seen from the last formula, the plot of the bending moment on the section AB is linear (x_1 in the first degree). In this case, $M^I(0) = 0$, $M^I(l) = -\frac{7}{8}ql^2$. Using these values of the function $M^I(x_1)$, we plot the moments on the plot AB (Fig. 19.5).

II. On the plot BC ($l \leqslant x_1' \leqslant 2l$) (Fig. 19.4), we have

$$M^{II}(x_1') = R_A x_1' + R_B(x_1' - l) = qlx_1' - \frac{15}{8}ql^2.$$

Analyzing the last formula, we come to the conclusion that the plot of BC is still linear. Calculating

$$M^{II}(l) = -\frac{7}{8}ql^2, \quad M^{II}(2l) = \frac{1}{8}ql^2.$$

Based on the values found in Fig. 19.5, a plot of bending moments is plotted on the BC section.

III. On the section ED (Fig. 19.4), we enter the coordinate x_2, counting it from the point E to the right. We have

$$M^{III}(x_2) = qx_2 \cdot \frac{x_2}{2} = \frac{1}{2}qx_2^2 \quad (0 \leqslant x_2 \leqslant l/2).$$

So, on the plot ED, the plot of moments is a square parabola. To construct it, we calculate the values of the moments at least at three points using the last formula.

$$\text{When } x_2 = 0 \ M^{II}(0) = 0;$$

$$\text{when } x_2 = l/4 \ M^{II}(l/4) = ql^2/32;$$

$$\text{when } x_2 = l/2 \ M^{II}(l/2) = ql^2/8.$$

Based on the values found, a parabola is constructed on the ED section (Fig. 19.5).

IV. On the section DC, the bending moment is constant, algebraically equal to the moment in the section D of the section ED (or the moment in the section C of the section BC), i.e.,

$$M^{IV} = M^{III}(l/2) = M^{II}(2l) = \frac{ql^2}{8},$$

as shown in Fig. 19.5.

Thus, the plot of the bending moments of the considered broken beam is completed. The ordinates of the plot which are everywhere in Fig. 19.5 are set aside on the side of the stretched fibers, so the signs on the plot are not specified.

Fig. 19.6 Plot of
longitudinal forces

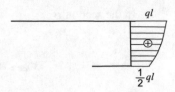

4. To determine the permissible speed of rotation of the shaft AB , we will
   need another plot of longitudinal forces. From the analysis of the calculation
   scheme (Fig. 19.4), it follows that there are no longitudinal forces in the sections
   AB, BC, *and* ED. To determine the longitudinal force ($N(x_3)$) in an arbitrary
   section of the section CD, we consider the equilibrium of the part of the broken
   bar located below the section with the coordinate x_3. We get

$$N(x_3) = q \cdot \frac{l}{2} + \frac{q + qx_3/l}{2} \cdot (l - x_3),$$

   i.e.,

$$N(x_3) = ql \left(1 - \frac{x_3^2}{2l^2}\right).$$

   To plot the plot $N(x_3)$, we calculate the values of the longitudinal force in
   three sections, namely, at the ends and in the middle of the section CD:

$$x_3 = 0: \ N(0) = ql;$$

$$x_3 = l/2: \ N(l/2) = \frac{7}{8} ql;$$

$$x_3 = l: \ N(l) = \frac{ql}{2}.$$

   The plot of the longitudinal forces, based on the values found, is shown in
   Fig. 19.6.
5. To determine the permissible number of revolutions ($[n]$) per minute, we find the
   most dangerous sections of the beam. These can be the cross-section of the shaft
   AC on the support B (Fig. 19.5), in which the maximum bending moment, or the
   cross-section of the section CD, is adjacent to the point C, since in this section
   the longitudinal force is maximum and the bending moment is large enough. In
   this situation, we will determine the permissible speed of rotation based on the
   strength conditions of each suspicious cross-section.
   We proceed from the strength condition: the maximum normal stress in the
   cross-section should not exceed the permissible value of $[\sigma]$:

$$\sigma_{max} \leqslant [\sigma].$$

In the section B

$$\sigma_{max} = \frac{|M^I(l)|}{W} = \frac{\frac{7}{8} q l^2}{\frac{\pi d^3}{32}} = \frac{28 q l^2}{\pi d^3}.$$

Using the formula (19.2), we can write

$$q = \frac{\pi F \omega^2 l}{g} = \frac{\gamma \pi d^2 \omega^2 l}{4g}.$$

Taking into account the last two formulas, the strength condition takes the form

$$\frac{28 l^2}{\pi d^3} \cdot \frac{\gamma \pi d^2 \omega^2 l}{4g} = [\sigma].$$

From here we find the square of the permissible angular velocity $[\omega]_B$ from the strength condition in the cross-section B:

$$[\omega]_B^2 = \frac{2d[\sigma]}{7 \gamma l^3} = \frac{1000 \cdot 980 \cdot 2}{7 \cdot 7, 8 \cdot 10^{-3} \cdot 10^3} = 3, 58 \cdot 10^4 \ c^{-2}.$$

In the section C

$$\sigma_{max} = \frac{N(0)}{F} + \frac{M^{IV}}{W} = \frac{ql}{\pi d^2/4} + \frac{q l^2/8}{\pi d^3/32} = \frac{\gamma \omega^2 l^2}{gd} \cdot (l + d).$$

Substituting this expression into the strength condition, we find

$$[\omega]_C^2 = \frac{2d[\sigma]}{\gamma l^2 (l + d)} = \frac{1000 \cdot 980 \cdot 2}{7, 8 \cdot 10^{-3} \cdot 10^2 \cdot (10 + 2)} = 21 \cdot 10^4 \ c^{-2}.$$

From the two found acceptable values of the square of the angular velocity, we take the smaller, i.e.,

$$[\omega]^2 = min\{[\omega]_B^2, [\omega]_C^2\} = 3, 58 \cdot 10^4 \ c^{-2}.$$

Hence we find that $[\omega] = 189 \ s^{-1}$. Then, using the relation between the angular velocity ω and the rotational speed n, rpm,

$$\omega = \frac{\pi n}{30}$$

find the permissible speed of rotation:

$$[n] = \frac{30 \cdot [\omega]}{\pi} = \frac{30 \cdot 189}{3,14} = 1800 \, (\text{rpm}).$$

Problem solved.

19.3 Natural Oscillations of Systems with One Degree of Freedom

The number of degrees of freedom of a mechanical system is the number of independent parameters that determine the position of the system. There are natural and forced oscillations of elastic systems. With its own (free) vibrations, the system is freed from the active external force influence and is left to itself. The motion of the elastic system in this case occurs as a result of the initial impulse communicated to the system.

Circular frequency ω_0 of natural oscillations of an elastic system with one degree of freedom is calculated by (Molotnikov 2006) using the formula

$$\omega^2 = \frac{1}{\delta_{11}m}. \tag{19.4}$$

Here m is the mass of the oscillating weight, and δ_{11} is the movement of this mass under the action of a statically applied unit force in the direction of the deviation of the center of mass from the static equilibrium position.

19.4 Forced Fluctuations

The oscillations of an elastic system are called *forced*, if they occur under the action of changing external forces, called *perturbing*. In engineering, the most interesting case is when the perturbing force (P) changes according to the harmonic law

$$P(t) = P_0 \sin \Omega t, \tag{19.5}$$

where P_0 is the maximum value of the force and Ω is the circular frequency of its change. In the presence of linear resistance forces, the amplitude A_{vn} of forced oscillations at a disturbing force (19.5) is expressed by the formula

$$A_{\text{вын}} = \frac{P_0 \delta_{11}}{\sqrt{\left(1 - \frac{\Omega^2}{\omega^2}\right)^2 + \frac{4n^2\Omega^2}{\omega^4}}}, \tag{19.6}$$

where $2n = \dfrac{\alpha}{m}$, α is the coefficient of proportionality between the drag force and the velocity of the oscillating mass m.

If there are no resistances, the formula (19.6) can be represented as

$$A_{\text{вын}} = P_0 \beta \delta_{11}, \quad \beta = \frac{1}{1 - \left(\dfrac{\Omega}{\omega}\right)^2}. \tag{19.7}$$

The β parameter is called the *oscillation rate*.

In the case under consideration, the voltage (σ_d) in elastic elements the system consists of voltage (σ_{St}) under static app power Q ($Q = mg$ g − the acceleration of gravity) and stresses (σ), the resulting displacement (19.6) power Q from the equilibrium position, i.e.,

$$\sigma_d = \sigma_{st} + \sigma = \sigma_{st}\left(1 + \frac{\sigma}{\sigma_{cst}}\right). \tag{19.8}$$

Since the stresses in elastic deformation are proportional to the displacements, the ratio σ/σ_{st} in the formula (19.8) can be replaced by the ratio of the corresponding displacements. The displacement from the statically applied force Q is equal to $Q\delta_{11}$, and the displacement causing the stresses σ is equal to the amplitude of the oscillations. Then the formula (19.8) can be represented as

$$\sigma_d = \sigma_{st}\left(1 + \frac{H\beta\delta_{11}}{Q\delta_{11}}\right) = \sigma_{st}\left(1 + \frac{H}{Q}\beta\right),$$

or

$$\sigma_d = k_d \sigma_{st}, \tag{19.9}$$

where the parameter is

$$k_d = 1 + \frac{P_0}{Q}\beta \tag{19.10}$$

it is called the *dynamic coefficient*.

If there is a linear resistance, the dynamic coefficient will be

$$k_d = \frac{A_{\text{вын}}}{P_0 \delta_{11}} = \frac{1}{\sqrt{\left(1 - \dfrac{\Omega^2}{\omega^2}\right)^2 + \dfrac{4n^2\Omega^2}{\omega^4}}}. \tag{19.11}$$

Hence, it can be seen that the dynamic coefficient for forced oscillations depends on the frequency ratio Ω/ω and the parameter $2n/\omega$, which characterizes the

Fig. 19.7 On the
phenomenon of resonance

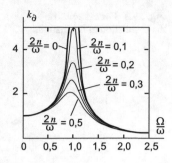

attenuation. Figure 19.7 shows a family of curves constructed from a dependency
(19.11). The graphs show that in the absence of attenuation ($n = 0$) and the
coincidence of the frequencies of natural and forced oscillations ($\Omega/\omega = 1$), the
dynamic coefficient increases indefinitely. For all other $n > 0$, the coefficient k_d is
a bounded value that has a maximum at $\Omega = \omega$.

   *The phenomenon of the increase in the amplitude when the frequencies of natural
and forced oscillations coincide is called resonance.*

   A sharp increase in the amplitude of the oscillations at resonance causes a
significant increase in the voltage, which is undesirable. Therefore, when designing
machines, they strive to ensure that the condition $\Omega/\omega \notin [0.7; 1.3]$ is met.

   Interval

$$0.7 \leqslant \Omega/\omega \leqslant 1.3$$

is called *resonant*. With the option $0.7 \leqslant \Omega/\omega$ it is most preferable, since in this
case the machine does not pass the resonant interval during the start-up and run-out.

   If the dynamic coefficient is found and the stresses σ_{mboxst} from the statically
applied force P_0 are determined, then the calculated voltage caused by the dynamic
action of this load will be

$$\sigma = \sigma_{st} k_d. \tag{19.12}$$

19.4.1 Example of Calculation of Stresses During Oscillations

On two beams of the I-beam section №16, an engine weighing Q (Fig. 19.8a) is
installed, making n revolutions per minute. The centrifugal force of inertia resulting
from the unbalance of the rotating parts of the engine is equal to H. The self-weight
of the beams and the drag forces can be ignored. You need to find:

(1) the natural oscillation frequency of the system ω_0;
(2) the frequency of change of the disturbing force ω;
(3) the coefficient of increase of fluctuations β;
(4) dynamic coefficient k_d;

Fig. 19.8 For example, on
the calculation of fluctuations

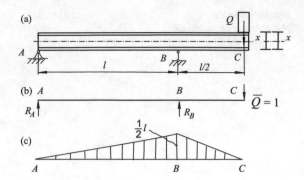

(5) the maximum normal stress in the beam is σ_d at the permissible stress $[\sigma] = 1000\,\text{kgf/cm}^2$.

If the dynamic coefficient is greater than the allowed one, then it is necessary to change the number of I-beams, confirming that the new profile was selected correctly by repeated calculation. When calculating, take (Fig. 19.8a): $l = 120\,\text{cm}$, $Q = 2000\,\text{kgf}$, $H = 1000\,\text{kgf}$, frequency $n = 600\,\text{rpm}$.

S o l u t i o n .

1. Calculate the natural oscillation frequency of the system by the formula (19.4). To calculate the displacement of δ_{11} , you need to plot the bending moments from the unit force $\overline{Q} = 1$, applied at the point C, (Fig. 19.8b). To this end, we determine the reactions of the supports R_A and R_B (Fig. 19.8b). We write the equations of equilibrium, equating to zero the sum of the moments of all forces relative to the points B and A:

$$\sum M_B = R_A \cdot l + 1 \cdot \frac{l}{2} = 0,$$

$$\sum M_A = R_B \cdot l - 1 \cdot \frac{3}{2} \cdot l = 0.$$

From here we find

$$R_A = -\frac{1}{2}; \ R_B = \frac{3}{2}.$$

With known reactions, you can begin to plot the bending moments. If there are only concentrated forces (R_A, R_B, $\overline{Q}$), the plot of moments is linear. Since the moments at points A and C (Fig. 2.1b) are zero, it is sufficient to calculate the moment M_B at point B to construct the plot. We have

$$M_B = \overline{Q} \cdot \frac{l}{2} = \frac{l}{2}.$$

The plot of moments from the force $\overline{Q} = 1$ is constructed in Fig. 19.8*in*.

Now we find δ_{11} by multiplying the unit plot by itself. To do this, we divide the plot into two triangles with bases AB and BC. Find the areas of these triangles S_1 and S_2:

$$S_1 = \frac{1}{2} \cdot l \cdot \frac{1}{2} \cdot l = \frac{1}{4} \cdot l^2; \quad S_2 = \frac{1}{2} \cdot \frac{l}{2} \cdot \frac{l}{2} = \frac{l^2}{8}.$$

Then we calculate the ordinates y_1 and y_2 of the unit plot under the centers of gravity of the specified triangles:

$$y_1 = y_2 = \frac{2}{3} \cdot \frac{1}{2} \cdot l = \frac{l}{3}.$$

Then

$$\delta_{11} = \frac{1}{2EJ_x} (S_1 \cdot y_1 + S_2 \cdot y_2) = \frac{1}{2EJ_x} \left(\frac{1}{4} \cdot l^2 \cdot \frac{l}{3} + \frac{l^2}{8} \cdot \frac{l}{3} \right) = \frac{l^2}{16EJ_x}.$$

Here E is the elastic modulus of the beam material; J_x is the moment of inertia of one I-beam relative to the $x - x$ axis (Fig. 19.8a). In the future, we will put $E = 2 \cdot 10^6$ kgf/cm^2. According to the tables of the rolled steel grade, we find for the I-beam №16: $J_x = 863$ cm^4, and the moment of resistance $W_x = 109$ cm^3. Substituting δ_{11} into the formula (19.4), we find

$$\omega_0 = \frac{1}{\sqrt{l^3/16EI_x \cdot Q/g}} = \frac{4}{l} \sqrt{\frac{EJ_x g}{Ql}}.$$

Using numeric values, we have

$$\omega_0 = \frac{4}{120} \sqrt{\frac{2 \cdot 10^6 \cdot 873 \cdot 980}{2000 \cdot 120}} = 75.5 \, \text{c}^{-1}.$$

2. The circular frequency of the disturbing force will be

$$\omega = \frac{\pi n}{30} = \frac{3,14 \cdot 600}{30} = 62,8 \, \text{c}^{-1}.$$

3. Calculating the coefficient rise of fluctuations. By the formula (19.7) we have

$$\beta = \frac{1}{1 - (62,8/75,5)^2} = 3,23.$$

4. According to the formula (19.10), the dynamic coefficient will be

$$k_{\text{д}} = 1 + \frac{1000}{2000} \cdot 3,23 = 2,62.$$

5. Using the formula (19.9), we calculate the maximum normal stress in the beam. To do this, we first calculate the voltage σ_{st} by dividing the greatest moment M_{max} from the statically applied force Q at the moment of resistance of the composite beam:

$$\sigma_{st} = \frac{M_{max}}{2W_x}.$$

Having a plot of moments from the unit force (Fig. 19.8c), the moment M_{max} is calculated by multiplying the maximum ordinate of this plot ($l/2$) by the magnitude of the force Q, i.e., $M_{max} = 0.5lQ$. Then

$$\sigma_{st} = \frac{0,5lQ}{2W_x} = \frac{Ql}{4W_x} = \frac{2000 \cdot 120}{4 \cdot 109} = 550\,\text{kgf/cm}^2;$$

$$\sigma_{d} = k_d\sigma_{st} = 2.62 \cdot 550 = 1441\ \text{kgf/cm}^2.$$

Thus, the greatest stress in the beam exceeds the permissible one ($1441 > 1000$). This result could have been predicted already when we calculated the frequencies ω_0 and ω. In fact, by calculating the ratio

$$\frac{\omega}{\omega_0} = \frac{62.8}{75.5} = 0.83,$$

we conclude that this ratio is in the so-called resonant range $[0.75; 1.25]$. Therefore, it was possible to expect the appearance of large dynamic stresses.

Changing the profile number. Take the I-beam №18; for this I-beam, we write out the necessary geometric characteristics from the assortment tables: $J_x = 1290\,\text{cm}^4$, $W_x = 143\,\text{cm}^3$. Repeat the calculation.

1. Find the circular frequency of natural oscillations of the load:

$$\omega_0 = \frac{4}{l}\sqrt{\frac{EI_xg}{Ql}} = \frac{4}{120}\sqrt{\frac{2 \cdot 10^6 \cdot 1290 \cdot 980}{2000 \cdot 120}} = 108\,\text{s}^{-1}.$$

2. The frequency of ω remains the same: $\omega = 62.8\,\text{c}^{-1}$.
3. Calculate the ratio

$$\frac{\omega}{\omega_0} = \frac{62.8}{108} = 0.58.$$

As we can see, this relation no longer belongs to the resonant interval. We have

$$\beta = \frac{1}{1 - (\omega/\omega_0)^2} = \frac{1}{1 - 0.58^2} = 1.51.$$

4. Calculate the dynamic coefficient

$$k_д = 1 + \frac{H}{Q}\beta = 1 + \frac{1000}{2000} \cdot 1,51 = 1,755.$$

5. Calculate

$$\sigma_{st} = \frac{Ql}{4w_x} = \frac{2000 \cdot 120}{4 \cdot 143} = 420\,\text{kgf/cm}^2;$$

$$\sigma_d = k_d \cdot \sigma_{st} = 1.755 \cdot 420 = 737.1\,\text{kgf/cm}^2.$$

Since now the strength condition is satisfied ($\sigma_d < [\sigma]$), we accept the selected profile number. Problem solved.

19.5   Impact Loads: Approximate Calculation of the Impact

We have already said (p. 105) that shock or pulse loads are called loads that act for a very short period of time (the duration of the load is small compared to the period of natural oscillations of the system).

Shock loading occurs when an elastic system is hit by a certain mass moving at a certain speed. Below we will consider only the case of inelastic impact, in which no rebound occurs after the contact of the striking and striking bodies.

The approximate calculation of the effect of shock loads is based on the assumption that the configuration of the system with free vibrations after impact is considered known and unchanged. For example, in the case of a transverse impact on a beam, it is assumed that after the impact, the configuration of the longitudinal axis of the beam will be the same as its curved axis from the static action of the force applied at the point of impact.

Stresses (σ_d) and displacements (Δ_d) of the impact load is determined by the following formulas:

$$\sigma_d = \sigma_{st} \cdot k_d,$$
$$\Delta_d = \Delta_{st} \cdot k_d,$$

(19.13)

Moreover, σ_{st} and Δ_{st} are the stresses and displacements caused by the static application of force at the point of impact, respectively, and k_d is the dynamic coefficient.

In cases where the inertial forces (the mass of the impacted system) can be neglected, the dynamic coefficient can be calculated by the formula

$$k_д = 1 + \sqrt{1 + \frac{2h}{\Delta_{cт}}},$$

(19.14)

where h is the height of the drop of the load hitting the system. From the structure of the last formula, it is clear that if the height of the fall of the load is many times greater than the static movement of Δ_{st}, an approximate dependence can be used to calculate the dynamic coefficient:

$$k_d = \sqrt{\frac{2h}{\Delta_{st}}}.$$

If the mass of the impacted system m_c is commensurate with the mass of the falling load m_p, then the dynamic coefficient is calculated by the formula

$$k_d = 1 + \sqrt{1 + \frac{2h}{\Delta_{st}(1 + \beta m_c/m_p)}}.$$

Here β is the coefficient of bringing the mass to the point of impact, calculated by the formula

$$\beta = \frac{\int\limits_{(m_c)} \delta_{st}^2(x) dm_c}{\Delta_{st}^2 m_c}, \tag{19.15}$$

Moreover, $\delta_{st}(x)$ is the movement of the mass element dm_c, defined by the coordinate x, from the statically applied load P at the point of impact. The integral in the formula (19.15) is calculated over all the masses of the impacted system.

Example On an I-beam lying freely on two rigid supports (Fig. 19.9a), from a height of h drops the load P. The following is required:

(1) find the largest normal stress in the beam;
(2) solve a similar problem, provided that the right support of the beam is replaced by a spring, the pliability of which (the draft from the weight of the load $\overline{P} = 1$) is equal to α;
(3) compare the results.

In the calculations, take I-beam №30; $l = 2.8\,\text{m}$, $a = 2.1\,\text{m}$; $P = 80\,\text{kgf}$; $h = 10\,\text{cm}$; $\alpha = 28\,\text{cm}\,/\,\text{t}$; $E = 2 \cdot 10^6\,\text{kgf/cm}^2$.

S o l u t i o n . Let's first solve the problem without taking into account its own weight beams.

1. Find the reference reactions from the statically applied force P. To do this, we make the beam equilibrium equations in the form of the sum of the moments of all forces with respect to the points A and B equal to zero:

$$\sum M_A = R_A \cdot l - P(l + a) = 0;$$

$$\sum M_B = -R_A \cdot l - Pa = 0.$$

Fig. 19.9 For example, the calculation of the impact load

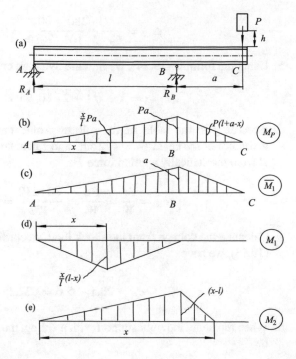

From here we have

$$R_B = P \cdot \frac{l+a}{l} = 80 \cdot \frac{2.8 + 2.1}{2.8} = 140\,\text{kgf};$$

$$R_A = -P \cdot \frac{a}{l} = -80 \cdot \frac{2.1}{2.8} = -60\,\text{kgf}.$$

2. Plot the load curve of the bending moments M_P (Fig. 19.9b).
3. To calculate the static deflection Δ_{st} under the point of application of the force P, we construct another plot of moments from the unit force $\overline{P} = 1$ applied at the point C (Fig. 19.9c). We get this plot from the previous one (Fig. 19.9b), assuming $P = 1$ in it.
4. Calculate the static deflection Δ_{st} by the formula

$$\Delta_{\text{st}} = \delta_{11} P.$$

The coefficient δ_{11} is found by multiplying the unit plot $\overline{M}_1$ (Fig. 19.9c) on itself

$$\delta_{11} = \frac{1}{EJ} \cdot \left(\frac{1}{2} \cdot a \cdot l \cdot \frac{2}{3} \cdot a + \frac{1}{2} \cdot a \cdot a \cdot a \cdot \frac{2}{3} \cdot a \right) = \frac{21 l^3}{64 EJ}.$$

where EJ is the bending stiffness. From the assortment tables, we find the maximum moment of inertia of the I-beam $J = 7080\,\text{cm}^4$. Next, we calculate

$$\Delta_{\text{ст}} = \frac{21 \cdot 280^3 \cdot 80}{64 \cdot 2 \cdot 10^6 \cdot 7080} = 0,041 \text{ cm.}$$

5. Using the formula (19.14), we find the dynamic coefficient:

$$k_{\text{d}} = 1 + \sqrt{1 + 2 \cdot 10/0.041} = 23.1.$$

6. According to the table of the rolling profile range, we find the moment of resistance of the beam $W = 472\,\text{cm}^3$ and calculate the stress in the cross-section B from the statically applied force P:

$$\sigma_{\text{st}} = \frac{M}{W} = \frac{Pa}{W} = \frac{80 \cdot 210}{472} = 35.6 \; \frac{\text{кгс}}{\text{см}^2}.$$

7. Calculate the voltage from the shock load. According to the first of the formulas (19.13), we have

$$\sigma_{\text{д}} = 23.1 \cdot 35.6 = 822.35 \; \frac{\text{kgf}}{\text{cm}^2}.$$

8. When replacing the right support with a spring, the movement of the point B will be

$$\Delta_B = R_B \cdot \alpha = 140 \cdot 10^{-3} \cdot 28 = 3.92 \text{ см.}$$

Given that the vertical displacement of the points of the beam axis at the spring draft is proportional to the distance from the support A, we find the displacement of the point of application of the statically acting force P in the presence of a spring:

$$\Delta_{\text{st}}^{II} = \Delta_{\text{ст}}^{I} + \Delta_B \cdot \frac{l+a}{l} = 0.041 + 3.92 \cdot \frac{280 + 210}{280} = 6.9 \text{ см.}$$

Here, the Roman numerals in the upper index indicate that the value belongs to the case of a rigid (I) and a spring support (II). The dynamic coefficient in the presence of a spring will be

$$k_{\text{d}}^{II} = 1 + \sqrt{1 + \frac{2 \cdot 10}{6.9}} = 2.97,$$

and the corresponding highest normal voltage is now equal to

$$\sigma_{\text{d}} = 35.6 \cdot 2.97 \approx 106 \,\text{kgf/cm}^2.$$

Comparing this result with the value found in point 7, we conclude that the installation of the spring reduced the stress from the impact by almost 8 times! Devices for this purpose are called *dampers*.

19.6 Example of Accounting for the Mass of the Impacted System

In the problem discussed in the previous paragraph, the mass of the beam:

$$m_b = \frac{q(l+a)}{g} = \frac{36.5 \cdot (2.8 + 2.1)}{g} = \frac{178.85}{g},$$

Moreover, q is the weight of one linear meter of the beam (given in the assortment tables). Mass of the striking body:

$$m_p = \frac{80}{g}.$$

The ratio of these masses will be

$$\frac{m_b}{m_p} = \frac{178.85}{80} = 2.23.$$

Hence, the interacting masses are of the same order, so for more reliable results, the mass should be taken into account the body being hit.

Find the function $\delta_{st}(x)$, which is included in the formula (19.15). In the problem under consideration, this function coincides with the equation of the elastic axis of a beam bent by a statically applied force P.

To calculate the function $\delta_{st}(x)$, we plot the moments from the unit force applied at a distance x from the support A. In this case, two cases should be considered:

(A) a single force is applied on the span AB; (B) a single force is applied on the console. The unit plots M_1 and M_2 corresponding to these two cases are plotted in Fig. 19.9d, e.

The function $\delta_{st}(x)$ for $0 \leqslant x \leqslant l$ is found by multiplying the plots M_P and M_1. We have

$$\delta_{cт}(x) = -\frac{1}{EI}\left[2 \cdot \frac{Pax}{l} \cdot \frac{l-x}{l} \cdot x \cdot \frac{x}{6} + \frac{l-x}{6}\right.$$
$$\left. \times \left(2 \cdot \frac{Pax}{l} \cdot \frac{l-x}{l} \cdot x + Pax \cdot \frac{l-x}{l}\right)\right].$$

After performing the simplest transformations, we get

$$\delta_{\text{ст}}(x) = -\frac{3P}{8EI}x(l^2 - x^2) \text{ if } 0 \leqslant x \leqslant l. \tag{19.16}$$

Similarly, by multiplying the plots M_P and M_2, we find

$$\delta_{\text{ст}}(x) = \frac{P(x-l)}{6EI}\left[\frac{17}{4}(x-l)l + \frac{7}{2}l^2 - (x^2 - l^2)\right] \text{ при } l \leqslant x \leqslant l+a. \tag{19.17}$$

We now calculate the integral in the numerator of the formula (19.15). Keeping in mind that

$$dm_c = \frac{q}{g}dx,$$

calculating

$$J = \int_{(m_c)} \delta_{\text{ст}}^2(x)dm_c = \frac{q}{g}\int_0^{l+a} \delta_{\text{ст}}^2(x)dx = \frac{q}{g}(J_1 + J_2),$$

where indicated

$$J_1 = \int_0^l \delta_{\text{ст}}^2(x)dx, \tag{19.18}$$

$$J_2 = \int_l^{l+a} \delta_{\text{ст}}^2(x)dx. \tag{19.19}$$

Substituting the functions (19.16) and (19.17), respectively, into the formulas (19.18) and (19.19), we find after calculating the quadratures:

$$J = \frac{3P^2l^7k}{196002E^2I^2},$$

where

$$k = (99\varsigma^5 + 231\varsigma^4 + 371\varsigma^3 + 260\varsigma^2 + 140\varsigma + 8), \quad \varsigma = \frac{3}{4}.$$

Substituting this integral into the formula (19.15), we find the coefficient mass reduction:

$$\beta = \frac{qP^2\varsigma^2l^7k}{3675gE^2I^2} : \frac{21^2l^6P^2ql}{64^2E^2I^2q} = \frac{1024}{416745} \cdot \frac{\varsigma^2k}{1+\varsigma}.$$

Numerical calculations give $\beta = 0.415$.
Then

$$k_d = 1 + \sqrt{1 + \frac{2 \cdot 10^6}{0.041 \cdot (1 + 0.415 \cdot 2, 23)}} = 16.9.$$

In this case, the highest normal stress will be

$$\sigma_d = 16, 9 \cdot 35, 6 \cong 502 \, \text{kgf/cm}^2.$$

Therefore, taking into account the mass of the impacted system allows you to detect the strength reserve of the impacted system. In other words, the neglect of the distributed mass goes to the safety margin of the system.

19.7 Impact Strength of Materials

At the beginning of Chap. 15 (p. 425), we talked about the fact that the behavior of the mother under dynamic loads in some cases can be very different from the behavior under static application of forces. To characterize the ability of a material to resist shock loads and its tendency to brittle fracture, a special mechanical value is used, which is called the *impact strength* of the material.

The impact strength is determined experimentally. A prismatic sample of square cross-section with an incision in the middle (Fig. 19.10) on a special copra is struck from the side opposite to the incision. A measure of the impact resistance of a material is the ratio of the work spent on the destruction of the sample to the cross-sectional area at the weakened location. Thus, in the SI system, the impact strength has the dimension $H \cdot m/m^2$. In the Russian scientific and technical literature, the

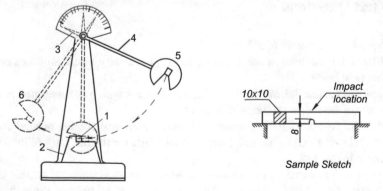

Fig. 19.10 Coper and test sample: 1, sample; 2, anvil; 3, arrow; 4, the initial position of the pendulum; 5, the striker; 6, the final deviation of the pendulum

Table 19.1 Impact strength of carbon steels

Carbon contin steel,%ent	Heat treatment			
	Annealing		Tempering and tempering	
	σ_{vr}, MPa	Impact viscosity $a_K \cdot 10^{-5}$ N/m	Ultimate strength σ_{vr}, MPa	Impact viscosity $a_K \cdot 10^{-5}$ N/m
$C < 0.15$	350...450	>25	360...500	>25
$C\ 0.15...0.20$	400...500	>22	450...650	>20
$C\ 0.20...0.30$	500...600	>20	550...750	>15
$C\ 0.30...0.40$	600...700	>16	700...850	>12
$C\ 0,40...0,50$	700...800	>12	800...950	>8
$C\ 0.50...0,60$	800...900	>10	900...1050	>5
$C\ 0.60...0.70$	850...950	>8	>1000	>3
$C > 0.7$	>950	>6	>1050	>2

impact strength is most often indicated by the symbols a_k. The work spent on the destruction of the sample is defined as the difference in the values of the potential energy of the pendulum in its initial and final position. The specified positions are fixed by the arrow 3 (Fig. 19.10).

Impact tests are usually carried out at room temperature by testing a batch of similar samples. The sample size of the samples is usually set to at least four. Such a number of samples is necessary because the random circumstances of their manufacture and conducting experiments can greatly affect the value of the impact strength. Table 19.1 shows the impact strength values for carbon steels with different carbon content.

The impact strength of most materials depends significantly on the temperature. For steels and many metals, a decrease in impact strength is found with a decrease in temperature. This phenomenon is called *cold breakage*.

Self-Test Questions

1. What is the dynamic load?
2. Why is the calculation for dynamic load significantly more complicated than for static load?
3. How are the inertial forces determined in systems with a concentrated mass?
4. How are the forces of inertia directed?
5. Plot the forces of inertia when a homogeneous rod rotates around an axis parallel to the axis of the rod. Derive the formula for the intensity of the inertial forces in this rod.
6. How are the inertial forces distributed when a homogeneous rod rotates around an axis drawn perpendicular to the axis of the rod and passing through one of its ends? Draw a diagram of the intensity of these forces.

7. How is the permissible speed of rotation of the shaft bearing the masses removed from the axis determined?
8. What can cause the natural oscillations of an elastic system?
9. What is called the dynamic coefficient?
10. Write down the formula for determining the circular frequency of natural oscillations of a system with one degree of freedom.
11. What are the vibrations of a mechanical system called forced?
12. Give an example of a real mechanical system in which forced oscillations take place.
13. What oscillations are called damped with linear resistance?
14. How are the stresses in the elements of a mechanical system determined with a known dynamic coefficient?
15. Name the condition for the occurrence of resonance.
16. What is the resonant interval?
17. Explain why it is desirable to choose the parameters of the machine that ensure its operation in all modes to the left of the resonant interval?
18. Give an example from the history of technology, when the structure (device) was destroyed due to resonance.
19. How are the displacements in the elements of a mechanical system determined with a known dynamic coefficient?
20. What determines the circular frequency of the disturbing force?
21. What is the number of natural vibration frequencies of a mechanical system?
22. What is called the main form of vibration of a mechanical system?
23. What, in your opinion, is the qualitative difference between the oscillation processes of systems with concentrated and distributed parameters?
24. What loads are called shock or pulse loads?
25. Formulate a hypothesis that is accepted in the approximate theory of impact.
26. How is the dynamic coefficient expressed when hitting a system whose mass can be neglected?
27. Write down the expression for the dynamic coefficient when hitting a system with a concentrated mass.
28. What is the dynamic coefficient in the case of a sudden load application?
29. How does the dynamic coefficient change when replacing a rigid support with a pliable one?
30. For what purpose are the dampers installed?
31. What is called the mass reduction coefficient?
32. What characterizes the impact strength of the material?
33. How is the impact strength determined?
34. What is the minimum number of tests for determining the impact strength?
35. Name the units of impact strength.
36. What is the manifestation of cold breakage?
37. Try to explain the reasons for the dependence of the impact strength on the body temperature.

References

V. Molotnikov, *Kurs soprotivleniya materialov [The Course of Strength of Materials]* (Krasnodar, Lan' Publ., Moscow, 2006)

V. Molotnikov, A. Molotnikova, *Dinamicheskie zadachi i raschyoty' na vy'noslivost' [Dynamic Tasks and Endurance Calculations]* (2011). https://v.molotnikov.de/dyn.pdf

V. Molotnikov, A Molotnikova, *Mexanika deformacij [Deformation Mechanics]* (RFNO Publ., Rostov-on-Don, 2015)

Y. Rabotnov, *Soprotivlenie materialov [Strength of Materials]* (Gosfiztexizdat Publ., Moscow, 1962)

Chapter 20
Fatigue Resistance

Abstract Fatigue resistance (or fatigue strength) is the property of a material that does not break down over time under the influence of changing workloads. In most cases, these are cyclic loads. If the level of alternating stresses exceeds a certain limit, then irreversible processes of damage accumulation occur in the material, which ultimately lead to the destruction of the structure. It is established that about 90% of all breakdowns of machine parts are the result of the development of fatigue cracks. This chapter introduces the main characteristics of variable voltages. The concept of the endurance limit is defined. The methods of practical calculation of machine parts for endurance are described. An example of calculating the shaft endurance is given.

Keywords Fatigue · Endurance · Microcracks · Macrocracks · Dislocation · Structural defects · Stress cycles · Endurance limit · Weller's experiments · Limit cycles · Fatigue strength margin · Stress concentration · Surface quality · Large-scale effect

20.1 The Concept of Fatigue Failure

Most machine parts are subject to stresses that change over time. The change in stress can be caused either by a change in the current loads or by a relative change in time of the loads and zones of the studied stresses. Experiments show that under the action of variable stresses, the destruction of materials occurs at stresses significantly lower than the ultimate strength or yield strength. Failure almost always occurs suddenly, and in this sense it is identical to brittle failure, even in cases where the material under static test reveals plastic properties.

The first researchers of the destruction of materials under variable stresses proceeded from the assumption that the material after a long period of operation is reborn; its crystal structure changes due to *fatigue*. At present, the hypothesis of a change in the crystal structure has been refuted by X-ray methods, but the term "fatigue" has been preserved in the literature, although it has acquired a different meaning.

449
V. Molotnikov, A. Molotnikova, *Theoretical and Applied Mechanics*,
https://doi.org/10.1007/978-3-031-09312-8_20

Metals and their alloys almost always have structural defects in the form of dislocations, vacancies, the introduction of foreign atoms, etc. These defects are zones of stress redistribution, but under static loading, the migration of defects does not occur. In the case of prolonged exposure to alternating stresses, on the contrary, there is a movement of defects in the structure of the material and their accumulation in limited areas. The accumulation of defects leads to the formation of microcracks, which can gradually grow and combine into macrocracks.

The process of damage accumulation under the influence of variable stresses is called material fatigue, and the property of the material to resist the action of variable stresses is called endurance.

The calculation of strength under the action of time-varying stresses is called the calculation of endurance or the calculation of fatigue strength.

Most often, the origin of macrocracks begins with the surface of the part in areas of increased stress. Such zones are acute risks after turning with the cutter, places of sharp changes in the size of the part, keyways, etc. However, there are cases when the formation of cracks begins inside the part. The tops of the formed cracks are zones of increased stress, which leads to the appearance of local inelastic deformations and the rupture of bonds. Cracks begin to grow. One of them becomes dominant. As a result of the development of the crack, the section of the part is weakened, and at some point, there is a sudden destruction.

The goal that the engineer sets for himself when calculating the endurance is, firstly, to determine the circumstances that cause the appearance of fatigue cracks and, secondly, to choose a design of the part that would guarantee it from destruction under variable stresses. The task of calculating endurance is extremely important. It is established that about 90% of all breakdowns of machine parts are the result of the development of fatigue cracks.

At present, no physical theory of fatigue failure has been created yet. Therefore, in solid-state mechanics, an experimental-phenomenological approach is used, which gives the engineer fairly reliable methods for evaluating fatigue strength.

20.2   Stress Cycles

The change in the stresses in the machine parts over time in the vast majority of cases is periodic. Let us first consider the case of an uniaxial stress state. We will put the time t on the abscissa axis and the voltage σ on the ordinate axis at some point in the part. Then the change in stress over time is represented by a certain periodic curve (Fig. 20.1, solid polyline).

The totality of all the stress values for a single period of its change is called the *cycle of variable stresses.* The highest cycle stress in the algebraical sense is called the *maximum stress* (σ_{max}) , and the smallest is *with the minimum stress* (σ_{min}) of the loop (Fig. 20.1).

Experimental studies have established that the fatigue strength is determined mainly by the values of the maximum and minimum stresses of the cycle and

Fig. 20.1 Stress cycles

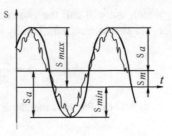

does not depend on the nature of the stress change in the interval between σ_{max} and σ_{min}. Therefore, it is assumed that the cycles shown in Fig. 20.1 by the solid and dashed lines are equivalent if values σ_{max} and σ_{min} are equal.. Using this property, we will further depict the cyclic change in the stresses of the sinusoidal curve. To characterize the cycles of alternating stresses, the following parameters are introduced (Fig. 20.1).

The algebraic half-sum of the maximum and minimum stresses is called the *average cycle stress*. Denoting this value with the symbols σ_m, by definition we have

$$\sigma_m = \frac{\sigma_{max} + \sigma_{min}}{2}. \tag{20.1}$$

Algebraic half-difference of maximum and minimum of the stresses is called the *amplitude of the cycle* of the variable stresses, i.e.,

$$\sigma_a = \frac{\sigma_{max} - \sigma_{min}}{2}. \tag{20.2}$$

From the formulas (20.1) and (20.2) follows

$$\sigma_{max} = \sigma_m + \sigma_a; \quad \sigma_{min} = \sigma_m - \sigma_a. \tag{20.3}$$

A cycle in which the maximum and minimum voltages are equal in magnitude and opposite in sign, i.e., $\sigma_{max} = -\sigma_{min}$, is called *symmetric*. All other loops are called *asymmetric*. Among the asymmetric ones, there are cycles in which either $\sigma_{min} = 0$ or $\sigma_{max} = 0$. Such cycles are called *pulsating* or *zero cycles*. Relationship

$$r = \frac{\sigma_{min}}{\sigma_{max}} \tag{20.4}$$

is called the cycle skewness coefficient . Sometimes, along with the skewness coefficient, the parameter is also used

$$\rho = \frac{\sigma_a}{\sigma_m}, \tag{20.5}$$

which we will call the *characteristic of the cycle*. This parameter is related to the cycle skewness coefficient by the dependency

$$\rho = \frac{\sigma_a}{\sigma_m} = \frac{\sigma_{max} - \sigma_{min}}{\sigma_{max} + \sigma_{min}} = \frac{1 - r}{1 + r}. \tag{20.6}$$

Stress cycles with the same parameters r or ρ are called *similar* stress cycles.

20.3   Endurance Limit

To calculate the endurance, in addition to the cycle parameters, it is necessary to know the properties of the material that characterize its ability to withstand the action of variable stresses. These properties are determined experimentally. The most common tests are those under symmetric cycle conditions.

A diagram of the simplest setup for performing such tests is shown in Fig. 20.2. Sample 1 of circular cross-section is fixed at one end in the chuck of the spindle 2 rotating at a certain speed. At the other end of the sample, bearing 3 is pressed, to the outer ring of which a kettlebell suspension with a weight P is attached. A special counter records the number of spindle revolutions at the moment of sample destruction.

For the fatigue test, at least ten identical samples are made. Each sample is tested to failure at a certain stress amplitude in the most loaded cross-section $A - A$, (Fig. 20.2). The magnitude of the amplitude σ_a for each sample is set by selecting the load P.

As a result of testing each sample, the researcher receives the maximum cycle voltage σ in the cross-section $A - A$ and the number of cycles N that the sample sustained before destruction. The data obtained during the tests are marked in the coordinate plane $N \sim \sigma$. As a result of numerous experiments on identical samples at different voltages, a line of the type indicated in Fig. 20.3 (Ustalost' materiala[Material Fatigue] 2021), which is called the *fatigue curve*. This kind of

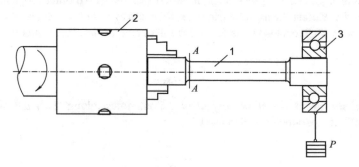

Fig. 20.2 Weller's scheme of experience

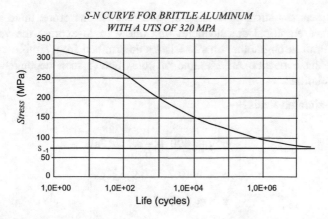

Fig. 20.3 Weller curve

mass experiments was performed by the German engineer A. Weller, so the fatigue curve is also called the Weller curve.

The fatigue curve for low- and medium-carbon steels, as well as for some grades of alloy steels and non-ferrous alloys, has a horizontal asymptote. This means that if the maximum cycle stress does not exceed a certain value, the material can withstand an unlimited number of loading cycles. The highest cycle stress at which there is no fatigue failure of the sample for an arbitrarily large number of stress cycles is called the *endurance limit* or *fatigue limit*.

This mechanical characteristic of the material is usually denoted by the symbol used to denote the stress under study, indicating in the subscript the value of the cycle asymmetry coefficient at which the endurance limit is determined, for example, σ_{-1}, σ_0. Sometimes the superscript also indicates the type of test in which the endurance limit was determined. So, σ_{-1}^b means the limit of endurance for a symmetric cycle, determined from experiments on bending fatigue, σ_0^t—the limit of endurance for a pulsating stress cycle under conditions of cyclic stretching, unloading, etc.

Since it is not possible to perform an infinitely large number of loading cycles during fatigue tests, it is limited to the so-called base number of cycles. For steels and cast irons, 10^7 cycles are taken as the base number. It is assumed that if a material can withstand a basic number of cycles, it will also withstand an infinitely large number of cycles. Conversely, if the sample can withstand the base number of cycles, then at no point in the material does the stress exceed the endurance limit.

For many non-ferrous metals, their alloys, and some alloy steels, the Weller curve has no horizontal asymptote. For such materials, the concept of *limit of limited endurance is introduced*, for which the highest absolute value of the cycle stress is taken at a certain base number of cycles of stress change. In general mechanical engineering, the base number for such materials is $5 \cdot 10^8$ cycles.

In fatigue tests, there is usually a large variation in the experimental data. To obtain reliable results, it is necessary to perform a large number of experiments with subsequent statistical processing of their results. Therefore, there is a natural desire to have empirical dependencies that allow us to express the value of the endurance limit through other characteristics determined from simple experiments with more stable results. An example of constructing such dependencies is the following relations:

- for high-strength steels

$$\sigma_{-1}^b \approx 400 + \frac{1}{6}\sigma_{vr} \text{ (MPa)};$$

$$\sigma_{-1}^t \approx (0.7\ldots0.9)\sigma_{-1}^b; \tag{20.7}$$

$$\tau_{-1} \approx 0.58\sigma_{-1}^b,$$

or for any steel $\sigma_{-1}^b \approx 0.4\sigma_{vr}$, $\sigma_{-1}^t \approx 0.28\sigma_{vr}$, $\tau_{-1} \approx 0.22\sigma_{vr}$;
- for non-ferrous metals and their alloys

$$\sigma_{-1} \approx (0.25\ldots0.5)\sigma_{vr}; \tag{20.8}$$

- for brittle materials

$$\tau_{-1} \approx 0.8\sigma_{-1}. \tag{20.9}$$

20.4 Limit Cycle Diagrams

In modern machines, many parts operate under unbalanced stress cycles. The calculation of such parts for fatigue strength leads to the need to know the limit of endurance of the material at different cycles of stress changes, and there is no other way except for experiments here so far. The results of endurance tests with different coefficients of cycle asymmetry are presented in the form of diagrams depicting the relationship between any two parameters of the cycle.

Let the fatigue tests of a series of samples be carried out at a given cycle skewness coefficient r, and the endurance limit σ_r is found. Determine the amplitude and average voltage of the cycle. Denoting these parameters by σ_a' and σ_m', respectively, and also bearing in mind that by definition $\sigma_r = \sigma_{max}'$, from the formulas (20.1)–(20.3), we get

$$\sigma_a' = \frac{\sigma_r}{2}(1 - r); \quad \sigma_m' = \frac{\sigma_r}{2}(1 + r). \tag{20.10}$$

A loop with parameters (20.10) is called *limit*.

Fig. 20.4 Limit amplitude
diagram (Hay diagram)

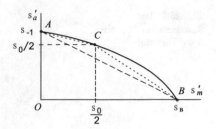

Performing fatigue tests at different cycle skewness coefficients r, we obtain in
each case the parameters of the limit cycles defined by the formulas (20.10). We
will plot the average stresses of the limit cycles along the abscissa axis and the
corresponding amplitudes along the ordinate axis. As a result, we get the ACB
curve (Fig. 20.4), which is called the *limit amplitude diagram,* or *Hay diagram.*

The point A of the curve ACB corresponds to the symmetric limit cycle ($\sigma'_m =
0$), and the point B represents the result of the static test ($\sigma'_a = 0$) at uniaxial tension
to the point of failure, which occurs when the ultimate strength is reached σ_b.

In the absence of the required number of experiments, the ACB curve can be
replaced by a straight line segment drawn in Fig. 20.4 with a dashed line. For a
more accurate approximation of the limit amplitude diagram, you need to have at
least one more point in addition to the points A and B. For these purposes, fatigue
tests with a pulsating cycle are most often used. With such a cycle, $r = 0$, and we
get the third point C of the limit cycle diagram. Connecting it to the points A and
B, we get the polyline ACB (Fig. 20.4, point line). The approximation of the curve
AB of the two-link polyline ACB is quite accurate.

For parts made of plastic materials, not only fatigue failure is dangerous but
also the occurrence of fluidity. Therefore, for this case, the limit amplitude diagram
must be changed. The greatest stress for plastic materials must not exceed the yield
strength, i.e., the condition must be met

$$\sigma_{max} = \sigma_m + \sigma_a \leqslant \sigma_t,$$

for limit cycles

$$\sigma'_m + \sigma'_a = \sigma_t. \tag{20.11}$$

The equation of the straight line (20.11) is represented on the diagram by the
line SDQ (Fig. 20.5), which in the section SD will be the boundary of the limit
amplitude diagram.

Let's now consider how to use the limit amplitude diagram in endurance
calculations. Let us assume that we have a certain working cycle of the sample,
and at the most loaded point of the material, there are periodically varying stresses
with an amplitude of σ_a at an average voltage of σ_m.

Fig. 20.5 The
Sorensen-Kinasoshvili
approximation

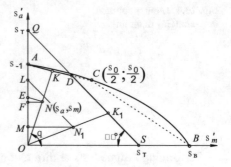

Let's draw in the plane $\sigma'_m \sim \sigma'_a$ a point N with the coordinates σ_m, σ_a (Fig. 20.5) and draw a ray through this point from the origin to the intersection with the diagram at the point K. There are two possible cases here: 1) the point K belongs to the segment AD, and 2) the point K belongs to the segment DS.

Let us first consider the first case. Denote by ϑ the angle of inclination of the beam ON to the abscissa axis. We have

$$\tan\vartheta = \frac{\sigma_a}{\sigma_m} = \rho,$$

where ρ is the cycle characteristic defined by the formula (20.5). The last result means that all the cycles represented by the points of the ray OK are similar to the given one. The cycle corresponding to the point K is the limit. For all other cycles for which the image point N is located below the point K, the sample is able to withstand an unlimited or at least basic number of cycles. If the image point of the working cycle is higher than the point K, the sample will collapse at some limited number of cycles less than the base number.

Let us call *the coefficient of fatigue strength* the ratio of the segments OK and ON in such cycles, i.e., the value of

$$n = \frac{OK}{ON}. \qquad (20.12)$$

The coefficient n characterizes the degree of proximity of the working and similar limit cycles of stresses.

Let's draw a straight line NE parallel to AC from the point N, as well as a horizontal line NF. From the similarity of the triangles OAK and OEN, we have

$$n = \frac{OK}{ON} = \frac{OA}{OE}. \qquad (20.13)$$

But

$$OA = \sigma_{-1}; \quad OE = OF + FE = \sigma_a + \sigma_m \tan\angle FNE. \qquad (20.14)$$

Denote

$$\tan \angle FNE = \psi_\sigma. \tag{20.15}$$

Given that the angle of inclination of the straight line NE to the abscissa axis is equal to the angle of inclination of the straight line AC to the same axis, we get

$$\psi_\sigma = \frac{\sigma_{-1} - \dfrac{\sigma_0}{2}}{\dfrac{\sigma_0}{2}} = \frac{2\sigma_{-1} - \sigma_0}{\sigma_0}. \tag{20.16}$$

Substituting the formulas (20.14) and (20.15) into the definition (20.13) gives the following formula for the fatigue strength factor of the standard sample:

$$n = \frac{\sigma_{-1}}{\sigma_a + \psi_\sigma \sigma_m}, \tag{20.17}$$

where the parameter ψ_σ is defined by the formula (20.16). Note that the formula for determining the parameter ψ_σ depends on how the limit cycle diagram is approximated.

Now consider the second case. Let the point representing the duty cycle be $N_1(\sigma_m, \sigma_a)$ such that the line ON_1 (Fig. 20.5) intersects the limit cycle diagram at the point K_1 belonging to the segment DS. Draw the straight lines $N_1L \parallel DS$ and N_1M from the point N_1 parallel to the abscissa axis. From the similarity of the triangles OK_1Q and ON_1L for the fatigue strength factor, we get

$$n = \frac{OK_1}{ON_1} = \frac{OQ}{OL} = \frac{OQ}{OM + ML} = \frac{\sigma_t}{\sigma_{max}} = n_t. \tag{20.18}$$

Thus, in the second case, the fatigue strength factor is the same as the usual safety factor with respect to the yield strength. This result could have been predicted, because by construction, the segment DS restricts the area of cycles that are dangerous in the sense of the occurrence of fluidity.

In engineering practice, another type of limit cycle diagrams is sometimes used, in which the average stress σ_m' and the maximum σ_{max}' and the minimum σ_{min}' of the cycle voltage are selected for the limit cycle parameters (see, e.g., (Molotnikov 2006)).

20.5 Fatigue Limit and Factors Affecting It

Experiments show that the endurance limit depends not only on the material properties of the part but also on its geometric dimensions, shape, manufacturing technology, and operating conditions. So, for example, rough turning of the surface of the part with a cutter creates a roughness that reduces the endurance limit by 20...50% compared to a polished sample. The presence of holes, keyways, sudden changes in size, press landings, etc. causes the appearance of local stresses, which also leads to a decrease in the endurance limit. The formation of corrosion can cause a sixfold reduction in the endurance limit!

There are factors that, on the contrary, increase the fatigue strength of materials. For example, when cold-working steel by rolling, drawing, or drawing, an incline is formed, which gives the effect of increasing the endurance limit. The increase in fatigue strength also occurs when heated. So, for carbon steels, heating to a temperature of 375 °C increases the endurance limit by about 40%. This effect is even more pronounced at very low temperatures. For steel 3 at a temperature of −185 °C, the endurance limit is almost twice as high as at room temperature.

When performing calculations on fatigue strength, it is not yet possible to take into account the influence of the entire set of geometric, technological, temperature, and other factors. Therefore, the main factors are chosen, the influence of which is predominant and well-studied. Let's move on to their consideration.

1. **Stress concentration**. The presence of recesses, holes, keyways, etc. creates zones of increased stress in the part, which are local in nature. The phenomenon of the occurrence of local zones of increased stress in the area of a sharp change in the shape of the part is called *stress concentration*. Geometric objects that cause stress concentration are called *stress concentrators*. An indicator of the degree of stress concentration is the so-called theoretical stress concentration coefficient, which is determined by the ratio

$$\alpha_\sigma = \frac{\sigma_{max}}{\sigma_{\text{ном}}}, \tag{20.19}$$

where σ_{max} is the highest local voltage and σ_{nom} is the rated ыекуыы determined by simple formulas of the resistance of materials without taking into account the effect of stress concentration. Similarly, if we are talking about tangential stresses, the theoretical coefficient of their concentration is calculated by the formula

$$\alpha_\tau = \frac{\tau_{max}}{\tau_{\text{nom}}}. \tag{20.20}$$

The highest local stresses σ_{max} (τ_{max}) are determined either analytically (when possible) or experimentally. For the main stress concentrators, the theoretical coefficients are determined and given in textbooks or reference literature in the form of formulas or graphs.

When calculating the fatigue strength, in contrast to the theoretical stress concentration coefficient, the so-called effective concentration coefficient k_σ (k_τ) is introduced, which is defined as the ratio

$$k_\sigma = \frac{\sigma_{-1}}{\sigma_{-1k}}, \quad k_\tau = \frac{\tau_{-1}}{\tau_{-1k}}, \tag{20.21}$$

Table 20.1 Effective stress concentration coefficients

Type of concentrator	k_σ σ_{vr}, MPa $\leqslant 700$	$\geqslant 1000$	k_τ $\leqslant 700$	$\geqslant 1000$
Fillet at				
$r/d = 0.02$	2.5	3,5	1.8	2.1
0.06	1.85	2.0	1.4	1.53
0.1 $(D/d = 1.25 \ldots 2)$	1.6	1.64	1.25	1.35
Undercut at $t{=}r$ and				
$r/d = 0.02$	1.9	2.35	1.4	1.7
0.06	1.8	2.0	1.35	1.65
0.10	1.7	1.85	1.25	1.5
Cross hole at $d/D = 0,05\ldots0,25$	1.9	2.0	1.75	2.0
Keyway	1.7	2.0	1.4	1.7
Slots	When calculating the internal diameter, take 1			
Press fit with compression not less than 20 MPa	2.4	3.6	1.8	2.5
Thread	1.8	2.4	1.2	1.5

where σ_{-1}, τ_{-1}—endurance limits for a symmetric cycle for a sample without a concentrator; σ_{-1k}, τ_{-1k}—the same, for the same sample with a concentrator. The effective stress concentration coefficients are established experimentally and shown in graphical or tabular form (see, e.g., Table 20.1).[1]

In the absence of experimental data, you can use the formulas

$$k_\sigma = 1 + q(\alpha_\sigma - 1); \quad k_\tau = 1 + q(\alpha_\tau - 1). \tag{20.22}$$

Here q is the coefficient of sensitivity of the material to the stress concentration. For high-strength alloy steels $q \approx 1$; for structural steels $q \approx 0.6 \ldots 0.8$; for gray cast iron $q = 0$. The insensitivity of cast iron to stress concentration is explained by the fact that in its structure large graphite grains are already concentrators in themselves, so that the influence of the geometric features of the part loses its significance.

---

[1] The data for building Tables 20.1, 20.2, and 20.3 of this paragraph is borrowed from the book: *Guzenkov P.I.* Machine parts.—M.: Higher School, 1982.

In non-responsible calculations, empirical formulas can also be used to determine the effective stress concentration coefficients:

– in the absence of sharp stress concentrators and a clean surface of the part

$$k_\sigma = 1, 2 + 0, 2 \cdot \frac{\sigma_{vr} - 4000}{11000} \quad (\sigma_{vr} - \text{in kgf/cm}^2);$$

– in the presence of sharp concentrators

$$k_\sigma = 1.5 + 1.5 \cdot \frac{\sigma_{vr} - 4000}{11000} \quad (\sigma_{vr} - \text{in kgf/cm}^2). \tag{20.23}$$

Note When using these formulas, you should not separately consider the impact of the quality of the surface treatment of the part.

2. **Scale effect**. If several batches of samples are made from the same material, so that in each batch the samples have the same diameters, but in different batches these diameters are different, and test each batch for fatigue, it turns out that with an increase in the diameter of the samples, the endurance limit decreases.

The phenomenon of a decrease in the endurance limit with an increase in the size of the part is called the *scaling effect*.

The scale effect is explained by the fact that the distribution of material structural defects, including fatigue microcracks, is statistical in nature. With an increase in the geometric dimensions of the part, the likelihood that the bulk of the material will contain a larger number of centers of possible crack initiation increases. In addition, when making samples of a smaller diameter, the work-hardening in the surface layer covers a larger percentage of the total material than in a sample of large dimensions. In calculations for fatigue strength, the scale effect is taken into account by introducing *the scale factor,* which is determined by the formulas

$$\beta_{\sigma sc} = \frac{\sigma_{-1}}{\sigma_{-1d}}; \quad \beta_{\tau sc} = \frac{\tau_{-1}}{\tau_{-1d}}. \tag{20.24}$$

Here σ_{-1}, τ_{-1} is the endurance limit of a standard sample with a diameter of 7.5 mm with a symmetric cycle; σ_{-1d}, τ_{-1d} is the endurance limit of a specimen of a given diameter d from the same material at a symmetric stress cycle.

The values scale factors for steels are given in Table 20.2.

3. **Influence of the surface condition of the part**. Fatigue cracks usually begin to develop from the surface of the part. Therefore, the cleanliness of the surface treatment has a significant impact on the strength at variable stresses. At the beginning of this section, we have already given an approximate quantitative assessment of this phenomenon.

Table 20.2 Scale factors for certain steels

Type loading material	Shaft diameter, mm							
	15	20	30	40	50	70	100	200
	Scale factor							
Bend, carbon steel	1.05	1.09	1.14	1.18	1,23	1.32	1.43	1.64
Bend, high-strength steel; torsion, for all steels	1.15	1.20	1.30	1.40	1.43	1.54	1.70	1.92

Table 20.3 Coefficients of surface sensitivity of axes and shafts

Type of processing and surface finish	Ultimate strength σ_{vr}, MPa		
	400	800	1200
	Coefficients $\beta_{\sigma s}$, $\beta_{\tau s}$		
Sanding R_z 16...0.4	1	1	1
Turning R_z 10...16	1.05	1.10	1.25
Roughing up R_z 80...10	1.20	1.25	1.50
Untreated surface with scale	1.35	1.50	2.20

In fatigue calculations, the effect of surface finish cleanliness on the endurance limit value is taken into account by the surface *quality coefficient,* which is also called the surface *sensitivity coefficient.* For cycles of normal and tangential stresses, it is defined by the relations

$$\beta_{\sigma s} = \frac{\sigma_{-1}}{\sigma_{-1s}}; \quad \beta_{\tau s} = \frac{\tau_{-1}}{\tau_{-1s}}, \tag{20.25}$$

where σ_{-1} (τ_{-1}) is the fatigue limit of the polished sample and σ_{-1s} (τ_{-1s}) fatigue limit when testing a series of samples with a given surface finish.

The values of the surface sensitivity coefficients for axles and shafts for some types of surface treatment are given in Table 20.3.

The combined effect of stress concentration, scale effect, and surface quality in endurance calculations is estimated *by the total coefficient of reduction of the endurance limit in a symmetric cycle,* which is taken as the product of the coefficients (20.23)–(20.25), i.e.,

$$k_{\sigma d} = k_\sigma \beta_{\sigma sc} \beta_{\sigma s}; \quad k_{\tau d} = k_\tau \beta_{\tau sc} \beta_{\tau s}. \tag{20.26}$$

In this case, the endurance limit for the part with a symmetric stress cycle will be

$$\sigma_{-1d} = \frac{\sigma_{-1}}{k_{\sigma d}}; \quad \tau_{-1d} = \frac{\tau_{-1}}{\tau_{-1d}}. \tag{20.27}$$

20.6   Practical Endurance Calculations

In the overwhelming majority of cases, fatigue analysis is performed as a verification. The design calculation of a part operating at alternating voltages is usually
performed in two stages. At the first stage, the calculation is carried out without
taking into account the variability of stresses, but at the same time lower values
permissible stresses are taken. After completing the first stage, they proceed to the
verification calculation taking into account the voltage variability. The calculation
consists in checking the condition

$$n \geqslant [n], \tag{20.28}$$

where n is the calculated coefficient of the fatigue strength margin and $[n]$ is the
permissible value of this value, which is assigned taking into account the experience of designing and operating the products under study. In general mechanical
engineering, take $[n] = 1.4 \ldots 3.0$.

If the condition (20.28) is not met, the size, shape, or material of the projected
structure is changed. The calculated coefficient of fatigue strength is most simply
determined for parts operating under a symmetrical cycle under conditions of
uniaxial stress (tension-compression, bending) and shear (torsion). In fact, for most
structural materials, the limit of endurance under a symmetric cycle is known. Using
the reference literature, we can also determine the total coefficient of reduction of
the endurance limit for the calculated part and then use the formulas (20.27) to
calculate the endurance limit of the part for a symmetric cycle. Dividing the result
by the maximum cycle stress, we get the calculated coefficient of fatigue strength.
Thus, with a symmetric working cycle of the part, we have:

- when bending

$$n_\sigma = \frac{\sigma^b_{-1d}}{\sigma_{max}} = \frac{\sigma^b_{-1}}{k_{\sigma d}\sigma_{max}}; \tag{20.29}$$

- when stretching-compressing

$$n_\sigma = \frac{\sigma^{st}_{-1d}}{\sigma_{max}} = \frac{\sigma^{st}_{-1}}{k_{\sigma d}\sigma_{max}}; \tag{20.30}$$

- when twisted

$$n_\tau = \frac{\tau_{-1d}}{\tau_{max}} = \frac{\tau_{-1}}{k_{\tau d}\tau_{max}}. \tag{20.31}$$

In the case of an asymmetric stress cycle in a part operating under tension-
compression, bending, or torsion, we can determine the calculated fatigue
strength factor based on the formula (20.27). Recall that this formula applies

Table 20.4 Coefficient values ψ_σ and ψ_τ

Coefficient ψ_σ and ψ_τ	Ultimate strength $\sigma_B \cdot 10^{-1}$, MPa				
	$35\ldots55$	$52\ldots75$	$70\ldots100$	$100\ldots120$	$120\ldots140$
ψ_σ (stretching, bend)	0	0.05	0.10	0.20	0.25
ψ_τ (torsion)	0	0	0.05	0.10	0.15

to a standard sample, but not to a part made of the same material. To apply it to the assessment of the endurance of the part, it is necessary to take into account the influence of the factors discussed above—the stress concentration, the scale effect, and the cleanliness of the surface of the part. Since the influence of these factors on the fatigue strength of the part manifests itself with an increase in the amplitude σ_a, and not the static component of the cycle σ_m, when calculating the coefficient n in the formula (20.17), the value of the amplitude is multiplied by the total coefficient of reduction of the endurance reserve.

Thus, with an asymmetric cycle, the calculated coefficient of the fatigue strength of the part is determined by the formulas:

- when bending

$$n_\sigma = \frac{\sigma_{-1}^b}{k_{\sigma d}\sigma_a + \psi_\sigma\sigma_m};$$ (20.32)

- when stretching-compressing

$$n_\sigma = \frac{\sigma_{-1}^{st}}{k_{\sigma d}\sigma_a + \psi_\sigma\sigma_m};$$ (20.33)

- when twisted

$$n_\tau = \frac{\tau_{-1}}{k_{\tau d}\tau_a + \psi_\tau\tau_m},$$ (20.34)

where, by analogy with the formula (20.16),

$$\psi_\tau = \frac{2\tau_{-1} - \tau_0}{\tau_0},$$ (20.35)

moreover, τ_{-1} and τ_0 are, respectively, the endurance limit of the standard sample for symmetric and pulsating cycles of tangential stress. In the absence of the necessary set of experimental data, the values of the coefficients ψ_σ and ψ_τ can be selected from Table 20.4.

In the case where the average stress of the normal stress cycle is negative, the presence of $\sigma_m < 0$ has the effect of «healing» fatigue cracks. Therefore, in the formulas (20.33) and (20.34), you should put $\psi_\sigma = 0$.

In the general case of a complex stress state, several fatigue strength hypotheses have now been created. For the case of simultaneous action of normal and tangential stress, which is most often encountered in the practice of engineering calculations, use the empirical formula *Gaf* and *Pollard*

$$\frac{1}{n^2} = \frac{1}{n_\sigma^2} + \frac{1}{n_\tau^2},$$ (20.36)

where n is the calculated fatigue strength coefficient of the part and n_σ, n_τ are the coefficients determined by the formulas (20.33) and (20.34).

The application of the formula (20.36) is not limited to the requirement of a common-mode change in the normal and tangential stresses. The extreme values of σ and τ may not be reached simultaneously, and the stress change frequencies and the asymmetry coefficients of the normal and tangential stress cycles may not coincide.

20.6.1 Example of Calculating the Endurance Shaft

Task. In the dangerous cross-section of the shaft with a diameter of $d = 32\,\text{mm}$, there is a torque $M_t = 250\,\text{N} \cdot \text{m}$ and a bending moment $M_b = 200\,\text{N} \cdot \text{m}$. The shaft is made of carbon steel, the ultimate strength of which is $\sigma_{vr} = 550\,\text{MPa}$ and has no sharp transitions, notches, and grooves; its surface is cleanly treated with a cutter.

Determine the calculated coefficient of fatigue strength in the dangerous cross-section of the shaft, taking the normal bending stresses varying in a symmetrical cycle and the tangential torsional stresses in a pulsating cycle. The concentration coefficients and the scale factor coefficients are considered the same for normal and tangential stresses.

S o l u t i o n .

1. Calculate the maximum normal and tangential stresses in the shaft body:

$$\sigma_{max} = \frac{M_b}{W} = \frac{M_b}{0.1 d^3} = \frac{200}{0.1 \cdot (32 \cdot 10^{-3})} \approx 61 \cdot 10^6 \, \text{Pa} \approx 61 \, \text{MPa};$$

$$\tau_{max} = \frac{M_t}{W_p} = \frac{M_t}{0.2 d^3} = \frac{250}{0.2 \cdot (32 \cdot 10^{-3})^3} \approx 38 \cdot 10^6 \, \text{Pa} \approx 38 \, \text{MPa}.$$

2. We will determine separately the coefficients of the fatigue strength reserve for normal and tangential stresses. According to the condition, the cycle is symmetric for normal stresses, so $\sigma_{max} = -\sigma_{min}$, $\sigma_a = \sigma_{max}$, $\sigma_m = 0$ and the formula (20.32) takes the form

$$n_\sigma = \frac{\sigma_{-1}^b}{k_{\sigma d} \sigma_a}.$$ (20.37)

According to the condition of the problem, the shaft has no other stress concentrators, except, perhaps, the ridges on the surface from the processing of the cutter. Therefore, the theoretical stress concentration coefficient is equal to one, and by the formula (20.22), we get

$$k_\sigma = k_\tau = 1 + q(1 - 1) = 1.$$

The coefficients of the scale factor are determined by Table 20.2. Using linear interpolation, for a 32-mm-diameter carbon steel shaft, we obtain

$$\beta_{\sigma sc} = \beta_{\tau sc} = 1.14 + (32 - 30) \cdot \frac{1.18 - 1.14}{40 - 30} \approx 1.15.$$

The surface sensitivity coefficients are found using Table 20.3. Let's use the linear interpolation method again:

$$\beta_{\sigma s} = \beta_{\tau s} = 1.05 + (550 - 400) \cdot \frac{1.1 - 1.05}{800 - 400} \approx 1.07.$$

Multiplying the obtained coefficients, using the formulas (20.26), we find the total coefficient of reduction of the endurance limit for a symmetric cycle:

$$k_{\sigma d} = k_{\tau d} = k_\sigma \beta_{\sigma sc} \sigma_{\sigma s} = 1 \cdot 1.15 \cdot 1.07 \approx 1.23.$$

Then, using the empirical formulas (20.7), we determine the limits of endurance for a symmetric cycle:

$$\sigma_{-1}^b \approx 0.4\sigma_{vr} = 0.4 \cdot 550 = 220 \,\text{MPa};$$

$$\tau_{-1} \approx 0.22\sigma_{vr} = 0.22 \cdot 550 = 121 \,\text{MPa}.$$

Using the formula (20.29), we find the coefficient of fatigue strength for normal stresses:

$$n_\sigma = \frac{\sigma_{-1}^b}{k_{\sigma d}\sigma_{max}} = \frac{220}{1.23 \cdot 61} = 2.93. \tag{20.38}$$

Let's check the margin of safety in relation to the yield strength. The yield strength of steel during bending can be approximately determined by the empirical formula $\sigma_t \approx (0.6 \ldots 0.7)\sigma_{vr}$. Put $\sigma_t \approx 0.65\sigma_b$. Calculate $\sigma_t = 0.65 \cdot 550 = 358 \,\text{MPa}$. Then by the formula (20.18)

$$n_{\sigma t} = \frac{\sigma_t}{\sigma_{max}} = \frac{358}{61} \approx 5.9. \tag{20.39}$$

As the calculated value, we take the smaller of the coefficients (20.38) and (20.39), i.e., $n_\sigma = 2.93$.

3. Let us now set the coefficient of the fatigue strength reserve for tangential stresses. For the zero cycle of tangential stresses, we have $\tau_{max} = 38\,\text{MPa}$; $\tau_{min} = 0$; $\tau_m = 0.5(\tau_{max} + \tau_{min}) = 0.5 \cdot 38 = 19\,\text{MPa}$; $\tau_a = 0.5(\tau_{max} - \tau_{min}) = 19\,\text{MPa}$. According to Table 20.4, we find that $\psi_\tau = 0$. Then, using the formula (20.34), we calculate

$$n_\tau = \frac{\tau_{-1}}{k_{\tau d}\tau_a + \psi_\tau \tau_m} = \frac{121}{1.23 \cdot 19 + 0 \cdot 19} \approx 5.18.$$

Let's check the tangential stresses with respect to the torsional yield strength. In the absence of experimental data, the value of the torsional yield strength can be determined in terms of the ultimate strength using the approximate formula $\tau_t \approx (0.34 \ldots 0, 36)\sigma_{vr}$. Let's say $\tau_t = 0.35\sigma_b = 0.35 \cdot 550 = 192\,\text{MPa}$. Then the margin of safety in relation to the yield strength will be

$$n_{\tau t} = \frac{\tau_t}{\tau_{max}} = \frac{192}{38} = 5.05.$$

Thus, for tangential stresses, the strength is limited not by fatigue failure, but by the occurrence of fluidity. Therefore, we choose the smaller of the last two values as the calculated value of the safety factor, i.e., we assume $n_\tau = 5.05$.

4. Using the Gaf and Pollard formula, we calculate the calculated coefficient of the fatigue strength reserve while taking into account both normal and tangential stresses. We have

$$\frac{1}{n^2} = \frac{1}{n_\sigma^2} + \frac{1}{n_\tau^2}; \quad n = \frac{n_\sigma n_\tau}{\sqrt{n_\sigma^2 + n_\tau^2}} = \frac{2.93 \cdot 5.05}{\sqrt{2.93^2 + 5.05^2}} \approx 2.53.$$

Problem solved.

Self-Test Questions

1. What is called material fatigue?
2. What physical processes underlie the phenomenon of fatigue?
3. What is called the endurance of the material?
4. Give the definition of the variable voltage cycle.
5. What cycles are called similar?
6. Which voltage cycle is called symmetric and which is called asymmetric?
7. Why is the average cycle voltage called its static component?
8. What is called the endurance limit of the material?
9. What is the base number of cycles and the limit of limited endurance?

10. For what purposes are limit cycle diagrams constructed?
11. List the main factors that affect the value of the endurance limit of the part.
12. What is called the scale effect? Explain the reason for this phenomenon.
13. What is the surface sensitivity coefficient?
14. How is the fatigue strength factor of the part determined with a symmetric stress cycle?
15. In which cases, the Gaf and Pollard formula is used to determine the endurance factor of the part? Write down this formula.

References

V. Molotnikov, *Kurs soprotivleniya materialov [The Course of Strength of Materials]* (Krasnodar, Lan' Publ., Moscow, 2006)
Ustalost' materiala[Material Fatigue] (2021). https://ru.wikipedia.org/wiki/Ustalost%60_materiala

Chapter 21
Stability of Compressed Rods

Abstract Earlier, in the course of theoretical mechanics (Chap. 7), the basic concepts of the theory of stability of equilibrium of mechanical systems were defined. Here, in particular, the stability of elastic compressed rods is considered, since in the structures of machines and structures, parts are widely used for which the design scheme of such a rod is ideally suited. The formulation and solution of the Euler problem on the stability of a rod with hinged ends are presented. Based on Euler's solution, solutions are written for cases with a different fixation of a compressed rod. The behavior of the rod outside the limits of proportionality is investigated.

Keywords Balance stability · Kernel · Elasticity · Inelasticity · Euler's problem · Flexibility · Ultimate flexibility · The power is critical · Reduced length

21.1 General Concepts of Rod Stability

In all the previous chapters of this section of mechanics, we investigated the strength of mechanical systems, by which we understood the ability of a system to withstand given loads without destroying its elements. However, the destruction of the system can occur not only because its strength will be violated (Molotnikov and Molotnikova 2021) but also because the system does not retain the shape that the designer defined for it. Let's take a look at the simplest example.

Take a thin wooden ruler, and subject it to longitudinal central compression by forces applied at the ends, as shown in Fig. 21.1a. As soon as the compressive forces reach a certain value, the ruler will bend and break.

At the moment preceding the beginning of the bend, the stresses in the ruler will be significantly less than those that would cause the destruction of the wood with a simple compression. However, with an additional small increase in the load, the ruler will collapse. This is because when the load exceeds a certain value, the ruler cannot maintain its rectilinear shape and bends, which causes the appearance of final additional stresses from bending. They say that the ruler has lost its stability.

Fig. 21.1 The phenomenon
of loss of stability of a
compressed rod

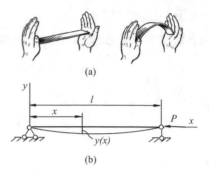

Having familiarized ourselves with the phenomenon of loss of stability, we now
accept the following definitions. *If the system has strength and at the same time does
not lose the ability to maintain to the desired extent the known law of its motion or
the specified limits of its deformation, then it is called sustainable. Otherwise, the
system is called unstable.*

From this definition, it follows that the necessary and sufficient conditions for the
stability of the system are its strength and some limitation of its movements.

*Load at the time of the transition of the system from a stable state to an unstable
one is called critical.*

21.2   The Euler Problem

Consider a rectilinear elastic rod of length l of constant cross-section with pivotally
supported ends (Fig. 21.1b). One of the supports is fixed, and the second allows the
longitudinal movement of the end of the rod. Load the rod with the compressive
force P. Requires determine the value of the critical force P_k.

This problem was first posed and solved by L. Euler in 1744 and is known in
the scientific and educational literature as the Euler problem. In the problem under
consideration, the rectilinear shape of the rod is the initial form of the equilibrium.
To judge its stability, consider the curved state of the rod, in which it is held by the
force $P = P_k$. The deflections of the rod $y(x)$ are small, which enables the use of
approximate differential equation of bending (16.28):

$$EJ\frac{d^2y}{dx^2} = M(x),$$

and J—minimum moment of inertia of the cross section.

In an arbitrary section with the coordinate x, the bending moment is

$$M(x) = -Py(x).$$

The «minus> > sign on the right side of this expression is due to the fact that, according to the sign rule adopted in Sect. 16.2, the bending moment in the section x is positive and the deflection $y(x) < 0$.

Let's introduce the notation

$$k^2 = \frac{P}{EJ}. \tag{21.1}$$

Then the differential equation of the bending of the rod takes the form

$$y'' + k^2 y = 0. \tag{21.2}$$

Let's add the boundary conditions here

$$y(0) = 0; \quad y(l) = 0. \tag{21.3}$$

The general integral of the Eq. (21.2) has the form

$$y(x) = A \sin kx + B \cos kx. \tag{21.4}$$

The first of the boundary conditions (21.3) gives $B = 0$, and the second implies that

$$A \sin kl = 0. \tag{21.5}$$

If $A = 0$, then it follows from the solution (21.4) that the deflection in any section of the rod is zero, i.e., the rod remains straight. But this contradicts the original assumption about the curvature of the rod. Therefore

$$\sin kl = 0, \text{ and } kl = \pi n \ (n = 1, 2, \ldots).$$

Hence, taking into account the notation (21.1), we get *Euler's formula* for determining the critical force

$$P_k = \frac{\pi^2 n^2 EJ}{l^2} \ (n = 1, 2, \ldots). \tag{21.6}$$

From the result obtained, it follows that the load holding a slightly curved rod in equilibrium can theoretically take many values, of which the smaller one will be at $n = 1$. In practice, in rods with hinged ends, this case is realized, i.e.

$$P_k = \frac{\pi^2 EJ}{l^2}. \tag{21.7}$$

This value of the critical force corresponds to the bending of the rod along a sinusoid with one half-wave

$$y(x) = A \sin \frac{\pi x}{l}. \tag{21.8}$$

Values of the critical force of the higher orders of magnitude at $n > 1$ can be practically obtained in the structure, for example, by installing intermediate supports that prevent the rod from deflecting. The installation of one intermediate support in the middle of the length of the rod leads to a bend with the formation of two half-waves of the sinusoid. With two intermediate supports, we have three half-waves, etc.

For a known critical force (21.7), we can determine the critical stress at the moment of loss of stability by dividing the critical force by the cross-sectional area of the rod, i.e.

$$\sigma_k = \frac{P_k}{F} = \frac{\pi^2 E J}{l^2 F} = \frac{\pi^2 E}{\left(\dfrac{l}{i}\right)^2} = \frac{\pi^2 E}{\lambda^2}, \tag{21.9}$$

where $i = \sqrt{J/F}$ is the minimum radius of inertia of the cross-section of the rod and λ denotes

$$\lambda = \frac{l}{i}. \tag{21.10}$$

This value is called the *flexibility* of the rod.

If the pinning of the rod is different from the two-pinned one discussed above, then the boundary conditions will change (21.3). With this in mind, the formula (21.6) can be written as:

$$P_k = \frac{\pi^2 E J}{(\mu l)^2}, \tag{21.11}$$

where μ—is the length reduction coefficient, depending on of the type of boundary conditions. The product of μl is called by the reduced length of the rod. Then in the general case

$$\lambda = \frac{\mu l}{i}. \tag{21.12}$$

For the most common cases of anchoring, the values of the length reduction coefficient are as follows:

- pivoting the ends $\mu = 1$;
- one end is pinched, the other is free $\mu = 2$;

- one end is immovably pinched, the pinching of the second has an axial displacement $\mu = 0.5$;
- one end is pinched, the support of the other is a movable hinge $\mu = 0.7$.

For other cases, the reader will find the values of μ, for example, in the reference (*Spravochnik metallista GNTIMP. T.2 [Metalworker's Handbook. Vol.2]* 1960).

21.3 Stability Beyond the Proportional Limit

The Euler formula (21.11) is applicable only when the stresses in the rod do not exceed the limit of proportionality σ_{pc}. Using the formula (21.9) for the critical voltage, we can write

$$\sigma_k = \frac{\pi^2 E}{\lambda^2} \leqslant \sigma_{pc}.$$

It follows that the Euler formula is applicable for rods whose flexibility index Flexibility is not less than the limit defined by the formula

$$\lambda_{pre} \geqslant \pi \sqrt{\frac{E}{\sigma_{pc}}}.$$

For example, for low-carbon steels, the proportionality limit is about 200 MPa, and the ultimate flexibility turns out to be approximately 100. Only long and thin rods have such flexibility. Shorter and more massive rods have significantly less flexibility and, consequently, lose stability at stresses above the limit of proportionality, i.e., in the plastic region.

The first statements of the problem of the stability of a compressed rod beyond the limit of proportionality belong to Karman and Yasinsky. The simplest and most reliable approximation of the dependence $\sigma_k \sim \lambda$ beyond the proportionality limit for steels is obtained by Yasinsky:

$$\sigma_k = \begin{cases} a - b\lambda & \text{at } 40 \leqslant \lambda \leqslant 100, \\ a - 40b & \text{at } 0 \leqslant \lambda \leqslant 40, \end{cases} \tag{21.13}$$

where the constants a and b for each material are determined from experiments. For steel 3, for example, the values of these constants are as follows: $a = 336$ MPa, $b = 1.47$ MPa. The dependence of the critical stress on the flexibility is shown in Fig. 21.2.

In engineering practice, the stability condition is most often used in the form of

$$\sigma \leqslant \varphi(\lambda)[\sigma]_c, \tag{21.14}$$

Fig. 21.2 Critical stresses of
an elastic and inelastic rod

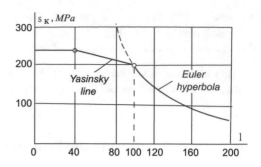

where σ is the absolute value of the calculated compressive stress, $[\sigma]_c$ is the main allowable stress when compressing the rod material, and $\varphi(\lambda)$ is the so-called coefficient of reduction of the main allowable stress, depending on the material and the flexibility of the rod.

To get the dependence $\varphi(\lambda)$, we write the stability condition by analogy with the strength condition in the form

$$\sigma \leqslant \frac{\sigma_{to}(\lambda)}{n_y}, \tag{21.15}$$

where n_y—is the stability margin factor. Comparing the right-hand sides of formulas (21.14) and (21.15), we find that

$$\varphi(\lambda) = \frac{\sigma_{to}(\lambda)}{n_y[\sigma]_c}. \tag{21.16}$$

The dependency (21.16) for different materials is usually set in tabular form. When compiling these tables, the coefficient of stability margin is assumed to be different for different flexibilities, and its greatest value is taken at the average flexibilities of the rod. The values of the φ coefficient for some structural materials are given in Table 21.1.

It was found that local weakening of the rod, such as small holes perpendicular to the axis, does not significantly affect its stability. Therefore, the calculated stress is calculated by dividing the compressive force by the gross area (without subtracting the area of attenuation). At the same time, of course, the strength condition must be met, written taking into account the area of weakening, i.e.,

$$\sigma = \frac{P}{F_{gross}} \leqslant \varphi[\sigma]_c; \quad \frac{P}{F_{net}} \leqslant [\sigma]_c. \tag{21.17}$$

In the case of a test calculation, when the load, material, length, shape, and dimensions of the cross-section of the rod are known, the calculation using the formulas (21.17) does not cause any difficulties. The situation is different in the design calculation. With the selected shape of the cross-section of the rod, we do

Table 21.1 On the calculation of critical stresses

Flexibility of the λ	Values φ for materials				
	Steel grades 2, 3, 4, OS	5 Steel	SPC steel	Cast iron	Wood
0	1.00	1.00	1.00	1.00	1.00
10	0.99	0.98	0.97	0.97	0.99
20	0.96	0.95	0.95	0.91	0.97
30	0.94	0.92	0.91	0.81	0.93
40	0.92	0.89	0.87	0.69	0.87
50	0.89	0.86	0.83	0.57	0.80
60	0.86	0.82	0.79	0.44	0.71
70	0.81	0.76	0.72	0.34	0.60
80	0.75	0.70	0.65	0.26	0.48
90	0.69	0.62	0.55	0.20	0.38
100	0.60	0.51	0.43	0.16	0.31
110	0.52	0.43	0.35	–	0.25
120	0.45	0.36	0.30	–	0.22
130	0.40	0.33	0.26	–	0.18
140	0.36	0.29	0.23	–	0.16
150	0.32	0.26	0.21	–	0.14
160	0.29	0.24	0.19	–	0.12
170	0.26	0.21	0.17	–	0.11
180	0.23	0.19	0.15	–	0.10
190	0.21	0.17	0.14	–	0.09
200	0.19	0.16	0.13	–	0.08

not know the size of the cross-section and therefore cannot determine the value of flexibility, as well as the coefficient φ. The selection of the cross-section in this case has to be carried out by successive approximations.

Self-Test Questions

1. What mechanical system is called stable?
2. Formulate the necessary and sufficient conditions for the stability of the system.
3. Provide a definition of the critical load.
4. Write down the Euler formula for determining the critical force.
5. What is called the reduced length of the rod?
6. Name the values of the length reduction coefficient for the main cases of fixing the ends of the rod.
7. What is called the flexibility of the rod?
8. What are the limits of applicability of the Euler formula?

9. Write down the Yasinsky formula for the critical voltage and name the limits of
   its applicability.
10. Write down the empirical formulas for calculating the stability of the com-
    pressed rods.

References

V. Molotnikov, A. Molotnikova, *The Theory of Elasticity and Plasticity*, (Springer Nature, Berlin,
   2021)
Spravochnik metallista GNTIMP. T.2[Metalworker's Handbook. Vol. 2] (1960)

Part IV
Machine Parts and Design Basics

Chapter 22
General Information About Machine Design

Abstract Design is the process of developing comprehensive technical documentation containing feasibility studies, calculations, drawings, layouts, estimates, explanatory notes, and other materials necessary for the production of the machine. According to the type of object image, there are drawing and three-dimensional design. Machine parts are characterized by a drawing design method. The set of design documents obtained as a result of the design is called the *project*. This chapter discusses the stages of project development, the requirements for parts, and the criteria for their performance. The principles of the theory of the reliability of technical systems are outlined. An overview of engineering materials is given. When presenting the material here, we were guided by the standards and standards of the Russian engineering industry, which to some extent may differ from European, American, or Japanese standards.

Keywords Detail · Design · Performance · Wear resistance · Reliability · Fail-safe resource · Structural materials · Steels · Heat treatment · Cast iron · Non-ferrous materials and alloys · Powder materials · Single crystals · Composite materials · Plastics · Rubber

22.1 Conditional Classification of Machines

Recall (see p. 286) that a part is a structural element made of a single brand of material without the use of assembly operations. All the variety of parts machines are usually classified into the following groups.

1. *Group of connections.* Many parts can be joined together after they are manufactured by itself constantly without subsequent disassembly. For example, there is no need to disassemble the steam boiler into separate sheets. The separation of such parts is impossible without their partial destruction or damage. Such connections are called *all-in-one.* Examples of all-in-one connections are welded, riveted, glued, etc. Another common type of machine

parts connections is *split connections*. These include threaded, keyed, wedge, pin, etc.

2. *Transmission mechanisms.* The details of this group are intended for transmission mechanical energy and the transformation of the parameters of mechanical motion from the motor machine to the technological machine that performs certain operations.

    The composition of the transmission mechanisms, as a rule, includes the actual transmission, shafts, and clutches. The most common transmission of rotational motion. They are mechanisms designed to transfer energy from one shaft to another with varying angular velocities and torques. There are friction gears and gearing gears. The former transfer energy through frictional forces. Varieties of such transmissions are belt and frictional transmissions. In gearing gears, the movement is transmitted by means of sequentially meshing elements—teeth. Among these gears are gear, worm, and chain.

3. *Shafts.* Earlier (p. 362), we have already said that this is the name of parts that serve to transmit torque along their axis. The same group often includes *axles,* designed to support rotating parts (gears, sprockets, pulleys, etc.).

4. *Couplings.* These parts are used to connect or disconnect the shafts.

5. *bearings.* Bearings serve as supports for shafts and axles.

6. *Translational motion guides.* Parts of this group are used to support translationally moving parts (e.g., a lathe caliper moves along the guides of the translational movement).

7. *Body parts and frames.* Parts of this group are designed to support (base) bearings and guides.

8. *Springs and springs.* Such parts are used to protect against vibration and shock, for accumulating the energy of elastic deformation, for making a reverse stroke in cam and other mechanisms, and for creating tension of belts, chains, etc.

9. *Flywheels, pendulums, weights, babas, and shab'ots.* Parts of this group are used to increase the uniformity of movement, balance mechanisms, or store energy due to their mass.

10. *Protection and lubrication devices.* The named devices serve to increase the durability of the machines.

11. *Details and control mechanisms.* These parts form machine or process control devices.

    In addition to these groups, specific parts of individual machines of groups A, B, and C are often distinguished. In:

    – energy machines: cylinders, pistons, valves, vanes and rotors of turbomachines, rotors and stators of electric machines, etc.;

    – for the transport of machines: wheels, tracks, rails, screws (water and air), buckets, etc.;

    – for machine tools: of the flask, the rolls, saboti, women, ammunition, shield, bobbins, shuttles, etc.

22.1.1 Fundamentals of Design and Development Stages of Machine Projects

Designing machines is a serious creative process that includes the following steps:

(1) *Task for a project or technical proposal;* the technical task is usually for a machine for mass production or serial production, and the technical task is for the manufacturer of special machines.

These documents define the main passport parameters of the machine and outline the general fundamental issues of the scheme of the future machine.

(2) Project *sketch* is a preliminary study of the general views of the main components of the machine.

(3) *Technical project* contains general views of the machine and its components, allowing you to proceed to detailing.

(4) *Working project*, which is a set of general views, working drawings of parts, specifications, programs for CNC machines, and other documents sufficient for the manufacture of the machine.

When designing, the optimal parameters of the product must be selected, which best meet the numerous requirements. The main criteria are efficiency, reliability, adaptability, economy, and ergonomics.

Operability is the property of the machine to perform the specified functions with the parameters set by the technical documentation.

Reliability is the property of the product to perform the specified functions without failures and failures in operation.

Adaptability is the ease of manufacturing, which means a minimum of labor, time, money, and other costs.

Cost-effectiveness implies a minimum of design costs, manufacture, operation, and repair of the product.

Ergonomics is the absence of harmful effects of the machine on the human body and beauty and perfection of forms, also referred to as *design.*

It is very difficult to meet all the above criteria at the same time. Therefore, multi-variant construction is used. In general, the optimal option is considered to be the one that provides the necessary performance indicators at the minimum cost of public labor. At the same time, theoretical calculations, design, and technological experience, as well as experiments, are involved.

22.2 Requirements for Parts, Performance Criteria, and Factors Affecting Them

As in the case of a machine, the main requirements for machine parts are operability and reliability. To meet these criteria, the designer has the ability to manage the shape, size, and material of the part. When selecting the shape of the part, the constructor often works together with the designer, and the size and material of the part are determined by calculation. The performance and reliability of machine parts are characterized by the following criteria: strength, rigidity, wear resistance, heat resistance, and vibration resistance. We will continue to get acquainted with concepts "strength" and "rigidity" which we have already met in the course of resistance of materials. And now let's move on to the criterion of wear resistance.

1. *Wear resistance*. Most often parts fail due to wear. Physically, wear and tear is a process destruction and separation of material particles from the surface of a solid body.

In this case, the process is also accompanied by the accumulation of residual deformations caused by friction in kinematic pairs. The consequence of the process is a change in the size and/or shape of the part.

Wear limits the durability of parts for the following reasons:
(a) loss of required accuracy (instruments, measuring tools, machine tools);   (b) reduced efficiency, loss of power or productivity (motors, pumps, etc.);
(c) reduced strength due to reduced cross-sections, uneven wear of supports, increasing dynamic loads;   (d) increased noise;   (e) complete abrasion (working bodies of earthmoving machines, etc.).

Types of wear are divided into the following groups.

1. Mechanical wear caused by abrasive particles.
2. Molecular-mechanical wear caused by the setting of materials of rubbing surfaces under the action of molecular forces. It is observed at low sliding speeds and high forces in the kinematic pair, causing the oil film to be squeezed out. At high speeds, seizure can occur due to heat, resulting in a reduction in viscosity of the lubricant.
3. Corrosion-mechanical wear is accompanied by chemical or electrolytic interaction of the part material with the medium. One of the types of such wear is *fretting corrosion*, which manifests itself in the destruction of constantly contacting surfaces under conditions of tangential micro-displacements without removing wear products. This kind of wear is typical for the seating surfaces of the rolling bearing rings and in the spline joints.

Another variety of corrosion-mechanical wear is the hydrogen wear associated with the release of hydrogen during the decomposition of water, oil, and petroleum products.

The calculation of wear resistance consists in providing the required thickness of the oil layer between the rubbing surfaces or, if necessary, limiting the pressure in the kinematic vapors. In this case, the results of experiments and the theory of similarity are used. The following dependence is taken as the initial one for the wear resistance characteristic:

$$p^m S = const.$$

where p is the pressure (contact stress) and S is the friction path. The indicator m is set based on the results of the tests ($m \in [1 \ldots 3]$).

2. *Heat resistance.* The operation of the machines is accompanied by heat release. As a result of heating, the following phenomena may occur, which worsen the working conditions of the machines.

- Lowering the load-bearing capacity of parts. For plastics, this phenomenon begins at a temperature of $100 \ldots 150\,°C$, for steels—at $300 \ldots 400\,°C$. When the temperature increases, creep occurs, the endurance limit decreases, and embrittlement of plastic materials occurs.
- Reduces the protective ability of the oil layer between the rubbing parts, which leads to increased wear and sometimes to jamming.
- The properties of the rubbing surfaces change, for example, the coefficient of friction in the brakes decreases.
- The accuracy of the machine decreases due to reversible temperature deformations. This is especially true for precision machines, such as precision metalworking machines.

The heat generation in mechanisms associated with friction in kinematic pairs is calculated directly or by the transmitted power and efficiency. The operability of the parts under heating conditions is ensured by joint thermal, hydrodynamic, and rheological calculations (for creep of deformations and stress relaxation).

3. *Vibration resistance.* This is the name given to the ability of the structure to operate in the desired range of modes without unacceptable vibrations. In real machines, there are two types of vibrations.

Forced oscillations, caused by external periodic forces. They are determined by calculation, and the parameters of the machine are selected in such a way as to exclude the operation of the machine in the resonant frequency range.

Self-oscillations, at which the disturbing forces are caused by the oscillations themselves. In case of danger of self-oscillation, the calculation of dynamic stability is necessary.

Vibration resistance is often associated with noise phenomena during machine operation. Noise is primarily associated with manufacturing errors of parts, but some parts are sources of noise, even with ideal precision manufacturing (e.g., gears when new teeth engage). The noise intensity is estimated in relative logarithmic units (decibels) and is limited by sanitary standards. The main ways to reduce noise are to improve the accuracy and quality of processing parts, reduce impact loads by structural methods, the use of materials with increased internal friction, and the use of special damping devices.

4. For many machines, *maintainability* is of great importance. The ratio of idle time in repair to working time time is one of the indicators of reliability. The design should provide easy access to the components and parts for inspection or replacement. Replacement parts must be interchangeable with spare parts.

In the design, it is desirable to allocate the so-called repair units. Replacing a damaged assembly with a pre-prepared one significantly reduces the machine's downtime.

22.3   Introduction to the Theory of Machine Reliability

Recall that reliability in mechanical engineering is understood as the property of a product to perform its functions for a given time without *failures* and *failures* in operation.

Reliability is one of the main indicators of product quality. The reliability of the product can be used to judge the quality of design work, production, and operation (Svetliczkij 2004; Kołłowrocki and Soszyňska-Budny 2011).

22.3.1   Failures and Reliability

Failure is called loss of the functional state of the product, divided between *full* and *partial failures*. In the first case, the operation of the product stops, and in the second case, the permissible limits of changing some parameters of the product are violated.

The main indicator of product uptime is the probability of $P(t)$ uptime for a given time (t) or operating time. The value $Q(t)$, equal to the difference

$$Q(t) = 1 - P(t),$$

determines the *probability failure*. It can be calculated using the formula

$$Q(t) = int_0^t f(\tau)d\tau,$$

where $f(\tau)$ is the probability density of failure at time τ.

The main indicators of durability of parts are:

– average resource, i.e., average operating time to the limit state;
– gamma-percentage resource that is provided by γ percent of identical parts.

For a product consisting of n parts, the probability of its failure-free operation $P_{iz}(t)$ is calculated by the multiplication theorem of the probabilities of failure-free operation of independent elements, i.e.

$$P_{iz}(t) = P_1(t)P_2(t)\ldots P_n(t). \tag{22.1}$$

When

$$P_1(t) = P_2(t) = \ldots = P_n(t)$$

we get

$$P_{iz}(t) = P_1^n(t).$$

For example, if $n = 10$ and $P_i(t) = 0.9$, we get the probability of failure-free operation of the system $P_{iz}(t) = 0.9^{10} = 0.35$.

During the *normal operation of the rm machine*, as a rule, gradual failures are not yet apparent, but *sudden failures* are not excluded. Such failures are caused by an unfavorable set of circumstances and have a constant intensity of λ independent of time. The probability of failure-free operation in this case calculated by the formula

$$P(t) = e^{-\lambda t},$$

where $\lambda = 1/\bar{t}$, and $\bar{t}$ is the average time to failure. Decomposing the exponent into a series, we have approximately

$$P(t) \approx 1 - \lambda t.$$

Example 1 Let the machine resource be 1000 hours. It is known that the intensity of sudden failures $\lambda = 10^{-6}$ 1/h. Evaluate the probability of failure-free operation during the operation of the machine.

We have

$$P(t) = 1 - \lambda t = 1 - 10^{-6} \cdot 1000 = 0.999.$$

For *gradual failures*, the law of normal distribution is valid. The probability density of such bounce is in the form

$$P(t) = \frac{1}{S\sqrt{2\pi}} exp\left[-\frac{(t - \bar{t})^2}{2S^2}\right],$$

where the mean time $\bar{t}$ and standard deviation S are calculated by the formulas:

$$\bar{t} = \frac{\sum t_i}{N_0}, \quad S = \sqrt{\frac{\sum(t_i - \bar{t})^2}{N_0 - 1}},$$

and N_0 is the total number of objects to be observed or tested.

The time t of uptime for a given probability P of uptime is determined by the dependence

$$t = \bar{t} + u_\gamma S,$$

Table 22.1 Normal
distribution quantiles

P	0.5	0.9	0.95	0.99	0.999	0.9999
u_γ	0	-1.28	-1.64	-2.33	-3.10	-3.72

where u_γ is a parameter called the *quantile of the normal distribution,* whose values
are taken from Table 22.1.

Example 2 The durability of the sliding transmission "screw-nut" is determined by
its wear resistance. The law of resource allocation over time in this case is close to
normal. Estimate 90%-th transmission resource, if $\bar{t} = 10{,}000$ h, $S = 2000$ h.

Calculating:

$$t_{90} = \bar{t} + u_\gamma S = 10{,}000 - 1.28 \cdot 2000 = 7440 \text{ h}.$$

Analyzing the formula (22.1), we can note the following:

- the reliability of a complex system is always less than the reliability of the most
  unreliable element, so it is important not to allow any weak element into the
  system;
- the more elements a system has, the less reliable it is.

If, for example, a system includes 100 elements with the same reliability $P_n(t) =$
0.99, then its reliability is $P(t) = 0.99^{100} = 0.37$.

Such a system, of course, cannot be considered workable, since it is doomed
to stand idle more than to work. This allows us to understand why the problem
of reliability has become particularly relevant in the modern period of technology
development along the way of creating complex automatic systems.

22.3.2 Ways to Improve Reliability at the Design Stage

It is known that many systems (automatic lines, rockets, airplanes, computers, etc.)
include tens and hundreds of thousands of elements. If these systems do not provide
sufficient reliability for each element, they become unusable or inefficient. par The
study of reliability is an independent branch of science and technology. The main
ways to improve reliability at the design stage are described below.

1. A reasonable approach to achieving high reliability is to design as simple as
   possible products with fewer parts. For each part , a sufficiently high reliability
   must be provided, equal to or close to the reliability of the other parts.
2. The simplest and most effective method of increasing reliability is to reduce the
   tension of the parts (increase the strength reserves). However, this requirement
   of reliability is in conflict with the requirements of reducing the size, weight, and
   cost of products. The compromise solution consists in the use of high-strength
   materials and strengthening technologies: alloy steels, thermal and chemical

treatment, surfacing of hard and antifriction alloys on the surface of parts, surface hardening by shot blasting or rolling in with rollers, etc.

For example, by heat treatment, it is possible to increase the load capacity of gears by 2...4 times. Chrome plating of the crankshaft necks of automobile engines increases wear life by 3...5 or more times. Shotblasting of gears, springs, springs, etc. increases the service life of the fatigue of the material in 2...3 times.

3. An effective measure to improve reliability is a good lubrication system: the right choice of oil grade, a rational system for supplying grease to rubbing surfaces, protecting rubbing surfaces from abrasive particles (dust and dirt) by placing products in closed cases, installing effective seals, etc.

4. Statically definable systems are more reliable. In these systems , the harmful effects of manufacturing defects on load distribution are less apparent.

5. If the operating conditions are such that accidental overloads are possible, then the design should provide safety devices (safety couplings or other devices of similar purpose).

6. The extensive use of standard components and parts, as well as standard structural elements (threads, fillets, etc.), increases reliability. This is due to the fact that standards are developed on the basis of extensive experience and standard components and parts are manufactured in specialized factories with automated production. At the same time, the quality and uniformity of products are improved.

7. In some products, mainly in electronic equipment, to to increase reliability, use parallel connection of elements and so-called redundancy, rather than serial connection. When the elements are connected in parallel, the reliability of the system is significantly increased, since the function of the failed element is assumed by the parallel or backup element. In mechanical engineering, parallel connection of elements and redundancy are rarely used, since in most cases they lead to a significant increase in the mass, size, and cost of products. An example of a justified application of a parallel connection can be service aircraft with two and four engines. An aircraft with four engines will not suffer an accident even if one or two engines fail.

22.4 Engineering Materials

We have already said (p. 481) that one of the ways to ensure the operability and reliability of machine parts is the choice of suitable materials. In modern mechanical engineering, both traditional structural materials (steel and cast iron, aluminum, magnesium, titanium, and copper alloys) and new materials (metal composite materials, polymers, silicate materials, powder materials, ceramic metal, and many others) are widely used. The day is not far off when graphene and other ultra-new materials will come to mechanical engineering (*Spravochnik metallista GNTIMP. T.3[Metalworker's Handbook. Vol.3]* 1960; Molotnikov 2017).

22.4.1   Steels

Steel is an alloy of iron with carbon and other elements, in which the carbon content does not exceed 2%. Depending on the carbon content, steel is divided into it low carbon ($C \leqslant 0.25\%$), medium carbon ($0.25 \leqslant 0.6\%$), and *high carbon* ($C > 0.6\%$). When marking steel, the average carbon content in hundredths of a percent indicates the number formed by the first two digits (e.g., steel 45 contains 0.45% carbon).

To improve the mechanical, corrosion, and other properties of steels, alloying additives are used. rm The presence of additives is reflected in the name of the steel grade by the letters: B, tungsten; Г, manganese; Д, copper; M, molybdenum; H, nickel; P, boron; C, silicon; T, titanium; X, chromium; Ф, vanadium; and Ю, aluminum. The numbers after these letters indicate the percentage of alloying additives in the steel composition. If the content of the alloying element is not more than 1%, then the number after the letter is not put. For example, 12X2VN4 steel contains an average of 0.12% carbon, 2% chromium, no more than 1% tungsten and 4% nickel.

According to the method of production, carbon steels are divided into ordinary-quality steels and high-quality structural steels. Carbon steels of ordinary quality, obtained by the open-hearth method, are designated in the order of increasing strength St0, St1,..., St7. They are used for the manufacture of non-responsible parts (housings, fasteners, etc.).

Alloy steels are divided into high-quality, high-quality (at the end of the brand designation, the letter A is added, e.g., 30XGSA) and especially high-quality. They are more expensive than carbon, have a high strength ($\sigma_{\text{in}} = 800\ldots 1400$ MPa), and are used for the manufacture of critical machine parts (gears, shafts, etc.).

Thermal and Thermochemical Treatment of Steels

To give the steel certain properties (hardness, strength, ductility, etc.), perform heat treatment of workpieces or finished parts. Heat treatment consists of three successive stages: heating to a certain temperature at a certain speed, holding at this temperature for a certain time, and cooling at the required speed.

The following four types of heat treatment are most commonly used: annealing, normalization, quenching, and tempering.

A n n e a l i n g—heating, holding at a certain temperature, and then slowly cooling (together with the oven or in the air). The purpose of annealing is to reduce hardness, improve workability by cutting, and remove residual stresses (after welding or roughing with a cutter). There are *full* and *low-temperature (high)* annealing. Parts made of carbon and carbon alloy steels are subjected to full annealing. The sequence of operations consists of heating to $800\ldots 900\,^{\circ}\mathrm{C}$, holding at this temperature, cooling together with the oven to $400\ldots 600\,^{\circ}\mathrm{C}$, and then cooling in air. In low-temperature annealing, high-alloy steel parts are heated to a lower temperature—$650\ldots 670\,^{\circ}\mathrm{C}$—and slowly cooled in air.

N o r m a l i z a t i o n differs from full annealing by the cooling rate. The parts are cooled in air after heating and holding. During normalization, a more uniform structure with a higher hardness and strength is obtained than after annealing.

Q u e n c h i n g differs from full annealing and normalization by a high cooling rate. Cooling is carried out in water, oil, aqueous solutions of salts, or alkalis (NaCl, NaOH, etc.). As a result, the hardness, strength, and wear resistance of the part increase, but the ductility of the material decreases, and the workability of cutting becomes more difficult. Several types of quenching are practiced, which differ in fast cooling modes. A wide application is found in *surface quenching*—heating of the surface layer of a steel part by high-frequency currents, laser beam, etc. above the temperature of phase transformations and subsequent rapid cooling to obtain a fine-grained structure in the surface layer. Surface quenching is applied to wheel teeth, cams, shafts, etc.

T e m p e r i n g—heating to a temperature below the phase transformations, holding, and subsequent cooling. As a result of tempering, the ductility of the material increases, and technological residual stresses decrease. Depending on the heating temperature, heating temperature $500 \ldots 670\,°C$) *receives a high release*, $250 \ldots 450\,°C$ *an average release*, and ($140 \ldots 230\,°C$) *a low release*. At a higher heating temperature, the ductility of the steel increases as a result of tempering.

T h e r m o c h e m i c a l t r e a t m e n t of steel parts is carried out in order to increase the hardness, strength, and wear resistance of the surface layers of the parts by saturating them with carbon (*cementation*), nitrogen (*nitriding*), and simultaneously saturating them with carbon and nitrogen (*cyanidation*) or boron (*boronation*). These elements form compounds with iron in the surface layer, which are characterized by high hardness.

Cementation is applied to parts made of low-carbon steels 15, 20X, 18X2H4MA, etc.; nitriding-medium-carbon alloy steels 38X2MYA, 38X2YU, etc.; and cyanidation-steel grades 15, 20, 45, 35X, 40X, etc.

22.4.2   Cast Iron

Iron-carbon alloys with a carbon content of more than 2% are called cast iron. They have high casting properties, so they are widely used in mechanical engineering. Depending on the structure, cast iron is divided into *gray, white*, and *malleable*.

Gray cast iron is marked with the letters SCH and a two-digit number, which approximately shows the tensile strength divided by 10 in MPa. For example, SCH15 means gray cast iron with a tensile strength of 150 MPa. The most commonly used cast irons are SCH15 and SCH20. They are used for the manufacture of body parts, pulleys, gear teeth, shafts, etc.

White cast iron has a high hardness and brittleness, it is processed by cutting only with a carbide tool. It is used for the manufacture of brake pads and parts subject to abrasive wear.

Ductile iron has high strength (tensile strength $\sigma_b^p \approx$ 300 MPa, for compression—$\sigma_{vr}^c \approx 630$ MPa). It is used for the manufacture of cast parts. Despite its name, ductile iron is not pressure-treated due to its low ductility.

22.4.3   Copper Alloys

Brass and *bronze* have main applications in mechanical engineering. Brass is divided into double alloys of copper and zinc and multicomponent, additionally containing lead, silicon, manganese, and other elements. Brasses have good technological properties. They are processed by casting, cutting, and pressure and have sufficient strength ($\sigma_{vr} \approx 250 \ldots 350$ MPa). In the marking of brass, there is a letter Л, for example, Л59, Л62, Л90. In mechanical engineering, its main use is found in multicomponent brasses such as ЛКС80-3-3, ЛМцС8-2-2 (new designation ЛЦ38Мц2С), etc., used mainly as antifriction materials.

Bronzes, in addition to the base metal copper, contain other elements that determine their name. In mechanical engineering, tin, lead, aluminum, beryllium, and other bronzes are used. The main advantages of these alloys are high antifiction properties, good corrosion resistance, and technological properties (among the bronzes, there are both foundry and pressure-treated). The greatest use of bronze is found in the device of sliding bearings, worm, and screw gears.

When marking bronzes, the letters Br are used, followed by letters denoting other (except copper) elements (A aluminum, B beryllium, W iron, K silicon, O tin, C zinc, F phosphorus, etc.) with a digital indication of their percentage content. For example, BRAZH9–4 means bronze with an average content of 9

22.4.4   Babbits

Alloys based on tin, lead , and calcium are called *babbits*. They are high-quality, well-developed anti-friction-bearing materials. They are indicated by the letter Б, followed by the percentage of tin, and there may also be letters indicating additional components, if any. For example, babbit Б88 contains $\approx 88\%$ of tin, and the rest is antimony and copper; Б83С is babbit containing $9 \ldots 11\%$ of antimony, $5 \ldots 6\%$ of copper, and the rest is tin.

Babbits are almost 20 times more expensive than high-quality steel, which hinders the expansion of their application in mechanical engineering.

22.4.5   Aluminum Alloys

These materials have a density almost 3 times less than the density of steel. At the same time, their strength is close to that of steel.

There are *foundry* and *deformable* aluminum alloys. Foundry alloys include aluminum with silicon—*silumins*. They are labeled as АЛ2, АЛ4, АЛ5, etc. Copper, zinc, and magnesium can be used as alloying additives. Silumins are well processed by cutting and after quenching can have a tensile strength of 170–250 MPa. Their main application is the casting of body parts.

Deformable aluminum alloys of the АМц, АМг grades are thermally non-hardening, with a content of manganese ($\approx.1.5\%$) or magnesium have good plastic properties and are used for the manufacture of housings, rivets, bearing separators, etc. Thermally hardened alloys D Д1, Д16, etc. *duralumins* (copper up to 5%, magnesium up to 2%, manganese up to 1%, the rest is aluminum) have a tensile strength of $\sigma_{vr} \approx 350 \ldots 430$ MPa and are used for the manufacture of pressure and cutting housings, transport parts, etc.

22.4.6 Magnesium Alloys

The main advantages of magnesium as a machine-building material are its low density, high specific strength (the ratio of strength to density), resistance to shock, and vibration loads. These properties allow the use of magnesium alloys for the manufacture of parts subject to dynamic loads (wheels of guns, aircraft, pistons, connecting rods, etc.).

Magnesium is almost 1.6 times lighter than aluminum, and its density is 1.7 g/cm^3. It is well processed with a cutting tool, easy to grind and polish. To remove the same volume of metal in the processing of magnesium requires a power of almost $6 \ldots 7$ times less than in the processing of steel. Magnesium-based alloys are the lightest structural materials.

Like aluminum, magnesium alloys are divided into foundry and deformable. The first have label МЛ: МЛ1, МЛ2, МЛ3, МЛ6, МЛ11, etc., while the second are labeled МА2, МА3, МА5, МА8, МА9, etc.

Cast alloys МЛ3, МЛ4, МЛ5 after heat treatment show the ultimate strength $\sigma_b = 200 \ldots 230$ MPa at yield strength $\sigma_t = 150 \ldots 180$ MPa. Alloys of the МЛ9, МЛ10, МЛ11, МЛ19 system are heat-resistant. For long-term operation, they can operate at temperatures up to $250 \ldots 300\,°C$ and for short-term operation up to $400\,°C$. The main alloying element in the alloys МЛ9, МЛ10, МЛ19 is neodymium, and in МЛ11 cerium mishmetal (75% Ce, the rest rare earth elements).

22.4.7 Titanium Alloys

In the domestic industry, titanium alloys are mainly used in chemical, heavy, energy, and transport engineering, mechanical engineering for light, food industry, and household appliances. Titanium is also widely used in products where rotating parts are used. As an example, we will point to the details of centrifugal machines (centrifuges, separators, dryers, compressors, etc.). When creating them, the designers developed a number of measures to improve the antifriction properties of alloys.

Titanium alloys with aluminum, copper, and other additives (BT3-1, BT5, BT9, BT16, BT22, etc.) are almost twice as light as steel and have a high strength after heat treatment ($\sigma_{vr} = 900\ldots1300\,\text{MPa}$). They are also characterized by high corrosion resistance.

The experience of using titanium alloys in Russia and abroad shows that it is most appropriate to use titanium alloys for parts of high-load engines, load-bearing structures, and chassis of cars.

As a result of scientific research and production practice, the following applications of titanium alloys are currently recommended:

For load-bearing structures of cars—medium-strength alloys of grades OT4-1; BT5-1, OT4, BT5, BT6;

For chassis of cars—medium-strength and high-strength alloys of grades BT6, BT3-1, BT8, AT6, BT5-1, BT14, BT15, BT16;

For engine parts—high-strength and heat-resistant alloys of the brands BT3-1, BT8, BT14, BT15. BT16, CT-1, CT-4, BT18.

22.4.8 Powder Materials

Powder metallurgy is a special type of technology for manufacturing parts by forming and sintering fine powders. It greatly simplifies or eliminates the subsequent machining of parts. The following types of powder metallurgy materials are used in domestic mechanical engineering: structural, heat-resistant, tool, porous, and highly porous. The field of application of powder materials is very extensive: parts of aircraft, brake assemblies of tractors and other machines, filters with a recovering filtering capacity, piston rings, and many other parts. The ultimate strength of powder materials based on iron-carbon powders ranges from 510 to 1280 MPa, on the basis of copper alloys from 140 to 300 MPa, and on the basis of aluminum from 70 to 100 MPa.

22.4.9 Single Crystals and Composite Materials

Single crystals of refractory and rare metals and alloys are ideal crystals, devoid of structural defects. The growth of single crystals can be carried out from the melt, metal vapor, or the solid phase by recrystallization. At first, the size of the grown single crystals was small, and they were called whiskers. In modern production facilities, the cultivation of single crystals reaching 50 mm or more in cross-section has been established.

Single crystals have extremely high plasticity, resistance to thermomechanical influences, and a number of other unique properties. The high plastic properties of single crystals have opened the way for their wide application in the production of so-called composite materials.

Metal composite materials are obtained on the basis of a metal matrix reinforced with metal or ceramic whiskers. Reinforced composite has significantly higher strength properties. The best result is obtained when reinforcing composites with ceramic whiskers with a diameter of 1 . . . 3 microns. The resulting materials have a tensile strength comparable to high-quality steel, and their Young's modulus is 5 . . . 10 times higher than that of steel.

Korean researchers have created composite materials using graphene that are 500 times stronger than the original ones. This is the first time that graphene has been successfully used to create strong composite materials. Due to the microscopic amount of graphene used, the weight of which in the total mass of the resulting material was 0.00004%, this breakthrough may lead to faster commercial use of graphene alloys than pure graphene, large-scale production of which is still problematic. China's huge investment in such technologies gives hope to expect that this is not far off.

22.4.10 Silicate, Ceramic, and Cermet Materials

Glass is widely used as a structural and technical material. The scope of its use is limited to structures in which shock loads and the appearance of tensile stresses are excluded or minimized.

The mechanical strength of glass is quite high. Thus, quartz glass has a compressive strength of 350 . . . 650 MPa and a bending strength of about 220 MPa. It is an indispensable material in the manufacture of chemically resistant equipment, pipelines, and other products operating in aggressive environments. Glass is also used in the manufacture of fiberglass and plastics. With a fiber diameter of 3 . . . 4 microns, the tensile strength of the glass fiber reaches a colossal value— 37,000 MPa! The industry also produces film and flake glass, which is mainly used in the production of plastics—glass fiberglass. The ultimate tensile strength of glass fiber composites reaches 250 MPa. Their important property is their translucency.

Under normal conditions, glass is an amorphous body. However, under certain conditions, its complete or partial crystallization occurs. The result is glass-crystal materials called sitals. These materials occupy an intermediate position between glass and ceramics. The bending strength of the sitals is in the range of 160 . . . 250 MPa. They are used for the manufacture of pipes, chemical equipment, bearings operating at high temperatures (about 500 °C) and without lubrication, pistons and cylinders of diesel engines, etc.

Technical ceramics are made either by sintering powdered materials or by forming products in the plastic state of the source material, followed by firing the product. Ceramics made of pure oxides of aluminum, beryllium, zirconium, and other metals, as well as their nitrides, borides, sulfides, etc., have the greatest strength properties. The ultimate strength of ceramic materials under compression varies widely, from 800 to 3000 MPa, and under tension, from 100 to 250 MPa.

22.4.11 Plastics

Plastic is obtained as a result of the joint processing of high-molecular organic compounds(synthetic resins), fillers, coloring agents, plasticizers, hardeners, and other additives. The main component of plastic, which determines its type and basic properties, is resin.

The filler significantly affects the characteristics of the plastic, changing its physical, mechanical, and electrical properties. In addition, the introduction of fillers reduces the cost of plastic, as it reduces the consumption of relatively expensive resin.

In mechanical engineering, the following types of plastics are most widely used:

(a) thermosetting laminated plastics: textolite (filler, cotton fabric), getinax (filler, sheet paper), asbotextolite, fiberglass, and wood plastics;
(b) thermosetting plastics (fiber, phenoplast, etc.) are used for the manufacture of pressing handles, joints, wheel hubs, and other parts;
(c) thermoplastic plastics (organic glass-plexiglass, viniplast, fluoroplast, etc.) are used for the manufacture of glasses, tubes, etc.;
(d) polyamides (nylon, etc.) are used for forming parts of complex shapes (belts, gears, etc.).

22.4.12   Rubber

Rubber is a material obtained on the basis of natural or artificial rubber. It has a high elastic pliability, well dampens vibrations, and resists abrasion.

Depending on the purpose, rubber is made soft (for tires), porous (for shock absorbers), and hard (ebonite, for the manufacture of electrical products).

To increase the strength of rubber products, they are reinforced with textile, polyvinyl alcohol, steel, or other elements. Such rubber is used for the manufacture of tires, drive belts, sleeves, etc.

Self-Test Questions

1. List the main groups of machine parts.
2. What is defined in the technical specification for the design of the machine (mechanism)?
3. What is a sketch project?
4. What are the main components of the technical project?

5. What are the main criteria that determine the optimality of the designed product?
6. What is the ergonomics of the machine?
7. What is meant by the wear resistance of the machine?
8. List the consequences of reducing the heat resistance of machine parts.
9. Define vibration resistance of engineering design.
10. What is meant by the maintainability of the machine?
11. What alloy is called steel?
12. What are alloying additives?
13. How are carbon steels divided according to the method of production?
14. How are alloy steels divided?
15. List the main stages of heat treatment of steel.
16. What are the main types of thermochemical treatment of steel?
17. What alloys are called cast iron?
18. Specify the scope of application of gray, white, and ductile iron.
19. Name the scope of application of brass and bronze.
20. Name the main components of babbits.
21. Name the scope of application of silumins and duralumins.
22. Specify the scope of application of titanium alloys.
23. What is the main idea of powder metallurgy?
24. What are single crystals? Where do they apply?
25. List the types of plastics used in mechanical engineering.
26. Name the technical applications of rubber and rubber products.
27. How is the number of system elements related to its reliability?
28. What is the relationship between the reliability of a complex system and the reliability of its most unreliable element?
29. List ways to improve reliability at the design stage. Name the conflicting trends in the design process.
30. What is called redundancy, used to improve reliability?
31. Which systems are more reliable—with serial or parallel connection of elements?
32. Which systems are preferable in terms of reliability—statically definable or statically indeterminate?

References

K. Kołłowrocki, J. Soszyňska-Budny, *Reliability and Safety of Complex Technical Systems and Processes* (Springer, Berlin, 2011)

V. Molotnikov, *Tekhnicheskaya mekhanika [Technical mechanics]* (Lan' Publication, Sankt-Peterburg, 2017)

Spravochnik metallista GNTIMP. T.3[Metalworker's Handbook. Vol.3] (1960).

V. Svetliczkij, *Statisticheskaya mexanika i teoriya nadyozhnosti [Statistical mechanics and reliability theory]* (Izd-vo MGTU im. Baumana., Moscow, 2004)

Chapter 23
Precision Manufacturing of Machine Parts

Abstract When manufacturing a part by any technological process, it is impossible to obtain a part of a certain predetermined size. This dimension can be obtained with a greater or lesser degree of approximation within the tolerances assigned by the designer of the machine and indicated in the drawing of the part. The basis of serial and mass production in mechanical engineering is the so-called interchangeability of parts and assemblies of machines. Interchangeability is ensured by the Unified System of Tolerances and Landings ESDP (GOST 25346, GOST 25347) and the requirements of the International Organization for Standardization ISO. This chapter discusses the tolerances for the maximum deviations from the nominal dimensions, the nature of the mating of parts depending on the specified tolerances, the accuracy of the geometric shape, the tolerances for the ultimate roughness, and waviness of the surface.

Keywords Interchangeability · Standardization · Tolerances · Fitments · Nominal size · Deviation · Non-straightness · Non-flatness · Roughness · Undulation

23.1 Interchangeability and Standardization

Interchangeability is a design principle, , which makes it possible to assemble independently manufactured parts into a node and nodes into a machine without additional processing and fitting operations.

To ensure the interchangeability of parts and assembly units, they must be manufactured with the specified accuracy, i.e., so that their dimensions, surface shape, and other parameters are within the limits set when designing the product (Molotnikov 2017).

Distinguish between *full* and *incomplete* interchangeability of parts assembled in assembly units. Complete interchangeability provides the ability to assemble (or replace during repair) any independently manufactured with a given accuracy of the same type of parts in an assembly unit. Complete interchangeability is required in, for example, bolts, nuts, washers, bushings, and gears.

© The Author(s), under exclusive license to Springer Nature Switzerland AG 2023
V. Molotnikov, A. Molotnikova, *Theoretical and Applied Mechanics*,
https://doi.org/10.1007/978-3-031-09312-8_23

Limited interchangeability refers to such parts, when assembling or changing which may require a group selection of parts (selective assembly), the use of compensators, adjusting the position of parts, fitting (e.g., the assembly of a gearbox, rolling bearings).

The interchangeability of the product production is characterized by the coefficient interchangeability, equal to the share of the labor intensity of manufacturing interchangeable parts to the total labor intensity of manufacturing the product.

There are also *external* and *internal* interchangeability. External interchangeability refers to purchased or co-operated products that are mounted in other, more complex products. For example, in electric motors, external interchangeability is provided by the shaft speed, power, and shaft diameter and in rolling bearings by the diameters of the outer and inner rings. Internal interchangeability applies to the parts, assembly units, and mechanisms included in the product. For example, in a rolling bearing, rolling bodies and rings have internal group interchangeability.

The basis for implementing interchangeability in modern industrial production is *standardization* (Münstermann and Weitzel 2008).

The largest international organization in the field of standardization is ISO. The purpose of ISO is stated in its charter: "... to promote the favorable development of standardization around the world in order to facilitate the international exchange of goods and to develop mutual cooperation in various fields of activity".

At the same time, standardization is understood as a planned activity to establish mandatory rules, norms, and requirements, the implementation of which increases the quality of products and labor productivity.

By definition, the *standard* is a normative and technical document that establishes requirements for groups of homogeneous products and rules that ensure their development, production, and application.

Along with the standards, the manufacturing companies often establish *technical specifications* (TU), which establish requirements for specific products, materials, their manufacture, and control. Unlike standards, technical specifications are primarily addressed to consumers (customers) of products and secondly to conformity assessment bodies.

To strengthen the role of standardization, a state system of standardization has been developed and put into effect. It defines the goals and objectives of standardization, the structure of standardization bodies and services, the procedure for the development, registration, approval, publication, and implementation of standards.

Depending on the scope, the following categories of standards are provided:

- international: ISO (International Organization for Standardization), IEC (International Standard in the Electrical Industry), ITU (International Telecommunication Union), UNECE (United Nations Economic Commission for Europe), and IAEA (International Atomic Energy Agency). International standards are not mandatory, but in the face of intense competition, manufacturers of products, in an effort to maintain the high competitiveness of their products, are forced to ensure that their products meet the requirements of international standards;

- regional—GOST (Russia and CIS), EU (EU);
- national—BSI (UK), NF (France), DIN (Germany), JISC (Japan), etc.;
- industry-specific—OST;
- enterprise standards, STP (or STO), Russia, are set by the enterprise (organization) itself, taking into account the requirements of Article 17 of the Federal Law "On Technical Regulation".

23.2 Tolerances

The geometric parameters of the parts are quantified by *dimensions*.

Nominal size is the common size for the connection parts, obtained as a result of the calculation and rounded in accordance with the series of normal linear dimensions established by GOST 6636-69 "Nominal linear dimensions".

Size limits—these are the two maximum allowed sizes between which the actual size of the usable part must be located.

The actual size is the size obtained as a result of processing the part and measured with an acceptable error. Its value may coincide with the specified one only by chance, since a large number of factors affect the *accuracy* of manufacturing and inevitably lead to errors. Here, the accuracy is understood as the proximity of the manufacturing results and the prescribed values. A measure of accuracy is an error equal to the difference between the obtained and prescribed values of a particular size. To ensure proper assembly and normal operation of the part, these errors are limited to the limit values. Economically feasible limit deviations of the dimensions of the parts are determined by a single tolerance system of the height and landings established by GOST 25347-82 and GOST 25346-82.

In the drawing, the size is most convenient to put down in the form of a nominal size with deviations. The difference between the largest and smallest limit dimensions is called the *tolerance*. As already mentioned, the tolerance is a measure of the accuracy of the size. It determines the labor intensity of manufacturing the part. The larger the tolerance, the easier and cheaper it is to manufacture the part.

The concepts of nominal size and deviations are simplified by the graphical representation of the tolerances in the form of diagrams of the *tolerance fields*. Figure 23.1 shows cylindrical holes (a) and shafts (b) combined in the formers, for which the nominal and limit values of the diameters are indicated. GOST 25346-82 establishes the following designations:

$$\text{size tolerance:} IT = d_{max} - d_{min};$$

$$\text{upper shaft deflection:} es = d_{max} - d;$$

$$\text{top hole deflection: } ES = d_{max} - d; \tag{23.1}$$

$$\text{lower shaft deflection: } ei = d_{min} - d;$$

$$\text{lower hole deflection: } EI = d_{min} - d,$$

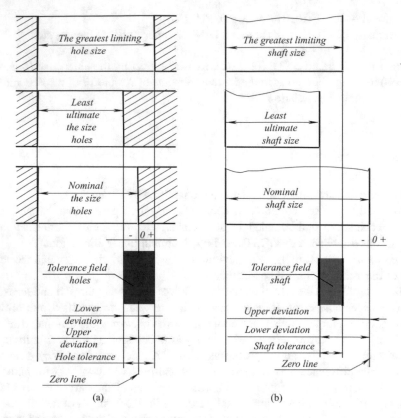

Fig. 23.1 Maximum dimensions of the hole (**a**) and the shaft (**b**)

where d is the nominal diameter of the shaft (hole).

It follows from the definitions (23.1) that the values of deviations can be either positive or negative. In the schematic representation of the tolerance field (Fig. 23.2), the first ones are put up and the second ones down from the zero line. The standard GOST 25346-82 defines the location of the tolerance field relative to the zero line as a letter (or two letters) of the Latin alphabet—uppercase for holes and lowercase for shafts (e.g., $H9$, J_s8, $h8$, j_s7, etc.).

The tolerance value IT is assigned depending on the diameter d, mm, as the product of

$$IT = ai,$$

where a is the number of tolerance units i defined by the formula

$$i = 0.45\sqrt[3]{d} + 0.001d, \text{ mkm.}$$

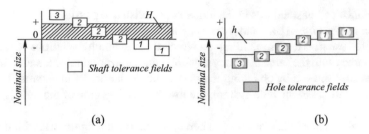

Fig. 23.2 Tolerance fields of shafts (**a**) and holes (**b**)

Depending on the number of a in the tolerance of IT, the standard sets 19 *qualities* (classes) of accuracy: 01, 0, 1, ..., 17. The tolerances in the 01, ..., 4 qualities are for end length measures, gauges, measuring tools, etc.; the 5, ..., 13 qualities are used for the tolerances of mating parts; and the 14, ..., 17 qualities are assigned to the tolerances of non-mating part sizes.

The limit values of the upper and lower deviations are indicated in the drawings in one of three ways.

1. For the nominal size, the numerical values of the limit deviations are indicated in millimeters in the upper or lower index, for example

$$15^{-0.036}_{-0.051}, \quad 22^{+0.016}, \quad 255^{+0.060}_{+0.080}.$$

2. The symbol of the tolerance field, expressed by a letter and a number denoting the quality, for example, $12e8$, $20h10$.
3. By simultaneously specifying the tolerance field, followed by numeric deviation values in parentheses, for example,

$$12e8^{-0.032}_{-0.059}, \quad 20h10_{(-0.084)}.$$

23.3 Landings

Fit is the nature of the coupling of two parts, which determines the greater or lesser freedom of their relative movement or the degree of resistance to their mutual displacement. Fit is formed by a combination of the shaft and hole tolerance fields.

If the diameter of the hole is larger than the diameter of the shaft, then a *gap* (positive diameter difference) is formed between them, creating the possibility of free axial and circumferential relative movement of the parts covering the hole and the parts covered by the shaft. If the hole size is smaller than the shaft size (negative diameter difference), then the connection is only possible with *tightness*.

Tolerance and fit standards in the industry establish two possible sets of fittings—*a system with an inner hole* and *a system with a shaft.*

A hole system is a set of landings in which the maximum deviations of the holes are the same (with the same accuracy class and the same nominal size), and different landings are achieved by changing the maximum deviations of the shafts (Fig. 23.2 it a rm). In all fits of the hole system, the lower limit deviation of the hole is always zero.

This hole is called the main hole. It can be seen from the figure that with the same nominal size (diameter) and constant tolerance of the main hole (the hole tolerance field is shaded in Fig. 23.2a), different landings can be obtained by changing limiting shaft dimensions. The landings can combine the tolerance fields of the hole and shaft of the same or different qualities. (Typically, the larger tolerance applies to the hole.) The base hole tolerance field is denoted as H.

Fits in the hole system are indicated by the nominal connection diameter followed by the hole tolerance fields and then the shaft, for example: $\varnothing 40H7/s6$ or $40H7 - s6$, or $40\frac{H7}{s6}$.

Shaft system is a set of landings in which the maximum deviations of the shafts are the same (with the same accuracy class and the same nominal size) and different landings are achieved by changing the maximum deviations of the holes. In all landings of the shaft system, the upper limit shaft deviation is always zero. Such a shaft is called a main shaft. The shaft tolerance field in this system is denoted by the letter h, and the landings in the drawings, by analogy with the hole system, are as follows: $\varnothing 40P7/h6$, or $40P7 - h6$ or $40\frac{P7}{h6}$.

The hole system is most common in mechanical engineering because it requires fewer tools to use for hole making than with a shaft system.

In both systems, different landings are divided into three groups: *mobile, fixed* (press or with interference), and *transitional.*

M o v i n g f i t s or landing with a clearance correspond to the tolerance fields marked in Fig. 23.2 by the numbers 1. Such fits are used in movable joints (e.g., in plain bearings, as well as in joints that are often assembled and disassembled). The most common landings are $H9/f9$; $H7/f7$; $H7/g6$; $H8/h6$, etc.

F i x e d f i t s, called also press or interference fits, are formed at the tolerance fields marked in Fig. 23.2 with the number 3. They are used for the fixed connection of parts without additional fastening (dowels, pins, etc.). In general mechanical engineering, landings of this type are most applicable with parameters $H7/p6$; $H7/r6$; $H7/s6$; $H8/e8$, etc.

T r a n s i t i o n a l l a n d i n g s are landings, which can be either clearance landings or interference fits, depending on the ratio of the actual dimensions of the hole and the shaft. Landings of this kind are formed by combining tolerance fields marked with 2 in Fig. 23.2. They are tried on for centering mating parts with a fixed connection with additional fastening with keys, screws, or pins. Examples of designation of transitional landings are $H7/j_s6$; $H7/k6$; $H7/n6$, etc.

23.4 Accuracy of the Geometric Shape of the Parts

Deviations from the specified geometric shape and the location of the surfaces of the manufactured part occur during machining as a result of inaccuracies and deformations of the machine and the device, wear of the cutting blade of the tool, deformations of the workpiece, as well as unevenness of the processing allowance.

The accuracy of the parts in terms of geometric parameters is characterized not only by size deviations but also by surface deviations. In this case, the deviation of the surfaces is determined by shape deviations, deviations in the location of the surfaces, waviness, and roughness.

Shape deviations of parts with flat mating surfaces include *non-straightness* and *non-flatness* (Fig. 23.3).

N o n - s t r a i g h t n e s s—deviation from the straight line of the profile of the cross-section of the surface by the plane normal to it in the specified direction (Fig. 23.3a).

N o n - f l a t n e s s is the deviation from straightness in any direction along the surface, for example, convexity and concavity (Fig. 23.3b).

Deviations in the shape of cylindrical surfaces are possible in the transverse and longitudinal sections. In the cross-section, it is possible that the contour of the surface deviates from the correct circle—*is non-circular*. In the longitudinal section of a cylindrical surface, there may be deviations from the straightness of its generators: barrel shape, concavity, curvature of the axis, and deviations from the parallelism of the generators—taper shape.

The difference between the largest and smallest diameters is taken as the value of deviations in the shape of a cylindrical surface. The maximum deviations in the shape of cylindrical surfaces are limited by the tolerance fields for the diameter.

The deviation of the location of surfaces (axes, profiles) is the discrepancy between the actual location of the surface, axis or profile, and the nominal location. The nominal position is determined by the nominal linear or angular dimensions between the surface under consideration (straight line, profile) and the base. The base is called the surface, axis, or point, with respect to which the location tolerances are set. The deviations of the relative position according to GOST 24642-81 include

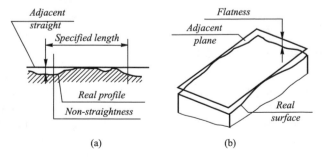

Fig. 23.3 Non-straightness (**a**) and non-flatness (**b**)

Table 23.1 Symbols of shape and location tolerances (GOST 2.308-79 ESKD)

Tolerance group	Tolerance	Sign
Tolerances shapes	Straightness	—
	Flatness	⌓
	Roundness	○
	Cylindricity	⌭
	Longitudinal profile	=
Tolerances locations	Parallelism	////
	Perpendicular	⊥
	Tilt	∠
	Alignments	◎◎
	Symmetries	≡
Total shape and location tolerances	Radial or face runout	↗↗
	Full radial or full end runout	⌰⌿
	Forms of the specified profile	⌒
	Shapes of the specified surface	⌓

non-parallelism, non-perpendicular (for planes); *misalignment, radial, and end runout* (for cylindrical planes); *misalignment of axes*, etc.

The maximum deviations of the shape and location of the surfaces are indicated in the drawings in the form of symbols (signs, symbols) and text entries (Table 23.1).

The shape and location tolerances are specified only if there are special requirements for the accuracy of the shape and location of the surfaces of the parts. The sign and the value of the deviation are entered in a rectangular frame divided into two or three parts. In the first part (on the left), indicate the deviation sign (Fig. 23.4), in the second the numerical value in millimeters, and in the third the letter designation of the base. The bases are indicated by a blackened triangle and connected to a frame in which the letter designation of the base is given (Fig. 23.4a) or the conditional designation of the tolerance (Fig. 23.4b). The direction of the deviation measurement line is indicated by a line segment with an arrow.

23.5 Roughness and Undulation

During the processing process, surface irregularities are formed on the surface of the part, which occur due to the oscillatory movement of the tool, copying defects in its cutting edge, and other factors. If the ratio of the pitch to the height of the irregularities is in the range from 50 to 1000, then such deviations of the surface are called undulation and if this deviation is less than 50, roughness.

Fig. 23.4 Designations of
tolerances for the shape and
location of surfaces

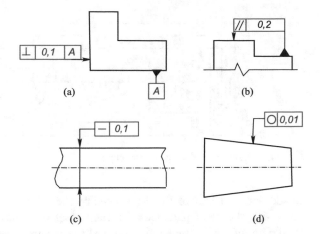

The height of the undulation W_z is the arithmetic mean of its five values
determined on the length of the measurement section.

The parameters and characteristics of the roughness are set by GOST 2789-73.
The base line (Fig. 23.5) for determining the values of the roughness parameters is
the middle line $m - m$.

The base length l is the length of the base line on which the numerical values
of the surface roughness parameters are determined. The middle line of the profile
$m - m$ is a base line that has the shape of a nominal profile and is drawn in such a
way that within l, the sum of the areas of the protrusions and depressions is equal to
each other. The protrusion and depression of the profile are the parts of the profile
bounded by the contour of the real surface and the midline. The irregularities of the
profile are formed by its protrusions and depressions.

The unevenness step S_{mi} is the length of the midline segment that intersects
the profile at three adjacent points. The pitch of the profile irregularities along
the vertices S_i is the length of the midline segment between the projections of
the two highest points of the neighboring projections on it. The average step of
the irregularities S_m and the average step of the irregularities along the vertices S,
respectively, the arithmetic mean of the step of the irregularities along the middle
line or along the vertices of the irregularities within the baseline:

$$S_m = \frac{1}{n} \sum_{i=1}^{n} S_{mi}, \quad S = \frac{1}{n} \sum_{i=1}^{n} S_i. \tag{23.2}$$

The arithmetic mean deviation of the profile R_a is defined as the arithmetic mean
of the absolute values y_i of the profile deviations from the midline within the base
length:

$$R_a \approx \frac{1}{n} \sum_{i=1}^{n} y_i. \tag{23.3}$$

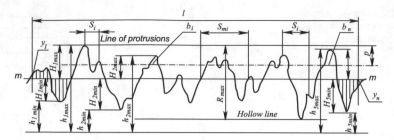

Fig. 23.5 Basic parameters of surface roughness

In addition to the value R_a, determined by the formula (23.3), according to GOST 25142-82, the main parameter for estimating roughness is the height R_z of the profile irregularities at ten points on the base length l (see Fig. 23.5):

$$R_z = \frac{1}{5} \left(\sum_{i=1}^{5} h_{i\,max} + \sum_{i=1}^{5} h_{i\,min} \right),$$ (23.4)

where $h_{i\,max}$, $h_{i\,min}$ are the distances to the specified profile points from a straight line parallel to the midline that does not intersect the profile (see Fig. 23.5).

Sometimes the relative reference length of the profile t_p (%) is used to characterize the roughness, which is the total length of the profile measured at a certain level of the profile section, related to the base length l:

$$t_p = \left(\sum_{j=1}^{n} b_j / l \right) \cdot 100\%.$$

In accordance with GOST 2.309-73, the signs are shown in Fig. 23.6. The sign of position (a) does not regulate the type of surface treatment. The sign shown at position (b) denote the surfaces formed by the removal of the matheriala layer. The sign in Fig. 23.6c indicates the surface that is not treated after casting and other types of previous processing.

The general case of roughness notation is shown in Fig. 23.6d. In place of frame 1, write the roughness parameter (s) (R_a, R_z, S_m, S, etc., for R_a—without the symbol (see Fig. 23.6d); instead of frame 2, write (if necessary) the type of processing (polish, scrape, etc.); in place of frames 3 and 4, respectively, write the base length (see Fig. 23.6g) and the symbol for the direction of the irregularities.

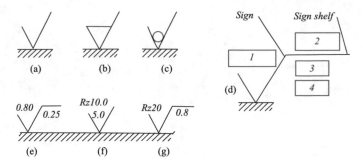

Fig. 23.6 Marking the surface roughness in the drawing

Self-Test Questions

1. What is called interchangeability?
2. What does the word "standardization" mean?
3. List the categories of standards, depending on their scope.
4. What is deviation and tolerance?
5. Define the concept of "landing".
6. Name the possible landing systems.
7. What is the reason for the preferential use of landings in the opening system?
8. Name the landing groups in the shaft system and the hole system.

References

V. Molotnikov, *Texnicheskaya mexanika [Technical mechanics]* (Lan' Publication, Sant-Peterburg, 2017)

B. Münstermann, T. Weitzel, What is process standardization? (2008). https://core.ac.uk/download/pdf/301346409.pdf

Chapter 24
Mechanical Transmission

Abstract In general case, *mechanical transmission* in a device involves the transfer and/or transformation of energy from some point in space to another point remote from the first. Devices consisting of a motor, connecting elements, and gears are called *drives*. This chapter describes the method of kinematic calculation of the drive, as well as the structure and design sequence of gears of various types: gear, chain, worm, and friction. As a rule, for each type of transmission, the calculated part is preceded by the necessary theoretical information.

Keywords Broadcast · Drive unit · Ratio · Gearing · Conical gearing · Chains · Worm gear · Reducers · Friction gears · Variators

24.1 Kinematic Calculation of the Drive

24.1.1 The Concepts of "Transmission" and "Drive": Kinematic Drive Scheme

Transmission is a mechanical device designed to transfer energy over a distance and coordinate the speeds of working machines and engines.

In technology, the following types of gears are most widespread: gear, worm, chain, belt, frictional gears "screw-nut", etc. Matching the operating speeds of engines and technological machines is not always possible to achieve a transmission device of any one type. On the contrary, it is more often necessary to use a set of sequentially connected gears. For the joint operation of these sets of gears, it becomes necessary to use devices that connect the shafts of different gears. These devices are called *couplers*.

Thus, to give movement to a technological machine, it is necessary to create a device consisting of an engine and gears connected by one or more clutches. Such devices are called *drives*. To depict all the drive units and the path of motion transmission from the engine to the process machine, *kinematic schemes* are used. Kinematic schemes are drawn according to GOST 2.770–68.

© The Author(s), under exclusive license to Springer Nature Switzerland AG 2023
V. Molotnikov, A. Molotnikova, *Theoretical and Applied Mechanics*,
https://doi.org/10.1007/978-3-031-09312-8_24

An example of a kinematic drive circuit is shown in Fig. 24.1. The caption below lists the devices connected to the drive.

24.1.2 Basic Power and Kinematic Ratios in Gears

From the course of theoretical mechanics, it is known that the linear velocity v of the points of a rotating body that are separated from the axis of rotation at a distance $d/2$ is determined by the formula

$$v = \omega d/2 = \pi dn/60, \qquad (24.1)$$

where d—in meters (m); v—in m/s; ω—rad/s; (c^{-1}); n—rpm. This value is called *circumferential velocity*.

The force that causes the rotation of bodies or resistance rotation, and directed tangentially to the trajectory of the point of its application, is called *circumferential force*. We agree to denote it with the symbol F_t. This force is related to the circumferential velocity v and the power P transmitted by the body of rotation, the dependence

$$P = F_t v, \qquad (24.2)$$

where P is in watts (W), F_t is in Newtons (H), v is in m/s, P is in kilowatts (kw), and v is still in m/s.

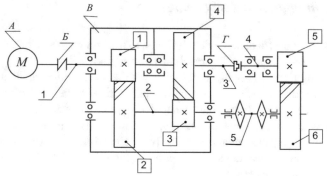

1 - driving (high-speed) shaft of the first stage of the gearbox; 2 - slow-moving (output) shaft of the first stage of the gearbox; 3 - gearbox output shaft;
4 - drive shaft of an open gear transmission; 5 - shaft of asterisks;
1 - drive gear of the first stage of the reducer;   2 - toothed wheel first stage reducer;   3 - second stage gear;   4 - second stage wheel;
5 - open gear gear; - low-speed;   6 -   open gear wheelю.

Fig. 24.1 Example of a kinematic drive circuit: A—electric motor; B—elastic coupling; C—gearbox; D—cam-disc coupling

The circumferential force is related to the torque T transmitted by the body of rotation, by the formula

$$F_t = 2T/d. \tag{24.3}$$

Let us agree to denote for the master and slave bodies of rotation (i.e., gears, pulleys, sprockets, etc.), respectively, the transmitted powers P_1 and P_2, the transmitted torques through T_1 and T_2, etc. Then, as is known from physics, the efficiency of the transmission can be determined by the formula

$$\eta = \frac{P_2}{P_1}. \tag{24.4}$$

From the formulas (24.1)–(24.3), we can get another useful dependency that will be useful in the future:

$$T = \frac{P}{\omega}. \tag{24.5}$$

When transmitting rotational motion from one body to another, the gear ratio (or gear ratio) is the ratio of the angular velocities of the master and slave bodies of rotation:

$$i = \frac{\omega_1}{\omega_2}. \tag{24.6}$$

We have already said that the drive can include various types of transmission. If the drive consists of several gears connected in series, then the efficiency of the drive is equal to the product of the efficiency of all its gears, i.e.

$$\eta = \eta_1 \eta_2 \ldots \eta_k, \tag{24.7}$$

and the gear ratio of the drive is equal to the product of the gear ratios of all the gears included in it:

$$i = i_1 i_2 \ldots i_k, \tag{24.8}$$

where k is the number of gears included in the drive.

The design of the drive usually begins with a kinematic calculation. The initial data for the calculation are:

- the nominal torque on the shaft of the machine being driven (it is also called *output torque* of the drive or simply *output torque*); we agree to denote it by T_{out};
- angular velocity ω_{out} or rotational speed n_{out} of the output shaft at—water;
- a schedule of changes in the load and speed for a certain period of operation of the drive.

Sometimes, instead of the torque T_{out}, the rated power P_{out} on the output shaft of the drive is set; in this case, the rated torque T_{out} can be determined by formula (24.5).

For conveyor drives, instead of the speed of rotation of the output shaft, set the speed of the belt or chain and instead of the torque on the drive shaft, the circumferential force F_t, which is called *by pulling force* on the belt or in the chain.

24.1.3 Selection of the Electric Motor

In modern production, most technological machines are driven by electric motors. Therefore, the next step in the design of the drive is the selection of the electric motor. It begins with determining the required power of the electric motor.

If the power P_{out} on the driven shaft of the drive is specified, the required (estimated) power P_p of the electric motor will be

$$P_p = \frac{P_{out}}{\eta}, \tag{24.9}$$

where η is the efficiency of the drive, determined by formula (24.8). To calculate it, use the reference data given in Table 24.1.

If the drive design assignment specifies the torque T_{out} (N·m) on the driven shaft and its angular velocity speed ω_{out} (rad/s), then the required motor power (W) will be

$$P_p = \frac{T_{out}\,\omega_{out}}{\eta}. \tag{24.10}$$

Table 24.1 Average efficiency of mechanical gears

Transfer	η_i	Transfer	η_i
Closed Gear		Chain	
With cylindrical		Closed	$0.95\ldots0.97$
With wheels	$0.97\ldots0.98$	Open	$0.90\ldots0.95$
With tapered wheels	$0.96\ldots0.97$	Belt	
Open Gear	$0.95\ldots0.96$	With flat belt	$0.96\ldots0.98$
Closed worm gear with number of worm visits		With wedge and Polyclinic	
$z_1 = 1$	$0.70\ldots0.75$	Belt	$0.95\ldots0.97$
$z_1 = 2$	$0.80\ldots0.85$		
$z_1 = 4$	$0.80\ldots0.95$		

1. The friction in the supports is taken into account by the multiplier $\eta_0 = 0.99\ldots0.995$
2. Losses in the elastic coupling are taken into account by the coefficient $\eta_m = 0.98\ldots0.99$

As already mentioned, when designing drives for conveyors and conveyors, they usually indicate the traction force on the belt is F (N), and its speed is v (m/s). In this case, the required engine power is determined by the formula:

$$P_p = \frac{Fv}{\eta}. \tag{24.11}$$

Based on the found value of the required engine power, calculated using a suitable formula from (24.9)–(24.11), the type and size of the engine are selected.

Most often, three-phase asynchronous electric motors are used in the drives. The types and characteristics of electric motors produced by the domestic industry are given in mechanical engineering reference books, as well as in many tutorials on machine parts (see, e.g., Chernavskij (1976)).

With the same design power P_r, you can choose an electric motor with different speeds. Therefore, it is necessary to consider several options and focus on the one that is most suitable for the operating conditions of the drive, bearing in mind that with an increase in the speed of rotation, the weight and dimensions, as well as the cost of the engine, decrease. However, the working resource is also reduced.

24.1.4 Drive Gear Ratio

After selecting the electric motor, they return to the kinematic calculation. To determine the gear ratio of the drive, you need to know the speed n of the motor at the rated operating mode (take from the same tables that are used when selecting the electric motor) and the speed n_{out} of the output shaft of the drive. If this frequency is not specified directly, it is determined from other source data. For example, if the speed v (m/s) of the conveyor belt and the diameter of the drum D (m) are given, then

$$n_{out} = \frac{60v}{\pi D}, \text{ rpm}. \tag{24.12}$$

The gear ratio of the drive is calculated by formula (24.8). To do this, you need to find the private gear ratios of the individual gears that are part of the drive from the reference books. To make your life easier, this data is given in Table 24.2.

For known partial gear ratios of the drive, the total gear ratio will be

$$i = \frac{n}{n_{out}} = i_1 i_2 \ldots i_k. \tag{24.13}$$

Remark In general, the choice of the kinematic scheme and the breakdown of the total gear ratio into partial multipliers allow for many solutions. For example, the installation of a two-stage gearbox allows you to get rid of the belt or chain transmission, and the use of a worm gearbox makes it possible to do without both.

Table 24.2 Average values
of partial gear ratios

Transfer	i
Toothed	
With cylindrical wheels	3...6
With tapered wheels	2...4
Worm	8...40 (up to 100)
Wave	50...1000
Chain	3...6
Belt	2...4

In projects designed for serial and mass production, the values of the gear ratios are taken from standard rows for each type of transmission

However, the efficiency of the worm gear is significantly lower than that of the gear, and the durability is the worm pair is much smaller than the toothed pair. Therefore, when designing the drive, it is necessary to take into account not only the dimensions but also reliability, cost-effectiveness, ease of installation, and operation. It is advisable to study the design in multiple ways in order to choose the optimal one from several options.

24.2   Gear Transmission

24.2.1   Concepts of Gearing Gears

Gears are machine elements in which the movement between the links is transmitted by means of successively engaged teeth. They are used in most machines and devices for transmitting and converting motion in a wide range of capacities (up to 150,000 tons) and speeds (up to 200 m/s).

To the *advantages of* gears the gearboxes include, along with those already mentioned, high reliability, high efficiency (0.97...0.98 for one pair of wheels), easy maintenance, and compact size.

The *disadvantages of* gears include high labor intensity of manufacturing, the possibility of noise in operation, etc.

The process of transmitting motion with the help of teeth is commonly called *gearing. The line of intersection of the lateral surface of the tooth with a given surface, such as a plane perpendicular to the wheel axis, is called the tooth profile.*

Gears are classified according to their geometric and functional features:

- according to the relative position of the axes—cylindrical (with parallel axes), conical (the axes of the wheels intersect), hyperboloid (axis mix), worm, helical, and hypoid;
- at the relative location of the surfaces—peaks and valleys of the teeth wheels, the transmission of external and internal gearing;

- at the nature of the motion axes—normal (fixed axis) and planetary (movable axes);
- in the direction of the teeth—straight-toothed and oblique-toothed;
- according to the profiles of the teeth—gears with involute engagement, with cycloidal engagement, and with Novikov engagement;
- according to the design—open and closed (in a separate closed case).

   The unit with a downshift is called a *gearbox,* and with an upshift *multiplier.* The involute gearing, invented by L. Euler, is the most widespread due to its adaptability. The line of engagement of the involute teeth is straight. Mated tooth profiles in general are rolled relative to each other with a slip. However, each wheel has one coaxial surface that touch each other and in which the relative velocity vector is zero at any point of contact. These surfaces are called *initial*, and the concentric circles perpendicular to them are called *initial circles.*

   Consider the transmission of rotation by two links 1 and 2 (Fig. 24.2). The links will rotate in opposite directions with angular velocities ω_1 and ω_2. At the point of contact $With$, the link velocities will be:

$$v_1 = \omega_1 C O_1, \quad v_2 = \omega_2 C O_2.$$

We draw the normal $N_1 N_2$ and the tangent t through the point C (Fig. 24.2) to the tooth profiles and decompose the velocities v_1 and v_2 into the normal and tangent components:

Fig. 24.2 Gearing kinematics

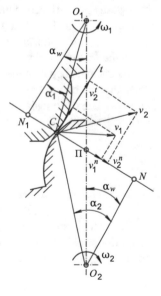

$$v_1^n = v_1 \cos\alpha_1 = \omega_1 O_1 C \cos\alpha_1 =$$
$$= \omega_1 O_1 N_1; \quad v_1^t = v_1 \sin\alpha_1;$$
$$v_2^n = v_2 \cos\alpha_2 = \omega_2 O_2 C \cos\alpha_2 =$$
$$= \omega_2 O_2 N_2; \quad v_2^t = v_2 \sin\alpha_2.$$

The condition for the contact of the links will be the equality of the normal components of the velocities of the points C belonging to links 1 and 2, i.e., $v_1^n = v_2^n$.

$$\frac{\omega_1}{\omega_2} = \frac{O_2 N_2}{O_1 N_1}.$$

Denote by N the intersection point of the center line $O_1 O_2$ and the normal $N_1 N_2$. From the similarity of triangles $O_1 N_1 N$ and $O_2 N_2 P$, it follows that $O_2 N_2 / O_1 N_1 = O_2\Pi/O_1\Pi$.

Then from the previous equality, we have

$$\frac{\omega_1}{\omega_2} = \frac{O_2\Pi}{O_1\Pi} = i_{12}.$$

This equality expresses *the basic law of engagement: the normal to the profiles at the point of contact divides the distance between the centers into segments inversely proportional to the angular velocities of the links.*

The point Π is called *the pole,* and the circles with centers O_1 and O_2 passing through the pole are called *the main ones.* They are the reamers of the involutes that form the tooth profiles. The segment of the normal $N_1 N_2$ is called the *linking line, and the angle α_w between the line of centers and the normal to the linking line is called the linking angle.*

The distance between the same (right or left) profiles of adjacent teeth, measured along the arc of the circle centered on the axis wheels, is called *circumferential pitch p_t.* The parameter m equally divided by π to the district step is called *module of gear:*

$$m = \frac{p_t}{\pi}.$$

Its values are standardized to form a series of (mm):
$$1, 1.25, 1.5, 1.75, 2, 2.25, 3, 3.5, 4, 4.5, 5, \text{ etc.}$$

Gear wheels are cut on gear milling or gear-cutting machines. In the latter case, the cutting tool is a rail that makes reciprocating motion, and the gear blank rotates slowly (Fig. 24.3a). The rail contour (Fig. 24.3b) has the following parameters:

$$h_a^* = 1; \quad c^* = 0.25; \quad \rho = 0.384m.$$

If the dividing line of the rail touches the initial circumference of the workpiece, then it is said that the teeth are cut *without offset* (*uncorrected* wheel). By shifting

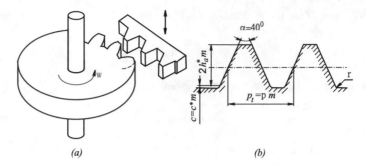

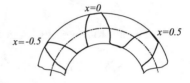

(a)   *(b)*

Fig. 24.3 Method of cutting teeth (**a**)—scheme of cutting teeth on a gear-cutting machine; (**b**)—tool rail

Fig. 24.4 Gear adjustment

the rail at a distance xm perpendicular to the wheel axis, you can get teeth of different widths (Fig. 24.4).

If z is the number of teeth of the i-th wheel, then its geometric parameters are calculated by the formulas:

– the diameter of the dividing circle

$$d_i = mz_i;$$

– the diameter of the main circle

$$d_{bi} = mz_i \cos \alpha;$$

– step along the main circle

$$p_b = \frac{\pi d_b}{z} = \pi m \cos \cos \alpha;$$

– the diameter of the circle of the depressions

$$d_{fi} = mz_i - 2(h_a^* + c^* - x_i)m;$$

– diameter of the circumference of the tops of the teeth

$$d_{a1} = 2a_w - d_{f2} - 2c^* m,$$
$$d_{a2} = 2a_w - d_{f1} - 2c^* m;$$

– the thickness of the tooth along the dividing circle

$$s = m \left(\frac{\pi}{2} + 2xtg\alpha \right);$$

– center distance in gears without displacement

$$a_w = 0,5m(z_1 + z_2),$$

in corrected transmissions

$$a_w = \frac{m(z_1 + z_2)}{2} \cdot \frac{\cos \alpha}{\cos \alpha_w}.$$

The center-to-center distances are standardized and form a series:

40; 50; 63; 80; 100; 125; 160; 180; 200; 225; 250; 280; 315; 355; 400; ... mm.

For non-standard transmissions the mismatch of the center distance to the standard value is allowed.

24.2.2 Strength Calculation of Gears

Gears rely on the *contact endurance* of the tooth material and on the *bending strength of the teeth.*

A. The calculation for *contact endurance* is the main one for *closed gears,* operating in conditions of heavy lubrication. The calculation is based on Hertz's solution of the problem of compressive stresses of contacting bodies. Using the solution for the stresses in the contact zone, the formula is obtained

$$\sigma_H = Z_M Z_H Z_Z \sqrt{\frac{2M_2 K_H (u + 1)}{d_2^2 b}}, \tag{24.14}$$

where

Z_M is a coefficient depending on the material of the gear wheel; for steel wheels, you can take $Z_M = 275 \text{ N}^{1/2}\text{mm}^{-1}$;

Z_H—a coefficient that takes into account the angle of engagement and the inclination of the teeth. For straight-toothed wheels at $\alpha = 20°$, $\beta = 0$ $Z_H = 1.76$;

Z_Z—takes into account the length of the contact line: for straight gears $Z_Z = 0.9$, for oblique gears $Z_Z = 0.8$;

M_2—torque on the wheel shaft (driven link), N·mm;

K_H—a coefficient that takes into account the unevenness of the load distribution between the teeth; it is represented as the product of three coefficients, the values of which are taken from the tables;

$u = z_2/z_1$—gear ratio;

d_2—wheel dividing circle diameter, mm;

b—tooth length, mm.

Replacing in formula (24.14), the diameter by the center distance is obtained for the *verification* calculation of the formula:

for straight gears

$$\sigma_H = \frac{310}{a_w} \sqrt{\frac{K_H M_2 (u+1)^3}{bu^2}} \leq [\sigma]_H, \qquad (24.15)$$

for skew gears

$$\sigma_H = \frac{270}{a_w} \sqrt{\frac{K_H M_2 (u+1)^3}{bu^2}} \leq [\sigma]_H, \qquad (24.16)$$

where $[\sigma]_H$ is the permissible voltage when calculated on contact endurance.

For the design calculation, it is convenient to rewrite these formulas in the following form:

for straight gears

$$a_w = (u+1) \sqrt[3]{\left(\frac{310}{u\sigma_H}\right)^2 \cdot \frac{M_2 K_H}{\psi_{ba}}}; \qquad (24.17)$$

for skew gears

$$a_w = (u+1) \sqrt[3]{\left(\frac{270}{u\sigma_H}\right)^2 \cdot \frac{M_2 K_H}{\psi_{ba}}}, \qquad (24.18)$$

and ψ_{ba} is the coefficient of the tooth width, taken according to the table recommendations.

B. The calculation of the teeth for *bending strength* is the main one for *open gears*. This calculation is also performed for closed gears with a high surface hardness of the teeth. The tooth is considered as a cantilever beam (Fig. 24.5). The force is considered to be applied to the top of the tooth in a normal way to its surface. Decomposing it into the transverse (bending) force P_t' and the longitudinal (compressive) P_r', we get the stress diagrams shown in Fig. 24.5.

According to GOST 21354-75 formula for the test calculation of straight teeth, the flexural endurance test has the form

Fig. 24.5 Tooth bend

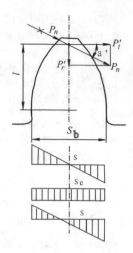

Table 24.3 Coefficient values Y_F

Z	17	20	25	30	40	50	60	80	100 and more
Y_F	4.28	4.09	3.09	3.80	3.70	3.66	3.62	3.61	3.6

$$\sigma_F = \frac{K_F P Y_F}{b_w m} \leq [\sigma]_F;$$

for the case of a design calculation, this formula is converted to the form

$$m = \sqrt[3]{\frac{2 K_F M_1 Y_F}{z_1 \psi_{bm} [\sigma]_F}},$$

where $\psi_{bm} = b_w/m$, $b_w = \psi_{bm} m$, M_1 is the torque on the gear shaft:

$$M_1 = 0,5 P m z_1;$$

K_F the load factor, which is represented by the product two tabular coefficients; Y_F the coefficient of the tooth shape, depending on the number of teeth. For wheels cut without offset, the values of this coefficient can be taken from Table 24.3:

For bevel gears and chevron gears, the calculation formulas are as follows:
– during the verification calculation:

$$\sigma_F = \frac{P K_F Y_F Y_\beta K_{F\alpha}}{b_a m_n} \leq [\sigma]_F, \tag{24.19}$$

where the coefficient K_F should be chosen by the equivalent number of teeth $z_v = z/\cos^3 \beta$ and the coefficient Y_β is calculated by the formula

$$Y_\beta = 1 - \frac{\beta}{140^0}.$$

In the design calculation of helical cylindrical wheels for bending endurance, the *normal engagement modulus* is calculated, determined by the dependence

$$m_n = \sqrt[3]{\frac{2M K_F K_{F\alpha} Y_F Y_\beta \cos \beta}{\psi_{bm} z [\sigma]_F}}.$$

Here, the value of M/z can be taken for any of the gearing wheels; the coefficient $\psi_{bm} = b/m_n$. The gear wheel for which the ratio $[\sigma]_F/Y_F$ is less should be calculated.

24.2.3   *Features of the Calculation of Bevel Gears*

We have already said (p. 514) that tapered wheels are used in gears with intersecting axes (Fig. 24.6). The angle Σ between the wheel axes can be in the range from 10 to 170 degrees, but most often $\Sigma = 90^\circ$. Table 24.4 shows the main geometric relations in the conic indexTransfer!toothed!tapered gears.

Fig. 24.6 Taper engagement

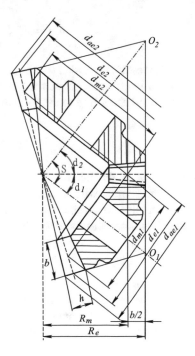

Table 24.4 Main parameters of the bevel gear

Parameters	Designation	Formula
Outer dividing diameter	d_{e2}	
Outer cone distance	R_e	$R_e = 0,5d_e/\sin\delta$
Gear Ring width	b	$b \le 0,3R_e$
Mean cone distance	R	$R = R_E - 0.5b$
Average Circumferential module	m	$m = m_e R/R_e$
Average dividing diameter	d	$d = mz$
Angle of the dividing cone	δ_2	$\delta_2 = \text{arctg}\, u$
		$\delta_1 = 90^0 - \delta_2$
Outer diameter of vertices teeth	d_{ae}	$d_{ae} = d_e + 2m_e \cos\delta$

The verification calculation for contact endurance is carried out according to the formula

$$\sigma_H = \frac{335}{R}\sqrt{\frac{K_H M_2 \sqrt{(u^2+1)^3}}{bu^2}} \le [\sigma]_H. \tag{24.20}$$

In the design calculation, the external dividing diameter of the wheel is determined by the formula

$$d_{e2} = 2\sqrt[3]{\left(\frac{335}{[\sigma]_H}\right)^2 \frac{K_H M_2 u}{(1-0,5\psi_b R_e)}}. \tag{24.21}$$

The result is rounded to the standard according to GOST 12289-76.

Checking the teeth of the bevel wheels for endurance by bending stresses is performed according to the formula

$$\sigma_F = \frac{P K_F Y_F}{bm} \le [\sigma]_F. \tag{24.22}$$

Here K_F is the bending load factor, which is chosen in the same way as for cylindrical spur wheels, and P is the circumferential force, which is considered to be applied tangentially to the middle dividing circle; based on the formula (24.3), the circumferential force will be

$$P = \frac{2M_2}{d_2}.$$

For wheels with a high surface hardness of the teeth, it may be that their dimensions will be determined by the bending strength. In this case, the design calculation is based on the bending stresses and for the average module is obtained

$$m = \sqrt[3]{\frac{2 K_F M Y_F}{[\sigma]_F \psi_{bm} z}}. \tag{24.23}$$

The width coefficient of the crown is taken by formula

$$\psi_{bm} = \frac{b}{m} = \frac{z_1}{6 \sin \delta_1}.$$

The calculation is carried out, as for cylindrical gears, on the wheel for which the ratio $[\sigma]_F / Y_F$ is less.

24.2.4   Design Calculation Sequence

The design calculation of the cylindrical gear train is recommended to be performed in the following sequence.

1. Determine the values included in the right part of one of the formulas (24.17), (24.18). For the gear ratio u, select a value from the standard series:

1st row: 1   1.25   1.6   2.0   2.5   3.15   4.0   5.0   6.3   8,0;
2nd row: 1.12   1.4   1.8   2.24   2.8   3.55   4.5   5.6   7.1   9.0.

The first row is preferable to the second.

When designing straight-toothed wheels, the crown width coefficient is $\psi_{ba} = 0.25$, and for oblique-toothed wheels $\psi_{ba} = (0.25 \ldots 0.63)$, checking that the condition is met

$$\psi_{ba} \geq \frac{2, 5 m_n}{a_w \sin \beta},$$

where β is the angle of inclination of the teeth.

2. The found center distance is rounded to the nearest value from the standard row (in mm):

1st row: 40 50 63 80   100   125   160   200   250   315   400   500   630   800   1000;
2nd row: 71   90 112 140 180   224   280   355   450   560   710   900.

3. Select the engagement module from the standard row:

1st row: 1   1.5   2   2.5   3   4   5   6   8   10   12   16   20
2nd row: 1.25   1.375   1.75   2.25   2.75   3.5 4.5 5.5 7 9   11   14   18.

For skew wheels, the standard is the normal module m_n, and for chevron wheels, the standard can be either the normal module m_n or the circumferential module m_t.

4. Determine the total number of teeth by the formula:
– for wheels with a standard circumferential module

$$z_\Sigma = z_1 + z_2 = \frac{2 a_w}{m_t};$$

– for bevel and chevron wheels with standard normal module

$$z_{\Sigma} = \frac{2a_w \cos \beta}{m_n}.$$

The angle β of the tilt of the teeth is assumed for skew wheels $8 \ldots 15°$, for chevron—$25 \ldots 40°$.

5. Determine the number of gear and wheel teeth:

$$z_1 = \frac{z_{\Sigma}}{u+1}, \qquad z_2 = z_{\Sigma} - z_1$$

and specify the gear ratio $u = z_2/z_1$. Deviation from the previously accepted one is allowed up to 4.5%. Then check the center distance:
for wheels with standard D. module

$$a_w = \frac{z_1 + z_2}{2} m_t;$$

to wheel with standard normal module:

$$a_w = \frac{(z_1 + z_2)m_n}{2 \cos \beta}.$$

If the obtained result does not correspond to the previously accepted standard value of the center distance, then the discrepancy is eliminated by changing the angle β:

$$\cos \beta = 0.5(z_1 + z_2)\frac{m_n}{a_w}.$$

The calculation should be carried out with an accuracy of five significant digits. Then you need to check the calculations by determining:

$$d_1 = z_1 \frac{m_n}{\cos \beta}, \quad d_2 = z_2 \frac{m_n}{\cos \beta}, \quad a_w = \frac{d_1 + d_2}{2}.$$

Figure 24.7 shows the appearance of a low-power conical gearbox (Gurin et al. 2019).

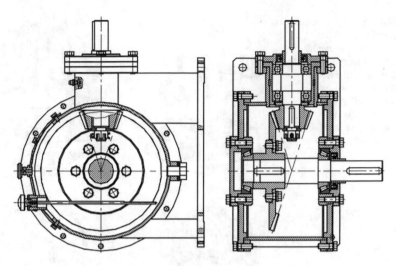

Fig. 24.7 The gearbox is conical with horizontal shafts

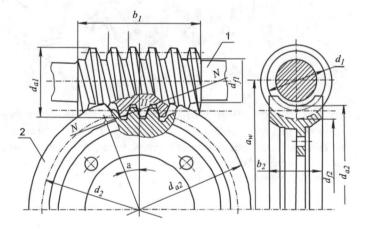

Fig. 24.8 Worm engagement

24.3 Worm Gears

24.3.1 Geometric Parameters of the Worm Engagement

Worm gear scheme is shown in Fig. 24.8, and Fig. 24.9 shows the appearance of the most common types of worm gearboxes. Such gears are used to lower the angular velocity of rotation with a constant gear ratio.

The leading link 1 is a worm, which represents as a cylindrical or globoid screw. The driven link 2 is a worm wheel, which is cut according to the rolling method using a worm cutter.

Fig. 24.9 Worm gearboxes:
(**a**) general view of the
gearbox with a detachable
housing; (**b**) general view of
the gearbox with a ribbed
detachable housing and
artificial blowing; (**c**) the
same, with the lid removed;
(**d**) general view of the
gearbox with an all-in-one
housing

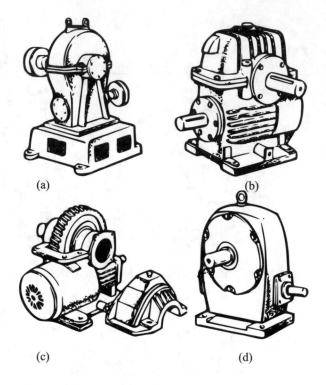

(a)   (b)

(c)   (d)

The advantages of worm gears are the ability to obtain large gear ratios
(10 . . . 100), smooth and quiet operation, self-braking. The disadvantages include
relatively low efficiency and heat generation, which requires finning of the housing
or special devices to remove heat.

We present the basic geometric relations in worm gears.

Worm Worm step:

$$p_1 = \pi m,$$

where m is the engagement module;
dividing diameter of the worm:

$$d_1 = qm,$$

and q is the coefficient of the worm diameter. With the growth of q, the rigidity of the
worm increases, but the angle of lifting of its thread decreases and falls Efficiency;
dividing angle of the turn lift: $\tan \gamma = \dfrac{z_1}{q}$, where z_1 is the number of worm passes;
diameter of the vertexes of the worm turns:

$$d_{a1} = d_1 + 2m = m(q + 1);$$

diameter of the depressions of the worm coils:

$$d_{f1} = d_1 - 2,4m = m(q - 2,4);$$

length of the sliced part of the worm:

$$\text{at } z_1 = 1 \text{ or } 2 \;\; b_1 \geqslant (11 + 0.06z_2)m,$$
$$\text{when } z_1 = 3 \text{ or } 4 \;\; b_1 \geqslant (12.5 + 0.09z_2)m.$$

Worm Wheel The dividing diameter of the worm wheel is determined by the engagement modulus and the number of teeth: $d_2 = z_2 m$;
diameter of the tooth tips:

$$d_{a2} = d_2 + 2m = m(z_2 + 2);$$

the diameter of the depressions with a radial clearance of $0.2m$:

$$d_{f2} = d_2 - 2.4m = m(z_2 - 2.4);$$

wheel rim width:

$$\text{at } z_1 = 1, \; 2 \text{ or } 3 \;\; b_2 \leqslant 0.75 d_{a1},$$
$$\text{when } z_1 = 3 \text{ or } 4 \;\; b_2 \leqslant 0.67 d_{a1}.$$

24.3.2 Calculation of Worm Gears

The calculation of the worm engagement on the contact endurance is carried out as a design, determining the required center distance by the formula

$$a_w = \left(\frac{z_2}{q} + 1\right) \sqrt[3]{\left(\frac{0.463}{[\sigma]_H z_2/q}\right)^2 M_{P_2} E_{\text{ol}}},$$

where M_{P_2} is the torque on the shaft of the worm wheel; E_{PR}, given the modulus of elasticity; and the index 1 refers to the worm and index 2 to the wheel.

The calculation of the teeth of the worm wheel by bending stresses is performed according to the formula

$$\sigma_F = \frac{1.2 M_{P_2} Y_F \xi}{m d_2 b_2},$$

where Y_F is the coefficient of tooth shape, which is taken from the tables depending on the number of teeth of the wheel, and ξ is the coefficient of tooth wear, taken according to the recommendations from the reference literature.

Thermal Calculation of the Worm Gear For worm gearboxes, the thermal calculation is mandatory. It boils down to checking the condition

$$\Delta t = t_{\mathrm{m}} - t_{\mathrm{in}} = \frac{N(1 - \eta)}{k_t F} \leqslant [\Delta t]$$

where t_{m} and t_{in} are, accordingly, the temperature of the oil in the gear case and ambient air; N, W, transmitted power; η gearbox efficiency; $k_t = 17$ W/m$^2 \cdot$C$^\circ$ heat transfer coefficient; $[\Delta t] = 60^\circ$ permissible temperature difference; and F, m^2, heat sink surface area (bottom area is not taken into account).

In the event that the specified condition is not met, provide for the device of fins, forced blowing, or another method of cooling.

24.4   Chain Transfers

24.4.1   Design, Scope of Application

Chain drives are used in drives to transmit rotary motion over long distances. The simplest gear consists of two sprockets 1 and 2 (Fig. 24.10a), connected by a chain 3. Often the design of the chain drive also includes a tensioner 4 (Fig. 24.10b).

The most widely used drive roller chains are marked with PR symbols. In Fig. 24.11, the drawings of such circuits are given in accordance with GOST 13568-97 (ISO 606-94).

Here it is indicated: 1-inner link; 2-outer link; 3-connecting link; 4-transition link; 5-double transition link. The parameters specified in the drawings (chain pitch t, etc.) are standardized.

The industry produces not only single-row but also multi-row chains (2PR, 3PR) (Fig. 24.10d). Multi-row chains are used at high loads and speeds. Chain drives are widely used in agricultural machines, robots, machine tools, etc. Most often, such transmissions are designed for low power (up to 1 kW), and the center distance a (Fig. 24.10) can reach 6 . . . 8 m.

The advantages of chain drives include a relatively high efficiency (0.96 . . . 0.98) and can be used in both high-speed (speed up to 25 m/s) and low-speed (speed up to 2 m/s) drive stages. Chain drives can be either downshift or upshift.

The disadvantages of chain gears include noise in operation, uneven running, pulling chains, and the need for tensioners.

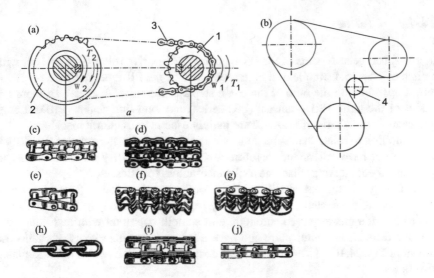

Fig. 24.10 Chain drive design and types of drive chains: (**a**) is the simplest three-link transmission; (**b**) is a chain transmission with a tension link; (**c**) is a single-row roller drive chain; (**d**) is the same, double-row; (**e**) is the same, with curved plates; (**f, g**) is a gear drive with guide plates; (**h**) is a round-link cargo chain; (**i**) is the same, plate-shaped; (**j**) same, traction bushing

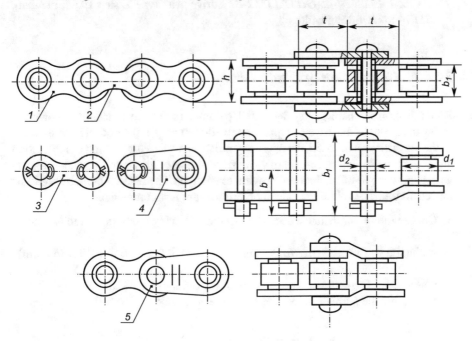

Fig. 24.11 Design of the PR chain and variants of its closing links

24.4.2 Chains

In low-speed gears, chains of the PR type are used—drive roller (Fig. 24.10c) with a pitch of $t > 25.4$ mm and in high speed also of the PR type, but with $t < 25.4$ mm. Depending on the ratio of the chain step t to the roller diameter D, there are chains of the light (PRL), normal (PR) series, and long-link chains (PRD). Long-link chains with a ratio of $t/D > 2$ are mainly used in agricultural machines.

In addition to roller chains, the industry also produces bushing chains (PV, 2PV). They do not have rollers and therefore are less durable. They are used mainly in mechanical engineering (machine tool construction, robotics, etc.).

Relatively limited use is still made of toothed chains, dialed from plates of a special shape, connected to each other by segmental prisms (Fig. 24.10e,f). Such chains provide greater speed, strength, and smooth operation with less noise, but they are difficult to manufacture and have a large mass. Other types of chains are shown in Fig. 24.10h–k. In the technical documentation, the chains are designated as follows.

Chain PR-25.4-6000 GOST 13568-75 (drive roller, load up to 6000 kgf);
Chain PV-9.525-1300 GOST 13568-75 (drive bushing, breaking load 1300 kgf);
Chain PRD 38.1-3180 GOST 13568-75 (long-link roller drive);
Chain PZ 1-19.05-74-45 GOST 13552-81 (drive gear, type 1, step 19.05, breaking load 74 kN, working width 45).

24.4.3 Calculation of Chain Gears

When designing chain gears, the initial parameters are the transmitted power N (kW), the speed of the drive n_1 (rpm) and the driven n_2 (rpm) sprockets, the location of the transmission in space, the operating mode, as well as the method of lubrication and chain tension. It is necessary to avoid steep slopes of the centerline and, for gears with a large angle of inclination, provide tensioners. The leading branch should be placed on top. The following calculation procedure is recommended.

A. Choose the center distance according to the following recommendations:

$$\text{optimal } a = (30\ldots 50)t; \quad a_{ma} = 80t; \quad a_{min} = 0.6(D_{e1} + D_{e2} + 30\ldots 50\,\text{mm}),$$

where D_{e1}, D_{e2}—the outer diameters of the circles of the stars.

B. Determine the number of teeth of the small sprocket:

$$\text{optimal } z_1 = 29 - 2u \leqslant \frac{3a}{t(u-1)}, \quad u = \frac{z_2}{z_1}.$$

At low speeds ($v < 1$ m/s), $z_1 = 11 \ldots 13$ is recommended.

C. Determine the number of teeth of the driven sprocket $z_2 = u z_1$. The number of teeth of the driven sprocket must be $z_2 < 120$, because otherwise the chain will jump off.

D. Find the length of the chain:

$$L = 2a + \frac{1}{2} z_c t + \frac{\Delta^2 t^2}{a}, \quad z_c = \frac{z_1 + z_2}{2}; \quad \Delta = \frac{z_2 - z_1}{2\pi}.$$

E. Count the number of chain links:

$$L_t = \frac{L}{t} = \frac{2a}{t} + \frac{1}{2} z_c + \frac{\Delta^2 t}{a}.$$

The resulting result is rounded to an integer (preferably even).

F. Recalculate the center distance expressed in chain steps:

$$a_t = 0.25 \left[L_t + 0.5 z_c + \sqrt{(L_t - 0.5 z_c)^2 - 8\Delta^2} \right] \quad \text{(do not round to whole!)}.$$

Then $a = a_t t$, mm. The idle branch of the chain should freely sag at $\approx 0.01a$, so it is necessary to provide for the possibility of reducing the center distance by $\approx 0.005a$ when installing the transmission.

Further, the calculated speed of the chain:

$$v = \frac{z_1 n_1 t}{60 \times 10^3} \quad (t - \text{mm} \ n - \text{rpm; get } v \ \text{m/s}).$$

In normal transmissions, there should be $v \leqslant 10$ m/s. From this condition, the recommended speed of the drive sprocket will be $[n_1] \leqslant 15 \cdot 10^3 / t$ (t – in mm). The number of chain strokes when running over the teeth asterisks and escapes from them:

$$w = \frac{4 z_1 n_1}{60 L_t}.$$

The condition must be met:

$$w \leqslant [w] = \frac{508}{t} \, c^1.$$

G. The durability of the chain depends mainly on the average pressure p in the joints. The strength condition is written as

$$p \leqslant [p],$$

where $[p]$ depends on the speed of the chain; the specified dependence is given in the following table form:

v, m/s	0.1	0.4	1.0	2.0	4.0	6.0	8.0	10
$[p]$, MPa	32	28	25	21	17	14	12	10

For the design calculation, the strength condition is converted to the form

$$t \geqslant 2.8 \sqrt[3]{\frac{T_1 k_e}{z_1 [p]}}, \tag{24.24}$$

where T_1 is the torque on the drive shaft,

$$k_e = k_1 k_2 \dots k_6.$$

The values of the coefficients $k_1, \dots, k_6$ are taken from the reference literature (see, e.g., (Chernavskij 1976, p. 385)).

Calculated by formula (24.24), the chains are rounded to the nearest standard value. After that, the center distance is recalculated.

H. Determine the safety margin of the chain by the formula

$$s = \frac{F_b}{F_t k_1 + F_c + F_f}, \tag{24.25}$$

where F_b is the load that breaks the chain (taken from the tables, for example, (Chernavskij 1976, Table 10.1)), in N;

F_t—circumferential force in N;
k_1—table coefficient (Chernavskij 1976, p. 385);
F_C—load from the centrifugal forces, determined by the formula:

$$F_C = m v^2,$$

and m the mass of 1 m of chain, kg, v chain speed, m/s;
F_t—force from chain sag, N:

$$F_t = 9.81 k_f m a,$$

where the coefficient k_f takes into account the slope of the centerline: for a vertical centerline, $k_f = 1$; for a horizontal centerline, $k_f = 6$.

After calculating the safety margin using the formula (24.25), the condition is checked

$$s \geqslant [s],$$

and the allowed values of the coefficient $[s]$ are taken from tables, for example, Table 12.11 (Chernavskij 1976).

24.5 Friction Transmissions

24.5.1 *Friction Gears and Variators*

Friction gears are friction gears with rigid links. They are used to convert rotational motion into rotational or translational motion at low power (up to 20 kW). The simplest is the friction transmission wheels with cylindrical rollers, the scheme of which is shown in Fig. 24.12. The transmission consists of a driving roller 1, the movable axis 3 of which presses it against the driven roller 2 with a fixed rack 4.

Thus, at the point of contact of the rollers, the normal pressure F_{21}^N is created, which is necessary for the appearance of the circumferential forces $F_{21}^T = F_{12}^T$.

Figure 24.13 shows a scheme of a friction transmission that converts the rotational motion of the leading link 1 into the translational motion of the link 3 at a speed of v.

Fig. 24.12 Friction gear "rotation–rotation"

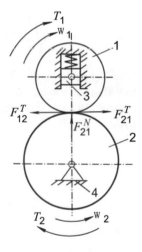

Fig. 24.13 Transmission "rotation–translational motion"

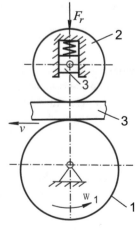

The condition for the normal operation of the friction transmission is

$$F_{21}^T > f F_{21}^N, \tag{24.26}$$

where f is the coefficient of friction in the absence of skidding.

The advantages of friction gears are simplicity of design, quiet operation, insensitivity to overloads and shocks, and the possibility of reversing. The disadvantages of friction gears include unstable gear ratio, rapid wear of the roller surfaces, heavy loads on the axles, and low efficiency.

24.5.2 Strength Calculation

In the case of cylindrical contact rollers, the calculated contact voltage is determined by the Hertz formula (see Sect. 17.6):

$$\sigma_H = \sqrt{\frac{E_1 E_2}{E_1(1 - \mu_2^2) + E_2(1 - \mu_1^2)} \cdot \frac{q}{\pi \rho_p}} \leqslant [\sigma]_H, \tag{24.27}$$

where $[\sigma]_H$ is permissible contact napojenie; E_1, E_2 and μ_1, μ_2, respectively, are the Young's moduli and Poisson's ratios of the materials rinks 1 and 2; q is the normal force per unit length of the line contact; and ρ_p is the radius of curvature of cylinders:

$$\frac{1}{\rho_p} = \frac{1}{R_1} \pm \frac{1}{R_2},$$

where R_1, R_2 are the radii of the rollers; sign "plus" or "minus" is taken depending on external or internal touch rollers.

In the design calculation, formula (24.27) is converted to the form

$$a = (u \pm 1) \sqrt[3]{\frac{T_2 K E_p}{\phi f} \cdot \left(\frac{0.418}{u[\sigma]_H}\right)^2}, \tag{24.28}$$

where $u = n_1/n_2$, n_1, n_2—frequency of rotation of the rollers; $a = R_1 + R_2$, $\phi = b/a$ the ratio of the width of the rink; T_2 torque on the axis of the driven roller; and E_p given modulus of elasticity:

$$E_p = \frac{2E_1 E_2}{E_1 + E_2}.$$

The permissible contact voltage $[\sigma]_H$ is assumed depending on the Brinell hardness of the roller surface: for a metal pair when working in oil, $[\sigma]_H = (2.5 \ldots 3.0)HB$ can be assumed, in the absence of lubrication, $[\sigma]_H = (1.2 \ldots 1.5)HB$.

24.5.3 Variators

In the absence of slippage, mechanisms of the type shown in Figs. 24.12 and 24.13 have constant gear ratios. However, with the help of rollers, it is not difficult to form mechanisms with a variable gear ratio, called *variators*.

Figure 24.14a shows a frontal variator in which the driving roller 1 can move along its shaft in the axial direction shown by the bidirectional arrow. In this case, the gear ratio continuously changes as the radius R changes. When moving the roller 1 to the left of the center of the roller 2, the rotation of the latter is reversed.

In the cone variator (Fig. 24.14b), the axial movement of the roller 1 at a constant r_1 causes a change in the contact radius R.

In engineering practice, other types of variators are also used (torus, V-belt, etc.). V-belt variators are widely used (Fig. 24.15), in which a stepless change in the gear ratio is achieved by sliding pulleys (Fig. 24.15).

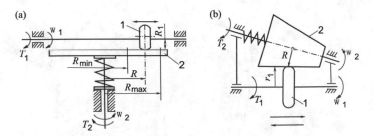

Fig. 24.14 Front (**a**) and cone (**b**) variators

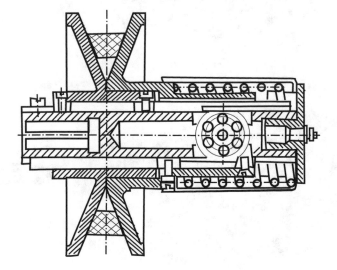

Fig. 24.15 V-belt variator

This type of variators is used in conveyor drives, agricultural machines, automobiles, and other devices. Figure 24.16 shows the appearance of industrial V-belt variators.

Note that the industrial use of V-belt variators was established in the USSR in the 1950s. Thus, the Pskov machine-building production and technical association developed by ENIMSA produced variators with a control range of four for transmitting power up to 10 kW. The material of the friction wheels must have

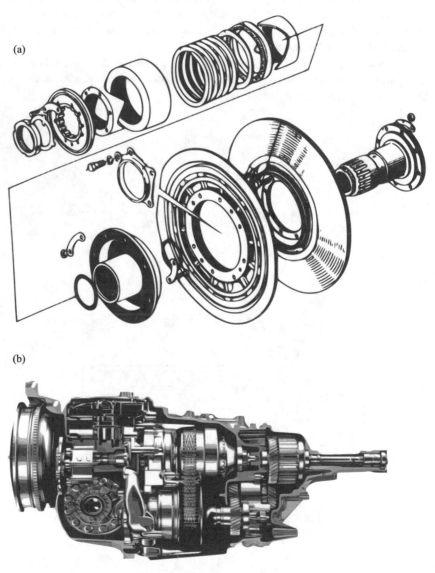

Fig. 24.16 Industrial Power Belt Variators: (**a**) harvester "Don 1500"; (**b**) car "Infiniti"

high wear resistance and surface strength, high coefficient of friction, and elastic modulus. These requirements are most fully met by the following combinations of friction wheel materials: hardened steel over hardened steel; steel over plastic; steel or cast iron over leather; pressed asbestos; or rubberized fabric.

24.5.4 Belt Drives

Belt drive scheme is shown in Fig. 24.17a. The transmission in the simplest case consists of the master 1 and slave 2 pulleys, as well as the tensioner 3 and the belt 4. The belt touches the surfaces of the pulleys on the arcs α_1 and α_2, which determine the girth angles of the master and slave pulleys, respectively. The distance a between the centerlines of the pulleys is called the centerline distance. Figure 24.17b–d shows the most commonly used belt profiles: rectangular (b), wedge (b), V-shaped, (d) circular (d), and toothed (e).

In flat-belt gears, rubberized belts of three types are used: (A) threaded with rubber layers (at speeds up to 30 m/s); (B) layer-wrapped (at speeds up to 20 m/s); and (C) spiral-wrapped without layers (speeds up to 15 m/s). In addition to rubberized, flat belts are cotton, solid-woven, and leather. They are used relatively rarely.

Our industry produces three types of V-belts: normal cross-section ($v \leqslant 30$ m/s), they are designated O, A, B, C, D, E; narrow (for increased speeds up to 40 m/s), their designations are UO, UA, UB, UV; wide (for continuously variable gears). Multi-ribbed belts are more flexible than V-belts, so they can be used with pulleys of small diameter. Sections of such belts are produced in three types: K, L, M.

Toothed belts are supplied, for example, by Habasit (Austria). They are designed for conveyors and linear transport systems that require high positioning requirements. The scope of application of these products is not limited to general-purpose conveyors. The flat gear belt is used in drives of gas distribution mechanisms of internal combustion engines, office equipment structures, door opening systems, and the food industry, for the transportation of fragile products and for special-purpose conveyors.

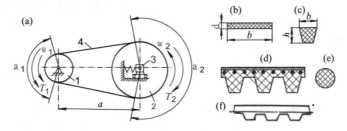

Fig. 24.17 Belt drives: (**a**) scheme; (**b–d**) types of drive belts

Belt drives are mainly used as downshifts. Their advantages are the ability to transmit movement over long distances (up to 8...10 m), simple design, low cost, high speed and noiselessness, and low sensitivity to overloads and shocks. Disadvantages of belt drives are low belt durability, large radial dimensions, significant loads on shafts and supports, and variable gear ratio.

Belt drives are designed for the traction capacity and durability of the belt. Let us consider the method of these calculations. In the calculations, empirical coefficients obtained from the experience of designing and operating gears are widely used.

24.5.5 Calculation of Flat-Time Transmission

The initial data for the calculation are the nominal transmitted power N, W (or kW); the speed of the drive pulley n_1, rpm; the gear ratio $i = n_1/n_2$; and the installation and operation conditions of the transmission. The following calculation sequence is recommended for Chernavskij (1976).

1. According to the Saverin formula, determine the diameter of the small pulley:

$$d_1 \approx 60\sqrt[3]{T_1}, \tag{24.29}$$

where T_1 (N · mm) is the torque on the shaft of the small pulley; d_1 is obtained by this formula in mm. The resulting value of d_1 is rounded to the nearest larger of the standard series (GOST 1783-73):

40, 45, 50, 56, 63, 71, 80, 90,100, 112, 125, 140, 160, 180, 200, 224, 250, 280, 315, 355, 400, 450, 500, 560, 630, 710, 800, 900, 1000, 1120, 1250, 1400, 1600, 1800, 2000.

2. Determine the diameter of the larger pulley according to the formula $d_2 = id_1$; the value d_2 are rounded according to GOST down, and clarify the ratio and frequency of rotation of the output shaft:

$$i = \frac{d_2}{d_1(1-s)}, \quad n_2 = n_1 \frac{d_1(1-s)}{d_2}, \tag{24.30}$$

where s is the relative slip ($s \approx 0.01$). The values obtained by formulas (24.30) should not differ from the original data by more than 3%.

3. Determine the belt speed:

$$v = \frac{\pi d_1 n_1}{60}. \tag{24.31}$$

If it turns out that $v \leqslant 10$ m/s, it is recommended to increase the pulley diameters.

4. Determine the center distance, the angle of girth of the small pulley, and the length of the belt according to the formulas:

$$a \approx 2(d_1 + d_2); \quad \alpha_1 = 180° - 60° \cdot \frac{d_2 - d_1}{a};$$

$$L = 2a + \frac{\pi}{2}(d_1 + d_2) + \frac{(d_2 - d_1)^2}{4a}.$$

(24.32)

5. Determine the circumferential force and the permissible useful stress $[k]$ per unit cross-sectional area of the belt according to the formulas:

$$F_t = \frac{N}{v}; \quad [k] = k_0 c_\theta C_\alpha C_v C_p,$$

(24.33)

where k_0 is a table value (see, e.g., (Chernavskij 1976, Table 9.2));
   C_θ—coefficient that takes into account the angle (θ) slope of the centerline: at $\theta \leqslant 60°$ $C_\theta = 1$; at $60° \leqslant \theta \leqslant 80°$ $C_\theta = 0.9$; at $\theta > 80°$ $C_\theta = 0,8$;
   C_α—the coefficient of the girth angle, calculated by the formula

$$C_\alpha = 1 - 0,003(180° - \alpha_1);$$

(24.34)

C_v—belt speed coefficient:

$$C_v = 1.04 - 0.0004v^2 \; (v - \text{in m/s});$$

(24.35)

   C_p—the ratio modes of operation: with a calm work with short-term inrush overload no above 120% $C_p = 1$; at moderate load variations (up to 150%) $C_p = 0,9$; in large fluctuations load (200%) $C_p = 0,8$; under shock loads and overloads of up to 300% $C_p = 0,7$.
6. Determine the required cross-sectional area of the belt and the number of gaskets in the belt:

$$b\delta = \frac{F_t}{[k]}.$$

(24.36)

At the same time, keep in mind that the belt thickness δ must satisfy the condition

$$\frac{\delta}{d_1} \leqslant \frac{1}{40},$$

(24.37)

and the width of the belt b should be assigned from the standard row:
20; 25; (30); 40; 50; (60); 63; (70); 71; (75); 80; (85); 90; 100; 112; (115); 125; 160; 180; 200; 224; 250; 280; (300); 355; 400.
7. Check the durability of the flat belt, computing resource t according to the formula:

$$t = \frac{\sigma_y^6}{\sigma_{max}^6} \cdot \frac{10^7 \cdot C_i}{3600 \cdot 2u} \text{ hours},$$

(24.38)

where σ_y is the fatigue limit; for rubberized belts $\sigma_y = 7$ N/mm^2;

$u = v/L$—number of belt runs per second;

C_i—coefficient depending on the gear ratio: in the range $i = 1 \ldots 4$ $C_i = 1 \ldots 2$;

$$\sigma_{max} = \sigma_1 + \sigma_{and} + \sigma_v;$$

$$\sigma_1 = \sigma_0 + \frac{F_t}{2b\delta}; \quad \sigma_{and} = E\frac{\delta}{d_1}; \quad \sigma_v = \rho v^2 \cdot 10^{-6} \text{ N / mm}^2, \tag{24.39}$$

and:

σ_0—the tension in the belt from the pre-tension (take from 1.6 to 2 N/mm^2, on average 1.8 N/mm^2);

σ_b—belt bending stress (when calculating it for rubberized belts, take the value of the Young's modulus $E = 200$ N/mm^2);

ρ—belt material density (for rubberized belts, take $\rho = 1100$ kg/m^3); v—belt speed, m/s.

For belts in light mode of operation, the resource calculated by formula (24.38) should not be less than 5000 h, for medium mode 2000 h, for heavy and very heavy modes at least 1000 and 500, respectively.

8. The calculation is completed by determining the forces in the transmission elements:

– pre-tensioning of each belt branch:

$$S_0 = \sigma_0 b\delta;$$

– tension of the master and slave branches of the belt:

$$S_{1,2} = S_0 \pm \frac{F_t}{2};$$

– shaft pressure:

$$Q = 2S_0 \sin\frac{\alpha_1}{2};$$

– maximum initial belt tension:

$$Q_{max} = 1.5Q.$$

24.5.6 Calculation of V-belt Gears

The initial data for the design calculation is the same as for the plane-time transmission. The calculation is recommended to be carried out in this order (Shigley and Mischke 1996).

A. Assign the center distance from the range

$$a_{min} = 0.55(d_1+d_2) + T_0,$$
$$a_{max} = 2(d_1 + d_2), \tag{24.40}$$

where T_0 is the height of the belt section (see Fig. 24.18); for the center distance, first take $a = 0.5(a_{min} + a_{max})$.

B. The last formula (24.32) is used to determine the length of the belt L. Round it to the nearest standard value L_p (e.g., according to Table 11.11 (Chernavskij 1976)), and specify the center distance using the formula

$$a = 0.25 \left[(L_p - w) + \sqrt{(L_p - w)^2 - 8y} \right], \tag{24.41}$$

where L_p is the estimated belt length measured by the neutral layer,

$$w = \frac{\pi}{2}(d_1 + d_2), \quad y = \frac{(d_2 - d_1)^2}{2}.$$

To install and replace the belt, it must be possible to reduce a by 2% at $L_p < 2$ m and by 1% at $L_p > 2$ m. To compensate for belt pulling, it is possible to increase the center distance a by 5.5% of the belt length.

C. Calculate the girth angle of the small pulley by the formula

$$\alpha_1 = 180° - 57° \cdot \frac{d_2 - d_1}{2}. \tag{24.42}$$

D. Determine the required number of belts by the formula

Fig. 24.18 Belt profile

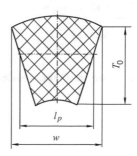

$$z = \frac{N}{p_p C_z}, \tag{24.43}$$

where

$$p_p = p_0 C_\alpha C_L / C_p, \tag{24.44}$$

and p_0 is the rated power transmitted by one belt, kW; it's taken from the table (e.g., (Chernavskij 1976), Table 7.8); C_L coefficient of influence of the length of the strap ((Chernavskij 1976), Table 7.9); C_p the ratio mode ((Chernavskij 1976), Table 7.10); C_z coefficient, taking into account the number of belts: if $z = 2$ or 3 $C_z = 0,95$; when $z = 4 \ldots 6$ $C_z = 0.9$; when $z > 6$ $C_z = 0.85$; C_α the ratio of the angle girth:

$\alpha°$	180	160	140	120	100	90	70
C_α	1.0	0.95	0.89	0.82	0.83	0.68	0.56

E. Using the nomogram (Chernavskij 1976, Fig. 7.3), the belt cross-section is selected, and then p_0 is determined using Table 7.8, p_p is found using formula (24.44), and p_p is found using formula (24.43)—the number of belts, rounding it to a larger integer.

F. Calculate the tension of the branch of one belt by the formula

$$S_0 = \frac{850 N C_p C_l}{z v C_\alpha} + \theta v^2, \tag{24.45}$$

where N is the transmitted power, kW; v m/s is the belt speed, and θ, $\text{H} \cdot c^2 / \text{m}^2$—coefficient that takes into account the influence of centrifugal forces:

Section	O	A	Б	B	Г	Д
θ	0.06	0.1	0.18	0.3	0.6	0.9

G. Determine the force acting on the shafts:

$$F_b = 2 S_0 z \sin \frac{\alpha_1}{2}. \tag{24.46}$$

H. Determine the working resource (in h):

$$H_0 = N_{\text{oc}} \frac{L_p}{60\pi d_1 n_1} \left(\frac{\sigma_{-1}}{\sigma_{\max}} \right)^8 C_i, \tag{24.47}$$

where L_p is the estimated length of the belt; N_{oc} is the number of cycles sustained by the belt (for cord-fabric belts section O, it is recommended to take $N_{OC} = 4.6 \cdot 10^6$; section B, B, G $N_{OC} = 4, 7 \cdot 10^6$; section D E, $N_{OC} = 2.5 \cdot 10^6$; for konsolowych belts all profiles $N_{OC} = 5, 7 \cdot 10^6$); σ_{-1} endurance limit (for V-belts, it is taken equal to 7 MPa); σ_{max} same as in the formula (24.39); C_i is the same, in formula (24.38).

The resource calculated by formula (24.47) must comply with the recommendations given on page 540 for flat belts.

Self-Test Questions

1. Give the definition of a mechanical transmission.
2. List the main types of transfers.
3. What is called a drive in mechanics?
4. What is called the gear ratio?
5. How to determine the efficiency of a drive consisting of several gears?
6. Name the initial data for the design of the drive.
7. What gears are called gears?
8. Name the advantages and disadvantages of gears.
9. List the classification of gears by geometric and functional features.
10. How does the gearbox differ from the multiplier?
11. Formulate the basic law of engagement.
12. Name the sequence of the design calculation of the gear train.
13. Name the advantages and disadvantages of worm gears.
14. What are the advantages and disadvantages of chain drives?
15. Specify the scope of application of chain gears.
16. List the main types of drive chains.
17. What are the friction gears?
18. What are the advantages and disadvantages of friction gears?
19. The design of the frontal and cone variators.
20. Name the scope of application of industrial variators.
21. List the main types of drive belts.
22. How do they compensate for belt pulling in the process of operation?

References

S. Chernavskij, *Proektirovanie mexanicheskix peredach* [*Mechanical Transmission Design*] (Mashinostroenie Publ., Moscow, 1976)

V. Gurin, A. Popov, V. Zamjatin, *Detali mashin. Kursovoe proektirovanie. Kniga 2. Uchebnik* [*Machine Parts. Course Design. Book 2. Textbook*] (YUrajt Publ., Moscow, 2019)

I. Shigley, C. Mischke, *Standard Handbook of Machine Design* (McGraw-Hall, 1996)

Chapter 25
Shaft and Axle

Abstract Axles and shafts are called bars of various sections, designed to accommodate rotating parts. The axles are rectilinear and only serve to support parts. They can be stationary and rotating. Shafts are always movable and, unlike axes, not only support parts but also transmit torque. This chapter describes the classification of shafts, shaft and axle loads, strength, and stiffness calculations of straight shafts. The calculation is given to resist fluctuations. A list of materials used in the manufacture of axles and shafts is given.

Keywords Axles · Shafts · Appointment of axles and shafts · Classification · Shafts are simple · Torsion shafts · Camshafts · Shafts are fast (quiet) running · Strength calculation · Stiffness calculation · Calculation for fluctuations

25.1 Purpose and Classification of Shafts

In technology, parts are often used that transmit torque in the axial direction or support rotating bodies. Parts for this purpose are called *shafts*. Most often, the shaft has the shape of a body of revolution and is installed in bearings. Depending on the perceived forces, there are simple shafts, torsion shafts, and axles (Shigley and Mischke 1996).

Simple shafts (or simply shafts) are used in gears. They transmit torque and perceive axial and radial loads. Hence, it is clear that a simple shaft experiences complex resistance: torsion, bending, and axial tension-compression.

Depending on the load distribution along the shaft axis, as well as on the assembly conditions, straight shafts are made smooth (Fig. 25.1a) or stepped (Fig. 25.1b). The step shaft is similar in shape to the beam *of equal bending resistance*, in which the moment of resistance of the cross section varies along the length of the beam, ensuring the constancy of the maximum stresses.

Smooth shafts are more technologically advanced and have a greater distribution. In reciprocating engines, compressors, and other devices, hollow shafts are used (Fig. 25.1 *in*). The purpose of the channel is twofold: it reduces the mass of the shaft

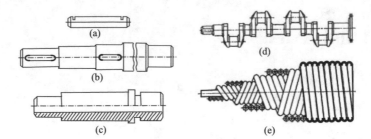

Fig. 25.1 Shaft and axle types

and is often used for the supply of oil or air and the placement of supports and the coaxial shaft.

Less common are shafts that are used only to support rotating parts and do not transmit torque. Such shafts work mainly on bending and are called axes. Another variety is shafts that transmit only torque. Such shafts are called torsion shafts.

In a number of agricultural, road, and other machines, long composite shafts (up to 20 m) are used, transmitting the moment to the working body. Such shafts are called *transmission shafts.* Transmission shafts have sufficient torsional rigidity and low bending stiffness.

In reciprocating engines and compressors, crankshafts with a polyline axis are used (Fig. 25.1d). To transfer the torque between units with the axes of the input and output shafts displaced in space, flexible shafts are used (Fig. 25.1e). They have a curved geometric axis.

Depending on the location, speed and purpose of the shafts are called input, intermediate and output, low—speed and high-speed, distribution, etc.

25.2   Calculation of Straight Shafts for Strength and Stiffness

Shafts are among the most important parts of the machine. Violation of the shape of the shaft due to high radial compliance or vibrations and even more so the destruction of the shaft entail the failure of the entire structure. Therefore, high demands are placed on the shafts, both in terms of manufacturing accuracy and in terms of strength, rigidity, stability, and limited vibrations.

25.2.1   Loads on the Shafts. Design Schemes of Shafts

To judge the strength, you need to know the stresses in the shaft sections from external loads and the nature of these loads (constant or variable). The loads on the shafts can be determined by calculation or experimentally.

The loads transferred to the shafts from the gear wheels mounted on them are shown in Fig. 25.2.

If the external loads are known, then a beam pivotally fixed in two rigid supports is taken as the design scheme of the shaft, as shown in Fig. 25.3 *and*. When installing

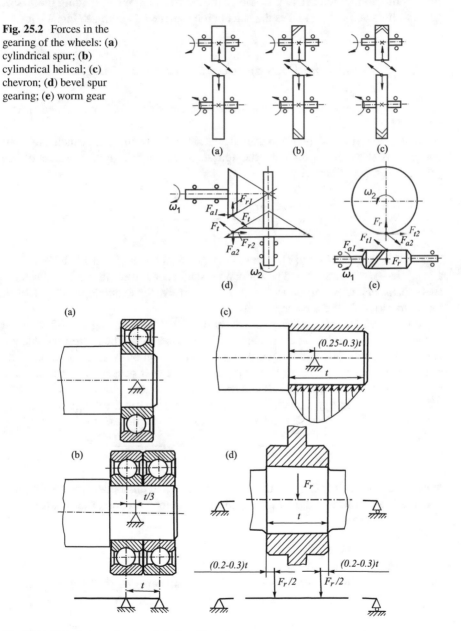

Fig. 25.2 Forces in the gearing of the wheels: (**a**) cylindrical spur; (**b**) cylindrical helical; (**c**) chevron; (**d**) bevel spur gearing; (**e**) worm gear

Fig. 25.3 Calculation schemes of shafts with various supports

two rolling bearings in one support, the conditional support is taken according to the scheme Fig. 25.3b. For shafts supported at the ends by plain bearings, the conditional supports are located at a distance of $(0.25\ldots 0.3)l$ from the inner end of the bearing (Fig. 25.3c). This takes into account the displacement toward the maximum contact stresses due to the deformation of the shaft and bearing. When landing pulleys or gears with a wide hub, (t) takes the calculation scheme according to Fig. 25.3d.

25.2.2 Calculation of Strength Under Stationary Loads

The strength calculation is the main one for drive shafts. It is performed in two stages.

At the first stage, a preliminary calculation is performed, in which the shaft diameter is determined only by the torque at an underestimated value of the permissible tangential stress:

$$d = \sqrt[3]{\frac{1000T}{0.2[\tau]_{\mathrm{K}}}} = \sqrt[3]{\frac{9554P}{0.2[\tau]_{\mathrm{K}}n}}, \tag{25.1}$$

where T is torque, N $\cdot$ m; $[\tau]_{\mathrm{k}}$ permissible torsional stress (for steel shafts $[\tau]_{\mathrm{k}} = 12\ldots 20$ MPa); P transmitted power, kW; n shaft speed, rpm. In the first stage, the shaft shank diameter can be assigned constructively, for example, $(0.8\ldots 1.0)$ of the motor drive shaft diameter.

At the second stage, the design of the shaft is developed, paying special attention to the manufacturability of its manufacture and assembly. Then a test calculation is carried out for static and fatigue strength, stiffness, stability, and vibrations. The static strength of the shafts is calculated by the greatest possible short-term load, the repeatability of which is small and cannot cause fatigue failure. Since the shafts work mainly on torsion and bending, the equivalent stress at the dangerous point of the most loaded section is determined by the third or fourth strength theory:

$$\sigma_{\mathrm{equ}}^{III} = \sqrt{\sigma^2 + 4\tau^2} \text{ or } \sigma_{\mathrm{equ}}^{IV} = \sqrt{\sigma^2 + 3\tau^2},$$

where σ and τ are, respectively, the greatest normal and tangential stresses from the bending moment M and the torque M_{k} in the dangerous section of the shaft:

$$\sigma = \frac{M}{W}, \ \tau = \frac{M_{\mathrm{K}}}{W_p},$$

moreover, W and W_p are the axial and polar moments of the resistance of the checked shaft section. Since for a circular shaft $W_p = 2W$, we can write

$$\sigma_{\mathrm{equ}}^{III} = \frac{32}{\pi d^3}\sqrt{M^2 + M_{\mathrm{K}}^2} \text{ or } \sigma_{\mathrm{equ}}^{IV} = \frac{32}{\pi d^3}\sqrt{M^2 + 0,75M_{\mathrm{K}}^2},$$

where d is the diameter of the shaft. Typically, the torque M_k is equal to the torque T transmitted by the shaft. The bending moment M is determined by the plot of the bending moments. At the same time, if the loads on the shaft do not lie in the same plane, then they are first projected into two orthogonal planes, and the bending moment in the two planes is found by geometric addition in the checked cross-section if the angle between the load planes does not exceed 30°, then for simplicity, it is allowed to count all loads in one plane:

$$M = \sqrt{M_x^2 + M_y^2}.$$

Then the safety factor is checked in relation to the yield point:

$$n_t = \frac{\sigma_t}{\sigma_{equ}} \leqslant [n]_t, \qquad (25.2)$$

where σ_t is the yield point and $[n]_t$ in general mechanical engineering is usually assumed to be $1.2 \ldots 1.8$.

Let us now consider the consideration of alternating voltages. Variable voltages arise from the action of both constant and varying loads. Let's analyze them separately.

Under constant loads, due to the rotation of the shafts, alternating cyclically varying bending stresses arise. They vary in a symmetrical cycle with the amplitude and average cycle voltage equal to:

$$\sigma_a = \sigma = \frac{M}{0,1d^3}; \ \sigma_m = 0. \qquad (25.3)$$

When the shafts are operating, the torque and therefore the shear torsional stresses change in a pulsating cycle. The amplitude and average stress of the cycle of shear stresses are determined by the formulas:

$$\tau_a = 0.5\tau = \frac{t}{0.4d^3}; \ \tau_m = \tau_a. \qquad (25.4)$$

Fatigue calculation is performed in the form of determining the endurance margin according to the Gough and Pollard formula:

$$n = \frac{n_\sigma n_\tau}{\sqrt{n_\sigma^2 + n_\tau^2}}, \qquad (25.5)$$

where n_σ, n_τ are the coefficients of the fatigue strength reserve for normal and tangential stresses. As is known from the section on the resistance of materials, these coefficients for the symmetric cycle σ and the pulsating cycle τ are determined by the formulas:

$$n_\sigma = \frac{\sigma^b_{-1}}{k_{\sigma d}\sigma_a}, \quad n_\tau = \frac{\tau_{-1}}{k_{\tau d}\tau_a}, \tag{25.6}$$

moreover, σ^b_{-1} and τ_{-1} are the limits of endurance in bending and torsion, respectively, and $k_{\sigma d}$ and $k_{\tau d}$ are the general coefficients of reducing the limit of endurance.

The safety margin calculated by formula (25.5) must be at least the allowed value $[n] = 1.5 \ldots 2, 0$. Fatigue strength increases are achieved by structural and technological methods. Firstly, they try to reduce the stress concentration as much as possible, and secondly, they use strengthening of the shaft surfaces by local plastic deformation (roller running, shot blasting, laser or plasma processing)

25.2.3   Calculation of Strength Under Non-stationary Loads

In this case, the strength calculation is carried out according to the equivalent stress:

$$\sigma_{equ} = \sqrt[m]{\frac{1}{N_0}\sum_{i=1}^{n}\sigma_i^m n_i} \leqslant \sigma_{max}, \tag{25.7}$$

where N_0 is the number of cycles corresponding to the inflection point on the fatigue curve; take $N_0 = (3\ldots5)\cdot10^6$ for small shafts and 10^7 for large shafts; n_i is the total number of cycles at the stress σ_i; i is the number of the loading stage on the cyclogram; m is the degree of the fatigue curve (for steel shafts $m = 9$); σ_{max} is the stress at the most loaded point of the shaft at the maximum long-acting load.

For a known value of σ_{equ}, the safety margin is determined by formula (25.2). At the same time, if it turns out that the maximum normal stress exceeds the equivalent stress according to a suitable strength theory, then take $\sigma_{eq} = \sigma_{max}$, since the shaft in this case operates in the zone of unlimited durability (in the zone of the horizontal section of the fatigue curve).

25.2.4   Calculation of Shaft Stiffness

The elastic movements of the shafts have an adverse effect on the gears, bearings, and other parts. Large movements from the bending of the shaft can cause the bearings to jam or the gears to break. Therefore, when designing, it is necessary to check the deflections and angles of rotation of the characteristic sections. To determine the deflections and angles of rotation of sections, use the Mohr integral or the ready-made results given, for example, in (Spravochnik metallista GNTIMP. T.2 1960).

The permissible values of deflections and angles of rotation depend on the requirements for the design. Usually, the permissible deflection is $0.0002\ldots0,0003$ from the span (the distance between the supports), and the permissible deflection under the gears is $0.01m$ for cylindrical and $0.005m$ for conical and hypoid gears (m—the engagement module). The angles of rotation of the cross-sections during bending are allowed in the range of 0.0001 to 0.05 radians and during torsion— $0.2\ldots1°$ per 1 m of the shaft length.

25.3 Calculation of Shafts for Vibrations

In shafts, two types of vibrations are possible: torsional and bending. Let us first consider the calculation for torsional vibrations. One of the main tasks of this calculation is to determine the natural frequencies of the system and to assess the possibility of the appearance of resonance phenomena.

The differential equation of torsional vibrations of a shaft of length l with one rigidly fixed end and a disk of mass m of radius R, placed at the free end of the shaft, has the form

$$J_m \frac{d^2\theta}{dt^2} = -M_K, \qquad (25.8)$$

where $J_m = mR^2/2$ is the moment of inertia of the disk.

The circular frequency of natural vibrations of such a system:

$$\omega_0 = \sqrt{c_\theta/J_m} = \sqrt{1/\lambda_\theta J_m},$$

where c_θ is the torsional stiffness of the shaft, i.e., the torque in H· mm required to rotate the disk by 1 rad; $\lambda_\theta = 1/c_\theta$ torsional malleability coefficient. The torsional stiffness is calculated by the formula

$$c_\theta = \frac{G J_K}{l}, \qquad (25.9)$$

where G is the shear modulus and J_k is the polar moment of inertia of the shaft cross-section.

The frequency of the forcing torsional load most often coincides with the gearing frequency of the gear teeth:

$$\omega = \frac{zn}{60}, \qquad (25.10)$$

where z and n are the number of teeth and the speed of the shaft, rpm.

To avoid the occurrence of resonance, when designing, try to satisfy the conditions: $0.7 \leqslant \omega/\omega_0$ or $\omega/\omega_0 \geqslant 1.3$.

Table 25.1 Coefficient of compliance with lateral vibrations of shafts

Sketch	λ	Sketch	λ
EJ m l	$\dfrac{l^3}{3EJ}$	EJ_1 m EJ_2 a b l	$\dfrac{a^3 b^2}{3EJ_1 l^2} + \dfrac{a^2 b^3}{3EJ_2 l^2}$
EJ m $l/2$ $l/2$	$\dfrac{l^3}{48EJ}$	EJ_1 EJ_2 m a b	$\dfrac{ab^2}{3EJ_1} + \dfrac{a^2 b}{3EJ_2}$

In addition to torsional vibrations, bending vibrations occur during the operation of the shafts. Vibrations of shafts of this type are associated with the deformation of the bending of the shafts from the impact of radial loads. In approximate calculations, a shaft with a gear wheel is taken for a two-hinged beam with a concentrated mass. The mass of the shaft is brought to the mass of the wheel by summing the masses, taking into account the mass reduction coefficient, depending on the location of the supports and the wheel.

The free oscillations of such a system are described by the equation

$$m\frac{d^2 y}{dt^2} + \frac{1}{\lambda}y = 0, \tag{25.11}$$

where m is the mass of the disk and λ is the compliance coefficient of the system, which depends on the location of the mass on the shaft. Formulas for its definition are given in Table 25.1.

The circular frequency of bending vibrations will be

$$\omega_0 = \sqrt{\frac{1}{\lambda m}}. \tag{25.12}$$

This relationship is valid for any single-mass system. When the design scheme changes (i.e., the position of the disk on the shaft), the shaft compliance coefficient λ will change. If the center of mass of the disk is located on the axis of the shaft, then the angular velocity (25.12) coincides with the critical angular velocity. In the case of displacement of the center of mass of the disk relative to the axis of the shaft by the amount e, the amplitude of vibration of the shaft can be calculated by the formula

$$A = \frac{e}{\left(\omega_{\text{кр}}/\omega\right)^2 - 1}. \tag{25.13}$$

From formula (25.13), it can be seen that when the frequencies coincide, there is a resonance. Therefore, for safety reasons, in shaft designs, they strive to ensure the condition $\omega \leqslant 0.7\omega_{\text{rot}}$, which was already mentioned earlier.

25.4 Materials for Making Shafts

The choice of material and heat treatment of the shafts is determined by the criteria of their performance (rigidity, bulk strength, and wear resistance at relative micro-displacements that cause corrosion), including the criteria for the performance of trunnions with supports or spline sections with hubs of the parts placed on them. The significance of the latter criteria in the case of sliding supports or movable spline (key) joints may even be decisive.

The main material for the shafts are carbon and alloy steels (rolled, forged, steel castings), since they have high strength, the ability to surface, and volume hardening; cylindrical workpieces are easily obtained by rolling and are well processed on machine tools, as well as high-strength modified cast iron and non-ferrous metal alloys (in instrument making). For shafts and axles that are subject to the criterion of rigidity and are not subjected to heat treatment, the following steels are used: Ст. 5; Ст. 6.[1] For most shafts, heat-treated steels 45, 40X are used. High-speed shafts that rotate in plain bearings require a very high hardness of the trunnions. They are made of cemented steels 12X, 12XН3A; 1XXГТ.

Only for the manufacture of heavy crankshafts, shafts with large flanges and longitudinal holes, high-strength (with spherical graphite) and modified cast iron are used.

In the automotive and tractor industries, engine crankshafts are made of ductile or high-strength cast iron.

Since the main criterion for the efficiency of the shafts is their fatigue strength (endurance), for the manufacture of most shafts, thermally improved medium-carbon steels 40; 45; 50 are used. They are used for the manufacture of shafts of stationary machines and mechanisms. The billet made of such steels is subjected to an improving heat treatment (HRC$\leqslant$36) before machining. The shafts are turned on a lathe, followed by grinding of the seats and trunnions on a grinding machine.

If the endurance of the thermally improved shafts is unsatisfactory or if they have wear areas (under the lip seals, spline, etc.), the shafts made of these steels are subjected to surface quenching in these places with HDPE heating and low tempering.

For non-responsible low-load structures of shafts and axles, carbon steels are used without heat treatment.

Responsible heavy-loaded shafts are made of alloy steel 40XHMA, 25XГТ, etc. These steels are used to make shafts for critical gears of mobile machines (shafts of gearboxes of tracked vehicles). Improving heat treatment (HRC?45) is most often subjected to the part after preliminary turning. Finally, the landing surfaces and trunnions are ground on grinding machines and in repair production sometimes on a lathe with the use of a special grinding head.

---

[1] Here the marking of materials is given in accordance with Russian standards.

For heavily loaded shafts of critical machines it is recommended to use expensive hardened alloyed steels 40KhN; 40KhN2MA; 30KhGT, etc.

High-speed shafts that rotate in sliding bearings require a very high hardness of the surfaces of their trunnions. In this regard, such shafts are made of cemented steels 15X; 20X; 18XГТ; 12XH3A, etc. or nitrided steels of the 38X2MЮA type. The shaft, manufactured with a minimum allowance for final processing, is subjected to surface chemical-thermal treatment (cementation, nitriding, etc.) and is hardened to a high surface hardness (HRC-55...65). The working surfaces of the slots, the landing surfaces, and the surfaces of the trunnions are ground after heat treatment in order to obtain the necessary accuracy.

Chrome-plated trunnions have high wear resistance. For example, from the experience of the automotive industry, it is known that chrome plating of the crankshaft necks increases their service life by 3...5 times.

Shafts operating in an aggressive environment are made of high-alloy stainless steels, and when using conventional steels, they are insulated with bronze or polymer jackets, rubberized or enameled.

For shafts whose dimensions are determined not by strength but by rigidity, carbon structural steels 20, 35, etc. are used without heat treatment. In this case, it is advisable to use heat-treated steels only when this is determined by the durability requirements of the trunnions, splines, and other wear surfaces of the shafts.

Self-Test Questions

1. What part is called a shaft?
2. Name the differences between the shaft and the axle.
3. What is the shaft-gear?
4. Draw the design of the shafts and axles.
5. The loads transmitted to the shafts from the gears.
6. What design scheme is used when calculating the shaft for the specified loads?
7. Calculation of straight shafts for strength and rigidity.
8. Calculation of shafts for strength under non-stationary loads.
9. What materials are used for the manufacture of shafts and axles?

References

V. Molotnikov, *Texnicheskaya mexanika [Technical Mechanics]* (Lan' Publ., Sant-Peterburg, 2017)

I. Shigley, C. Mischke, *Standard Handbook of Machine Design* (McGraw-Hall, 1996)

Spravochnik metallista GNTIMP. T.2 [*Metalworker's Handbook*, vol.2] (1960)

Chapter 26
Shaft and Axle Supports

Abstract Supports of shafts and axles are designed to support the rotary or rocking motion of the shafts and axles and transfer forces from them to the housing. The accuracy of the action and the reliability of the mechanism as a whole largely depend on the design of the supports. Supports designed for the perception of radial or combined (radial and axial) loads are commonly called bearings, and supports that take only axial loads are called thrust bearings. This chapter describes the types of bearings and their design features. Particular attention is paid to the design of bearing assemblies. Attention is paid to the correct choice of lubricants.

Keywords Shafts · Axles · Supports · Bearings · Pivot · Strength calculation · Shaft stiffness · Vibrations of shafts · Bearing capacity · Resonance · Shaft materials

26.1 Rolling Bearings

Bearings are used as bearings for axles and shafts. By the type of friction, bearings are differentiated into rolling bearings and plain bearings. We'll start by looking at rolling bearings. They are assembly units consisting of the following parts (Fig. 26.1a): (1) outer and inner rings with tracks, (2) rolling bodies (balls or rollers), and (3) separators, which serve to separate bodies rolling. In combined supports, one or both rings may be missing. In this case, the rolling elements roll directly on the grooves of the shaft or housing.

Rolling bearings are a group of parts that are most widely standardized internationally, interchangeable, and mass-produced. Compared to plain bearings, rolling bearings have the following advantages :

- lower moments of friction forces and heat dissipation;
- low dependence of the friction forces on the speed;
- significantly smaller (5 . . . 10 times!) starting moments;
- lower maintenance requirements and lower lubricant consumption;

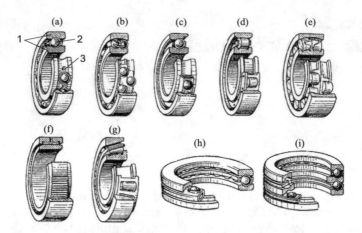

Fig. 26.1 Main types of rolling bearings: (**a**) ball radial single row; (**b**) ball radial double row spherical; (**c**) radial-thrust ball roller; (**d**) radial roller with short cylindrical rollers; (**e**) radial spherical double-row roller with barrel-shaped rollers; (**f**) needle roller; (**g**) radial-thrust roller; (**h**) single-row ball thrust; (**i**) double-row ball thrust

- large bearing capacity per unit bearing width;
- significantly lower consumption of non-ferrous metals.

The *disadvantages of* rolling bearings include:

- increased diametrical dimensions;
- high contact voltages that limit their service life;
- less ability to dampen vibrations;
- noise at work.

Bearings in the range of internal diameters 3 . . . 10 mm are standardized through 1 mm, up to 20 mm through 2 . . . 3 mm, up to 110 mm through 5 mm, up to 200 mm through 10 mm, up to 500 mm through 20 mm, etc.

In accordance with GOST 3395-75, in the direction of perceived loads, rolling bearings are divided into:

- radial, designed for purely radial loading; they are also able to fix the shafts in the axial direction and take small axial loads;
- radial-thrust, receiving both radial and axial loads;
- thrust bearings designed for axial loading;
- thrust-radial, bearing an axial and small radial load.

According to the shape of the rolling elements, bearings are divided into ball bearings (Fig. 26.1a–b, h, i) and roller bearings (Fig. 26.1d–g). A roller bearing with thin and long rollers is called a needle bearing (Fig. 26.1f).

Roller bearings perform the following types:

- cylindrical with short rollers (Fig. 26.1e);

- radial spherical double-row with barrel-shaped rollers (Fig. 26.1d);
- conical radial-thrust (Fig. 26.1g).

Ball bearings are faster than roller bearings, but roller bearings have a higher (by 50 . . . 70%) load capacity.

According to the number of rows of rolling elements: bearings are divided into single-row (main distribution), two-row, and multi-row.

According to the ability to self-align: distinguish between self-aligning bearings (Fig. 26.1b, e) and non-self-aligning.

By overall dimensions: bearings are divided into size series—by radial dimensions (ultralight (two series), extra light (two series), light, medium, and heavy (seven series in total)) and by width (narrow, normal, wide, and extra wide). The main distribution is especially light, light, and medium series of bearings.

Bearings have symbols made up of numbers and letters. The number formed by the last two digits, multiplied by 5, gives the inner diameter of the bearing. The third digit on the right indicates the bearing series: extra light 1, light 2, medium 3, heavy 4, light wide 5, medium wide 6, etc. The fourth digit on the right indicates the type of bearing:

0. single-row radial ball;
1. radial ball double-row spherical;
2. radial with short cylindrical rollers;
3. radial roller double-row spherical;
4. roller with long cylindrical rollers or needles;
5. roller with twisted rollers;
6. angular contact ball bearing;
7. tapered roller;
8. thrust ball bearing;
9. thrust roller.

The fifth or fifth and sixth digits on the right, which are not entered for all bearings, indicate design features, for example, the contact angle of the balls in angular contact bearings, the presence of a locking groove on the outer ring, the presence of integrated seals, etc.

The numbers 6, 5, 4, and 2, separated by a dash before the symbol of the bearing, indicate its accuracy class (in ascending order). Class 0 is not specified.

Examples of designations of bearings of accuracy class 0: single-row ball bearings with an internal diameter of 50 mm of the light series, 210; medium series, 310; and heavy, 410.

Roller bearings with an inner diameter of 80 mm, with short cylindrical rollers and sides on the inner ring: light series 2216, medium 2316, heavy 2416; tapered light series 7216, light wide 7516, medium 7316, medium wide 7616.

Bearing 5-210—single row ball bearing of the light series of precision class 5.

Foreign and some Russian bearing manufacturers use a different system of designations, which can be found, for example, in the brochure (Kacz 2007). The system given in this source is used by the world's leading bearing manufacturers, for example, Sweden, "SKF"; Germany, "FAG," "INA"; USA, "Timken"; Japan, "Koyo," "NSK"; etc. The foreign system of designations for most types of manufactured bearings is also used by CJSC "Vologda Bearing Plant" (and not only when marking products but also in all documentation).

26.1.1 Design Procedure for Rolling Bearings

We recommend that you design the supports in the following order.

1. According to the sketch layout, the distances between the supports and the parts fixed to the shaft are outlined.
2. Determine the loads on the supports.
3. Pre-select the type of bearing, taking into account the design of the rolling support, operating conditions, and installation.
4. Determine the estimated durability of the bearing, assigning its approximate size, and compare it with the recommended one.
5. Depending on the requirements for the quality of the rolling support, assign the accuracy class of the bearing, fit on the inner and outer rings, and choose the method of fixing the bearing rings.
6. Select the type of lubricant, brand of lubricant, and seal design.
7. Finally formalize the design of the rolling support.

26.1.2 Calculation for Durability

The method of calculation of rolling bearings is defined by the interstate standards 18854-91 (ISO76-87) and 18555-94 (ISO281-89). The design calculation is performed according to two criteria:

– by the basic dynamic radial C_r and axial load capacity C_a;
– according to the basic static radial C_{0r} and axial C_{0a} load capacity.

Basic dynamic radial load capacity C_r is a constant stationary radial load that the bearing can theoretically handle with a basic design life of 1 million revolutions.

Basic dynamic axial design load capacity C_a is the constant central axial load that the bearing can theoretically withstand for 1 million revolutions of the shaft.

Basic static radial C_{0r} and axial C_{0a} load capacities are the loads that the bearing can withstand without rotating the rings.

Bearings with a ring speed of $n > 1$ rpm are calculated for a given resource by dynamic load capacity and at a speed of $n < 1$ rpm-by static load capacity.

26.1.3 Dynamic Load Capacity Calculation

This calculation is performed at a frequency of $n \geqslant 10$ rpm. If $n \in (1; 10)$ rpm, then assume $n = 10$ rpm. In this case, according to ISO 281-89, the basic design resource L_{10} is used as a criterion for the bearing's operability. This resource corresponds to 90% reliability for a particular bearing (or a group of identical bearings).

The nominal durability (resource, million revolutions) is calculated by the formula

$$L = \left(\frac{C}{P}\right)^m,$$ (26.1)

where P is the equivalent load; C dynamic load capacity of the bearing according to the catalog; m exponent:

$$m = \begin{cases} 3 \text{ for ball bearings,} \\ 10/3 \text{ for roller bearings;} \end{cases}$$

Rated endurance in hours:

$$L_h = \frac{10^6 L}{60n} = \frac{10^6}{60n}\left(\frac{C}{P}\right)^m.$$ (26.2)

When using formulas (26.1), (26.2), you must strictly ensure that the dimensions in C and P coincide.

For single-row and double-row spherical deep groove ball bearings, single-row angular contact ball, and roller bearings, the equivalent load is calculated by the formulas:

$$P = \begin{cases} (VXF_r + YF_a)K_6K_\text{T} & \text{if } \dfrac{F_a}{VF_r} > e; \\ VF_rK_6K_\text{T} & \text{if } \dfrac{F_a}{VF_r} < e, \end{cases}$$ (26.3)

where V is a coefficient that takes into account the effect of the ring rotation: when the inner ring rotates, $V = 1$; when the outer ring rotates, $V = 1.2$; F_r and F_a are radial and axial loads, respectively, N; K_T is the temperature coefficient depending on the temperature of the working medium (see (Chernavskij 1976), Table 9.20); K_6—safety factor, taking into account the nature of the load (taken from tables, see, for example, (Chernavskij 1976), Table 9.19); for reducers $K_6 \in (1.3; 1.5)$; axial load factor e and bearing parameters X, Y are taken from catalogs or reference books (for example, (Chernavskij 1976), Table 9.18).

The equivalent load for single-row and double-row bearings with short cylindrical rollers (without flanges on the outer or inner rings) is determined by the lower formula (26.3) and for thrust (ball and roller) bearings by the formula

$$P = F_\alpha K_6 K_T. \tag{26.4}$$

If necessary, the bearings are calculated, taking into account the operating conditions, variability of operating modes, seals, etc.). Limiting the axial movement of the shaft is achieved by attaching the bearings to the shafts and fixing them in the housing.

26.1.4 Fitting of Rolling Bearings

Landing of the inner rings on the shaft is carried out in the hole system and the outer rings in the housing in the shaft system. Bearing landings differ from those generally accepted in mechanical engineering by the location and values of the tolerance fields on the landing surfaces of the rings. According to GOST 3325-85*, tolerance fields for the bearing bore diameter depending on the bearing accuracy class (0, 6, 5, 4 and 2) denote $L0$, $L6$, $L5$, $L4$, and $L2$. The tolerance fields for the outer diameter of the bearing are $l0$, $l6$, $l5$, $l4$ and $l2$, respectively. Examples of designations of bearing landings on the shaft are $\varnothing 60L0/k6$; in the housing—$\varnothing 100H7/l0$.

In assembly drawings of bearing assemblies, it is allowed to specify only the tolerance field for the diameter of the part mated to the bearing without specifying the tolerance field for the landing diameters of the bearing rings, for example: $\varnothing 60K6$; $\varnothing 100H7$. The choice of the tolerance field depends on the *mode of operation of the bearing* and its type and dimensions. Depending on the ratio of the equivalent load P to the basic dynamic load capacity C, the following operating modes are distinguished:

$$
\begin{array}{lll}
\text{easy} - & \text{for} & \dfrac{P}{C} \leqslant 0.07; \\[2mm]
\text{normal} - & \text{for} & 0.07 \leqslant \dfrac{P}{C} \leqslant 0.15; \\[2mm]
\text{heavy} - & \text{for} & \dfrac{P}{C} > 0.15.
\end{array}
$$

The choice of landing is carried out using the recommendations given in all textbooks on machine parts and reference books of a mechanical designer.

26.1.5 Securing the Bearings in the Housing and on the Shaft

Fixing supports limit the axial movement of the shaft in one (Fig. 26.2a, b) or both (Fig. 26.2c, d) sides and perceive both radial and axial loads, respectively, in one or both directions.

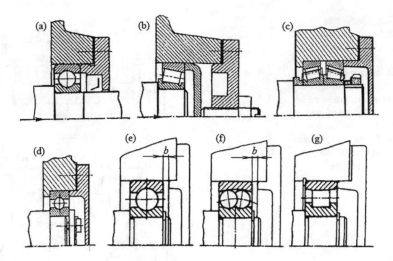

Fig. 26.2 Bearing fixing methods

Floating supports (Fig. 26.2c–h) do not restrict axial movement of the shaft and can only take a radial load. For floating bearings, single-row radial ball bearings (Fig. 26.2e) or double-row bearings (Fig. 26.2f) are used, but bearings with short cylindrical rollers are more often used (Fig. 26.2g). In the supports made according to the schemes of Fig. 26.2e, f, there must be a gap between the ends of the outer ring and the bearing cover $b \geqslant 0.01l$, where l is the distance between the ends of the bearing rings.

26.1.6   Bearing Unit Designs

The shafts are installed according to the following schemes: with two floating supports; with one fixing and one floating support; with the installation of vraspor bearings; and with the installation of bearings in a stretch.

The design of the shaft installation in two floating supports is shown in Fig. 26.3. In the first version of the design, the shaft is mounted on two roller bearings (Fig. 26.3a). The axial movement of the shaft is ensured by the fact that the inner rings of bearings with a set of rollers can move axially relative to the fixed outer rings. The application of this scheme requires high precision of the L and l dimensions. In the case of insufficient accuracy of the specified dimensions, a significant initial axial displacement of the rings S is possible after assembly. In addition to the L and l errors, the S offset is also affected by the b_1 and b_2 size errors.

In the second variant (Fig. 26.3b, the movement inside the housing is limited by the sides of both bearing rings and toward the bearing covers by the gap z.

Fig. 26.3 Shafts on two floating bearings: (**a**) the inner rings are fixed on the shaft, and the outer rings are fixed in the body; (**b**) the outer rings have some freedom of axial movement; (**c**) the inner rings of the bearings are fixed on the shaft, and the outer ones are free and can move along the holes in the housing

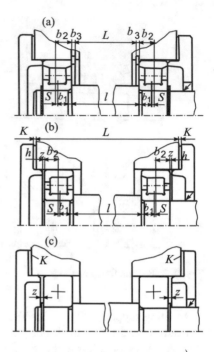

When the shaft is axially floating, the inner rings of bearings with roller sets are displaced relative to the outer rings. Possible errors in the dimensions of l, L, and h are eliminated by compensatory gaskets K.

In the third variant (Fig. 26.3b), radial single-row ball, ball, or roller double-row spherical bearings are used in the supports. The value of the axial movement is limited by the gaps z, which are set during assembly by selecting the compensator gaskets K. The main advantage of this scheme is the possibility of its use with non-rigid shafts and a low degree of alignment of the landing surfaces of the shaft and the housing.

This type of design is used when it is necessary to self-install the shaft in the axial direction (e.g., in chevron gears).

The diagram of the shaft with one fixing support is shown in Fig. 26.4a. In this design, the shaft is fixed by a single radial bearing, in which the inner ring is fixed to the shaft and the outer ring is fixed in the housing. The inner ring of the second support is fixed on the shaft, and the outer ring is not fixed in the housing (floating support).

In the supports made according to the schemes of Fig. 26.4, a gap is provided between the cover and the end of the outer ring of the bearing $b \geqslant 0.01l$, where l is the distance between the ends of the bearing rings (Fig. 26.4a, b). With the temperature extension of the shaft, the outer floating ring moves, which compensates for the thermal expansion of the shaft. Similar designs are widely used in gearboxes, gearboxes with cylindrical gears, and other mechanisms where there are no heavy loads and the axial movement of the shaft does not affect the accuracy of the work.

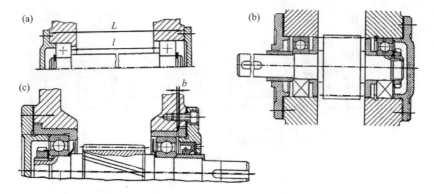

Fig. 26.4 Shafts with one fixing support

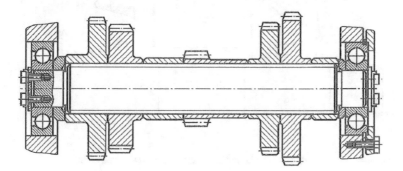

Fig. 26.5 Supports fixing (right) and floating (left)

At high axial loads, two angular contact ball or roller bearings are installed in the fixing support. This increases the rigidity of the system and almost completely eliminates the axial "play" of the shaft.

Figure 26.5 shows a shaft with one floating (left) and one fixing (right) support. Such designs are used for relatively long shafts with unbalanced axial forces.

The design schemes of mounting bearings on the shaft vraspor (Fig. 26.6a) and vrastyazhku (Fig. 26.6b) are used. In the first case, the ends of the inner rings of the bearings rest on the shaft collars, and the outer ends of the outer rings on the ends of the covers. It turns out that each support restricts the displacement of the shaft in one direction. This scheme is the simplest and cheapest, since the boring of the bearing sockets in the housing is performed in one pass. In addition, the design has a minimum of details. To compensate for the thermal expansion, a gap is provided between the cover and the outer end of one of the outer rings of the bearings:

$$a = \Delta L + 0.15 \ldots 0.2 \text{ mm},$$

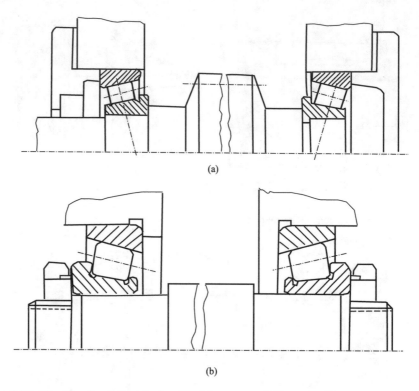

(a)

(b)

Fig. 26.6 Mounting the shaft on the bearings (**a**) vraspor and (**b**) vrastyazhku

where ΔL is the thermal elongation of the shaft, which usually ranges from 0.2 ... 1.0 mm.

The vraspor scheme is recommended for short shafts. For the same shafts, use also the installation of bearings vrastyazhku (Fig. 26.6b). Here, the inner ends of the outer rings of the bearings rest against the body collars and the outer ends of the inner rings in the nuts, which regulate the axial "game" during installation. The inner ring is set here by the fit $L0/js6$, which provides its axial displacement. The vrastyazhku assembly eliminates pinching of the bearing, which is an important advantage of such a design solution.

As an example, Fig. 26.7 shows the granulator press.[1] Here, the pinion shaft 1 is designed on vraspor bearings, and the gear shaft 5 is designed on vrastyazhku bearings.

---

[1] The sketch of the granulator is borrowed from the book by M. N. Erokhin (Eroxin et al. 2005). The unit is designed for the production of pellets from grass flour. Despite the fact that grass flour and grass pellets belong to coarse feeds, they are close to concentrates in terms of their energy value, which gives a high economic effect from the use in feeding farm animals.

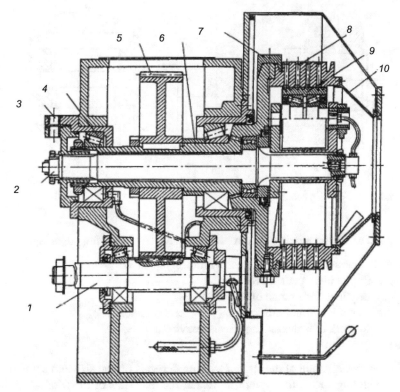

1 - pinion shaft;   2 - axle;   3 - cup roller taper bearing;   4 - bearing;
5 - gear wheel; 6 - main shaft; 7 - segments; 8 - matrix; 9 - pressing rollers;
10 - conical receiver.

Fig. 26.7 Granulator press for making pellets from grass flour

26.2   Shaft Supports on Plain Bearings

26.2.1   *Designs and Operating Modes of Plain Bearings*

Plain bearing is a rotation pair. It consists of a support section of the shaft 1
(Fig. 26.8a) and the actual bearing 2 in which the shaft slides. If the support section
of the shaft is located at its end, then in the case of a radial load F_r, it is called a
pivot (Fig. 26.8a) and under an axial load F_a (Fig. 26.8b) *is a thrust bearing.* If the
support section of the shaft is located on the span, then it is called the *neck*.

According to the type of perceived load, plain bearings are divided into:

radial—only accepts radial loads (Fig. 26.8a);
thrust—perceive axial forces (Fig. 26.8b, c);
radial-thrust—for radial and axial loads (Fig. 26.8b).

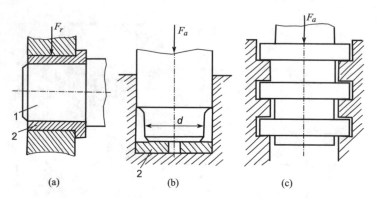

Fig. 26.8 Types of plain bearings

Compared to rolling bearings, plain bearings have the following advantages:

- reliable operation under shock and vibration loads;
- operability at high shaft speed;
- operability in a wide range of temperatures;
- ability to work without lubrication (with certain materials);
- ability to work in water and aggressive environments;
- silent operation.

The operating mode of the sliding bearing depends on the size and nature of the load, the sliding speed, the type of lubricant, the materials of the friction surfaces, etc. At very low sliding speeds (up to 0.01 m/s), the friction surfaces in the bearing touch each other, and in the absence of lubrication, the effect of *dry friction* occurs. The resistance to rotation in this case is determined only by the coefficient of friction of the rubbing surfaces.

If the load is such that not all the grease is squeezed out and in the contact zone the working surfaces of the bearing are covered with the thinnest layer (up to 0.1 microns) of oil, the friction is called *boundary*. With such friction, in the areas where the shaft micro-roughness touches the inner surface of the bearing, the materials are set and the microparticles are separated. This leads to wear of the rubbing surfaces, but much less than with dry friction.

With an increase in the sliding speed in the presence of lubrication, the rotating shaft carries the lubricant with it into the wedge gap between the rubbing surfaces. A hydrodynamic lifting force is created to reduce the contact stresses, but the shaft is not yet fully weighted in the lubricant. This type of friction is called *semi-liquid*, since both liquid and boundary friction exist simultaneously. The resistance to rotation of the shaft in this case is less than in the case of boundary or dry friction and depends not only on the materials of the rubbing surfaces but also on the quality of the lubricant. The semi-liquid friction coefficient for common bearing materials is $0.008\ldots0.1$.

A further increase in the sliding speed leads to an increase in the hydrodynamic lifting force, and the shaft neck floats up. The thickness of the oil layer increases, and the working surfaces of the shaft no longer touch one another. This type of friction is called *liquid*. It provides high wear resistance, jamming resistance, and high bearing efficiency. In such conditions, for example, the bearings of the crankshaft of an internal combustion engine work.

In addition to these types of friction, gas-lubricated plain bearings are also used.

The instrument supports work, as a rule, at low loads and low speeds. Therefore, holes in housings and boards are often used as bearings. The axles are usually installed in two bearings (Fig. 26.9a), and in rigid plates the axles are often placed in one bearing (Fig. 26.9b). Instead of a hole in the board, a more complex support structure is used, shown in Fig. 26.9c. In such a support, the bearing length is usually set equal to 5...6 axle diameters, and a groove is made in the middle part.

The sliding bearings of most agricultural machines often work with insufficient lubrication, low sliding speed ($\approx$0.05 m/s), and a large specific load. Their modes correspond to semi-liquid, boundary, or even dry friction. Bearings with self-contained all-in-one (Fig. 26.10a) and detachable (Fig. 26.10b) supports are widely used in agricultural machinery. The split bearing consists of a housing 1, a cover 2, a liner 3, mounting bolts with nuts 4 and an oilcan 5. The insert connector is made according to its diameter, and the housing connector is stepped to prevent the cover from shifting relative to the bearing housing. The housings of autonomous bearings are made of cast iron SCH15, steel, etc. For the manufacture of inserts, antifriction materials with good hardness, thermal conductivity, workability, and wettability

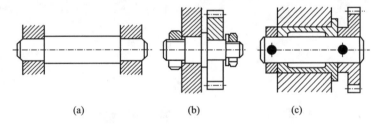

Fig. 26.9 Structures of support units of devices

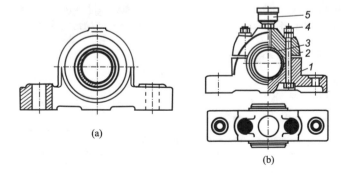

Fig. 26.10 Autonomous plain bearings: (**a**) one-piece; (**b**) detachable

with lubricants are used. Most often, in the manufacture of inserts, high-grained babbits Б16 and Б83, bronzes БрОФ10-1, БрЩ6Ц6С3, БрАЖ9-4, antifriction cast irons АСЧ-2, etc. are used.

26.2.2   Calculation of Sleeve Bearings

As mentioned above, sleeve bearings of shaft bearings in agricultural machines operate under conditions of semi-fluid, boundary, and dry friction. To ensure their durability, two types of (Eroxin et al. 2005) calculations are performed:

calculation for the permissible specific pressure p, which ensures the wear resistance of the bearing, consists in checking the condition

$$p = \frac{F_r}{ld} \leqslant [p], \tag{26.5}$$

where F_r is the radial load on the bearing, H; l journal length, mm; d shaft journal diameter, mm; and $[p]$ permissible specific pressure, the value of which can be taken from Table 26.1;

calculation of the product of specific pressure and sliding speed, providing the permissible thermal regime of the sliding pair, is performed according to the formula

$$pv \leqslant [pv], \tag{26.6}$$

with

$$v = \frac{\pi dn}{60 \cdot 10^3} \text{ m/s},$$

where n is the speed of the shaft, rpm, and the permissible values of $[pv]$ are given in Table. 26.1. This calculation is usually performed as a test, since the diameter of the shaft neck is found from the calculation of endurance and the length of the bearing (l) is assigned structurally. But you can also not set the length of the trunnion (or neck) and, with the selected bearing material, determine l from formula (26.5):

$$l \geqslant \frac{F_r}{d[p]}.$$

Table 26.1   Allowed values $[p]$ and $[pv]$ for sleeve bearings

Parameters	Parameter values for liner materials					
	СЧ20	АСЧ-2	БрОФ10-1	БрАЖ9-4	Б16	Nylon
v (less)	0.5	1	10	4	12	4
$[p]$, MPa	4	12	15	15	15	15
$[pv]$, MPa·m/s	–	12	15	12	10	15

Note. The v values are maximum

26.2.3 Lubricants

For lubrication of sleeve bearings operating at a specific pressure of more than 10 MPa and a sliding speed of less than 3 m/s, antifriction grease lubricants are used, given in Table 26.2.

According to the composition, there are calcium-based lubricants (solidols, characterized by water resistance, but low heat resistance); sodium-based or sodium-calcium-based lubricants (constalins, heat-resistant, but soluble in water); and complex calcium lubricants (uniols, litols, etc., thermo-and water-resistant).

For bearings operating in any mode of friction, from liquid to boundary, liquid mineral, and synthetic industrial oils can be used, having the letter "И" in the designation.

According to their intended purpose, industrial oils are divided into four groups, denoted by the second letter:

Л for lightly loaded bearings;
Г for hydraulic machines;
Н for sliding guide bearings;
Т for heavy-duty bearings (gears).

According to their performance properties and composition, industrial oils are divided into five subgroups, denoted by the third letter:

А without additives;
В with antioxidation and anticorrosion additives;
С, Д, Е additionally with anti-wear, extreme pressure, and anti-pump additives.

Kinematic viscosity classes of oils (in centistokes) at a temperature of 40 °C are the following: 2, 3, 5, 7, 10, 15, 22, 32, 46, 68, 100, 150, 220, 320, 460, 680, 1000, 1500.

The following industrial oils are most applicable in (viscosity classes): И-Л-А (7; 10; 22); И-Г-А (32; 46; 68); И-Л-С (3; 5; 10; 22); И-Г-С (32; 46; 68; 100; 150;

Table 26.2 Lubricants

Brand, GOST (TU)	Interval operating temperatures	Characteristics
Solid oil: synthetic, GOST 4366-76 *;	−20 . . . 65	General purpose, gradually replaced
fatty, GOST 1033-79	−25 . . . 65	
Litol-24, TU 21150-75	−40 . . . 130	Multipurpose (main) grease
Uniol-1, TU 201150-78	−40 . . . 150	Heat-resistant, complex, wide application
CIATIM-202, GOST 6267-74 *	−60 . . . 90	Frost resistant, general purpose

220); И-Г-В (46; 68); И-Н-Е (68; 100; 220); И-Г-Н-Е (32; 68); И-Т-С (320); И-Т-Д (68; 100; 220; 460; 680).

Along with liquid and plastic lubricants, solid lubricants are also used. Examples of such substances are colloidal graphite, molybdenum disulfide, fluoride compounds, etc. They are used to lubricate parts operating in a vacuum, in conditions of very low (below $-100\,°C$) or very high ($>300\,°C$) temperatures, when working in aggressive environments, when the presence of a minimum amount of oils or even their vapors is unacceptable.

Self-Test Questions

1. Classification of bearings by type of friction.
2. List the advantages of rolling bearings over plain bearings.
3. List the types of rolling bearings.
4. Classification of rolling bearings according to the direction of perceived loads.
5. Classification of rolling bearings according to the shape of rolling bodies.
6. Classification of rolling bearings by overall dimensions.
7. Symbols of rolling bearings.
8. List the design procedure for the rolling bearings.
9. What is the basic dynamic radial load capacity of a rolling bearing?
10. What is the calculation of the dynamic load capacity of the bearing?
11. Landing of rolling bearings.
12. What bearings are used for floating supports?
13. Draw a diagram of the installation of bearings vraspor and vrastyazhku.
14. Design of the sliding bearing.
15. Advantages of plain bearings over rolling bearings.
16. List the operating modes of the sliding bearings.
17. Draw a diagram of an autonomous sliding support.
18. Calculation of sliding bearings according to the permissible specific pressure.
19. What determines the purpose of the type of lubricant for the sliding bearing?
20. For which sliding bearings are liquid industrial oils used?

References

S. Chernavskij, *Proektirovanie mexanicheskix peredach [Mechanical Transmission Design]* (Mashinostroenie Publ., Moscow, 1976)

M. Eroxin et al., *Detali mashin i osnovy' konstruirovaniya [Machine Parts and Design Basics]* (KolosS Publ., Moscow, 2005)

M. Kacz, *Sistema uslovny'x oboznachenij podshipnikov kacheniya sharikovy'x podshipnikov, sharikov i rolikov [Legend System for Rolling Bearings, Ball Bearings, Balls and Rollers]* (ZAO "Sfera" Publ., Moscow, 2007)

Chapter 27
Couplings

Abstract The definition of a coupling as a device for connecting (disconnecting) the shafts of jointly operating machine components is given. The classification of couplings with a detailed description of the design of each type of device is given. Attention is drawn to the following. When selecting a coupling, you have to consider all the system's requirements. It is not enough to know what the driver and load are and how big the shaft is. You must also know how the two halves are assembled and whether there is misalignment, as well as the system's operating range and the operating temperature.

Keywords Couplings · Blind couplings · Compensating couplings · Elastic couplings · Coupling clutches · Overrunning clutch · Controlled clutches · Self-driving couplings

27.1 Purpose and Classification of Couplings

Couplings are devices for connecting shafts of jointly operating units (aggregates) of machines, parts of composite shafts, as well as for connecting shafts with parts located on them (gears, sprockets, etc.) (Schwerdlin 1996).

In addition to connecting functions, couplings often perform other functions at the same time:

- control the operation—turn on and off the actuator when the engine is running, make it easier to start the machine, etc. These functions are performed by the so-called *controlled* clutches;
- adjust the parameters—limit the speed of rotation, protect the parts from accidental (unacceptable) overloads, *safety* couplings.

Couplings that perform several functions simultaneously must periodically connect and disconnect the connected shafts. Such couplings are called *coupling*. In contrast, couplings that perform only connecting functions are called *constants*.

The couplings must have compensating and damping capabilities. The first of them should to some extent compensate for the deviations of the connected shafts

V. Molotnikov, A. Molotnikova, *Theoretical and Applied Mechanics*,
https://doi.org/10.1007/978-3-031-09312-8_27

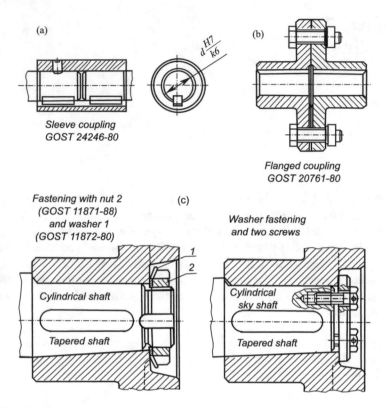

Fig. 27.1 Blind couplings

from the coaxial position. The damping capacity is necessary to mitigate the adverse effect of external load fluctuations. Damping refers to the ability to dampen and calm vibrations caused by shocks or shocks. These properties are necessary especially for reverse loading. The couplings are diverse in design. Consider the most common coupling designs.

The simplest in design are the so-called blind couplings (Fig. 27.1), forming a rigid connection of the shafts. Here, at the position (a), the bushing coupling on the prismatic keys is shown. Along with prismatic keys, segmental dowels are also used. Taper pins are sometimes used instead of dowels. In the latter case, locking screws are not required to prevent axial displacement of the bushing (Molotnikov 2017).

Blind couplings are used for shafts with a diameter of up to 70 mm. When using them, the offset of the axes should not exceed 3 . . . 5 microns. To facilitate installation, often blind couplings are made in the form of two half-couplings, connected by bolts or other means (Fig. 27.1b). This type of coupling is called flanged. The methods of attaching the half-couplings to the shaft are shown in Fig. 27.1c. Flanged couplings are standardized (GOST 20761-80), designed to connect shafts with a diameter of 12 . . . 200 mm when transmitting torque from 8 to 45000 N·m.

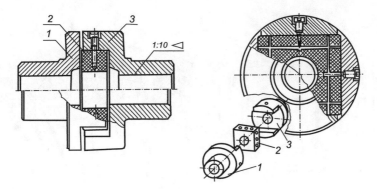

Fig. 27.2 Compensating coupling

Couplings in the form of autonomous devices (units) produced by specialized enterprises are common. This makes it easier to standardize the couplings. The most common are *compensating* and *elastic* permanent couplings. When installing the coupling, they strive to ensure that the contribution of shaft and support deformations to compensate for deviations from the coaxial position is minimal. The coupling must perform the functions of the compensator of the system not only during installation but also during operation of the drive.

The design of the compensating coupling is shown in Fig. 27.2. Such a coupling allows the connection of shafts with increased mutual displacements of the shafts. It consists of two half-couplings 1 and 3 with grooves that act as guides for the intermediate link 2.

Another type of compensating couplings are gear couplings (Fig. 27.3a). They are used in high-load structures for shafts with a diameter of $40 \ldots 200$ mm and allow an angular displacement of up to $0.5°$ with a radial displacement of the order of $0.008L$, with L being the distance between the gear joints of the coupling. The coupling consists of two bushings 1 and two clips 2, respectively, with external and internal involute teeth.

The connection of the bushings to the shaft is carried out by a key and a fit with tension or slots. The large number of simultaneously engaged teeth ensures the compactness and high load capacity of the gear couplings. When rotating shafts mounted with a skew, there is a cyclic radial and axial displacement of the teeth of the bushings relative to the clips. This leads to tooth wear. To increase their wear resistance, the tooth surfaces are hardened to a hardness of $45 \ldots 55\ HRC$, and oil is poured into the coupling. if

Chain couplings (Fig. 27.3b) are used for shaft connections with diameters of $20 \ldots 140$ mm when transmitting torque from 63 to 8000 H · m at a speed of 500 to 1600 rpm. The torque is transmitted by means of sprockets and a chain. The couplings compensate for the radial displacement of the shafts up to 1.2 mm and the angular displacement up to $1°$. The connection of the coupling halves to the shafts can be carried out using dowels or splines. To protect against contamination

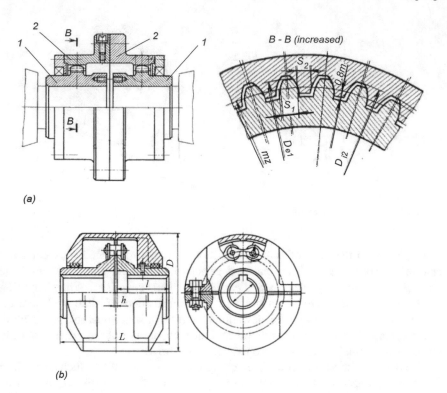

Fig. 27.3 (**a**) Toothed clutch type $M3$ according to GOST 5006 - 83 and chain (**b**) Chain coupling GOST 20742 - 81

and better lubrication, it is advisable to close the coupling with a casing with sealing elements. Chain couplings have a simple design, are easy to maintain and technologically advanced in manufacturing, have small dimensions and weight, and are easy to install. Their disadvantage is the presence of "dead move." This limits their use in reverse gears.

27.2   Flexible Couplings

Flexible couplings are the most common. They are able not only to compensate for radial and angular displacements but also to damp vibrations, shock absorbers, and shocks.

The following types of standard flexible couplings are used.

Here in position (a): 1 corrugated rubber sheath, 2 steel pin. For a clutch with a toroidal shell (Fig. 27.4b: 1 shell, 2 pressure ring, 3 conical surface, 4 bolt, 5 coupling half. *Elastic sleeve-finger* MUVP type couplings (Fig. 27.4a). They are used in electric motor drives and in other cases for shafts with a diameter of

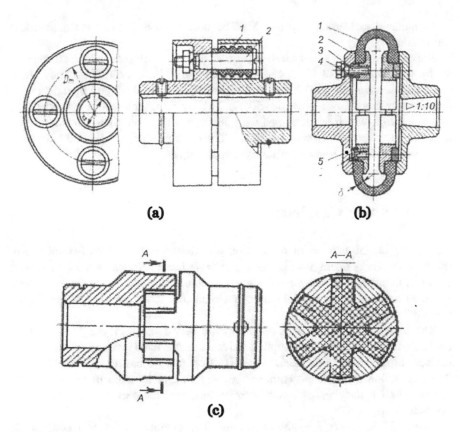

Fig. 27.4 Elastic couplings: (**a**) sleeve-finger; (**b**) with a toroidal shell; (**c**) with an asterisk

10 . . . 180 mm at torques of 6.3 . . . 6000 H · m. The standard provides couplings of type I (with a cylindrical hole for the shaft) and type II (with a conical hole for the shaft). They can be made in two versions—for long and short shaft ends. The torque between the coupling halves is transmitted through the rubber corrugated bushings 1, put on the fingers 2. Depending on the size, the couplings allow a radial displacement of the shaft axes by 0.2 . . . 0.4 mm, a longitudinal displacement of the shafts by 10 . . . 15 mm, and an angular displacement by 1°. Operating temperature ranges from −40 to +50 °C.

Elastic couplings with a torus-shaped shell (Fig. 27.4b) of a convex or concave profile have good compensating properties. The coupling consists of a rubber elastic element (shell) 1 and a half-coupling 2. The clamping half-rings 5 are attracted to the half-couplings by screws 3 through the rings 4. When assembling the coupling, the semi-rings are connected to the ring by 4 screws 6, located between the screws 3. The coupling can connect shafts with diameters of 14 . . . 240 mm and transmit torques from 200 to 40000 N ·. Depending on the size, the coupling can compensate for longitudinal displacements up to 11 mm, radial displacements up to 5 mm, and angular displacements up to 1.5°. The disadvantage is the large diametrical dimensions.

Elastic couplings with an asterisk (Fig. 27.4c) are recommended for connecting shafts with diameters of 12 ... 45 mm at a torque of 2.5 ... 400 N·m. They allow shaft displacement up to 0.4 mm and skew angle up to 1.5° and operate in the temperature range from −40 to +50 °C. The coupling consists of two identical half-couplings, which are equipped with end cams. Between these cams is placed a rubber sprocket with four or six petals. The couplings are simple and have high working qualities but transmit relatively small torques.

27.3   Coupling Couplings

Depending on the method of connection and disconnection of half-couplings in the process works distinguish between *controlled* and *self-controlled* (automatic) *clutches.* Controlled clutches are designed to connect or disconnect shafts or parts mounted on them using special control mechanisms. They are used in gear boxes and similar mechanisms.

The designs of the controlled couplings are very diverse. Figure 27.5a shows a controlled cam coupling. Its coupling halves 1 and 2 have protrusions on the end surface cams, shown in Section $A - A$. Half coupling 1 has no axial movements. The connection and disconnection of the shafts occur by axial movement of the coupling half 2, the cams of which enter (or exit) into the sockets at the end of the opposite coupling half.

Figure 27.5b shows a toothed coupling. The operation of a gear coupling is similar to that of a cam coupling. Its closing and opening are carried out by axial movement of the bushing 1 of the drive shaft along the key 3. To reduce the wear of the teeth, a thick lubricant is poured into the coupling housing. The tightness of the housing is ensured by the seal 4. The axial movement of the gear sleeve to the right restricts the projection on the drive shaft.

To activate the cam or gear coupling, the shafts must be stopped. Clutches of such structures fail due to wear of the cams or teeth. Therefore, their calculation is carried out in the form of limiting the average pressure on the cams and teeth.

In addition to controlled coupling couplings, self-controlled couplings are often used. They perform one of the following functions:

- torque transmission in one direction only—overrunning clutches;
- switching on and off at a given speed—centrifugal clutches;
- limitation of the transmitted load—in this case, the coupling is called a safety one.

Let's look at examples of the designs of such devices.

Figure 27.6a shows a roller overrunning clutch. When the sprocket 1 rotates clockwise, the rollers 2 are wedged between the sprocket and the outer ring 3. The

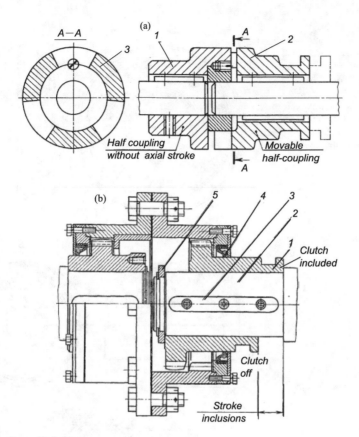

Fig. 27.5 Controlled clutches

clamping devices 4 reduce the "dead" stroke and contribute to an even distribution of the load between the rollers. When the sprocket rotates in the opposite direction, the rollers will not jam, and the rotation is not transmitted. This effect is used in the gears of bicycles, motorcycles, machine tools, etc.

Figure 27.6b shows a centrifugal six-cylinder coupling. Here, the driven part of the coupling is located on the sliding bearings. Such couplings are used for smooth starting of conveyor drives, lifting machines, etc. They allow the electric motor to easily accelerate and, upon reaching a certain speed, begin a smooth acceleration of the working machine. The contact between the pads and the drum will occur when the centrifugal force reaches the value of

$$F = m\omega^2 r,$$

where m is the mass of the pad, ω is the angular velocity of the drive shaft, and r is the distance from the center of mass of the pad to the axis of rotation.

Fig. 27.6 Self-driving
clutches: (**a**) roller
overrunning: 1 sprocket, 2
roller, 3 outer ring, 4
clamping device; (**b**)
centrifugal six-core: 1–2
coupling halves, 3 pad; it in
safety ball: 1 driving coupling
half, 2 pressure spring, 3 ball,
4 driven coupling half, 5 seal

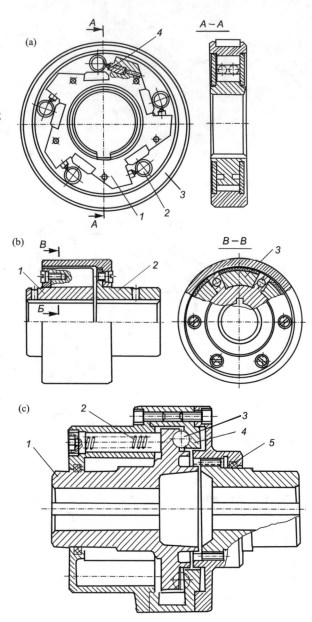

Figure 27.6c shows a safety coupling. The clutch is triggered when the torque
exceeds a certain set value. The value of this moment depends on the stiffness of the
pressure springs 2.

In conclusion of this chapter, we note that in the vast majority of cases of drive
design, the designer chooses couplings from standard products. The initial data in

this case are the transmitted torque, the purpose and features of the drive, the landing diameters of the coupling halves and shafts, etc.

The interested reader can find more information about couplings in the Schwerdlin (1996) article.

Self-Test Questions

1. Name the purpose of mechanical couplings.
2. Give the classification of couplings according to the principle of their operation.
3. What errors in the arrangement of shafts connected by couplings are possible when installing agricultural machinery units?
4. What parameters are used to choose a standard coupling?
5. Name the scope of application of blind couplings.
6. What advantages of the chain coupling make it the most versatile among compensating couplings?
7. What are the main types of elastic couplings?
8. What is the purpose of safety couplings?
9. Draw a sketch of the overtaking clutch. Name the technical devices that use overtaking clutches.
10. Why is the elastic sleeve-finger coupling most applicable in the electric drive?
11. What types of couplings used in agricultural machinery are known to you from experience?

References

V. Molotnikov, *Texnicheskaya mexanika [Technical Mechanics]* (Lan' Publ., Sant-Peterburg, 2017)
H. Schwerdlin, *Standard Handbook of Machine Design*, Chapter 29. Couplings (McGraw-Hall Publ., 1996) , pp. 941–974

Chapter 28
Connections of Parts and Units of Machines

Abstract The parts that make up the machine are interconnected in one way or another. The most typical types of these links are considered here. First, the design, technological characteristics, advantages, and disadvantages, as well as strength calculations of permanent joints, welded, glued, riveted, and tight joints, are given. For detachable connections—threaded, keyed, splined, and pinned—multivariate designs, areas of their application, and rational methods of strength calculations are considered. An example of calculating a multi-bolt connection is given.

Keywords Details · Nodes · Communication · Connections · Strength · Constructions · Welds · Bolts · Rivets · Dowels · Slots · Pins

28.1 General Characteristics of Connections

Machines are assembled from parts by connecting them and fixing the relative position. The mating parts of the parts together with the connections form connections, the names of which are usually determined by the type of connection or connecting element, for example, bolted, welded, keyed, etc. (Chernavskij and Reshchikov 1976).

Depending on the design, technological, operational, and other requirements, the connections can be *detachable* or *all in one*. Detachable connections are disassembled without damaging the parts, and non-removable ones can be disassembled only by destroying the connections or parts. In this case, the landing surfaces or the parts themselves are often damaged.

Detachable connections are made both movable and stationary. In moving joints under load, the relative movement of parts provided for by the functional purpose is possible. The main use is for fixed joints, in which the parts do not perform relative movement during operation. In such joints, under load, the mutual displacement of the contact points of the parts occurs only due to their deformations.

Joint parts form the most common class of machine parts. The quality of these parts and the performance of the assembly often determine the reliability of the entire structure.

28.2   All-in-One Connections

28.2.1   Welded Joints

Depending on the method of forming interatomic connections of parts, distinguish the following classes of welding:

- thermal class of welding, in which welding is carried out by general or local heating of the connected parts;
- mechanical class of welding is carried out by plastic deformation of parts in the joint zone;
- thermomechanical class of welding performed during joint heating and plastic deformation of the parts to be welded.

In practice, more than 60 welding methods are used, in which the material is melted (arc, gas, electron beam, etc.), deformed without heating (cold, explosion, etc.), or heated and plastically deformed (contact, gas-press, high-frequency, etc.).

Advantages of welded joints:

- they are better than other types of connection to bring the composite parts to the whole;
- easier to provide equal strength condition;
- reduce the cost of the product due to the relatively low labor intensity of the welding process;
- contribute to a relatively small weight of the structure;
- ensure tightness and tightness of the joint.

The main *disadvantages of* welded joints are:

- presence of residual stresses due to non-uniform heating and cooling;
- possibility of warping of parts during welding (especially thin-walled parts);
- the possibility of formation of hidden defects (cracks, non-vapors, slag inclusions);
- sensitivity to vibration and shock loads.

Quality control of welded joints using destructive or non-destructive methods is usually mandatory for detecting defects.

Of the various welding methods, electric arc welding is the most common. There are three types of arc welding: automatic, semi-automatic, and manual.

Automatic submerged and shielded gas welding allows you to weld a wide variety of materials, including high-strength, stainless, and heat-resistant steels, aluminum alloys, etc. In this case, the thickness of the welded parts can be from fractions to several tens of millimeters. It provides a high quality of the seam, not inferior in strength to solid metal. Such welding is high performance and economical.

Semi-automatic welding is used to perform long seams in places that are inaccessible to automatic welding, as well as in the case of short or intermittent seams.

Manual welding is used when other welding methods are not rational (difficult to reach seam location and other reasons). It has low productivity, and the quality of the seam depends on the qualification of the welder.

28.2.2 Structures of Welded Joints and Their Strength Calculation

Depending on the location of the parts to be welded, the following types of joints obtained by arc and gas welding are distinguished: butt, lap, T-bar, and corner. Butt joints (or butt joints) (Fig. 28.1) are the simplest and most reliable. They give a strength close to that of the base metal and provide the least addition of mass to the joint and a minimum stress concentration.

Depending on the thickness of the δ metal, welding is performed with one-sided (Fig. 28.1a, b) or two-sided (Fig. 28.1c–e) seams. For medium and large thicknesses, the edges are processed to ensure penetration by performing straight (Fig. 28.1b–d) or curved (Fig. 28.1e) bevels.

In automatic welding, the thickness of parts that are welded without cutting edges is doubled, and the bevel angle is halved ($30\ldots35°$). In addition to sheet metal, pipes, channels, corners, and other shaped profiles are also butt-welded.

With high-quality welding, the joint is destroyed mainly in the zone of thermal influence. Therefore, the calculation of the butt joint is usually performed by the size of the section of the part in this zone. Calculation of the butt joint working on the:

Fig. 28.1 Butt welded joints: **(a)** $\delta \leqslant 8$ mm; **(b)** $\delta \leqslant 16$ mm; **(c)** $\delta \leqslant 30$ mm; **(d)** $\delta \leqslant 40$ mm

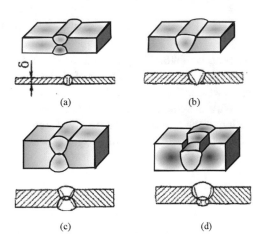

(a)   (b)

(c)   (d)

Fig. 28.2 Lap joint welds:
(**a**) frontal seams; (**b**) flank
seams; (**c**) oblique seam; (**d**)
combined seam

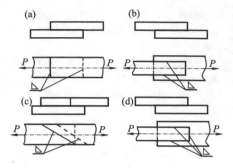

$$stretching : \sigma_p = \frac{P}{F} \leqslant [\sigma]_p;$$

$$compression : \sigma_c = \frac{P}{F} \leqslant [\sigma]_c; \ bend : \sigma_{\text{ГЁ}} = \frac{M}{W} \leqslant [\sigma]_p; \qquad (28.1)$$

$$tensile \ bend : \sigma = \frac{M}{W} + \frac{P}{F} \leqslant [\sigma]_p,$$

where P is the tensile or compressive force, N; F is the area of the calculated cross-section, mm^2; M is the bending moment, H·mm; W is the moment of resistance of the calculated cross-section, mm^3; $[\sigma]_p$ and $[\sigma]_c$, respectively, are the permissible tensile and compressive stresses, MPa.

Lap joints (Fig. 28.2) are made with corner seams. Cutting the edges in this process is not required. Depending on the location of the seam relative to the load, there are frontal (Fig. 28.2a), flank (Fig. 28.2b), oblique (Fig. 28.2b), and combined (Fig. 28.2d) seams.

For flank and frontal sutures, the most likely destruction is along the plane of the bisector of the right angle formed by the suture catheters. In this case, the area of the calculated cross-section will be

$$F = Lk_p \cos 45°,$$

where L is the total length of the weld and k_p is the calculated seam length; usually take $k_p = (0.9 \ldots 1, 2)\delta$, where δ is the smallest of the thicknesses of the elements to be welded $(k_{p\,min}) = 3$ mm at $\delta \geqslant 3$ mm).

With a symmetrical arrangement of the corner joints relative to the line of action of the tensile or compressive force, it is assumed that the stresses are evenly distributed along the entire length of the perimeter of the joint, and the strength condition is written as

$$\tau = \frac{P}{0.7k_pL} \leqslant [\tau]_w, \qquad (28.2)$$

where $[\tau]_w$ is the allowable shear stress of the seam material. In fact, the stress distribution along the length of the flank seam is uneven. Therefore, the length of the flank seam is limited to the value $60k_p$.

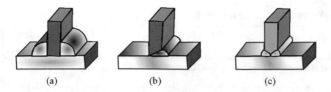

Fig. 28.3 T-bar welds: (**a**) without cutting edges; (**b**) with one-sided cutting edges; (**c**) with two-sided cutting edges

In the case of an asymmetric arrangement of the seams relative to the line of action of the external force, the welded joint is calculated, taking into account the load that falls on each seam. For example, in the case of the connection of the corner and the nodal shape, the seams at the rim and at the shelf perceive different loads even in the case of the action of the force P along the axis of the corner. Therefore, the total length L of the flank seams is distributed between the length at the rim (L_o) and the length at the pen (shelf)(L_p) inversely proportional to the distance from the centerline:

$$L_o = \frac{b - z_0}{b} L, \quad L_p = \frac{z_0}{b} L, \tag{28.3}$$

where b is the width of the angle flange to be welded and z_0 is the distance from the center of gravity of the angle section to the non-welded flange.

T-joints (Fig. 28.3) are used if the connected elements are located perpendicular to each other. Such connections are performed both without cutting edges (Fig. 28.3a) and one-way (Fig. 28.3b) or two-way (Fig. 28.3b) cutting edges. When performing a corner seam joint without cutting the edges, the seam is calculated for the cut using the formula (28.2) where L is the length of each of the corner seams. T-joints with cutting edges are calculated as butt joints (depending on the current load) according to the formulas (28.1).

28.2.3   Permissible Stresses for Welded Joints

The value of the allowable stress of the weld material depends on the following factors:

- weldability of the material (low-and medium-carbon steels are well welded);
- of the technological process type;
- of the electrode type;
- the nature of the active loads.

The values of permissible stresses for low-and medium-carbon steels, as well as low-alloy steels under static load, are given in Table 28.1.

Table 28.1 Permissible stresses for welds

Welding	Permissible voltages at		
	Stretch	Compress	Slice
Automatic submerged, as well as manual electrodes E42A, E50A	$[\sigma]_p$	$[\sigma]_c$	$0.65[\sigma]_p$
Manual electrodes E42, E50	$0.9[\sigma]_p$	$[\sigma]_c$	$0.6[\sigma]_p$

28.2.4   Soldered Connections

Such compounds are formed by local heating of a low-melting filler material-solder, which spreads over the heated surfaces and forms a solder seam when cooled. The joint material is diffusively and chemically bound to the material of the parts to be joined.

Heating of the solder and parts during soldering is carried out with a soldering iron, a gas burner, in furnaces, etc. When soldering in furnaces, the solder is laid in the form of wire and tape contours, pastes, etc. To weaken the harmful effect of oxidation of the surfaces of parts, special fluxes based on rosin, borax, etc. are used. Soldering is also used in a neutral gas environment or in a vacuum.

The design of the solder joint includes three main points: the choice of the base metal, the choice of the solder, and the soldering method. In this case, the main condition is the solderability of the base metal and ensuring the required strength of the joint. The sensitivity of the base metal to heat and its tendency to crack under the influence of molten solders are also taken into account. Therefore, for example, brazing steel with brass is of limited use, since copper causes embrittlement of the joint.

The solder should wet the fat-free surfaces of the parts well. As solders, sometimes, pure metals are used but mainly alloys based on tin, copper, silver, etc. For soldering steel products of non-responsible purpose, copper-based solder of the Γ<63 brand is used. For internal seams of medical equipment, as well as in electrical engineering and instrumentation, tin-lead solders ПОС 90, ПОС 61 and POS 50 are used.

In solder joints, butt joints and lap joints are used, as well as combinations of these connections. The calculation of butt and lap solder joints is similar to the calculation of welded joints, but in the lap joint, the area of the calculated cross-section is assumed to be equal to the contact area of the parts: $F = bl$, where b and l are the width and length of the contact area. The shear strength of joints made with tin-lead solders, as well as solders based on copper or silver, reaches $(0.8 \ldots 0.9)\sigma_{\mathrm{vr}}$, where σ_{vr} is the ultimate strength of the solder.

28.2.5 *Adhesive Joints*

Glue is an all-in-one joint, a substance that can connect materials and hold them together by bonding surfaces. Adhesive joints are widely used in many branches of mechanical engineering. Synthetic adhesives connect almost all materials of industrial significance (steel, alloys, copper, silver, wood, plastics, etc.). In some cases, bonding is the only way to connect dissimilar materials.

In metal structures, adhesives reliably and firmly connect heterogeneous materials of different thicknesses. Adhesive joints are cheaper than welded, riveted, or bolted ones. In addition, these connections do not weaken the base metal, since there is no need to drill holes for bolts or rivets. Often these connections are sealed without additional sealing. The most widely used joints in gluing are lap joints and telescopic joints. For telescopic joints, liquid glue is used (more often cold solidification). For lap joints, a high-strength adhesive, such as a film adhesive, is required. Common designs of adhesive joints are shown in Fig. 28.4.

The condition of the shear strength of the lap joint has the form

$$\tau = \frac{P}{bl} \leqslant [\tau]_{sh}, \tag{28.4}$$

where b and l are the width and length of the overlap and $[\tau]_{sh}$ is the permissible shear stress.

For phenolonitrile rubber adhesives $[\tau]_{sh} = 33 \ldots 40$ MPa; for epoxy ≈ 20 MPa; for polyurethane $10 \ldots 16$ MPa. The strength of the adhesive bond depends on the thickness of the adhesive layer. Typically this thickness is $0.05 \ldots 0.15$ mm. Quality control of joints is carried out by destructive and non-destructive methods (X-rays, infrared rays, etc.).

Fig. 28.4 Of adhesive joints: (**a**) butt; (**b, c**) overlapping; (**d**) mustache; (**e**) overlapping with undercut; (**f**) butt plate; (**g**) the same, with a double overlay; (**h**) butt joint with recessed double pad; (**i**) half-spike; (**k**) butt with beveled pads

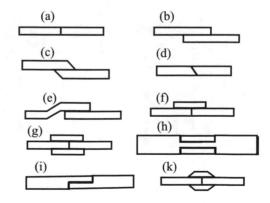

28.2.6 Riveted Connections

Riveted connections are carried out using a rivet—a short solid or hollow cylindrical rod with a locking head (Fig. 28.5). The use of this or that head shape in the design is mainly determined by the operational requirements (aerodynamic, etc.). Connections of this type are non-detachable.

During assembly, the rivet is installed in a pre-prepared hole in the parts, and a second closing head is formed with a special tool with a draft (riveting). In this case, the package of connected sheets is tightened due to the transverse elastic-plastic deformation of the rivet body, and the initial gap between the rod and the walls of the holes is filled.

Rivet joints are used mainly in aircraft structures, metal structures, and other products in which external loads act parallel to the joint plane, and the use of welding, soldering, or gluing is difficult or impossible for structural or technological reasons (non-welded materials, the inadmissibility of heating, etc.).

The sheets are overlapped (Fig. 28.6a) or butt-to-butt (Fig. 28.6) with one (b) or two (c) overlays. In this case, the rivets in the joint are arranged in simple (Fig. 28.7a) or staggered rows (Fig. 28.7b).

Fig. 28.5 The main types of rivets: (**a**) with a semicircular head; (**b**) with a low semicircular head; (**c**) hollow with a rounded head; (**d**) with a half-countersunk head; (**e**) with countersunk head; (**f**) flat head

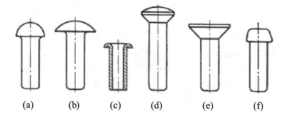

(a)   (b)   (c)   (d)   (e)   (f)

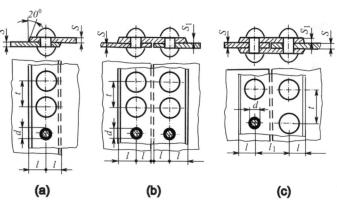

(a)   (b)   (c)

Fig. 28.6 Types of rivet joints: (**a**) lap-up ($d = S + 8$ mm, $t = 2d + 8$ mm, $l = 1.35d$); (**b**) butt joint with one pad [$d = S + 8$ mm, $t = 2d + 8$ mm, $l = (1.35\ldots1.5)d$, $S_1 = (1\ldots1.25)S$]; (**c**) butt joint with two pads [$d = S + (5\ldots6)$ mm], $S_1 = (0.6\ldots0.7)S$, $t = 2.6d + 10$ mm, $l = 1.35d$, $l_1 = 3d$

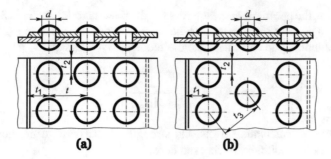

Fig. 28.7   Rivet layout diagrams: (**a**) simple rows ($t \geqslant 3d$); (**b**) chess rows ($t_3 \geqslant 3d$)

Riveted joints have a relatively high manufacturability but at the same time high labor intensity of manufacture and quality control.

28.2.7   Calculation of Riveted Joints

Operating experience has shown that the violation of the integrity of riveted joints occurs due to breakage of heads and destruction of rivet rods or due to crushing of the walls of the holes of the parts to be joined.

The calculation of the rivets for the cut is performed according to the formula

$$\tau_c = \frac{P}{F} = \frac{4P}{n\pi d^2} \leqslant [\tau]_{sh}, \tag{28.5}$$

where P is the force acting on the joint; F total cross-sectional area n of rivets with rod diameter d; and $[\tau]_{sh}$ permissible shear stress of the rivet body. In the design calculation, the rivet diameter is determined:

$$d \geqslant \sqrt{\frac{4P}{\pi n [\tau]_{sh}}}. \tag{28.6}$$

A connection with two overlays (Fig. 28.6c) can fail by cutting the rivets along two planes. Such connections are called *double-cut.* When calculating them, it is assumed that the effort per section is half the total effort.

If the rivets are made of a material that is less durable than the parts to be joined, then the calculation of the rivets for crushing is also performed:

$$\sigma_{\text{cr}} = \frac{P}{sdn} \leqslant [\sigma]_{\text{cr}}, \tag{28.7}$$

where $[\sigma]_{cr}$ is the permissible shear stress for the rivet material and s is the smallest of the thicknesses of the elements to be joined. From the condition (28.7) in the design calculation, you can determine the required rivet diameter:

$$d \geqslant \sqrt{\frac{P}{ns[\sigma]_{cr}}}. \qquad (28.8)$$

In addition to calculating rivets, the strength of the parts to be joined is also calculated. Two conditions are checked here.

First, the normal stresses in the cross-section weakened by rivets should not exceed the permissible ones:

$$\sigma = \frac{P}{F_{net}} = \frac{P}{s(b - dz)} \leqslant [\sigma]_p, \qquad (28.9)$$

where s is the thickness of the sheet, b its width, d is the diameter of the rivet hole, and z is the number of rivets in one row.

Second, the strength of the joined sheets is calculated for crushing the walls of the holes according to the formula

$$\sigma = \frac{P}{2s(t - 0.5d)z} \leqslant [\sigma]_{cr}, \qquad (28.10)$$

where $(t - 0.5d)$ is the length of the dangerous section and $[sigma]_{cr}$ the permissible shear stress for the material of the part.

The permissible stresses in the calculation of riveted connections are taken as follows:

$$[\tau]_{sh} = (0.5 \ldots 0.6)[\sigma]_{ten}, \quad [\sigma]_{cr} = (2 \ldots 2.5)[\sigma]_{ten}, \quad [\sigma]_p = (0.4 \ldots 0.5)\sigma_t,$$

$$(28.11)$$

where σ_t is the yield strength of the material of the corresponding element of the connection.

28.2.8   Interference Joints

In tight joints, the spanning part 1 (Fig. 28.8) has a hole diameter d_o less than the diameter d_b of the shaft 2. After assembly, the resistance to the mutual displacement of the parts is created by the elastic deformation of the stretching of the enclosing part and the compression of the shaft. A strong connection is provided by the tightness—the difference $\delta = d_b - d_o$ of the shaft and hole diameters.

Connections with guaranteed tension are conditionally divided into longitudinal-press and transverse-press. Assembly of longitudinal press joints is performed using manual, pneumatic, or hydraulic presses by applying an axial force F (Fig. 28.8a)

Fig. 28.8 Scheme of longitudinal-press connection in the process (**a**) and after (**b**) pressing

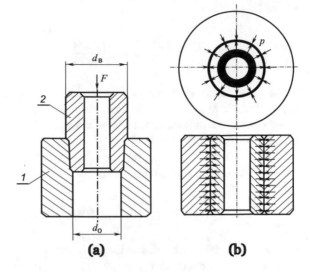

(a)　　　　　(b)

to one of the parts. On the contact surface of the parts, there are significant normal (perpendicular to the surface) pressures p (Fig. 28.8b) and friction forces that prevent the parts from shifting.

Before pressing, the axes of the parts must be combined. Usually, for this purpose, a groove is made on the shaft or in the hole; as a result, the shaft is freely inserted into the covering part for a certain length. The bearing surfaces of the parts must be strictly perpendicular to their axes to avoid distortions during pressing.

In cross-press joints, the convergence of the mating surfaces occurs radially to the surface. To obtain such connections, the enclosing part is usually heated before assembly. With this method, the strength of the joint increases by 2 … 3 times, since during assembly, there is no wear of the surfaces, and the micro-roughnesses seem to adhere to each other. After cooling, the part tightly covers the shaft.

The assembly of connections with cooling of the covered part is also used. When using carbon dioxide (dry ice) for this purpose, the temperature of the part is reduced to 195 K; when using liquid nitrogen, temperature is reduced to 78 K. For very large tightness, a combined method is used: the covering part is heated, and the covered part is cooled.

28.2.9 Calculation of Guaranteed Tightness Landings

The minimum pressure on the cylindrical seating surface of the parts (Fig. 28.8b), providing adhesion strength, is calculated by the formula ((Rabinovich 1965), pp. 260–261):

$$p_{min} = \frac{T}{f \pi D L},$$

(28.12)

where

$$T = \sqrt{\left(\frac{2M_\kappa}{D}\right)^2 + F^2}$$

is the shear force when the joint is loaded with a torsional moment M_κ and an axial force F; D and L are the diameter and length of the landing surface; and f is the coefficient of adhesion, taken when assembling by pressing $f = 0.08$, with temperature assembly $f = 0.14$.

With a known minimum pressure (28.12), the required design relative interference on the landing surface is determined:

$$\delta_{min} = k\frac{2p_{min}}{E_1}, \qquad (28.13)$$

where E_1 is Young's modulus of the shaft material.

Then the required tabular (measured) minimum interference will be

$$\Delta_{min} = \delta_{min}D + u, \qquad (28.14)$$

where u is the correction for the unevenness of the landing surfaces, determined by the formula

$$u = 1.4(l_1 R_{a1} + l_2 R_{a2}), \qquad (28.15)$$

where (28.15) R_{a1} and R_{a2} are the arithmetic mean deviations of the profile of the landing surfaces (see, e.g., (Rabinovich 1965)) and l_1 and l_2 are coefficients depending on the cleanliness of the landing surfaces (see Rabinovich (1965), p. 261).

According to the value of Δ_{min}, determined by formula (28.14), a standard fit is selected, in which the smallest interference value is equal to or slightly greater than Δ_{min}.

28.3   Detachable Connections

28.3.1   Threaded Connections

28.3.2   Threaded Connections

Among all types of connections, threaded are the most common. This is due to their most important advantages:

Fig. 28.9 Threaded connections: (**a**) bolted; (**b**) screw; (**c**) hairpin

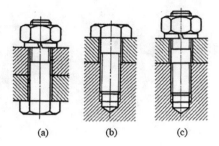

(a)   (b)   (c)

- high load capacity and reliability;
- ease of use;
- high degree of standardization;
- easy to manufacture;
- small size and weight;
- low cost.

The main disadvantage of threaded connections is the concentration of stresses in the thread, which reduces their strength, especially under cyclic loads. This disadvantage almost always causes the failure of the threaded connection, and the consequence is the destruction of the bolt rod or thread.

Mounting screws, depending on the type of threaded connection, are used in the following versions.

- Bolts are screws with nuts (Fig. 28.9a). They are used to fasten parts of not very large thickness, as well as to fasten parts made of materials that do not provide sufficient strength and durability of the thread. Bolts are also used when frequent screwing and unscrewing is necessary. Bolts do not require threading into parts, but they are not always easy to assemble and do not contribute to giving the machine perfect shapes.
- The screws (Fig. 28.9b) are screwed into one of the fastened parts. They are used in the case of sufficient strength of the material of the part and its sufficient thickness, in the absence of a place for the location of the nut, with strict requirements for the weight of the product.
- Pins (Fig. 28.9c) are used in the same cases as screws, but when the material of the connected parts does not provide sufficient thread durability during frequent assembly and disassembly of the connection.

Standard threaded connections are calculated based on the least durable element—the bolt (screw, stud) rod (Fig. 28.9). The exception applies in the case of using low nuts. According to the found diameter of the bolt d, other details of the threaded connection are selected: nuts, washers, etc. The length of the bolt is assigned depending on the thickness of the connected parts. The calculation of bolts depends on the nature of loading and technological features of the assembly of threaded connections (tightened, un-tightened, with a gap between the bolt and the hole, or without a gap). Let's consider the most common calculation cases.

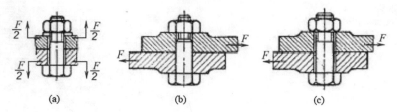

Fig. 28.10 Basic design cases

1. The bolt (stud) is installed in the hole of the body parts with a gap and tightened. The joint is loaded with an external axial force F (Fig. 28.10a). It is assumed that the entire external load is absorbed by the bolt, and the strength condition is written as

$$\sigma = \frac{F}{S_1} = \frac{4F}{\pi d_1^2} \leqslant [\sigma]_{st}), \tag{28.16}$$

where S_1 is the cross-sectional area of the bolt along the inner thread diameter d_1 and $[\sigma]_{st}$ the permissible tensile stress of the bolt material.

To prevent parts shifting in this case, the joint must be tightened so that the friction force F_{fr} in the joint is not less than the shear force, i.e.

$$F_{\text{fr}} = fF \geqslant Q,$$

where F is the required tightening force and f is the coefficient of friction (for dry surfaces made of steel and cast iron, $f \approx 0.15\ldots0.2$). In calculations they take

$$F = \frac{kQ}{f}, \quad k = \begin{vmatrix} 1.2\ldots1.5 \text{ under static load,} \\ 1.8\ldots2 \text{ at variable load.} \end{vmatrix} \tag{28.17}$$

When designing calculations from the condition (28.16), you can determine the required internal thread diameter:

$$d_1 \geqslant \sqrt{\frac{4F}{\pi[\sigma]_{st}}}. \tag{28.18}$$

Using the result (28.18) and GOST 8724-81, determine the required outer diameter and thread pitch.

2. The bolt is installed in the holes of the parts without clearance (Fig. 28.10b), and the joint is loaded with a lateral force. In this case, destruction can occur as a result of shearing the bolt along the plane of the joint (as in a riveted joint). The

condition for the strength of the bolt in terms of the permissible shear stresses has the form

$$\tau = \frac{4F}{\pi d^2} \leqslant [\tau]_{sh},\qquad(28.19)$$

where d is the bolt shank diameter and $[\tau]_{sh}$ is the allowable shear stress. From the condition (28.19), the required diameter of the bolt bar is determined:

$$d \geqslant \sqrt{\frac{4F}{\pi [\tau]_{sh}}}.\qquad(28.20)$$

The scope of application of connections according to the scheme (Fig. 28.10b) is limited to structures made of thin-sheet elements (aircraft and shipbuilding, etc.) for technological reasons. The complexity of making a gas-free joint in production conditions forces you to install bolts with a small tension, which significantly increases the cost of assembly work.

3. The bolt is installed in the hole of the connected parts with a gap, and the joint is loaded by a transverse force. This case corresponds to the connection shown in Fig. 28.10c. The absence of displacements in the joint here can be ensured by friction forces. The condition of mutual immobility of the connection parts will be

$$F \leqslant F_{fr},\qquad(28.21)$$

where F_{fr} is the friction force, which is determined by the product of the friction coefficient f and the tightening force of the connection F_0:

$$F_{fr} = f F_0.$$

The tightening force F_0 causes tensile and torsional stresses in the section along the thread ID. The strength condition is recorded by the equivalent stress:

$$\sigma_{equ} = \frac{1,3 \cdot 4F_0}{\pi d_1^2} \leqslant [\sigma]_{st}.\qquad(28.22)$$

In the case of a design calculation, the condition (28.22) gives

$$d_1 = \sqrt{\frac{5.2F}{\pi f [\sigma]_{st}}}.\qquad(28.23)$$

28.3.3 Design of Threaded Connections Under Variable Loads

The reason for the malfunction of threaded connections at alternating stresses is the fatigue failure of bolt rods (screws, studs) at the transition of a smooth rod to the threaded part. The bolt, the diameter of which is determined by formula (28.18), is additionally checked for fatigue strength, which is estimated by the safety factor for the amplitude of the cycle:

$$n_a = \frac{\sigma_{an}}{\sigma_a} \geqslant [n]_a, \qquad (28.24)$$

where σ_a is the amplitude of the stress cycle,

$$\sigma_{an} = \frac{\sigma_{-1}}{k_\sigma},$$

with

$$k_\sigma = \begin{vmatrix} 3.5 \ldots 4.5 \text{ for carbon steels,} \\ 4 \ldots 5.5 \text{ for alloy steels,} \end{vmatrix}$$

and tightened connection. For knurled threads, the given values of k_σ are reduced by $20 \ldots 30\,\%$. The amplitude of the cycle in formula (28.24) is calculated by the formula

$$\sigma_a = \frac{\chi F}{2S_1},$$

where χ is the axial load coefficient, which is usually taken equal to $0.2 \ldots 0.4$, and the permissible cycle amplitude coefficient is taken as $[n_a] = 2.5$.

28.3.4 Features of the Calculation of Group Threaded Connections

Consider first a connection with equally loaded bolts. This is, for example, the fastening of the vessel lid shown in Fig. 28.11a. In this connection, an axial force acts on the bolt, tending to open the joint. The bolts are pre-loaded and loaded with an additional external axial force. One bolt has a force $F = R/z$, where R is the resultant of external axial forces and z is the number of bolts in the joint. With a known force F, the design stress in the bolt is determined by formula (28.16), and then the bolt diameter is calculated using formula (28.18).

Otherwise, the group connection works, which is subject to a system of forces that shift the parts at the joint (Fig. 28.11b). In such joints, the most loaded bolt

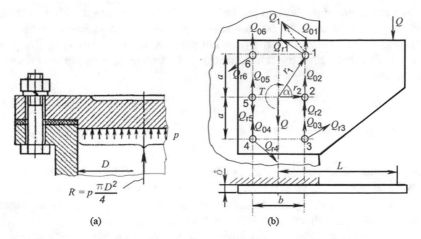

Fig. 28.11 Group bolted connections

(screw, stud) is first determined, which is then checked for strength by the methods discussed above. The remaining (less loaded) bolts take the same diameter as the calculated one.

So, in the case of the connection shown in Fig. 28.11b, the cantilever force Q in the design scheme is transferred to the center of gravity of the joint section, adding the moment T (shown by the dotted line). It is assumed that the force Q is distributed evenly between the bolts. One bolt has a cutting force $Q_{0i} = Q/z$, $(i = 1, \ldots, 6)$, where z is the number of bolts. The reactions of Q_{0i} are directed parallel to Q, but in the opposite direction.

From the moment T, the bolt that is as far away from the center of gravity of the joint as possible will be the most loaded. The cutting force from the moment T in this bolt will be

$$Q_{Tmax} = \frac{Tr_{max}}{\sum\limits_{i=1}^{z} r_i^2}, \tag{28.25}$$

where r_i is the distance from the axis of the i-th bolt to the center of gravity of the joint section and r_{max} is the largest of these distances. The reactions Q_{Ti} are perpendicular to the radii r_i and are directed opposite to the moment T.

The geometric sum of forces (28.25) and Q_0 is taken as the design load, i.e.

$$F = |\mathbf{Q} + \mathbf{Q}_{Tmax}|. \tag{28.26}$$

In the mounting bracket diagram shown in Fig. 28.11b, the most loaded bolts will be the 1st and 3rd, not the 4th and 6th (explain why).

The mounting of the bracket can be made with bolts installed with or without a gap. In the first version, the external load is perceived by the friction forces at the joint and in the second directly by the coupling of the bolts with the holes of the bracket. The specifics of the calculation in each of these options are considered by an example.

28.3.5 Example

Calculate the mounting bolts of the bracket shown in Fig. 28.11b, with the following data: $Q = 3000$ N, $L = 400$ mm, $a = 250$ mm, $b = 400$ mm. Material of bolts and bracket—St. 3 with yield strength $\sigma_t = 200$ MPa. The thickness of the bracket $\delta = 100$ mm. The tightening of the bolts is uncontrolled.

The calculation is performed in two versions: when installing bolts with a gap and without a gap.

Solution The calculation is performed in this order.

1. Bring the force Q to the center of gravity of the joint and calculate the force Q_0:

$$Q_0 = \frac{Q}{z} = \frac{3000}{6} = 500 \text{ N}.$$

2. We calculate the moment:

$$T = QL = 3000 \cdot 400 = 1.2 \cdot 10^6 \text{ N} \cdot \text{mm}.$$

3. Using formula (28.25), we determine the shearing force on bolt 1 from the moment T:

$$r_1 = \sqrt{a^2 + (b/2)^2} = \sqrt{250^2 + (400/2)^2} = 320 \text{ mm}.$$

$$Q_{Tmax} = \frac{Tr_1}{\sum r_i^2} = \frac{1.2 \cdot 10^6 \cdot 320}{4 \cdot 320^2 + 2 \cdot 200^2} = 674 \text{ N}.$$

4. We calculate the loads from the moment on bolts 2 and 5:

$$Q_{T2} = Q_{T5} = \frac{Tr_2}{\sum r_i^2} = \frac{1,2 \cdot 10^6 \cdot 200}{5696 \cdot 10^2} = 421 \text{ N}.$$

5. Calculate the estimated load Q_1:

$$\alpha = \text{arctg} \frac{a}{(b/2)} = 51.35°; \quad \beta = \frac{\pi}{2} - \alpha = 38.65°;$$

$$Q_1 = \sqrt{Q_0^2 + Q_{Tmax}^2 + 2Q_0 Q_{Tmax} \cos \beta} =$$
$$= \sqrt{500^2 + 674^2 + 2 \cdot 500 \cdot 674 \cdot \cos(38.65°)} = 1110 \text{ N}.$$

A. Bolts installed with clearance. Taking the friction coefficient at the joint $f = 0.15$ and the safety coefficient $k = 1.3$, we determine the required bolt tightening force by formula (28.17):

$$F = \frac{kQ_1}{f} = \frac{1.3 \cdot 1110}{0.15} = 9620 \text{ N}.$$

In case of uncontrolled tightening, the safety factor is assumed to be $n = 4$. By condition, the yield strength of steel is $\sigma_t = 200\text{MPa}$. Then

$$[\sigma] = \frac{\sigma_t}{n} = \frac{200}{4} = 50 \text{ MPa}.$$

Now, according to the formula (28.23), we find

$$d_1 = \sqrt{\frac{4 \cdot 1.3 \cdot F}{\pi [\sigma]}} = \sqrt{\frac{4 \cdot 1.3 \cdot 9629}{3.14 \cdot 50}} = 17.85 \text{ mm}.$$

We accept the M20 bolt.

B. Bolts installed without play. Determine the bolt diameters from the shear strength condition using formula (28.20):

$$d \geqslant \sqrt{\frac{4F}{\pi [\tau]_{sh}}},$$

where $[\tau]_{sh} = 0.3\sigma_t = 0.3 \cdot 200 = 60\text{MPa}$. We have

$$d = \sqrt{\frac{4 \cdot 9620}{3.14 \cdot 60}} = 14.29 \text{ mm}.$$

The connection can be made with M14 bolts. Let's check for crushing of the connection «bolts - bracket holes»:

$$\sigma_{cr} = \frac{Q_1}{d\delta} = \frac{1110}{14 \cdot 10} \approx 8 \text{ MPa}.$$

Assuming that the allowable stress of the material of the connected parts on the crumple $[\sigma]_{cr} = 0.8\sigma_t = 0.8 \cdot 200 = 160$ MPa, we make sure that the condition of the crumple strength is met.

Thus, both options are acceptable: both the setting of the M20 bolts with a gap and the setting of the M14 bolts without a gap. However, for technological reasons, the second option is more time-consuming and expensive.

28.3.6   Keyed Joint

Keyed connections are called the connections of coaxial parts by means of a special part—the key. The mating surfaces can be either cylindrical or conical.

In mechanical engineering, non-stressed connections with prismatic or segmental dowels (Fig. 28.13a), as well as stressed connections (Fig. 28.13b) with wedge dowels, are used. The dowels of the listed types are standardized. The variety of keyways used is shown in Fig. 28.12.

The advantages of keyed connections are simplicity of design, low manufacturing cost, ease of assembly, and disassembly. Disadvantages are inapplicability in fast-rotating joints and the need for manual fitting of joint elements.

Key connections (Fig. 28.14a) are the most commonly used. They have a relatively small plunge-in depth into the shaft and are easy to assemble and dismantle. In many cases, they are also used in interference fittings.

The prismatic key has a rectangular cross-section with a height-to-width ratio of $b/h = 1$ for shafts up to 22 mm in diameter and $b/h = 0.5$ for shafts of large

Fig. 28.12 Connections: (**a**) key; (**b**) taper key

Fig. 28.13 Connections: (**a**) square or rectangular key; (**b**) square or rectangular key with one end rounded; also available with both ends rounded; (**c**) square or rectangular key with gib head; (**d**) woodruff key; also available with flattened bottom; (**e**) tapered rectangular key

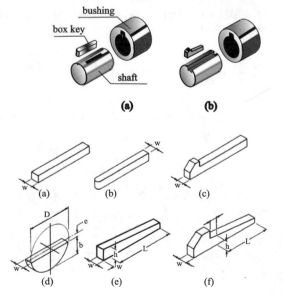

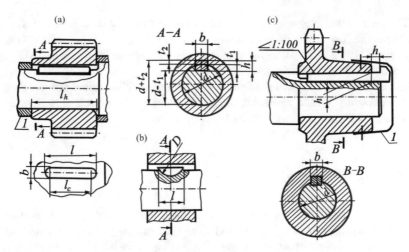

Fig. 28.14 Sketches of the main types of keyed connections

diameters. Working prismatic dowels are the side narrow faces. A gap is provided in the radial direction. The key material is pure drawn steel with a yield strength of at least 600 MPa.

Segment dowels (Fig. 28.14b) have a deep fit and are not prone to skewing under load. They are interchangeable. However, a deep groove significantly weakens the shaft. Therefore, segment dowels are used mainly on low-loaded sections of the shaft.

Keyway connections fail due to crumpling of the working surfaces of the faces. A key cut is also possible. The calculation of the joint for crumpling is performed according to the formula

$$\sigma_{cr} = \frac{2T}{dl_p t_2} \leqslant [\sigma]_{cr}, \tag{28.27}$$

where T is the torque; l_p key working length (Fig. 28.14a); $t_2 = 0, 4h$ the depth of the key cut into the hub; and $[\sigma]_{cr}$ permissible collapse stress. From the condition (28.27), you can determine the working length of the key:

$$l_p = \frac{2T}{dt_2[\sigma]_{cr}}. \tag{28.28}$$

The shear strength of the keys is usually not checked, since this condition is met when using standard keys and recommended values.

If the strength condition (28.27) is not met, then the joint is formed using two dowels set at an angle of 120° or 180°.

Segment key connections are also calculated using formula (28.27). In this case, take $t_2 = h - t_1$ (see section $A - A$ in Fig. 28.14a).

The permissible crushing stresses at constant load in the connection of the steel shaft and the key made of pure-drawn steel take $[\sigma]_{cr} = 500\ldots600$ MPa, for other steels $150\ldots180$ MPa, for cast iron and aluminum $80\ dotsc100$ MPa, and for textolite and woodplastic $15\ldots25$ MPa.

Large values are assigned for light operation, and smaller values are assigned for heavy operation. With a reversible load, the values of the permissible stresses are reduced by 1.5 times and with shock loads by 2 times.

28.3.7 Spline Joints

The spline connection of the shaft (male surface) and the hole (female surface) is carried out using splines (grooves) and teeth (protrusions) radially located on the surface. The most important advantages of spline joints are high strength and smaller overall dimensions than keyed joints, ensuring the alignment of the shaft and the hole in combination with the possibility of axial movement of the part along the axis.

According to the shape of the cross-sections of the teeth, *straight-line* (Fig. 28.15a), *involute* (Fig. 28.15b), and *triangular* (Fig. 28.15c) connections are used.

Connections with straight teeth are most common in domestic mechanical engineering. Depending on the number of teeth ($z = 6\ldots20$) and their height, the standard sets three series of connections for shafts with a diameter from $d = 23$ mm to $d = 125$ mm—light, medium, and heavy. A larger number of teeth have connections of the heavy series.

Involute teeth are more technologically advanced compared to straight teeth and have higher accuracy and strength. To increase the wear resistance of the tooth surfaces of straight and involute joints, they are cemented to give a hardness of about $50\ldots60HRC$.

Connections with a triangular tooth profile are used mainly in instrument engineering and in fixed landings with small overall dimensions.

The main dimensions of the spline joint (outer diameter D and length l) are set by the designer during the design, and then a check calculation of the spline joint for crumpling is performed using the formula

Fig. 28.15 Types of spline connections

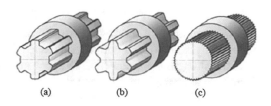

(a)   (b)   (c)

$$\sigma_{cr} = \frac{2T}{zd_m lh\psi} \leqslant [\sigma]_{cr}, \tag{28.29}$$

where d_m is the average diameter of the joint; z is the number of teeth; h and l are the height and length of the contact surface of the teeth, respectively; and $\psi = 0.7 \ldots 0.8$ is a coefficient that takes into account the stress concentration at the ends of the teeth. To set the height and length of the contact surface, you can use the following recommendations:

– For straight teeth:

$$h = \frac{D-d}{2} - 2f; \quad d_m = \frac{D+d}{2};$$

– For involute teeth:

$$h = m; \quad d_m = mz,$$

where m is the engagement modulus.

The number of teeth and the diameters are taken from the standards depending on the shaft diameter. The length of l is taken from the spanning part (hub).

28.3.8 Pin Connections

Pin connections are used to connect axles and shafts with parts mounted on them when transmitting small torques. In Fig. 28.16a, gear 1 is mounted on shaft 2 and fixed with a pin 3. The pins are also used to relieve threaded connections from lateral forces (Fig. 28.16b) and exact fixing details and for other purposes (Shigley 1996).

The advantages of pin connections are simplicity of design; ease of installation and disassembly, and precise centering of parts due to the tight fit; they work as a safety guard, especially when attaching wheels to the shaft.

The disadvantage of pin connections is the loosening of the connected parts by the hole.

The main types of pins used in mechanical engineering are shown in Fig. 28.16c.

The most widely used are tapered smooth pins. They are installed in through holes. Taper pins with threads are placed in the blind holes. In the joints that experience shocks and shocks, put adjustable pins. The same conical pins are used in the joints of fast-rotating parts. The tapered pins can be removed repeatedly and put back in place without compromising the reliability of the connection. Structurally, they are performed with a taper of 1:50.

Cylindrical pins are placed in the holes with tension. In moving joints, cylindrical pins are placed with the ends riveted. A big disadvantage of cylindrical pins is the loosening of the fit during reassembly and disassembly.

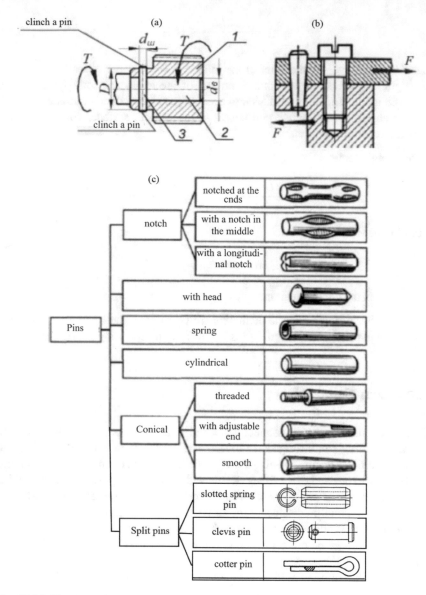

Fig. 28.16 Pin connections

In general, cylindrical pins are used as mounting parts for increased accuracy of fixing the connected parts relative to each other and in cases where there is a need to protect the connected parts from lateral displacement forces acting in opposite directions. Cylindrical and conical pins are made of structural steels. The sizes of pins of the listed types are specified in the corresponding GOST.

Some designs of notched pins are shown in Fig. 28.16c. They differ from smooth pins in that they have grooves of various shapes on the surface. When such pins are driven into the holes, the material previously squeezed out of the grooves is elastically deformed in the opposite direction. This position provides increased grip strength. It is important to note that the notched pins allow multiple mounting and dismounting without weakening the coupling force. These pins are made of spring steel.

The spring pins resemble a cylindrical tube cut along the generatrix. They are made of spring steel with subsequent heat treatment. Spring pins are inserted into holes that are smaller in diameter than the diameter of the pin. A reliable connection is made due to the elastic forces of the pin material. Repeated assembly and disassembly do not lead to any noticeable weakening of the coupling force.

Pins with heads, as well as notched ones, have three longitudinal grooves (along the axis of the rod), which create a spring action when they are installed in the hole. They are mainly used for fixing radio and electrical equipment parts with clamps on panels, shields, etc.

Smooth pins are made of steel 45 and A12, while pins with grooves and spring pins are made of spring steel.

When the wheels are attached to the shaft, the pins transmit both the torque and the axial force. Like rivets, pins work to cut and crumple. The corresponding calculations are usually performed as a test:

$$\tau_{cp} = \frac{8T}{\pi d_s^2 d_p} \leqslant [\tau]_{cp},$$

$$\sigma_{cr} = \frac{2T}{d_s d_p (D - d_s)} \leqslant [\sigma]_{cr}, \tag{28.30}$$

where T is the torque in the joint; d_s and d_p, respectively, are the diameter of the shaft and pin (see Fig. 28.16a); D is the hub diameter; and $[\tau]_{sh}$ $mbox{and}$ $[\sigma]_{cr}$ are the allowable shear and collapse stresses for the pin material, respectively.

Grooved pins are calculated in the same way as smooth, but allowable material stresses are reduced by 50%.

Self-Test Questions

1. Types of joints of machine parts.
2. Welding classes.
3. Advantages and disadvantages of welded joints.
4. Quality control of welded joints.
5. Designs and calculation of butt welded joints.
6. Lap welded joints (design, calculation).
7. T-bar welded joints.

8. Permissible stresses for welded joints.
9. Soldered and glued joints (structures, scope of application, calculation).
10. Rivet joints (structures, calculation).
11. Tight joints. Calculation of landings with guaranteed tightness.
12. Threaded connections (designs, advantages and disadvantages, calculation).
13. Calculation of threaded connections at variable loads.
14. Features of the calculation of group threaded connections.
15. Keyway connections (designs, advantages, disadvantages, scope of application, calculation).
16. Spline connections as multi-key connections.
17. Pin connections (scope, advantages, disadvantages).

References

S. Chernavskij, V. Reshchikov, *Spravochnik metallista [Metalist's Handbook]* (Mashinostroenie Publ., Moscow, 1976)

S. Rabinovich, *Raschyot posadok s garantirovanny'm natyagom. Kratkij spravochnik mashinostroitelya [Calculation of Landings with Guaranteed Interference. A Short Reference Book of a Mechanical Engineer]* (Mashinostroenie Publ., Moscow, 1965)

J. Shigley, *Standard Handbook of Machine Design*, chapter 22. Unthreaded Fasteners (McGraw-Hill Publ., 1996), pp. 676–704

Chapter 29
Body Parts of Machines and Mechanisms

Abstract The body parts of the machines are the basic elements. Various parts and assembly units are installed on them, the accuracy of the relative position of which must be ensured both in statics and during the operation of the machine under load. In accordance with this, the body parts must have the required accuracy, the necessary rigidity, and vibration resistance, which ensures the correct operation of the mechanisms and the absence of vibration. In this chapter, the definition of a group of parts that can be conditionally assigned to the group of body parts is given. Depending on the purpose, the body parts are divided into three groups, of which the most capacious is the group of body parts of machine components. The main technological requirements for cast body parts are outlined. Some software products used in the design of housing parts are listed.

Keywords Body · Plate · Box · Bed · Frame · Bearing body · Node · Rack · Table · Support · Finite elements

29.1 General Characteristics of Connections

The frames serve as a support and a connecting element of the main units of the machines. They ensure their correct mutual placement and perceive the main forces arising in the machine. *Plates* support entire machines and machine drives or their aggregates. *Boxes* and similar body parts enclose the mechanisms of the machines. All parts of these three groups are united by a common name—b o d y p a r t s.

Body parts are largely responsible for the performance of machines according to the criteria of vibration resistance, accuracy of work under load, and durability. In stationary machines, they make up from 75 to 85% of the mass of the machine. It follows that material savings are most likely to be obtained by improving the design of body parts.

The body parts are very specific to each group of machines and therefore very diverse in design. In the course of machine parts, only general questions of the design of body parts are considered.

© The Author(s), under exclusive license to Springer Nature Switzerland AG 2023
V. Molotnikov, A. Molotnikova, *Theoretical and Applied Mechanics*,
https://doi.org/10.1007/978-3-031-09312-8_29

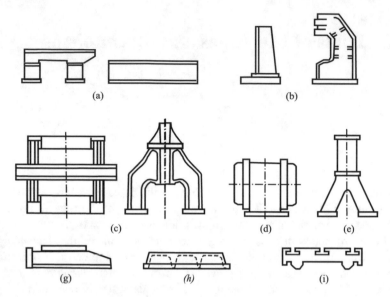

Fig. 29.1 Body parts: (**a**) frames; (**b**) racks; (**c**) portal frames; (**d**) ring frame; (**e**) cylinder blocks; (**f**) slider; (**g**) plate; (**h**) table

By purpose, all body parts can be divided into the following groups:

(1) frames, frames, and load-bearing bodies;
(2) bases and foundation slabs;
(3) body parts of the units.

The latter, in turn, can be divided into:

(a) housings, boxes, and cylinders;
(b) racks, brackets, suspensions, and other fixed supporting parts;
(c) tables, calipers, sliders, and other movable body parts;
(d) Casings and covers.

Body parts are also distinguished by the following features:

(a) with two overall dimensions, significantly smaller than the third,—long machine frames, crossbars, and sliders;
(b) with one overall size significantly smaller than the other two—plates and flat tables;
(c) with overall dimensions of the same order—boxes.

Typical body parts are shown in Fig. 29.1.

The main criteria for the operability and reliability of housing parts are strength, rigidity, and durability. Strength is the main criterion for housing parts operating in heavy conditions under shock and variable loads. Rigidity is also very significant for most body parts, since large elastic deformations disrupt the proper operation of mechanisms, reduce kinematic accuracy, and contribute to the occurrence of vibrations.

Durability in terms of wear is of great importance for body parts with guides or with cylinders made in one piece, without linings and sleeves. The service life of the main body parts is usually longer than the service life of the machines due to their wear and tear.

The stiffness requirements are best met by materials with a high modulus of elasticity. In this case, the material must allow the production of parts of perfect shape and without heat treatment. These requirements are met by cast iron and steel. Most of the body parts are cast in cast iron. However, in the presence of large mass forces, light alloys are used. Welded body parts are used to reduce weight and size mainly in single and small-scale production. In large-scale production, stamped and bent elements are used. In the absence of significant loads, plastic body parts (portable handheld machines and devices, covers, casings, etc.) are used.

29.2 Construction of Cast Parts

With body parts subject to bending and index Parts! Cast torsion, it is advisable to make them in the form of thin-walled structures, in which the wall thickness is determined by technological conditions. In this case, the twisted parts must have closed sections and bent ones with the maximum material separation from the neutral axis. If it is necessary to install viewing windows, the weakening of the housing must be compensated for by flanges or rigid covers.

It should be borne in mind that reducing the wall thickness is the most effective way to reduce the material consumption of body parts. By reducing the wall thickness by a factor of k while maintaining constant stiffness and similarity of the contour, one can reduce the mass by a factor of $k^{2/3}$. In this case, the required stiffness of the walls is provided by ribbing. The recommended thicknesses δ of the outer walls of cast iron castings are chosen depending on the given size of the casting N:

N, м	0.05	0.15	0.3	0.7	1.0	1.5	2.0	3.0
δ, mm	4	5	6	8	10	12	15	20

where the reduced size N is calculated by the formula

$$N = \frac{2L + B + H}{3}, \qquad (29.1)$$

where L, B, and H are the length, width, and height of the casting, respectively.

Fig. 29.2 Conjugation of
walls of castings

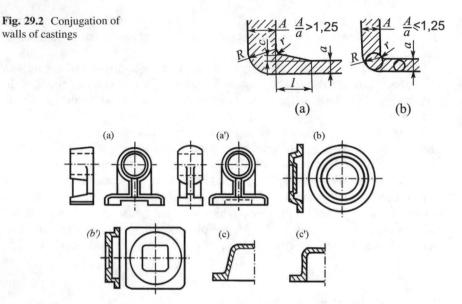

Fig. 29.3 Examples of constructions of cast parts, technological (**a–c**) and low-tech (**a′–c′**) for molding

The inner walls and fins cool more slowly than the outer ones. Therefore, it is recommended to choose their thickness on the order of 0.8 of the thickness of the outer edges. The height of the edges is recommended to assign no more than five times their thickness. At the same time, the walls of steel castings are chosen thicker than cast iron, by $20\ldots40\%$. Non-ferrous cast alloys allow for much smaller wall thicknesses than cast iron.

The wall should be as constant as possible in thickness. If this cannot be sustained, then the change in thickness should be smooth. Thus, when the walls with a thickness of A and a are conjugated at an angle, if $A/a > 1.25$, it is desirable, and at $A/a \geqslant 1.5$, it is mandatory to perform a wedge thickening of the thin wall (Fig. 29.2).

It is recommended to choose the radii of the rounded castings from the normalized series and minimize the number of radii in each cast. The correctness of the conjugation forms can be checked by entering the section of the walls of the circles into the contour (Fig. 29.2b). If possible, the diameters of the inscribed circles should not differ by more than 25% (up to 50% is allowed).

For the manufacturability of the models, the geometric shapes of the castings should allow machine-tool processing, i.e., the walls should be outlined with flat, cylindrical, or conical surfaces. We should also strive to ensure that the molding of simple castings is possible in a single semi-mold (Fig. 29.3a). To facilitate the removal of models from the forms, it is desirable to provide small structural slopes of the side walls (Fig. 29.3b, c).

29.3 Contents of Calculations

Body parts that are close in shape to the bars are usually subject to bending and torsion. Parts of this type with a closed circuit and the effect of loads on the partitions work as a whole, and they are calculated according to the corresponding formulas of the resistance of materials. Housing parts of the same type, but consisting of two walls with perpendicular or diagonal partitions (such as lathes), are calculated as thin-walled statically indeterminate systems. The portal frames are subjected to the same calculation.

Generally speaking, the calculation of body parts is preferably performed using computer programs such as Ansys, COSMOSWorks, etc., which use the finite element method (Molotnikov 2012, ch. 22). For an example, the results of this calculation in the COSMOSWorks program (Alyamovskij 2010) are shown in Fig. 29.4. Here, in the left position, the analyzed part is shown in the form of a box body with a finite element grid applied. The position on the right shows the (exaggerated) deformed state of the box, colored in tones of different saturation, and the scale of correspondence of these tones (colors of color) to the equivalent voltage levels (in Pa).[1]

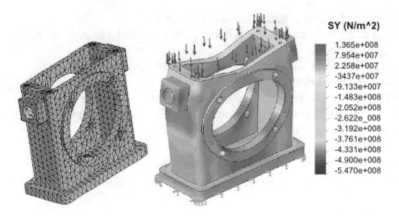

Fig. 29.4 Calculation of the box by the finite element method in the COSMOSWorks package

---

[1] The results of the COSMOSWorks program are the most visible and suitable for analyzing the stress-strain state of the part under study in color printing.

29.4   Foundations of Beds and Slabs

Horizontal frames and slabs are designed taking into account their joint work with the foundation, i.e., as elements located on an elastic base. In this case, the base is modeled according to Winkler, Nicolai, or Love (Lur'e 1970).

Machines subject to high dynamic loads, heavy machines, and high-precision machines are installed on individual foundations, while the rest of the machines are installed on a common concrete floor of the workshop with a thickness of 200...250 mm. In this case, the machines are installed on gaskets, wedges, or shoes, attracted by anchor bolts and poured with cement mortar.

High-precision machines, for which vibrations are unacceptable, as well as machines that are a source of strong vibrations in the shop, are installed on elastic vibration-proofing supports or gaskets, in particular critical cases-on vibration-proofing foundations. Such foundations are supported by springs or rubber shock absorbers.

Self-Test Questions

1. List the body parts of machines and mechanisms.
2. Why is the metal economy most likely to be obtained by improving the design of body parts?
3. List the groups of body parts for their intended purpose.
4. Name the main criteria for the performance of body parts.
5. What are the basic rules for designing cast parts?
6. The content of calculations of body parts. The concept of the method of finite (boundary) elements.
7. Foundations of walls and slabs.
8. Which machines are installed on individual foundations?
9. In what cases are the machines installed on vibration-proofing foundations?

References

A. Alyamovskij, *Inzhenerny'e raschety' v SolidWorks Simulation [Engineering Calculations in SolidWorks Simulation]* (DMK Press Publ., Moscow, 2010)
A. Lur'e, *Teoriya uprugosti [Elasticity theory]* (Nauka Publ., Moscow, 1970)
V. Molotnikov, *Mekhanika konstruktsii [Mechanics of Structures]* (Krasnodar, Lan' Publ., SPb., Moscow, 2012)

Part V
Introduction to CAD Based on AutoCad ~ AutoLisp

Chapter 30
Application System AutoCAD

Abstract This chapter introduces the concept of a computer-aided design system—CAD. A brief history of the development and current state of CAD is presented. The purpose and structure of the system, the user interface of the simplest system AutoCad $\sim$ Autolisp, and the technique of creating a drawing are described. Exercises for building drawings of machine parts are given.

Keywords Design; Automation; An outline of the history of CAD; AutoCAD system; Purpose and composition of the system; System interface; Drawing elements; Temporary tracking point; Binding; Hatching

30.1 A Brief Outline of the History of CAD

Modern scientific and technological progress and the rapid development of high technologies have created the need to create computer-aided design systems, CAD, which can significantly reduce the time for developing engineering projects and performing design documentation. The current level of technical and software tools of electronic computer technology allows us to talk about the automation of the engineer's work. The use of CAD frees it from time-consuming, often the same type of calculation and drawing work, which the computer performs faster, more efficiently, and more accurately (Molotnikov and Molotnikova 2006).

The first versions of various CAD systems contained mainly tools for simple two-dimensional drawing and were essentially a very convenient "electronic kulman." But already at this stage of evolution, CAD tools differed favorably from an ordinary kulman with such great advantages as the ability to form an electronic archive of drawings, create analog drawings based on prototype drawings, edit drawings, etc. At the turn of the century, the apotheosis of achievements in CAD development was the creation of parametric solid-state modeling systems, in which powerful graphical capabilities are combined with computational service tools. Examples of such integrated programs are the mechanical desktop applications of Autodesk, "Compass" of the domestic company "Askon," APM development of Bauman

Moscow State Technical University, Cadmech Desktop, AVS, Rotation of Intermech (Belarus), architectural applications of ARCO (Russia, APIO Center), Archicad (Softdesk), etc.

Impressive success in the production of professionally oriented CAD systems creates the illusion that basic general technical disciplines (descriptive geometry, engineering graphics, resistance of materials, etc.) are "going out of fashion." Such a judgment can satisfy or even please only an unqualified user. There is no need to prove that in the labor market, a higher quotation will be used by a specialist for whom computer applications are not a "thing in themselves" but a powerful tool that can be supplemented or changed by the user himself, depending on the class of tasks being solved. And this requires not only knowledge of basic disciplines but also elements of algorithm and programming, as well as skills in working with graphics packages. In our book, we aimed to show that reaching this level of skill is a feasible task.

We do not claim that we managed to say our own word when describing the structure and components of AutoCAD ~ AutoLISP. It is not at all easy to do this nowadays, since the system has long been established, expanded, and deepened. And we did not set ourselves the task of writing another dissertation. The main goal that the authors set for themselves was to popularize knowledge in the field of using AutoCAD and programming in AutoLISP, as well as familiarization with the techniques of creating parametric images.

30.2   An Outline of Modern Computer-Aided Design (CAD) Systems

According to the standards adopted in the 1980s, CAD is not just some kind of program installed on a computer, it is an information complex consisting of hardware (computer), software, descriptions of methods used in working with the system, data storage rules, and much more.

There are a large number of CAD systems on the modern market that solve different problems. In this review, we will consider the main computer-aided design systems in the field of mechanical engineering. The reader can find the CAD classification, for example, in the overview (*Obzor populyarny'kh sistem avtomatizirovannogo proektirovaniya (CAD)* [An overview of popular computer-aided design (CAD) systems] 2021).

30.2.1   Basic and Light CAD

Lightweight CAD systems are designed for 2D design and drafting, as well as for creating separate 3D models without the ability to work with assembly units.

The undisputed leader among basic CAD systems is AutoCAD. AutoCAD is a basic CAD system developed and supplied by Autodesk. AutoCAD is the most widely used CAD system in the world that allows you to design in both 2D and 3D environments. With AutoCAD, you can build 3D models, create and design drawings, and much more. AutoCAD is a platform-based CAD system, i.e., this system does not have a clear focus on a specific project area, and it is possible to carry out at least construction projects, even machine-building projects, and work with surveys, electrical engineering, and much more.

The computer-aided design system AutoCAD has the following distinctive features:

- the de facto standard in the CAD world;
- wide possibilities of customization and adaptation;
- tools for creating applications in embedded languages (AutoLISP, etc.) and using the API;
- Abundance of third-party programs.

In addition, Autodesk develops vertical versions of AutoCAD—AutoCAD Mechanical, AutoCAD Electrical, and others, which are intended for professionals in the relevant field.

30.2.2 Mid-Level CAD

Medium CAD systems are programs for 3D product modeling, calculations, and design automation of electrical, hydraulic, and other auxiliary systems. Data in such systems can be stored both in a conventional file system and in a unified electronic document management and data management environment (PDM and PLM systems). Often in middle-class systems, there are programs for preparing control programs for CNC machines (CAM systems) and other programs for technological design.

From mid-level CAD systems, the largest number of users received applications: Autodesk Inventor from Autodesk, SOLIDWORKS from Dassault Systemes (Alyamovskij 2010), Solid Edge from Siemens PLM Software (Tickoo 2020), and Compass-3D from the Russian company Ascon (Askon 2017).

Midrange CAD systems are the most popular systems on the market. They successfully combine the price/functionality ratio, are able to solve the overwhelming number of design problems, and satisfy the needs of most of the customers.

30.2.3 "Heavy" CAD

"Heavy" CAD systems are designed to work with complex products (large assemblies in aircraft construction, shipbuilding, etc.). Functionally, they do everything the same as medium-sized systems, but they have a completely different architecture and work algorithms.

Among the "heavy" systems, NX is the flagship CAD system from Siemens PLM Software. It is used for the development of complex products, including elements with a complex shape and dense arrangement of a large number of component parts.

Here are the main features of the NX system:

- support for various operating systems, including UNIX, Linux, Mac OS X, and Windows;
- simultaneous work of a large number of users within one project;
- full-featured simulation solution;
- advanced industrial design tools (freeforms, parametric surfaces, dynamic rendering);
- tools for modeling the behavior of mechatronic systems;
- deep integration with the Teamcenter PLM system.

The CATIA system from Dassault Systemes has similar advantages.

PTC Creo from PTC compares favorably with these "heavy" CAD systems. Let's name the following differences of the Creo system from competing solutions:

- efficient work with large and very large assemblies;
- history-based modeling and direct modeling tools;
- work with complex surfaces;
- the ability to scale the functionality of the system depending on the needs of the user;
- different views of a single, centralized model developed in the system;
- tight integration with PLM system PTC Windchill.

30.2.4   "Cloud" CAD Systems

Recently, "cloud" CAD systems have been actively developing, which work in a virtual computing environment, and not on a local computer. Access to these CAD systems is carried out either through a special application or through a regular browser. The indisputable advantage of such systems is the possibility of using them on weak computers, since all the work takes place in the вЂҕcloudвЂҝ.

Cloud CAD systems are actively developing, and if a few years ago they could be classified as light CAD systems, now they are firmly established in the category of medium CAD systems.

Let's name the two most popular cloud-type systems.

CAD Fusion 360 is focused on solving a wide range of tasks, from simple modeling to complex calculations. The system developer is Autodesk. Fusion 360 Features:

- advanced user interface;
- combination of different modeling methods;

- advanced tools for working with assemblies;
- ability to work in online and offline modes (with and without a permanent connection to the Internet);
- affordable purchase and maintenance cost;
- calculations, optimization, and visualization of models;
- built-in CAM system;
- possibilities of direct output of models for 3D printing.

Onshape is a fully cloud-based CAD system developed by Onshape. Things to look out for when choosing Onshape:

- access to the program through a browser or mobile applications;
- work only online;
- narrow focus on engineering design;
- complete set of functions for modeling mechanical engineering products;
- version control of created projects;
- support for the FeatureScript language for creating your own applications based on Onshape.

In conclusion of this section, we note the following. When choosing a CAD system, it is necessary to focus on the needs of the enterprise, the tasks that the users face, the cost of acquiring and maintaining the system, and many other factors.

30.3   AutoCAD Applied Base System

30.3.1   Purpose and Composition of the System

The first versions of the system (the 1980s) were mainly tools for simple two-dimensional drawing. Gradually, from version to version, the system was supplemented and developed, turning into a powerful, dynamic means of automating the design of a wide variety of objects. Starting from the tenth version, the system began to allow performing quite complex three-dimensional constructions in any plane of space. But the problem of obtaining a drawing of orthogonal projections based on a three-dimensional model of an object or structure was still problematic.

The twelfth version of the system turned out to be a turning point. It provided the user with the ability to work with extended memory, introduced dialog boxes into the interface, and found a mechanism for obtaining projections from a three-dimensional model. Since the thirteenth version, the system has learned вЂк to work under Windows, and since the appearance in 1999 of the fifteenth version, the system has been assigned the year of launch of the next version (the 15th version— AutoCAD 2000, etc.).

The system consists of three main components: the AutoCAD graphic editor, the high-level programming language AutoLISP, and tools for creating a user graphical interface. The latest versions of the system require at least 16 MB of RAM for installation and 150–200 free MB on the hard disk, and for working with three-dimensional models, it is also desirable to have an 8–16 MB video adapter with a graphics accelerator. Menu

30.3.2   User Interface

Select the shortcut of the AutoCAD program with the mouse pointer on the display screen, and launch the program by double-clicking on the icon with the left mouse button. After loading the program into the machine's RAM, the Getting Started window appears, shown in Fig. 30.1. You need to move the mouse pointer to the button with one of the options for getting started and click Ok, after which the desktop design window opens (Fig. 30.2), and you can start creating a drawing. We will not describe the menu composition here, give a description of a set of tools, and interpret a verbal description of the system's responses when a particular button is pressed. The reader will find these descriptions in any self-help book (e.g., Poleshchuk 2001). It is best to sit at the computer and experiment with buttons and tools.

The user's dialogue with the AutoCAD system is carried out in the command language. The system is endowed with the ability to receive commands in several ways.

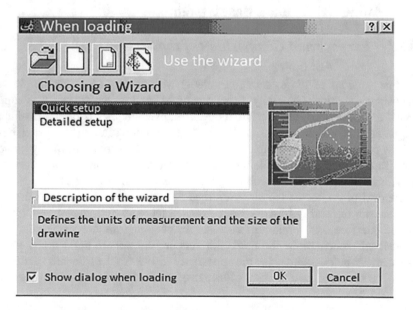

Fig. 30.1 System settings windowСҌ

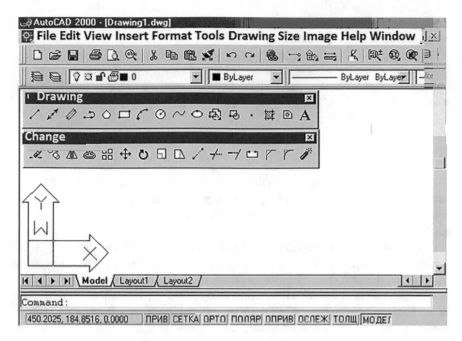

Fig. 30.2   Main menu and toolbars

The easiest way to enter commands is to type them on the keyboard in response to the **Command:** prompt. Here and further, we give commands, service words, etc. both in the Russian and English versions. If you have the Russian version of the system installed, then the command set in English should be preceded by an underscore, for example, **_line**. When typing, the machine does not distinguish between lowercase and uppercase letters. The command is completed by pressing the "Enter" key.

Another way to enter commands is to select the appropriate menu items or toolbar buttons. If no command is entered, then pressing the "Enter" key repeats the call of the previous command. The execution of any command is interrupted by pressing the "Esc" key.

30.3.3   Creating Drawing Elements

Just as any complex substance consists of atoms of chemical elements, any arbitrarily complex drawing consists of the simplest elements. In the AutoCAD system, these elements are called primitives. Primitives can be simple and complex.

Simple primitives include a point, a segment, a ray, a construction line, a circle, an arc, an ellipse, a spline, or a text. The group of complex primitives includes polyline, multiline, multitext, dimension line, callout, tolerance, hatching, block or external link occurrence, attribute, or bitmap image. Spatial primitives and viewports are also used less often.

Below we will look at a small part of the primitives that are most often encountered when developing drawings. After working with these primitives, the reader will acquire the necessary skills and will easily be able to independently master the handling of other primitives that we do not mention here.

Point

To construct this primitive, the POINT command is used, which, in addition to entering from the keyboard, can be called from the Draw panel. The corresponding button is marked in Fig. 30.3 with number 13. In this case, the following messages are displayed:

Current point modes: PDMODE=0 PDSIZE=0.0000 Specify point:

Specify the required point on the screen with the mouse. A point is formed at the specified location. Then the system issues the above request again, and you can enter the next point or interrupt the command by pressing the Esc key. Let us explain the meaning of point modes. If PDMODE = 0 and PDSIZE = 0, then the point is displayed as a pixel, which is not always convenient. Therefore, use the Point Style item in the Format drop-down menu, and select the display type and point size in the dialog box that appears (Fig. 30.4).

Drawing Segments

There are several ways to get a straight-line segment. Place the cursor in the command line, and enter the LINE command from the keyboard. This command can also be called by clicking on the button 1 of the DRAWING panel (Fig. 30.3) or from the DRAWING drop-down menu (Draw).

After pressing the Enter key, the AutoCAD system will ask a number of questions in the following sequence:

Specify first point:

The easiest way to set the first point of a line segment is to point it with the mouse on the visible part of the monitor screen. In this case, it is useful to use the

Fig. 30.3 The toolbar for drawing primitives

Fig. 30.4 The POINT
STYLE dialog box

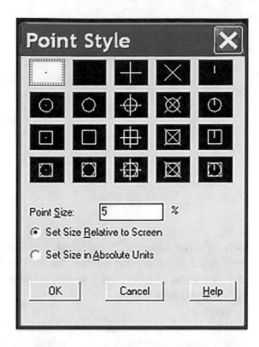

coordinate counter, which is located in the lower-left corner. After specifying the first point and pressing the Enter key, the system sets the following query:

Specify next point or [Close/Undo]:

As you can see, here, in addition to the *Undo* option, there is also the *Close* option. To build the closing segment, enter *U (C)* from the keyboard, and press Enter. If you do not need to build a bar, then just press Enter. Then the third segment will not be built. Another option for completing the *Line* command is to click the right mouse button. At the same time, a context menu appears in the position of the mouse pointer, shown in Fig. 30.5. Depending on the command being executed at the moment, you can select the desired option from this menu. Let's explain the last two options.

The *Pan* and *Zoom* items allow you to change the size and scale of the display of the required drawing area on the screen. If you select INPUT instead of specifying the second point of the segment, the LINE command will end without building any object. But if you press Enter without even selecting the first point, the end point of the last constructed object will be taken as the first point. If there is none (the sheet is empty), an error message will be displayed:

No line or arc to continue. Specify first point:

When drawing drawings, the position of the points of the beginning or end of the segment is usually selected not by using the mouse pointer but by entering the coordinates of the point from the keyboard, for example:

75, 123.31

This means that a point with coordinates X=75 mm, Y=123.31 mm is selected. It should be remembered that when entering from the keyboard, the comma is used

Fig. 30.5 Context menu the
command INTERCEPT
(Line)

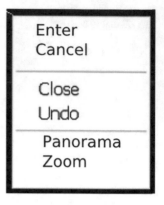

as a separator between the values of the abscissa and the ordinate of the point. The
dot is used as a separator between the integer and fractional parts of a number in its
representation as a decimal fraction. Another way to enter points is to use relative
coordinates, for example:

@60, 24

This record means that a point is selected with a shift relative to the previous
point along the abscissa axis by +60 mm (i.e., to the right of the previous point) and
along the ordinate axis (up) by +24 mm. The entered numbers can be both integers
and real (positive, negative, and zero).

The third way to enter points is to set the position of the entered point relative to
the last point by relative polar coordinates, for example:

@30.4<60

This entry means that the new point is separated from the previous one at a
distance of 30.4 mm, which is deposited on the ray coming from the last point at an
angle of 60° with the positive direction of the abscissa axis.

Finally, the fourth way to enter points is through the object snapping functions.
These functions are accessed either through the group button on the standard panel
or through the Object Snap panel. The specified group button is a button with a
black triangle in the lower-right corner. You will find it on the second line of the
AutoCAD panel under the DRAWING item of the main menu (Fig. 30.6). Position
the mouse pointer over this button, and click without releasing the left mouse button.
This will open a set of buttons for the tools of this group (Fig. 30.6). Move the
mouse pointer down to the button you need, and only then release the left mouse
button.

However, it is better to have the Object Snap panel always вЋ‌њat handвЋ‌к—
on the monitor screen. To do this, select the View item in the main menu and the
PANELSвЋ‌‌ option in the drop-down menu (Fig. 30.7, left). A list of Customize
panels will appear (Fig. 30.7, on the right). Select Object Snap here, and the
OBJECT SNAP panel (Fig. 30.7, bottom) will appear on the screen, just like
Figs. 30.7 and 30.6.

Fig. 30.6 Object Snap
buttons

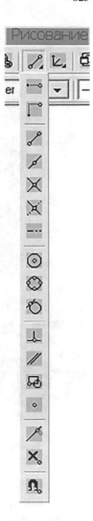

This panel contains the following buttons (from left to right in Fig. 30.7 or from top to bottom in Fig. 30.6):

Temporary tracking point—used to track with an intermediate point;

Snap From—snap to another (auxiliary) point;

Snap to Endpoint—Snap to an endpoint;

Snap to Midpoint—snap to midpoint;

Snap to Intersection—Snap to the intersection point;

Snap to Apparent Intersection—snap to the intersection point of the continuation of two objects;

Snap to Extension—Snap to the continuation point;

Snap to Center—Snap to the center of a circle or arc;

Snap to Quadrant—Snap to a point on the quadrant of a circle or arc defined by a polar angle of 0, 90, 180, or 270 degrees;

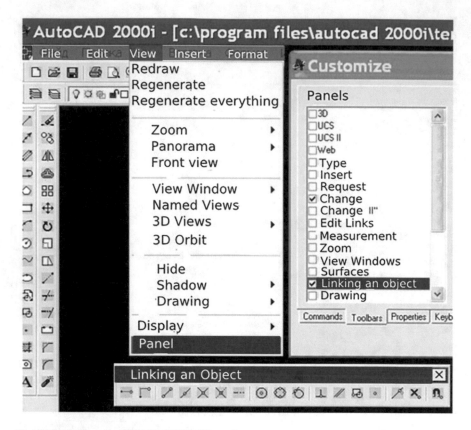

Fig. 30.7 Calling the OBJECT BINDING panel

Snap to Tangent—touch point;

Snap to Perpendicular—normal to the object;

Snap to Parallel—parallel to the object;

Snap to Insert—snap to the insertion point of text, block, or xref;

Snap to Node—snap to a node point;

Snap to Nearest—snap to the point nearest to the object;

nap to None—do not use object snapping;

Object Snap Settings—configures parameters for object snap modes.

Let's use an example to explain the use of the object snap buttons. Let two segments be drawn on the screen and you want to build a third segment, which goes from the end of the first segment to the beginning of the second. If we perform this operation without using object snaps, choosing the ends of the segments "by eye," then we will always have an error of coincidence of the endpoints of the third segment with the endpoints of the first two segments. This significantly degrades the quality of the drawing. To avoid this, we use object snapping. To draw the third segment, call the *Line* command, and in response to the *Specify first point:*

query, use the left mouse button to press the *Snap to Endpoint* button on the object snapping panel. Now, when you move the mouse pointer to the desired end of the first segment, a yellow square appears, signaling that the function of snap to the end point is enabled. When the left mouse button is pressed, "capture" occurs—the starting point of the third segment is aligned with the end point of the first. Similarly, for the request *Specify next point or [Undo]:* use the left mouse button to select the *Snap to Endpoint* button on the object snapping panel, and then move the mouse pointer to the desired end of the second segment and press the left mouse button. The next request *Specify next point or [Undo]:* will follow. Press the *Enter* key, and the *Line* command will be terminated.

Other object snaps are used in a similar way. In order to get used to handling them, you need to be patient and practice using them.

Ray

To build this primitive, the RAY command is used, which can be entered into the command line from the keyboard or called using the RAY item of the Draw drop-down menu. This will be followed by a request:

Starting point:

Specify start point:

After specifying the first point, the system cyclically requests other points and builds rays outgoing from the first point and passing through the rest:

Through point:

Specify through point :

To end the command, press the Enter key, or end the command by clicking the right mouse button as described above.

Construction Line (Straight)

The construction line is an endless straight line in both directions. To build it, use the *XLine* command, which can be entered into the command line from the keyboard or called from the DRAW drop-down menu or from the toolbar (Fig. 30.3, button 2). A request will follow:

Specify a point or [Hor / Ver / Angle / Bisect / Offset]: *Specify a point or [Hor / Ver / Ang / Bisect / Offset];*

In response to this request, you should specify a period using any of the methods discussed above. The system will build a bundle of straight lines passing through the selected point. To fix the position of a straight line, it is enough to indicate the second point in response to the request:

Through point: *Specify through point:*

By specifying several points, we get a bundle of straight lines passing through the first specified point. Using the options enclosed in square brackets, we get horizontal (vertical) lines, lines tilted at a given angle, etc. Experiment.

Circle

This primitive is drawn by the CIRCLE command, which can be entered from the keyboard into the command line, or called from the Draw panel, Fig. 1.6, or by choosing the CIRCLE item from the drop-down menu of the same name. When the command is called, the system issues a request:

Specify center point for circle or [3P / 2P / Ttr (tan tan radius)]:

If, in response to this request, you specify a point, then it is taken as the center of the future circle, and the next action will be the request:

Specify radius'of circle or [Diameter]:

In response, you need to enter a number that will be the radius of the circle. It can also be specified with a dot. If, in response to the last request, you type the character D (D), then the request is issued:

Specify diameter of circle:

If the 3T (3P) option is selected instead of the center of the circle, then AutoCAD will draw a circle based on three points of the plane. In this case, requests for input (or pointing with the mouse) of three points will be alternately displayed, and after correctly specifying them, a circle will appear on the screen. If option 2T (2P) is selected, only two points are requested. In this case, it is considered that they belong to the same diameter.

The Ttr option is designed to draw a circle tangent to two other objects and having a given radius. Selecting this option prompts you:

Pick a point on the object that defines the first tangent:

Specify point on object for first tangent of circle:

The second would be the request:

Specify point on object for second tangent of circle:

It will be followed by a request:

Specify radius of circle:

The radius can be specified by a number or a pair of points, the distance between which is taken as the value of the radius of the circle.

For example, in Fig. 30.8a, circle and a tangent are constructed. To do this, first, using the *Snap to Perpendicular* object snap function, the centerlines are constructed. To remind you, we have placed the OBJECT BINDING panel at the top position of Fig. 30.8. Next, using the *Snap to Intersection* function, a circle is built with the center at the intersection of the center lines. Then the *Line* command is called, the starting point is selected, and after requesting the second point, the *Tangent* button on the OBJECT BINDING panel is pressed with the mouse pointer. Then bringing the mouse pointer to the circle, we will find a yellow square indicating the point of intersection of the straight line and the circle. By clicking the left mouse button, we get a tangent. Then you can either finish the construction or build the next segment. In Fig. 30.8b, two tangents are constructed. Try to repeat the construction of these drawings.

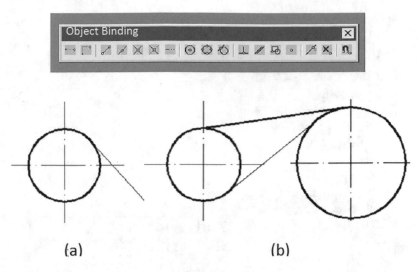

Fig. 30.8 Examples of constructing circles and tangents

Arc

The ARC command is used to construct this primitive. It can be entered into the command line from the keyboard or called from the DRAW toolbar by clicking on the button marked in Fig. 30.3 with the number 7.

The third way to draw an arc is to select ARC from the Draw drop-down menu. Here this item is detailed with a set of ten sub-items (Fig. 30.9).

After entering the ARC command, AutoCAD will prompt you:

Arc start point or [Center]:

Specify start point of arc or [CEnter]

When prompted, you can either specify the starting point of the arc or select the CENTER (CE) option. If you skip these steps and just press the Enter key, then the end point of the last object created in the drawing from the number of lines and arcs will be taken as the starting point of the arc. The system then draws an arc tangent to the last object. The end point of the arc is requested:

End point of arc:

Specify end point of arc:

After selecting the end point, an arc is drawn, which is a continuation of the previous object. If, in response to the request for the starting point, you enter this point in one of the previously considered ways, the system will request the second point of the arc:

Second point of arc or [Center / End]:

Specify second point of arc or [CEnter / End] :

After specifying the second point, you will be prompted:

End point of arc:

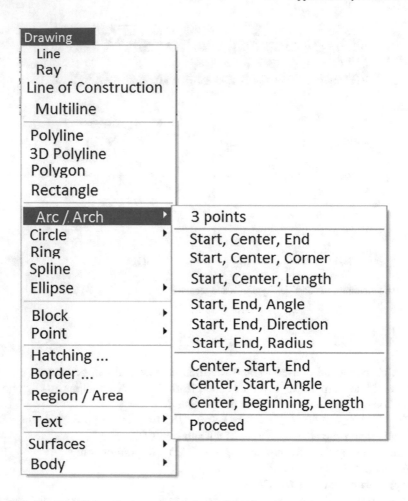

Fig. 30.9 The DRAW drop-down menu and the ARC/ARC option sub-items

Specify end point of arc:

As a result, an arc is constructed along three of its points. If, in response to the request of the second point, you select the CENTER option, then the request will follow:

Center of the arc:

Specify center point of arc:

After specifying the center, the following request will be:

End point of arc or [Angle/Chord Length]:

Specify end point of arc or [Angle/chord Length]:

When specifying the end point of an arc, the system calculates the radius of the arc based on the first point and center. As you can see from the last request, in this case, there are two more options for terminating the ARC command. Selecting the Angle option prompts you:

Center angle:

Specify included angle:

The angle can be entered from the keyboard (taking into account the sign) or defined by the mouse pointer. In the event that instead of the option ANGLE (ARC) select CHORD LENGTH (chord Length), you will receive a request:

Chord length:

Specify length of chord:

The chord length can be entered from the keyboard as a signed number, which affects the direction of the arc traversal. Another way to specify the chord length is to select a point. In the latter case, the distance from the starting point of the arc to the last entered point will be taken as the chord length.

On this we will restrict ourselves to the considered options for constructing an arc. The reader is encouraged to experiment with the other options listed in the ARC option sub-clauses.

From the group of complex primitives, we will consider only two that are most often used when creating mechanical engineering drawings.

Polyline

This complex primitive consists of several lines and arc segments, combined into one object. To draw a polyline, use the PLINE command, which is entered from the keyboard or called from the Draw toolbar (Fig. 30.3, button 4). The command can also be selected as the Polyline item in the Draw drop-down menu. After entering the command in one of the named ways, the system issues a request:

Starting point:

Specify start point:

You need to select the starting point of the polyline. After specifying the starting point, the following query will look like this:

The current width of the polyline is 0.0000

Next point or [Arc / Close / Half Width / Length / Undo / Width]:

Current line-width is 0.0000 Next point or [Arc / Close / Halfwidth / Length / Undo / Width] :

The primitive in question can have a variable line width (width), which is set by the user. To change the current line width, enter W followed by Enter and the ending line width. An example of constructing polylines of different widths is shown in Fig. 30.10. At position (a), the polyline consists of three straight-line segments of varying width, and at position (b), it consists of three arc segments.

After building the first link of the polyline, the system issues a request:

Next point or [Arc / Close / Half Width / Length / Cancel / Bus]:

Next point or [Arc / Close / Halfwidth / Length / Undo / Width] :

The next point is selected, and the next link of the required width is built. In addition to specifying points, you can select the option indicated in square brackets:

Fig. 30.10 Variable width polylines

Close—closes the polyline with a straight link;

Width—sets the width for the next polyline segment;

Halfwidth—specifies the width in terms of half-width (e.g., the required width of 10 can be specified as a half-width equal to 5);

Length—construction of a link, which is a continuation of the previous section with a given length; the length is specified by a number or a dot;

Undo—undo the last operation of the PLINE command;

Arc—switch to the mode of drawing an arc. When this option is selected, the system offers the following options:

End point of arc or [Angle / Center / Close / Direction / Half-width / Linear / Radius / Second / Undo / Width]:

Specify endpoint of arc or [Angle / CEnter / Close / Direction / Halfwidth / Line / Radius / Second pt / Undo / Width] :

After specifying the end point of the arc, the system draws an arc segment that touches the previous link in the polyline. The options listed in square brackets provide the following options:

Angle вЂ"—setting the value of the central angle of the arc segment; Center вЂ"—setting the center of the arc; Direction вЂ"—setting the direction for building an arc segment; Half-width вЂ"—setting the half-width for constructing an arc segment; Linear вЂ" Line)—switching to the mode of drawing straight links of a polyline; Radius—setting the radius of the arc segment; The second (Second pt)—setting the second point for constructing an arc segment based on three points; Undo вЂ"—undo the last action; Width—sets the width of the next segment. Being a complex primitive, a polyline can be divided into separate links at any time. This can be done using the EXPLODE command, which is located in the MODIFY panel or in the Modify drop-down menu, the EXPLODE item. However, the Width property disappears. In AutoCAD, it is also possible to combine previously drawn segments and arc segments that are adjacent to each other in series into a polyline. To do this, use the PEDIT command. When you enter this command, the system issues a request:

Specify the polyline:

Select polyline:

Specify the first element to be combined into a polyline. The system will display a message:

The selected object is not a polyline.

Make it a polyline? <D>:

Object selected is not polyline Do you want to turn it into one? <Y>:

In response, press the Enter key, or enter *Y*. Then the selected object is transformed into a polyline link, and the request will follow:

Set the option [Close / Add / Width / Top /

Smooth / Spline / Unsmooth / Tiplin / Undo]:

Enter an option [Close / Join / Width / Edit vertex / Fit / Spline / Decurve / Ltype gen / Undo] :

Select *J* to add the next link to the polyline. The following request will be repeated cyclically:

Select objects :

You must specify the object to be attached each time and complete the selection by pressing the *Enter* key.

Hatching

This primitive is used for hatching closed areas and is called by the BHATCH command, which is called either from the Draw panel by button 14 (Fig. 1.3) or using the Hatch item of the Draw drop-down menu, after which the dialog box shown in Fig. 30.11.

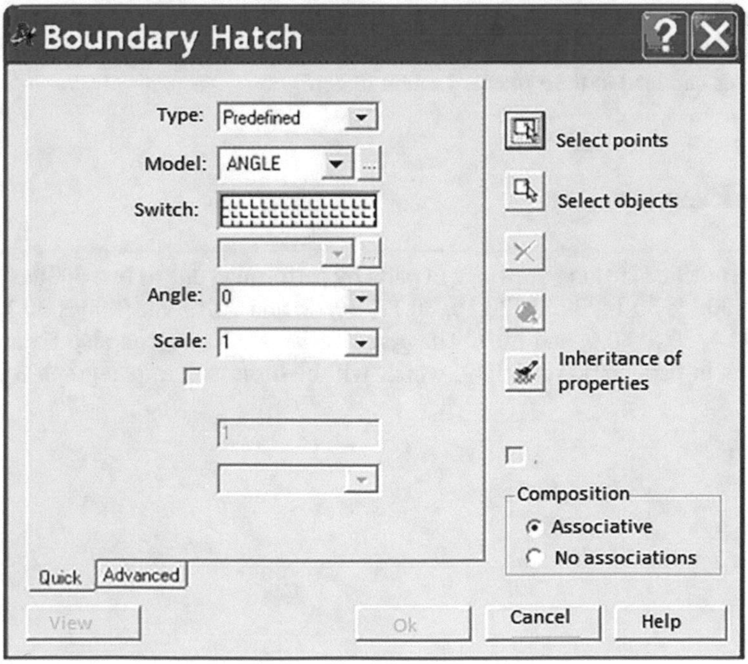

Fig. 30.11 HATCH dialog box

Fig. 30.12 Object editing panel

The hatching area can be selected by simply specifying points inside a closed contour or by selecting objects that limit the closed area. The system offers a rich set of standard hatches. The desired hatching is selected by name or from the palette of hatching samples. If you select the option BY LINE TYPE (User defined) in the Type field, the hatching pattern is built using the current line type, angle, and distance between the hatching lines.

Using the right part of the HATCHING dialog box, you can specify the Pick Points of the internal areas to be hatched. The Select Objects button allows you to mark those objects whose intersection gives the area to be filled with hatching.

Hatching is created as a single primitive. If you need to decompose it into its component segments, you should use the Explode command. In conclusion, a brief description of the primitives will indicate that any of the constructed objects can then be edited. To do this, use the CHANGE toolset (Fig. 30.12), as well as the drop-down menu of the same name.

The reader can find a description of the editing tools in any tutorial on the Auto-CAD system. However, this is unlikely to be necessary, since direct experimentation with tools can lead to their mastery much faster.

30.4   Exercises

You can practice creating drawings of parts by performing one or two options shown in Figs. 30.13, 30.14, 30.15, 30.16, 30.17, 30.18, and 30.19 and Tables 30.1, 30.2, 30.3, 30.4, 30.5, 30.6, and 30.7. The given variants of tasks can also be used for exercises in parametric modeling, which will be discussed in paragraph 3 of this manual.

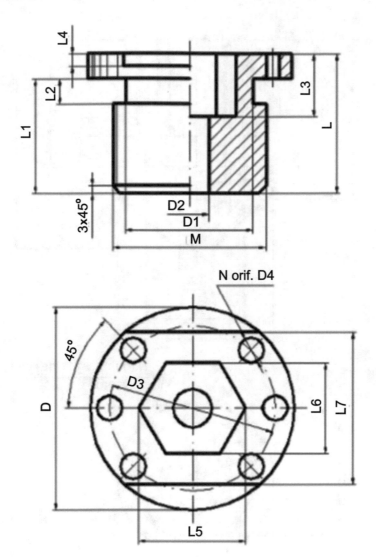

Fig. 30.13 Variant 1

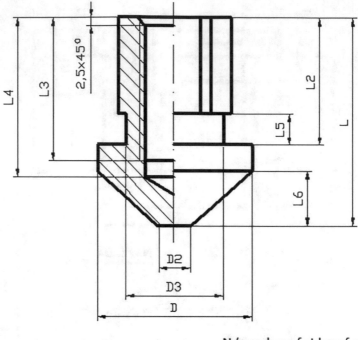

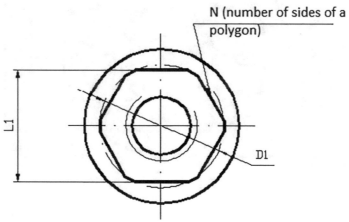

Fig. 30.14 Variant 2

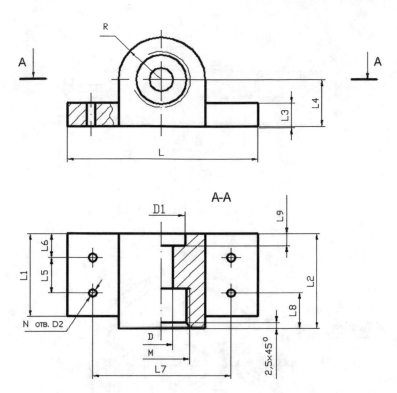

Fig. 30.15 Variant 3

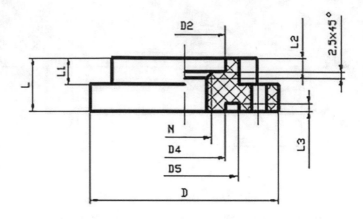

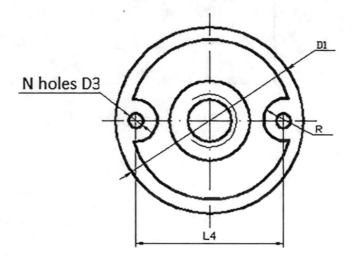

Fig. 30.16 Variant 4

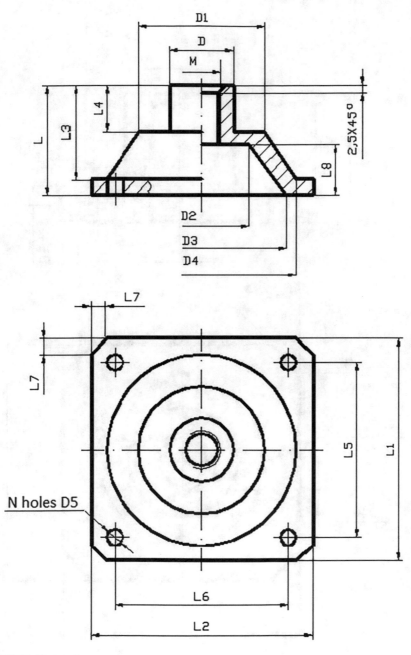

Fig. 30.17 Variant 5

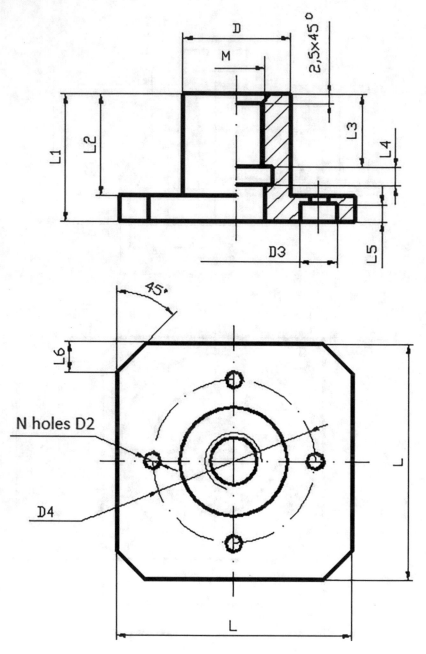

Fig. 30.18 Variant 6

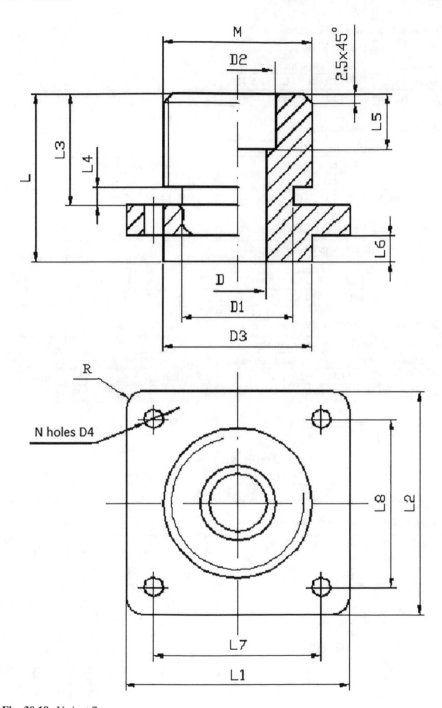

Fig. 30.19 Variant 7

Table 30.1 Source data of option 1

Data for building the program and images														
Options	Part base options						Optional parameters							
Input	D	D1	L	L1	L2	M	D2	D3	D4	L3	L4	L5	L6	L7	N
For debugging	80	50	55	45	10	60	15	65	5	25	5	36	31	60	6

Table 30.2 Source data of option 2

Data for building the program and images													
Options	Part base options						Optional parameters						
Input	D	D1	L	L1	L2	N	D2	D3	L3	L4	L5	L6	M
For debugging	48	35	65	32	40	6	10	30	45	50	10	17	20

Table 30.3 Source data of option 3

Data for building the program and images														
Part base options							Optional parameters							
D	L	L1	L2	L3	L4	R	D1	D2	L5	L6	L7	L8	L9	M	N
10	80	35	40	10	20	18	20	3.5	15	10	60	20	5	24	4

Table 30.4 Source data of option 4

Data for building the program and images													
Part base options				Optional parameters									
D	D1	L	L1	D2	D3	D4	D5	L2	L3	L4	M	R	N
70	60	20	10	30	5.5	30	40	5	3	55	20	8	2

Table 30.5 Source data of option 5

Data for building the program and images														
Part base options					Optional parameters									
D	L	L1	L2	L4	D1	D2	D3	D4	D5	L3	L5	L6	L7	L8	M	N
20	35	70	70	15	40	30	54	60	5	30	55	55	5	16	12	4

Table 30.6 Source data of option 6

Data for building the program and images													
Part base options				Optional parameters									
D	L	L1	L2	D1	D2	D3	D4	D5	L3	L4	L5	L6	M	N
30	65	35	28	15	4.5	10	45	20	20	5	5	8	16	4

Table 30.7 Source data of option 7

Data for building the program and images														
Part base options								Optional parameters							
D	D1	L	L1	L2	L3	L4	M	D2	D3	D4	L5	L6	L7	L8	N	R
15	30	45	60	60	30	5	40	20	40	5	15	7	45	45	4	5

References

A. Alyamovskij, *Inzhenerny'e raschety' v SolidWorks Simulation* [Engineering Calculations in SolidWorks Simulation] (DMK Press Publication, Moscow, 2010)

L. Askon, *Kompas-3D v17 Rukovodstvo pol'zovatelya* [Compass-3D v17 User Manual] (SPb, Ascon Press Publication, Fukuyama, 2017)

V. Molotnikov, A. Molotnikova, *Vvedenie v SAPR na osnove Autocad Autolisp* [Introduction to CAD based on AutoCAD Autolisp] (Rostov-on-Don, RIATM Publication, Philippines, 2006)

Obzor populyarny'kh sistem avtomatizirovannogo proektirovaniya (CAD) [An overview of popular computer-aided design (CAD) systems] (2021). https://www.pointcad.ru/novosti/obzor-sistem-avtomatizirovannogo-proektirovaniya

S. Tickoo, *Solid Edge 2020 for Designers*, 17th edn. (Purdue University Northwest Publication, Indiana, USA, 2020)

N. Poleshchuk, *Visual LISP i sekrety adaptatsii AutoCAD Seriya: Master* [Visual LISP and the secrets of adapting AutoCAD Series: Master] (Publishing house: BHV-Petersburg, 2001)

Chapter 31
Elements of Programming in the AutoLISP language

Abstract The chapter opens with general information about the Lisp language. The purpose and capabilities of the language are specified. It is noted that in Lisp, the form of writing programs and the data processed by them have the same structure. Both are presented in the form of lists. Due to this advantage, programs can process other programs and even themselves, which is very important in creating self-learning programs (expert systems, artificial intelligence systems, etc.). Operations on data in Lisp are carried out in the form of functions. Functions in AutoLISP have the same structure as lists. The built-in functions of the language, data manipulation functions, functions for organizing computational processes, functions for accessing primitives, and AutoCAD tools are described.

Keywords Programming · Functional · List · Artificial intelligence · Parameterization · Expression · Pseudo-function · Operation · Result · Objects

31.1 General Information About the Lisp Language

AutoLISP refers to one of the dialects of the COMMON LISP high-level programming language, developed in 1986 at the Massachusetts Institute of Technology on the basis of the fundamental version of Lisp created earlier (1962) by J. McCarthy. In the literal meaning of the word, Lisp is translated as lisp (lisping). The most acceptable semantic translation of the phrase LISting Processing is list processing. The fact is that in Lisp, the form of writing programs and the data processed by them have the same structure. Both are presented in the form of lists. Due to this advantage, programs can process other programs and even themselves, which is very important in creating self-learning programs (expert systems, artificial intelligence systems, etc.). According to the accepted classification, Lisp refers to functional programming languages of the declarative type. As applied to engineering graphics, Lisp provides one of the possible ways to solve the problem of automatically constructing an object based on a prototype. This technique is called parameterization. The basis of the Lisp language is working with lists. A list is an ordered sequence of elements enclosed in parentheses. The following data types can be elements of the sequence:

- *numeric constants*, for example,

231   – integer. 16-bit numbers from the interval [- 32768; + 32767];

6.28

real number—defined as a double-precision floating point number. Real numbers are passed to AutoCAD as 32-bit values;

$13.025E5$ – number represented by mantissa and order $(13.025 \cdot 10^5)$;

- *text constants* are any sequence of characters and spaces enclosed in quotation marks. Control characters can be used inside text constants:

$\backslash n$ – switching to a new line;

$\backslash t$ – tab;

$\backslash\backslash$ – sign $\backslash$;

$\backslash nnn$ – a character with the octal code nnn;

- *logical constants* – (T (true) – true, NIL (false) – false);
- *characters* (or *identifiers*) are names consisting of letters, numbers, and special characters with the exception of parentheses, a period, an apostrophe, quotation marks, and a semicolon. The symbol starts with a letter, and initially it is not assigned a specific value. Uppercase and lowercase letters in names are not distinguished by the interpreter;
- *lists*.

When writing a structure of the list type, an apostrophe is placed before the opening parenthesis, and the list items inside parentheses are separated by spaces. Examples of lists are:

'(3.14' "Hello!" '(a b c) -3.0E-2);

'("Wednesday" "Thursday" 1 2 '("cat" "mouse" '("January" 31))).

From the above definitions, it follows that a list is a multi-level hierarchical data structure in which the number of opening parentheses must match the number of closing ones. If there are no elements in the list, then it is called empty and denoted by the name NIL or '().

31.2   Functions in AutoLisp

Operations on data in Lisp are performed in the form of functions. In this case, you can use programming based on the traditional sequential execution of operators, the transfer of control, or the execution of cyclic procedures. However, Lisp is focused on a different programming ideology, the main means of which is the so-called functional programming. Here, programs are built from logically separate function definitions consisting of calculations, control structures, and nested (recursive) function calls. Due to this structure of the program, the execution of each action gives a value, which then becomes an argument for performing the next action, etc.

Functions in AutoLISP have the same structure as lists. A distinctive feature of such a list is that it is written in the form of an expression. An expression is a list, the first element of which is the name of the function. The expression in AutoLISP has the form:

(<function name> [<argument1>] [<argument2>] [<>]...)

Here and further in such constructions, square brackets mean that an element in the list may be but may also be absent. Arguments (or parameters) are variables, constants, or expressions that serve as a means of passing values (data) to a function. The number of arguments can be variable, fixed, or zero. The arguments are separated by at least one space. A list of the expression type indicates in AutoLISP that the enclosed in parentheses should be treated as a calculated expression. Any argument of the function can be another evaluated expression, limited by parentheses. For example, the entry

(−(+ 12.04 50E-1) (− 231.5 16.4)) means the following: "-" the name of a function (built-in arithmetic function subtraction) that has two arguments, each of which is a calculated expression. The first argument is the sum of the numbers 12.04 and 5.0, since the arithmetic addition function "+" is used here, and the second is the difference between the real numbers 231.5 and 16.4. The return value of this expression is the negative real number -202.56. It can be stored in a variable with a certain name (e.g., ALPHA,) using the SETQ pseudo-function, which acts as an assignment operator, and has the following syntax:

(SETQ <variable1> <expression1> [<variable2> <expression2> ... <variablen> <expressionn>] ...]) that is, we can, for example, write

(SETQ Alpha (− (+ 12.04 50E-1) (− 231.5 16.4))) .

The SETQ pseudo-function essentially allocates named memory sections for storing data and is the main means of storing values returned by other expressions.

There are three types of functions in AutoLISP:

- *built-in functions*,
- *user functions*,
- *functions developed by third-party users.*

31.3 Built-in Functions of the AutoLISP Language

Functions of this type can be classified according to their purpose:

* data entry functions;
* functions for data manipulation;
* functions for working with numeric data; функции functions for organizing computing processes;
* data output functions;
* functions for accessing primitives and AutoCAD tools;
* other functions.

Here are the most frequently used built-in functions. Below we will adhere to the following sequence: function-assignment-example of a call-return value. Phrases placed after the ";" to the end of the line mean AutoLISP comments. The reader will find a complete list of built-in AutoLISP functions in the related tutorials at the end of this chapter (LispWorks for the Windows Operating System. IDE User Guide 2021; Titarenko and Maly'shenko 2002; Poleshhuk 2001).

31.3.1   Data Entry Functions

▶ **(PROMPT <"Explanatory message">) – enter a message.**
   (PROMPT "Enter the main data of the task")
   Enter the main data of the task.
▶ **(GETINT <"Explanatory message">) – enter an integer.**
   (GETINT "\n Enter the number of shaft steps:")
   Enter the number of shaft steps: 3
   ;the number 3 is entered after the message appears.
▶ **(GETREAL <"Explanatory message">) – input of a real number.**
   (GETREAL "\n Diameter of the first stage of the shaft: D1=")
   The diameter of the first stage of the shaft: D1= 22.4
   ;the number is entered after the message appears.
▶ **(GETSTRING [<CR>] [<"Explanatory message">]) - waiting for input lines with spaces, if there is <CR>, otherwise entering a line before the 1st the space bar.**
   (GETSTRING "\n Enter a name: ")
   Enter a name: Yuri
   ; entering a name on the request.
▶ **(GETDIST [<T1>] [<"Explanatory message">]) – waiting for input points T2;**
   if there is <T1>, then the distance R=T2-T1, otherwise enter
   the distance R as an integer or real number.
   (GETDIST '(2.5 6.0) " \n Enter the second point:")
   Enter the second point: 8.5 14.0; Enter the coordinates of the second point.
   10.0; The result is the distance between the points.
▶ **(GETANGLE [<T1>] [<"Explanatory message">]) – waiting for input**
   points T2; if there is <T1>, then the angle is equal to the angle between the x-axis and the segment [T1 T2], otherwise the angle value in radians is entered as an integer or a real number.
   (GETANGLE ' (2.0 2.0) "\n Line slope angle ")
   Line slope angle 4.0 4.0; Enter the coordinates of the second point.
   0.7854; The result is the line slope angle.
▶ **(GETPOINT [<T1>] [<"Explanatory message">]) – waiting for input**
   a list of the values of the x and y coordinates of the point, if there is no T1,
   otherwise it draws a "rubber" line from point T1 to the cursor position.

(GETPOINT "Enter the base point:")
Enter the base point: 12, 32 ; Enter the point on request.

▶ **(GETCORNER [<T1>] [<"Explanatory message">]) – image a "rubber" rectangle from point T1 to the cursor position.**

▶ **(GETKWORD [<"Explanatory message">]) – waiting for input the keyword described by the function (INITGET...).**

▶ **(INITGET [<sum of numbers>] [<"keywords...">]) – installation data entry mode.** The sum of the numbers is a bit flag that should be an integer from 0 to 255, which is the sum of the bits with the corresponding values for each of them: Numbers:

1	– no empty input is allowed (i.e., pressing the <Enter>key);
2	– does not allow entering zero;
4	– does not allow entering negative numbers;
8	– allows specifying points outside the limits;
16	– this bit is not involved;
32	– forces AutoCAD to use a dotted line for the image a "rubber" line or rectangle;
128	– returns an arbitrary code from the keyboard.

Examples of using the bit flag:

(initget 1)	– empty input is not allowed;
(initget 3)	– empty input and zero input are prohibited $(3 = 1 + 2)$;
(initget 7)	– empty input, zero input, and negative input are not allowed numbers $(7 = 1 + 2 + 4)$;
(initget 6)	– empty input is allowed, but zero input and input are prohibited negative numbers $(6 = 2 + 4)$.

The "keywords" argument of the initget function is a string bounded by double quotes on both sides, which specifies keywords that limit the allowed input options. Keywords are listed separated by a space.

Calling the initget function can precede accessing the input functions getint, getreal, getdist, getangle, getpoint, etc. in the program. Example:

(initget 1 "Yes No"); Function call.

(getkword " Change the coordinates of a point? (Yes/No)")

Change the coordinates of a point? (Yes/No) Yes; Enter a keyword.

▶ **(READ-LINE [<handle>]) – reading a line from a file with the specified a handle;** if the handle is omitted, then a line is expected to be entered from the keyboard; the input ends by pressing the <Enter> key.

(Read-line)   ;Function call.
Enter the text of the technical requirements;   ;The text entered from the keyboard.

▶ **(READ-CHAR [<descriptor>]) - reading a character from a file with the**
   specified a handle; if a handle is not specified, then a character is expected to be entered from the
   keyboard; returns the code of the read character; pressing the key

                              (READ-CHAR) ;   Function call.
   <Enter> returns code 10.   A;   Entering the A character from the keyboard.
                              65;   The ASCII code of the A character is returned.

If the <handle> argument of the read-line and read-char functions is not empty, the file named <handle> must be previously opened for reading, which can be done, for example, in one of the following ways:
   (READ-LINE (OPEN "C:
DATA
FileData.txt" "R")) or
   (SETQ F (OPEN "C:
DATA
FileData.txt" "R")) ; Assign to the variable F

                    ; name of the procedure for opening the file for reading
                    ; FileData.txt, saved in the DATA folder of the device with.

   (READ-LINE F)
After opening and using the file, it is recommended to close it:
   (CLOSE F)   ;Closing file <F>.

31.3.2   *Data Manipulation Functions*

▶ **(SET '<variable> <expression>) – assigning**
   a value to a variable expressions (list, function, etc.).
   The (SET ...) function is similar to pseudo-functions (SETQ ...).
   Both functions expect a character as its first parameter.
   The only difference is that (SET ...) accepts an expression that returns a character, while (SETQ ...) does not.
   (SET 'A 17.0)
   (SET (READ ?A?) 17.0).

▶ **(CAR <list>) – retrieves the first element of the list.**

    (CAR '(5.0 21.3 0.17)) ; Function call.
    5.0   ; Returns the first element of the list.

▶ **(CDR <list>) – return a list without the first element.**

    (CDR '(A B C D)) ; Function call.
    (B C D)   ; The result is a list without the first element.

▶ **(CAAR <list>) – equivalent to (CAR (CAR <list>)).**
▶ **(CDAR <list>) – equivalent to (CDR (CAR <list>)).**
▶ **(CADR <list>) – equivalent to (CAR (CDR <list>)).**
▶ **(CDDR <list>) – equivalent to (CDR (CDR <list>)).**
▶ **(CADAR <list>) – equivalent to (CAR (CDR (CAR <list>)))), etc.,**
 up to four levels of nesting.
▶ **(CONS <new first element> <list>) – return a list**
 with a new element at its beginning (the added element
 can be list).

    (CONS '(2.0 5.16) '(3.12 4.01 14.1)) ; Function call.
    ((2.0 5.16) 3.12 4.01 14.1)   ; The result is a list with
                    ; a new one the first element.

▶ **(LAST <list>) – returns the last element of the list.**

    (LAST '(A B C D E)) ; Function call.
    E   ; The result is the last item in the list.

▶ **(LIST <element> <element>...) – creating a list of elements.**
 Its main function to is create a list. As arguments
 <elementN>, from which the list is formed, can be any
 objects operated by AutoLISP.

    (LIST 6.12 0.25 1.82)   ; Function call.
    (6.12 0.25 1.82)   ; The result is a list, the elements of which
                 ; are three real numbers.
    (LIST 1 "Hello!" (LIST 2 4 6)) ; Function call.
    (1 "Hello!" (2 4 6))   ; The result is a list.

► **(REVERSE <list>) – reversal of the list.**

```
(REVERSE '(15.6 0.25 1.98))    ; Function call.
(1.98 0.25 15.5)               ; The returned list.
```

► **(APPEND [<list1> [<list2>... [<listN>...]]]) – merge lists into one.**

```
(APPEND '(10 20) '(0.21 1.76 4.0))    ; Function call.
(10 20 0.21 1.76 4.0)                 ; Returned result.
```

► **(NTH <number> <list>) – extracting an element from the list by serial number.** (List items are numbered starting at zero from left to right).

```
(nth 1 '(0.21 1.76 4.0))    ; Function call.
1.76                        ; Returned result.
```

► **(MEMBER <elementN> <list>) – selection of a list, starting with element N.**

```
(member 6 '(2 3 6 8 10))    ; Function call.
(6 8 10)                    ; Returned result.
(MEMBER 0 '(2 3 6 8 10))    ; Function call.
NIL                         ; Returned result.
```

► **(ASSOC <key> <list>) – search for an item in the list by key.**

```
                                       ;Defines variable X
(SETQ X (LIST '(1 A) '(2 B) '(3 C) '(4 D)))  ; as a list whose elements are two
                                       ; object - an integer and a symbol.
(ASSOC 2 X)                            ; Function call.
(2 B)                                  ; Keyed sublist.
```

► **(SUBST <'new element> <'old element> <list>) – replacement in the list of the old item to the new one.**

```
(SETQ Z '(A B C D))    ; Defining Z/ Variable
(SUBST 'Y 'C Z)        ; Function call.
(A B Y D)              ; The replacement of an item in the list is complete.
```

▶ **(FIX <real number>) – converting a real number to whole.**

        (FIX 8.25)   ; Function call.
        eight   ; The result is an integer.

▶ **(FLOAT <integer>) – converting an integer to real.**

        (FLOAT 8)   ; Function call.
        8.0   ; The result is a real number.

▶ **(CHR <number>) – converting a number to an ASCII character code.**

        (CHR 77)   ; Function call.
        "M"   ; Received the ASCII character code of the number 77.

▶ **(ASCII <?ASCII character?>) is the inverse function (CIIR....).**

        (ASCII "G")   ; Function call.
        71   ; The result is the ASCII numeric code of the G character.

▶ **(ATOF <"number">) – conversion of a**
  numeric text constant to a real number.

        (ATOF "14.2")   ; Function call.
        14.20000   ; Returns a real number.

▶ **(ATOI <число number?>) – converts a numeric text constant to an integer.**

        (ATOI "85")   ; Function call.
        85   ; The result is an integer.

▶ **(ITOA <integer>) – converts an integer to a text constant.**

        (ITOA 85)   ; Function call.
        "85"   ; The result is a text constant.

▶ **(RTOS <real number> [<mode>] [<precision>]) – converts a real number to a text constant.**

    (RTOS 63.5 1 4)   ; Converts 63.5 to "6.3500 E+01" – scientific mode.
    (RTOS 63.5 2 4)   ; The result is 63 63.5000? – decimal mode.
    (RTOS 34.5 3 4)   ; Result – "2'-10.5000" – technical mode;(feet and inches).
    (RTOS 34.5 4 4)   ; Result – "2'-10 1/2 " – architectural mode; (feet and inches).
    (RTOS 63.5 5 4)   ; The result is "63 1/2 " – the correct fraction mode.

▶ **(DISTOF <text constant> <mode>) – inverse function (RTOS...).**

    (DISTOF "34 1/2" 5)   ; Function call.
    34.5   ; The result is a real number.

▶ **(STRLEN <строка string?>) – returns the number of characters in the string.**

    (STRLEN "AutoCAD")   ; Function call.
    7   ; The result is the number of characters in the word AutoCAD.

▶ **(LENGTH <list>) – defines the length of the list as the number of elements of the first nesting level.**

    (LENGTH '('(A B C) D E))   ; Function call.
    3   ; Result – there are three elements of the first one in the list
                         ; nesting level.

31.3.3   *Functions for Working with Numeric Data and Expressions*

▶ **(+ <N1> <N2> <N3>...) – addition of numbers <N1>, <N2>,> ...**

    (+ 16 2.5 0.51)   ; Function call.
    19.01   ; The result of the operation.

▶ **(- <N1> <N2> <N3>...) – subtraction of numbers <N2>, <N3>... from the number <N1>.**

(- 50.5 45.1 3 2.3)   ; Function call.
0.1   ; The result of the operation.

▶ **(*<N1> <N2> <N3>...) – multiplication of numbers <N1>, <N2>, <N3>, etc.**

(*2.2 4 5)   ; Function call.
44   ; The result of multiplying numbers.

▶ **(/ <N1> <N2> <N3>...) – dividing the number <N1> by <N2>, <N3>, etc.**

(/ 70 5.5 2)   ; Function call.
6.36364   ; The result of dividing 70 by 5.5 and then dividing
               ; quotient from dividing thse numbers by 2.

▶ **(ABS <N>) – determination of the absolute value (modulus) of the number <N>.**

(SETQ A (- 16.2 19.4))   ; Save the difference between 16.2 and 19.4
                         ; in variable A (A = -3.2).
(SETQ A (ABS A))   ; Variable A takes on the value 3.2.

▶ **(1+ <N>) – number <N> is increased by one.**

(1 + 10)   ; Function call.
11   ; Result.

▶ **(1- <N>) – number <N> decreases by one.**

(1 − 48)   ; Function call.
47   ; Execution result.

▶ **(SIN <angle>) – calculates the sine of the angle given in radians.**

> (sin 0.54)   ; Function call.
> 0.514136   ; The result of calculating the sine.

▶ **(COS <angle>) – calculates the cosine of the angle, given in radians.**

> (cos 1.4)   ; Function call.
> 0.169967   ; The result of calculating the cosine.

▶ **(ATAN <N>) – definition of arctangent <N>, the result is in radians.**

> (ATAN 1)   ; Function call.
> 0.785398   ; Result ($\pi/4$).

▶ **(EXP <N>) – calculation.**

> (EXP 2)   ; Function call.
> 7.38906   ; The result of calculating the exponent.

▶ **(EXPT <N1> <N2>) – calculation.**

> (EXPT 6 2)   ; Function call.
> 36.0000   ; Calculation result.

▶ **(LOG <N>) – calculation of the natural logarithm of the number N.**

> (LOG 7.38906)   ; Function call.
> 2.0   ; Calculation result ln(7.38906)=2.0.

▶ **(MAX <N1> <N2> <N3>...) – search for the largest of the numbers <N1>, <N2>,... .**

> (MAX 3/2 5.21 0.6 7 1.02)   ; Function call.
> 7.0   ; Search result.

► **(MIN <N1> <N2> <N3>...) – search for the algebraically smaller of the list arguments.**

► **(REM <N1> <N2>) – returns the remainder of the division of <N1> by <N2>.**

        (REM 72 10)   ; Function call.

        2   ; Result.

► **(GCD <N1> <N2>) – calculation of the greatest common divisor.**

        (GCD 64 112)   ; Function call.

        16   ; The returned result.

► **(ANGLE <T1> <T2> – determination of the angle (in radians) between the axis**
   X and the line defined by the points <T1>, <T2>.

        (ANGLE?(2.0 2.0) ?(3.0 8 0))   ; Function call.

        1.40565   ; The returned result.

► **(DISTANCE <T1> <T2>) – returns the distance between the points <T1> and <T2>.**

        (DISTANCE '(2.0 2.0) '(3.0 8.0))   ; Function call.

        6.08276   ; Result.

► **(INTERS <T1> <T2> <T3> <T4> [<ON>]) – point definition**
   intersections of two segments passing through points <T1>,<T2> and <T3>,<T4>.
   If there is no <ON>, then the intersection point it must belong to segments. Otherwise the intersection point of infinite lines is determined, defined by segments.

      (INTERS'(1 1) '(5 5) '(1 5) '(5 1))   ; Function call.

      (3.0 3.0)   ; The result is the coordinates of the intersection
                                    point of the segments.

▶ **(SQRT <N>) – returns the arithmetic square root of the number <N>.**

(SDRT 25)	; Function call.
5.0	; The returned result.

31.3.4   Functions for the Organization of Computing Processes

▶ **(COND (<(condition 1)> <(function 1,1)> [(function 1,2)]...)**

....

(<condition N)> <(a function of N,1)> [(a function of N,2)]...)) – conditions are calculated sequentially until one of them becomes true, after which the corresponding functions (expressions) are calculated. The value of the last evaluated expression is returned. If none of the conditions is true, NIL is returned.

(SETQ i 5 j 9)	; The variables i and j are assigned the
	values 5 and 9.
(COND ((= i j) (MAX 5 10 3))	; Determination of the maximum number (not
	; executed, because the condition is false).
((< i j) (MIN 5 6 20))	; Determination of the minimum of the numbers
	; is completed, the condition gives T).
((> i j) (* 5 7 78 9)))	; Calculation of the product of numbers (not used).
5	; The result of the execution.

▶ **(IF (<condition 1)> <(function 1)>**
   [(function 2)]]) - if <(condition 1)> is satisfied, then
   <(function 1)> is calculated, otherwise - <(function 2)>.

(SETQ I 7)	; Set variable I to 7.
(IF (= I 3) (MAX 55 100 13)	; If I = 3, then find the maximum number,
(MIN 15 64 2))	; otherwise find the minimum number.
2	; Calculation result.

▶ **(REPEAT <integer> <expression> – <expression>**
   is evaluated <integer> times.

```
(SETQ G 4)            ; Set variable G to 4.
(REPEAT 4             ; Cycle organization.
(SETQ G (* G 2)))     ; The variable G is assigned the value 64.
```

▶ **(PROGN <expression> ... – combines several (at least two!)
expressions into one, when, according to the syntax of the AutoLISP
language, use only one (as in the if... function).**

```
(IF (<A B) (PROGN (SETQ A (*A 3))    ; If A <B, then assign A to
(SETQ B (*B 5))))                    ; value 3A, otherwise assign B to 5V.
```

▶ **(APPLY <'function name> <'(list of arguments)>) – Applies a
function to the list of arguments.**

```
(APPLY '+' (2 11.25 5))     ; Function call.
18.25                       ; Result.
```

▶ **(MAPCAR <'function name> <'(list of arguments)>...) – execution of this
function over the first elements of the lists, then over the second eye elements,
etc.**

```
(MAPCAR '*' (1 2 3) '(4 5 6)' (7 8 9))    ; Function call.
(28 80 162)                               ; Result.
```

▶ **(FOREACH <variable> <list> <(function 1)> <(function 2)>...) – assigns
<variable> the value of the first element of the <list> and evaluates functions 1,
2, etc., then the second element of the list, and so on.**

```
(FOREACH Fi '(0.5 1.0 1.5)    ; Function call.
(PRINC (COS Fi)))             ; Calculating the cosine for each element
0.877583                      ; This list and display the values on the screen.
0.540302
0.070737
```

▶ **(WHILE <(condition)> <(function 1)> <(function 2)>...) – calculation
of functions in a loop as long as <condition>is met. (SETQ I 1**

Factorial 1)	; Assign the variables I and factorial the values 1.
(WHILE (< i n)	; If the condition is true, then
(Setq i (1+ i))	; i is incremented by one and factorial is multiplied
(Setq factorial (* factorial i))	; to (i+1). If the condition is false,
then)	; The end of the while loop.
	; end of the cycle. Therefore, n! is calculated

▶ **(TRANS <dot> <from> <to> [<T>]) – converts a point or distance from one coordinate system to another.**
< Point> – a list of three real numbers that can be
interpreted as a three-dimensional point or a three-dimensional vector;
<from> – the code of the coordinate system in which the point is located:
0 – world (WCS), 1 – user, 2 – screen.
<T> – if not nil, then the point should be perceived as a vector.

31.3.5   Data Output Functions

▶ **(WRITE-CHAR <integer> [<F1>]) – print an ASCII character.**
If [<F1>] is not nil, then printing the text to the file <F1> without quotes, and return in quotes.

(WRITE-CHAR 70)	; Function call.
F70	; The screen displays the letter F and echo of command 70.

▶ **(PRIN1 <expression> [<F1>]) – printing and returning the value of <expression>; if <expression> is text, the text is output in quotation marks. If [<F1>] is not nil, then printing <expression> to the <F1> file and returning the value of <expression>; text in quotation marks.**

(PRIN1 (/(cos Pi) 2))	; Function call.
-0.5 -0.5	; Execution result.
(PRIN1 "Happy New Year!")	; Function call.
"Happy New Year!" "Happy New Year!"	; Result of execution.

▶ **(PRINC <expression> [<F1>]) – printing the value of <expression>;**
if this is the text, then the output of the text in the same line and without quotes, then

echo – <expression> in quotation marks. If [<F1>] is not nil,
then the <F1> file is printed and the <expression>value is returned;
text without quotes in the last line of the file.

(PRINC (cos (/ pi 3)))	; Function call.
0.5 0.5	; Result.
(PRINC "Happy New Year", 2004!)	; Function call.
"Happy New Year", 2004! "Happy New Year, 2004!".	

▶ **(PRINT <expression> [<F1>] –**
printing the value of <expression> from the first
line followed by a space ; if <expression> is text, then
output the text with quotes, then echo – <expression> in quotes.

(PRINT (sin (/ pi 2)))	; Function call.
1.0 1.0	; Result.
(PRINT "Signature")	; Function call.
"Signature" "Signature"	; Result.

31.3.6 *Functions for Accessing Primitives and AutoCAD Tools*

▶ **(ENTDEL <primitive name>) – deletes or restores**
previously deleted primitive. It should be borne in mind that after drawing
saving operations all primitives marked as deleted are erased from
the drawing and can no longer be restored otherwise than painted again.
▶ **(ENTNEXT <primitive name>) – selects the next primitive from the database**
data.
▶ **(ENTLAST) – returns the name of the last non-deleted primitive**
in the drawing.
▶ **(ENTSEL [<hint>]) – calls the name of the primitive at the specified**
point.
Command: (SETQ A (ENTSEL "Enter the coordinates of the point
belonging to corresponding to the selected primitive: ")) ; Selection of the
primitive.
Enter the coordinates of the point belonging to the selected primitive: 2,5.
▶ **(SSGET [<method of selecting primitives>] [<point1> [<point2>]]) –**
creates a set of primitives.

Fig. 31.1 ALERT function
dialog box

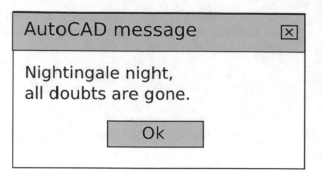

(SSGET)	; A method for selecting primitives is requested.
(SSGET "T")	; Wipes the current set of primitives.
(SSGET "P")	; Selects the last primitive.
(SSGET '(2 3))	; Selects a primitive passing through the specified ; dot.
(SSGET "P" '(1 1) '(5 7))	; The primitive in the frame is selected.
(SSGET "C" '(1 3) '(12 23))	; Selects primitives that intersect ; a frame.

► **(SSGETFIRST) – determines which objects are selected and captured.**
► **(SSADD [<primitive name>] [<set>]) – adds the primitive
name to set.**
► **(SSSETFIRST <objects to capture> [<objects to select>]) –
sets a set of selected (captured) objects. The returned
values are a set of objects.**
► **(ALERT <message>) – displays a dialog box
with the message of the AutoCAD system. Returns nil. To output
messages in several lines, put "\\n" at the end of the line.**
(SETQ LINES (STRCAT "Nightingale night, \\n"
"All doubts are away"))
(ALERT LINES) ; A dialog box is displayed
; Exit the dialog box – click on the
; OK button.
In Fig. 31.1 of the ALERT function. As indicated above, the dialog box is exited
by clicking the Ok button.
► **(INTERS <point1> <point2> <point3> <point4> [attribute>] – calculates the intersection of two segments. If the <attribute> argument is
omitted or is not equal to NIL, then the intersection of two segments**
textbfdefined by the points <point1> <point2> and <point3> <point4>is calculated.
**If <attribute> is set or is equal to NIL, then the intersection of
two straight lines passing through the specified segments**

is calculated.
(INTERS '(2.25 4.50)'(-2.25 6.335) '(1.27 0.50) '(-8.65 -2.30))
        ; Returns NIL.
(INTERS'(2.25 4.50) '(-2.25 6.335) '(1.27 0.50) '(-8.65 -2.30) NIL);
        ; Whorotates (7.64593 2.29966).

▶ **(POLAR <point> <angle> <distance>) – returns the coordinates of the point that is separated from the original <point> by the specified <distance>. The vector going from the initial point to the calculated new point must form a given < angle> in radians with the abscissa axis.**

    (POLAR '(16.32 4.782) 0.345 4.79)   ; Returns (20.8278 6.40196).

▶ **(OSNAP <dot> <mode>) – returns the result of applying an object binding function to <point> with setting the mode specified by the <mode>argument. Snapping modes are specified using strings in which no spaces comma-separated lists the names of the modes: "end", "mid","int","nea", "nod" etc.**

                                ; Returns the coordinates
                                  of the point,
(OSNAP '(16.32 4.782) "end, mid")   ; refined with the
              help of simultaneously applied functions.–
                     ; object binding to the
             end (end) and middle (mid) points.

31.3.7   User Functions

The user can create their own functions and use them in parallel with the built-in functions of the AutoCAD system. A special function DEFUN is used to create user functions:

▶ **(DEFUN <name> ([<formal parameters>] [/ <local arguments>]) <body of the function being defined>).** It is not recommended to use reserved names as function names, since this can lead to incorrect operation of individual commands of the AutoCAD system. A list of AutoLISP reserved names can be obtained using the ATOMS-FAMILY function, which has the syntax:

▶ **(ATOMS-FAMILY <form> [<list>]).** If the <list> argument is specified, the function returns the names of those characters that are reserved in this version. If the <list> argument is not specified, the function returns a complete list of reserved characters. The <form> argument can take the values 0 or 1, which affects the form of the returned list.

Formal parameters are a set of arguments of a defined function, which is a sequence of characters separated by spaces.

Local arguments are symbols used to denote other AutuLISP objects only inside the defined function.

The body of the defined function is one or more expressions. The value of the function is the value of the last evaluated expression. When developing custom functions, the recursion method is widely used; when during the definition of a function, it calls itself. Let's look at some examples of defining user-defined functions.

Example 1 Calculation of the exponential function of the integer argument $Y(A, N) = A^N$. Let's represent the defined function in the form:

$$Y(A, N) = A * Y(A, N - 1) = A * A^{N-1} = A^N.$$

In the AutoLISP language, we define this function as IND, and the program fragment by its definition will be:

$(DEFU NIND(AN)$; Determination of exponential function.

$$(IF(= N1)A(*A(IND A(-N1))))$$

)   ; End of the fragment.

The call to the described function looks like $(IND\ A\ N)$. For example, writing $(IND\ 3\ 6)$ in the text of the program or on the command line of AutoCAD gives the result of the calculation 729, which corresponds to 36.

Example 2 Calculation (Molotnikov and Molotnikova 2006) of the factorial $n!$. Let's name the function being defined FACT. A fragment of the program will look like:

$(DEFU N FACT(N)$; Definition of the function for calculating the factorial.

$$(IF(= N0)1(*N(FACT(-N1))))$$

)   ; End of function description.
(FACT 5)   ; Function call.
120   ; Calculation result 5!

In order to use the created user function, it must be saved in a text file with the lsp extension. You can create such a file by choosing the AutoLISP item from the drop-down menu TOOLS (Fig. 31.2), in which you should select the Visual LISP EDITOR item. The previously created LISP file can also be loaded using the Load item of the specified drop-down menu.

After typing the program or downloading it, the program needs to be compiled. To do this, a special vlisp-compile function is used and—even easier—the toolbar button of the Visual LISP main menu, marked with an asterisk at the bottom in Fig. 31.3.

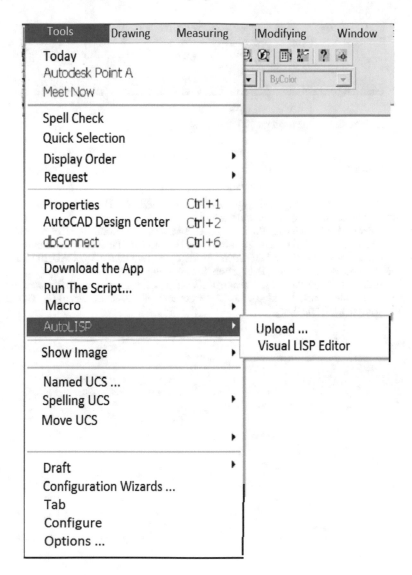

Fig. 31.2 Calling the Visual LISP editor

At the end of the compilation of the program, the user function can be called from the AutoCad command line by typing the function name and its informal parameters in brackets, for example, (FACT 4). The system will return the result: 24.

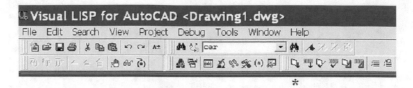

Fig. 31.3   Fragment of the main menu Visual LISP

References

LispWorks for the Windows Operating System. IDE User Guide (2021). http://www.lispworks.com/
documentation/lw60/IDE-W/html/ide-w.htm

V. Molotnikov, A. Molotnikova, *Vvedenie v SAPR na osnove Autocad Autolisp* [Introduction to
CAD based on AutoCAD Autolisp] (Rostov-on-Don, RIATM Publication, Philippines, 2006)

N. Poleshhuk, *Samouchitel' AutoCAD 2000 i Visual LISP* [Tutorial for AutoCAD 2000 and Visual
LISP] (SPb., BKhV-Peterburg Publication, St Petersburg, 2001)

N. Titarenko, A. Maly'shenko, *Mechanical Desktop 4,5,6. Iskusstvo tryokhmernogo proektirovaniya* [Mechanical Desktop 4,5,6. The art of three-dimensional design] (≪ TID ≪ DS≫
Publication, Kiev, 2002)

Chapter 32
Parametric Image of Objects

Abstract The concept of parametrization is formulated. A variant method of parametric image of objects based on the representative model is considered. The main advantages of the variant parameterization method are indicated. Two simple examples of the program description of parametrically specified graphic images-ball and roller bearings-are considered. The results of the construction of dimensional and assembly drawings of a cylindrical single-stage gearbox are presented.

Keywords Parameterization · Variant method · Generating method · Execution · Macro command · Model · Relational base · Bearing · The reducer is cylindrical

32.1   Methods of Parametric Image of Objects

Earlier (Sect. 31.1), we noted a remarkable feature of the Lisp language, namely, the ability to provide one of the possible ways to solve the problem of automatically constructing an object from a prototype. As we already know, this technique is called *parameterization.*

When parameterizing technical objects, two main methods are currently used: *variant* and *generating.* In this brochure, we will consider only the first of these methods (Ugural 2015).

In the variant method of creating parametric images of objects from a group of geometrically similar products, a representative model is compiled with all the geometric features of the products of this group. With the help of the representative model, all the geometric shapes of the modeled product are then obtained. At the same time, a representative of a class of products is also called a standard or *complex model,* and the forms obtained from it are *variants* or *versions.* In the simplest case, each instance of execution is determined only by the specified parameters. Zeroing some parameters leads to the disappearance of individual components of the graphic image. This type of construction is called principal construction. It is used in cases when the choice of geometry for the designed product has already been made. The areas of application of basic design can be the design of individual parts (e.g., fasteners, springs, bearing assembly covers, etc.), as well as complex functional components (e.g., bearings, seals, lubrication devices, etc.).

The variant method turns out to be extremely convenient for creating a custom relational database. Having at its disposal user functions describing standard models of products, the designer gets the opportunity to use them in the development of assembly drawings, when assigning the technology of processing parts, etc. In addition, once described, standard models can be used to describe other standard models as macros.

Let's consider two examples of a program description of parametrically specified graphic images.

32.2   Parametric Image of a Ball Bearing

The rules for drawing the internal structure of bearings are given in the book by Dunaev and Lelikov (1998), quoted in the list of sources. The initial data are the parameters of the internal geometry of the bearing, which are given in GOST. The specified data is placed in a text file. For GOST-8338 ball bearings, this file is named in the program below ball-GOST8338.txt. The following is the content of the text file ball-GOST8338.txt (Molotnikov and Molotnikova 2006). The data for its creation is borrowed from the reference book of the designer V. I. Anuryev (1992). Here, the bearing number, inner diameter, outer diameter, width, and radius of rounding of the edges are written sequentially separated by a space. This is followed by an entry for another bearing, etc. No other delimiters other than a space are allowed.

200 10 30 9 1 201 12 32 10 1 202 15 35 11 1 203 17 40 12 1 204 20 47 14 1.5 205 25 52 15 1.5 206 30 62 16 1.5 207 35 72 17 2 208 40 80 18 2 209 45 85 19 2 210 50 90 20 2 211 55 100 21 2.5 212 60 110 22 2.5 213 65 120 23 2.5 214 70 125 24 2.5 215 75 130 25 2.5 216 80 140 26 3 217 85 150 28 3 218 90 160 30 3 219 95 170 32 3.5 220 100 180 34 3.5 300 10 35 11 1.0 301 12 37 12 1.5 302 15 42 13 1.5 303 17 47 14 1.5 304 20 52 15 2.0 305 25 62 17 2.0 306 30 72 19 2.0 307 35 80 21 2.5 308 40 90 23 2.5 309 45 100 25 2.5 310 50 110 27 3.0 311 55 120 29 3.0 312 60 130 31 3.5 313 65 140 33 3.5 314 70 150 35 3.5 315 75 160 37 3.5 316 80 170 39 3.5 317 85 180 41 4.0 318 90 190 43 4.0 319 95 200 45 4.0 320 100 215 47 4.0

The following program contains detailed comments that make additional explanations unnecessary. Most of the points defined in the program are shown in Fig. 32.1, which makes it much easier to read it.

In the text of the program, we did not strive for compactness and elegant constructions that make it difficult to read but built the program in the way that a student usually does when he first ventured to create a software product in the AutoLISP environment. Don't let it bother you. The understanding of beauty and elegance will come to you a little later.

Fig. 32.1 To build a ball
bearing model

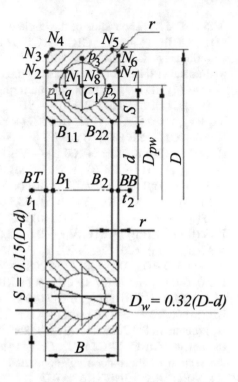

Here we present only a fragment of the program. The interested reader will find
the full listing in our book (Molotnikov and Molotnikova 2006).

;Ball Bearings (DEFUN BALL (XB YB N) ; XB – abscissa of the base point
W, ; YB-ordinate of the base point W, ; N-bearing number. (SETQ F1 (EVAL (List
OPEN "C:
AUTOCAD2000i
ball-GOST8338.txt" "r"))) ; In a text file C:
AUTOCAD2000i
ball-GOST8338.txt recorded; geometric dimensions of standard bearings according
to GOST-8338.

(setq SS (READ (STRCAT " ("(READ-LINE F1) ")"))); Reading ;data from
the file (CLOSE F1); Closing the data file (SETQ SS1 (member N SS)) ;N - bearing
number (SETQ dv (car (cdr SS1)) ; dv - the inner diameter of the bearing D (NTH
2 SS1); D is the outer diameter of the bearing B (NTH 3 SS1); B is the width of the
bearing ring r (NTH 4 SS1); r is the radius of rounding of contour faces)

(setq BT (LIST XB YB) ;Base point t1 (list (- XB 3) YB) ; Auxiliary points for
t2 (list (+ XB B 3) YB) ; construction of the bearing centerline

RD (/ D 2.0); Bearing outer radius RV (/dv 2.0); Bearing inner radius) (setq rs
(* 0.16 (- D dv)) ;The radius of the ball s (* 0.15 (- D dv)) ;Ring thickness Rpw (*

0.25 (+ D dv)) ;The radius of the circle of the centers of the balls C1C2 (- Rpw Rv s) tt (SQRT (- (* rs rs) (* C1C2 C1C2))) ; Length of the segment C1q)
　　(command "LAYER" "SET" "OSI" ""); Transition to the centerline drawing layer (command "LINE" t1 t2 ""); Drawing the centerline of the bearing (setq
　　cc (- (/(- RD RV) 2.0) rs); minimum ring thickness)
　　(setq C1 (list (+ XB (/B 2.0)) (+ YB Rpw)); point C1 and others K1 (list XB (+ YB RV)) ; (see Fig. 32.1) K2 (list XB (+ YB RV s)) K3 (list XB (+ YB RD (- 0 s))) K4 (list XB (+ YB RD (- 0 r))) K5 (list (+ XB r) (+ YB RD)) K6 (list (+ XB B (- 0 r)) (+ YB RD)) K7 (list (+ XB B) (+ YB RD (- 0 r))) K8 (list (+ XB B) (+ YB RD (- 0 s))) K9 (list (+ XB B) (+ YB RV s)) K10 (list (+ XB B) (+ YB RV)) K11 (list (+ XB B (- 0 r)) (+ YB RV)) K12 (list (+ XB B) (+ YB RV r)) K13 (list (+ XB r) (+ YB RV)) K14 (list XB (+ YB RV r)) K15 (list XB (+ YB RV s (- 0 (/ r 2.0))))) (setq K16 (list (+ XB (/ r 2)) (+ YB RV s)) K17 (list (+ XB B (- 0 (/ r 2.0))) (+ YB RV s)) K18 (list (+ XB b) (+ YB RV s (- 0 (/ r 2.0)))) K19 (list (+ XB B (- 0 (/ r 2.0))) (+ YB RD (- 0 s))) K20 (list (+ XB B) (+ YB RD (- 0 s) (/ r 2.0))) K21 (list (+ XB (/ r 2.0)) (+ YB RD (- 0 s))) K22 (list XB (+ YB RD (- 0 s) (/ r 2.0))) C2 (list (+ XB (/ B 2.0) (- 0 tt)) (+ YB RV s)) C3 (list (+ XB (/ B 2.0) (- 0 tt)) (+ YB RD (- 0 s))) C4 (list (+ XB (/ B 2.0) tt) (+ YB RD (- 0 s))) C5 (list (+ XB (/ B 2.0) tt) (+ YB RV s)) BB (list (+ XB B) YB) B1 (list (+ XB r) YB) B2 (list (+ XB B (- r 0)) YB)
) (command "LAYER" "SET" "OSN" "") ;Transition layer main ;drawing lines (command "CIRCLE" C1 rs ""); Drawing a ball; the following command - ; drawing the section of the outer ring (command "PLINE" C3 K21 "ARC" K22 "LINE" K4 "ARC" K5 "LINE" K6 "ARC" K7 "LINE" K20 "ARC" K19 "LINE" C4 "") (command "PLINE" K16 C2 "ARC" K15 "LINE" K14 "ARC" K13 "LINE" K11 "ARC" K12 "LINE" K18 "ARC" K17 "LINE" C5 "") (command "LINE" K15 K22 "") (command "LINE" K14 BT "") (command "LINE" K12 BB "") (command "LINE" K18 K20 "") (command "LINE" K11 B2 "") (command "LINE" K13 B1 "") (command "LAYER" "SET" "OSI" "") ; Transition layer centerline (setq p1 (polar C1 (pi* 1.1 rs)) p2 (polar C1 0 (* 1.1 rs)) p3 (polar C1 (/ pi 2) (* 1.1 rs)) p4 (polar C1 (* pi 1.5) (* 1.1 rs))) (command "line" p1 p2 "") (command "LINE" p3 p4 "") (command "LAYER" "SET" "VSP" ""); Transition to the layer of auxiliary lines to perform ; hatching (command "LINE" K21 K16 "") (command "LINE" K19 K17 "") (command "LAYER" " SET ""OSN" ""); Switch to the main lines layer setq r1 (list (- XB 4) (- YB 1)) ; Area selection r2 (list (+ XB B 4) (+ YB RD 1)) ;drawing subject to
　　Example of accessing the function

　　　　(BALL 170 250 212)　　; Example of accessing the function (the first two numbers are
　　　　　　　　　　　　　　　　; coordinates of the base point W,
　　　　　　　　　　　　　　　　; the third number is the bearing number.

　　The result of the execution is shown in Fig. 32.2.

32.3   Program of Construction of a Roller Conical Bearing

The contents of the text file ROLLER-GOST27365.txt are borrowed from the mentioned reference book by V. I. Anuryev (1992) and look like this:

7204 20 47 15.5 14 12 1.5 0.5 7205 25 52 16.5 15 13 1.5 0.5 7206 30 62 17.5 16 14 1.5 0.5 7207 35 72 18.5 17 15 2 0.8 7208 40 80 20 18 16 2 0.8 7209 45 85 21 19 16 2 0.8 7210 50 90 22 20 17 2 0.8 7211 55 100 23 21 18 2.5 0.8 7212 60 110 24 22 19 2.5 0.8 7213 65 120 25 23 20 2.5 0.8 7214 70 125 26.5 24 21 2.5 0.8 7215 75 130 27.5 25 22 2.5 0.8 7216 80 140 28.5 26 22 3 1

As in the case of a ball bearing, the bearing number, inner diameter, outer diameter, bearing width, inner ring width, outer ring width, and rounding radii of the edges r1 and r2 are indicated here, separated by a space (Fig. 32.2). Then the bearing data of the same name is listed for the next number. The use of separators other than a space is prohibited. When reading data files, the handle must contain the file access path, for example, C:\\ AUTOCAD2000\\. Note that when writing a descriptor, a double backslash is used as a separator in AutoLISP. The main points used to build the bearing are shown in Fig. 32.2.

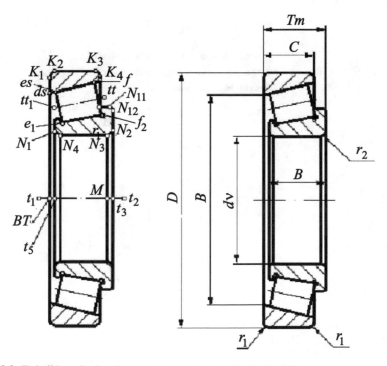

Fig. 32.2 To build a roller bearing model according to GOST 27365-87

Unlike the case with a ball bearing, we present here a complete listing of the program and the result of its execution (Fig. 32.2). The reference to the program is given at the end of the listing.

```
; Roller bearings GOST 27365-87
(DEFUN ROLLER (XB YB N)
(SETQ F1 (EVAL (List OPEN
"C:
AUTOCAD2000i
ROLLER-GOST27365.txt" "r")))
(setq SS (READ (STRCAT " ("(READ-LINE F1 ) ")") ))
(CLOSE F1)
(SETQ SS1 (member N SS))
(SETQ dv (car (cdr SS1))
D (NTH 2 SS1); Outer diameter
Tm (NTH 3 SS1); Bearing width
B (NTH 4 SS1); Width of the inner ring
c (NTH 5 SS1); Width of the outer ring
r (NTH 6 SS1); Radii of rounding edges
r1 (NTH 7 SS1)
  )


            (princ "\ n")    ; Display of the entered parameters on the screen
            (PRINC dv)       ; for the purpose of checking the correctness of the values.
            (princ "\ n")    ;15 lines can be commented out


(princ D)
(PRINC "\ n")
(PRINC Tm)
(PRINC "\ n")
(PRINC B)
(princ "\ n")
(princ c)
(princ "\ n")
(princ r)
(princ "\n")
(princ r1)
(princ "\ n")
(setq BT (LIST XB YB) ;Base point
t1 (list (- XB 3) YB) ;Bearing centerline points
t2 (list (+ XB Tm 3) YB)
)
    (setq rs (list (+ XB (/ Tm 2)) (+ YB (/ dv 2) (/ (- D dv) 4))) tt (polar rs (/ (* 14 pi)
180) (/ c 2.3)) ; Axial points of the roller tt1 (polar rs (+ pi (/ (* 14 pi) 180)) (* 0.4
Tm)) )
```

(COMMAND "LAYER" "SET" "OSI" "")
; Transition to the axial layer
(command "LINE" t1 t2 ""); lines and drawing them (command "LINE" tt tt1 "")
(setq r3 (polar rs (/pi 2) (/(- D dv) 8))
alpha (/(*16 pi) 180); The angle of inclination of the external generator
ccos (cos alpha) ; of the roller and its cosine
ds (polar r3 (+ pi alpha) (/ (* 0.5 Tm) ccos))
es (polar ds alpha (* 0.05 (- D dv)))
f (polar ds alpha (+ (/ c ccos) (- 0 (* 0.05 (- D dv)))))
K (polar ds alpha (/ c ccos))
RD (/ D 2.0)
RV (/ dv 2.0)
K1 (list XB (+ YB RD (- 0 r)))
K2 (list (+ XB r) (+ YB RD))
ccos (+ XB c)
K3 (list (- ccos r1) (+ YB RD))
K4 (ccos list (+ (- RD r1) YB))
alpha1 (/ (* 14 pi) 180) ;the angle of inclination of the axis of the roller
alpha2 (/ (* 12 pi) 180); and the angle of inclination of the inner forming
s1 (polar rs (* 1.5 pi) (/ (- D dv) 8))
s11 (polar s1 (+ pi alpha2) Tm)
es1 (polar es (+ (* pi 1.5) alpha1) Tm)
e1 (s1 s11 Inters es es1)
s12 (polar s1 alpha2 Tm)
f1 (polar f (+ alpha2 (* 1.5 pi)) Tm)
f2 (inters f f1 s1 s12)
Ll (/ (Distance f f2) 4)
l (polar f2 (+ (* 0.5 pi) alpha2) Ll)
l1 (polar l 0 Tm)
t3 (polar BT 0 Tm)
t4 (polar t3 (/ pi 2) RD)
N1 (inters t3 t4 l l1)
N11 (polar N1 pi r1)
N12 (polar N1 (* 1.5 pi) r1)
e2 (polar e1 alpha2 r1)
e3 (polar e1 (+ (* 1.5 pi) alpha1) r1)
e4 (polar e2 (+ alpha2 pi) (* 2 r1))
t5 (polar BT 0 (- Tm B))
t6 (polar t5 (/ pi 2) RD)
e5 (inters s11 e1 t5 t6)
t7 (polar e5 (/ pi 2) (* 0.124 (Distance f f2)))
e7 (polar t7 0 (/ (Distance e4 e5) 2))
k5 (polar k (* 1.5 pi) RD)
k6 (k5 inters k l l1)
f3 (f2 polar (+ (/ pi 2) alpha1) r1)

```
f4 (polar f2 (+ pi alpha2) r1)
N2 (polar t3 (/ pi 2) (+ Rv r))
N3 (list (+ XB Tm (- r 0)) (+ YB Rv))
N4 (N3 polar pi (- B r r1))
N5 (list (+ XB (- Tm B)) (+ YB Rv r1))
M3 (polar N3 (* 1.5 pi) Rv)
M4 (polar N4 (* 1.5 pi) Rv)
)
(command "LAYER" "SET" "OSN" "")
;Transition layer main
(command "Line" es e1 "") ;lines
(command "line" f f2 "")
(command "pline" l N11 "ARC" N12 "Line" t3 "")
(command "PLINE" K2 K3 "ARC" K4 "Line" K ds k1 "ARC" K2 "")
(command "ARC" e4 e3 e2 "")
(command "ARC" e4 e7 e5 "")
(command "Line" e5 t5 "")
(command "Line" e1 f2 "")
(command "Line" k k6 "")
(command "ARC" f4 "C" f2-f3 "")
(command "Line" ds BT "")
(command "Pline" N12 N2 "ARC" N3 "Line" N4 "ARC" N5 "Line" e5 "")
(command "Line" N3 M3 "")
(command "Line" N4 M4 "")
(setq rr1 (list (- XB 4) (- YB 1)))
rr2 (list (+ XB Tm 4) (+ YB RD 1))
a (ssget "–C" rr1 rr2); the selection pane to reflect
H1 (polar es (/ pi 2) r)
H3 (list (+ XB (* r 2)) (- YB RD (- 0 r)))
H2 (polar N12 pi r)
H4 (polar N3 (* 1.5 pi) (+ dv r))
)
(command "–MIRROR" a "" t1 t2 "N" ""); Reflection
(command "LAYER" "SET" "VSP" "")
(command "–BHATCH" "–P" "ANSI31" 1.0 0.0 H1 "");
    Hatching
(command "–BHATCH" H3 "")
(command "–BHATCH" "–P" "ANSI31" 1.0 90.0 H2 "")
(command "–BHATCH" H4 "")
)
    (ROLLER 170 250 7216); Example of a function call.
```

Note. Before accessing the BALL and ROLLER functions, create the necessary layers with the names used in the programs and also disable object binding.

32.4   Gearbox Assembly Drawing

For the parametric image of the gearbox, the source data from the text file is used
"C:
AutoCAD2000i
reduktor.txt". Here are the contents of this file: 4 140 340 219 105 100 70 0 0 0 200
270 112 25 30 16 18 30 35 151 93 141 156 32 100 9.367 1.5

Here are sequentially recorded the number of holes for foundation bolts, the
distance from the wheel axis to the left side of the upper belt, etc. A full explanation
of the entered parameters is given in the comments of the program listing. A space
is used as a separator between two adjacent parameters. The use of other separators
is prohibited.

The program uses a combined image construction technique. Many drawing
elements are described by sequential commands. For the image of fasteners and
some other elements (Molotnikov and Molotnikova 2006), user functions are built.

When constructing a horizontal projection of the gearbox, some assembly units
are designed as blocks.

The reader will find the full listing of the program in the book (Molotnikov and
Molotnikova 2006).

In conclusion, we present the constructed drawings of the gearbox (Fig. 32.3).

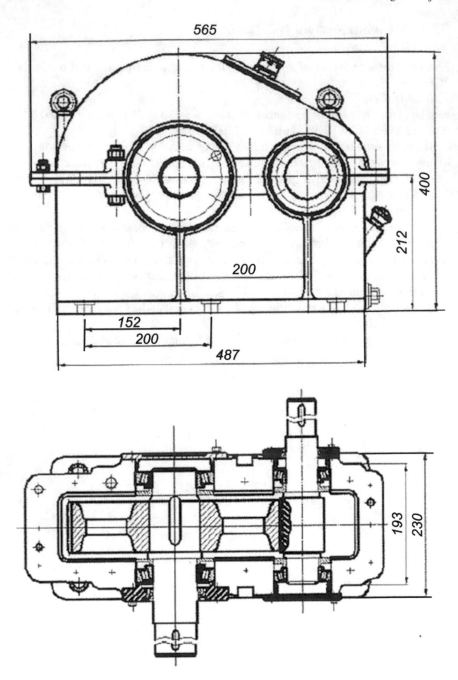

Fig. 32.3 Single-stage cylindrical reducer

References

V. Anur'ev, *Spravochnik konstruktora-mashinostroitelya* [Handbook of a Designer-Machine Builder] (Mashinostroenie Publication, Moscow, 1992)

P. Dunaev, O. Lelikov, *Konstruirovanie uzlov i detalej mashin* [Designing of Machine Components and Parts] (Vysshaya shkola Publication, Moscow, 1998)

V. Molotnikov, A. Molotnikova, *Vvedenie v SAPR na osnove Autocad Autolisp* [Introduction to CAD based on AutoCAD Autolisp] (RIATM Publication, Rostov-on-Don, 2006)

A.C. Ugural, *Mechanical Design of Machine Components* (CRC Press, New Jersey, 2015)

Index

Printed in the United States
by Baker & Taylor Publisher Services